ETHOLOGICAL DICTIONARY

ARMIN HEYMER

GERMAN – ENGLISH – FRENCH

with a foreword by Irenäus Eibl-Eibesfeldt

GARLAND PUBLISHING, INC.
NEW YORK & LONDON

VERLAG PAUL PAREY
BERLIN & HAMBURG

Drawings by KATHARINA MUNOZ-CUEVAS, Brunoy, and
HERMANN KACHER, Seewiesen

American Edition Published by
Garland Publishing, Inc., New York.

Library of Congress Cataloging in Publication Data

Heymer, Armin.
Ethological dictionary.

English, French, or German.
Bibliography: p.
Includes indexes.
1. Animals, Habits and behavior of—Dictionaries—Polyglot. 2. Dictionaries, Polyglot. I. Title.
QL750.3.H49 591.5 77-78418
ISBN 0-8240-7005-4

Printed in Germany

I hate definitions

BENJAMIN DISRAELI,

1804–1881

Foreword

Die vergleichende Verhaltensforschung – Ethologie – darf mittlerweile auf eine mehr als 50jährige Geschichte zurückblicken, nimmt man die ersten Veröffentlichungen von WHITMAN und HEINROTH als eigentliche Geburtsstunde des Faches, gefolgt von LORENZ mit seinen Arbeiten zur Soziologie der Corviden. Sie hat hauptsächlich als kontinentaleuropäische Disziplin ihr Begriffssystem erarbeitet. Distanzierte Beobachtung, stammesgeschichtliche Ausrichtung und die damit verbundene, vergleichend morphologische Betrachtungsweise und schließlich die physiologische, kausalanalytische Fragestellung bestimmten ihren Werdegang, der mit den Namen KONRAD LORENZ, NICOLAAS TINBERGEN, ERICH VON HOLST und KARL VON FRISCH untrennbar verbunden ist. Bereits im Jahre 1937 wurde als erstes ethologisches Veröffentlichungsorgan die Zeitschrift für Tierpsychologie gegründet.
Nach dem Kriege begann dann die sehr lebhafte Auseinandersetzung mit einigen stark milieutheoretisch orientierten Behavioristenschulen Amerikas. Die Diskussion führte zu Begriffsklärungen und zu einer Annäherung der Standpunkte. Daß es im tierischen Verhalten Vorprogrammierungen in feststellbaren Bereichen gibt, wurde schließlich als Faktum akzeptiert, und daß es im menschlichen Verhalten ähnlich sein dürfte, beginnt sich herumzusprechen. Ein neues Fach, die Humanethologie, entwickelt sich als interdisziplinär ausgerichtete Verhaltensbiologie des Menschen.
In einer Reihe von Lehrbüchern – TINBERGENS »Study of Instinct« eröffnete den Reigen – hat man mittlerweile die Ergebnisse der Verhaltensforschung publik gemacht. Dabei kam es fast unvermeidlich zu einem terminologischen Wirrwarr. Die Begriffe wurden ja erst neu geschaffen und definiert, und für die Übersetzungen mußten wieder neue Wörter gefunden werden. Erst allmählich kam es zu Übereinkünften und zu einer Nomenklatur des Faches.
ARMIN HEYMER hat sich in Zusammenarbeit mit einer Reihe von Experten die große Mühe gemacht, die Begriffe dieses Faches mit kurzen Erklärungen in deutsch, englisch und französisch zusammenzustellen. Das ist eine höchst verdienstvolle Aufgabe, geeignet, die Kommunikation zu erleichtern. Es muß allerdings darauf hingewiesen werden, daß viele der Begriffsbestimmungen noch vorläufigen Charakter tragen. Die hier vorgelegten Erklärungen sollen nicht als Korsett das Fach zum Erstarren bringen. Die Begriffe sind als vorläufiges Ordnungsgerüst zu verstehen, das System soll für Verbesserungen offenbleiben. Auch ist die Auswahl noch keineswegs vollständig. Viele Fachkollegen werden finden, daß ihr Begriffssystem zu kurz kam, jeder Experte, glaube ich, wird da und dort was zu bemäkeln haben und dennoch wird er ARMIN HEYMER und seinen Beratern für die geleistete Arbeit danken. Das Buch ist eine große Hilfe für Forschung und Unterricht. Ich wünsche ihm viele Auflagen, damit es an konstruktiver Kritik weiterwachse.

Ethology – the biology of behaviour – can soon look back upon a history of more then 50 years if one takes WITHMAN's and HEINROTH's publications as the beginning of modern ethology, followed by LORENZ' publications on the social behaviour of the Corvidae. The discipline developed mainly in continental Europe with respect to its early conceptualization. Objective observation, an evolutionary orientation, along with a comparative morphological point of view, and finally the physiological causal analysis determined its course. The names of KONRAD LORENZ, NICO TINBERGEN, ERICH VON HOLST and KARL VON FRISCH are inseparably tied with respect to this development and in 1937 the Zeitschrift für Tierpsychologie – Journal of Comparative Ethology has been founded as the first periodical for ethological publications.
After WW II a vigorous discussion began with several behavioristically oriented schools of thought in American psychology. The exchanges led to the clarification of concepts and some rapproachment between theoretical positions. That the behaviour of animals is made up of preprogrammed behaviour patterns in specific functional systems was finally accepted as fact, and that the same thinking applies to human behaviour is now being talked about. A new branch, human ethology, is developing as an interdisciplinary biology of human behaviour.

In a number of textbooks – TINBERGEN's »Study of Instinct« led the way – the results of ethological research have been publicised. It was unavoidable that some confusion arose in the rapidly expanding terminology. Concepts were stated and defined for the first time, and new words had to be coined for their translations. Agreement on a nomenclature of behaviour was gradual.
ARMIN HEYMER, in collaboration with a number of experts, undertook the formidable task of compiling short explanations of concepts in German, English and French. This is a most worthy undertaking that should greatly facilitate communication. However, it should be pointed out that many concepts are still in a preliminary stage of definition. The explanations that are presented here are not meant to force ethology into a rigid framework. These concepts should be seen as a preliminary framework, and leave the system open for improvements. The selection itself is not complete. Many ethologists will find that their areas may have been somewhat short-changed. Indeed each expert will find some room for improvement here and there. Yet they will at the same time thank ARMIN HEYMER and his consultants for the work they have produced. This book will be of great help to researchers and students. I hope that this work will have many editions, and that it may grow as a result of constructive criticism.

L'étude comparée du comportement – l'Ethologie – possède à l'heure actuelle une histoire de plus d'un demi-siècle, si l'on considère les publications de WHITMAN et de HEINROTH comme véritable date de naissance de cette science, suivies de celles de LORENZ sur la sociologie des Corvidae. L'éthologie, qui était à cette époque une discipline principalement européenne continentale, commença alors a élaborer son système conceptuel. Son developpement a été guidé et déterminé par l'observation objective, une orientation phylogénétique étroitement liée à la morphologie comparée, puis par l'approche physiologique et analytique causale. Les noms de KONRAD LORENZ, NICOLAAS TINBERGEN, ERICH VON HOLST et KARL VON FRISCH y sont inséparablement liés. C'est alors qu'en 1937 fut créée la Zeitschrift für Tierpsychologie – Revue d'Ethologie Comparée, comme premier organe de publications dans le domaine de l'Ethologie.
Après la dernière guerre, une vive discussion s'engagea avec les écoles behavioristes de l'Amérique du Nord, partisanes de la théorie du milieu. Ces discussions aboutirent à un certain éclaircissement conceptuel et à un rapprochement des positions. Le fait qu'il existe une programmation dans le comportement animal, a été finalement accepté comme tel et on commence à reconnaître qu'il pourrait en être de même dans le comportement humain. Un domaine nouveau s'est ainsi développé, l'éthologie humaine, qui se veut interdisciplinaire au niveau de la biologie du comportement de l'Homme.
Dans une série de traités – »L'étude de l'Instinct« de TINBERGEN étant le premier parmi eux – les résultats des recherches sur le comportement ont été plus largement diffusés. Ceci entraîna presque inévitablement un désordre terminologique, étant donné que les différentes notions venaient à peine d'être créées et avaient besoin d'être définies. Par ailleurs, pour les différentes traductions, il était nécessaire d'inventer de nouveaux termes. Ce n'est que petit à petit que des accords sur la nomenclature de notre discipline sont intervenus.
ARMIN HEYMER, en collaboration avec un certain nombre d'experts, a entrepris cette lourde tâche de rassembler les différentes notions et concepts éthologiques, accompagnés d'explications en allemand, anglais et français. Ceci est un effort hautement méritoire qui facilitera la communication. Il est évidemment nécessaire d'attirer l'attention sur le fait que nombre de ces conceptualisations ne peuvent avoir qu'un caractère provisoire. Les explications présentées dans cet ouvrage ne devraient pas provoquer un engourdissement de la discipline. Les notions doivent être comprises comme un cadre préliminaire et le système doit rester ouvert à toute sémantisation. Le choix n'est évidemment pas complet et de nombreux collègues trouveront que leur système conceptuel spécialisé a été trop brièvement traité. Chaque expert, je crois, trouvera çà et là de quoi critiquer et pourtant, il sera profondément reconnaissant à ARMIN HEYMER et ses collaborateurs pour ce travail accompli. Ce livre est d'un grand secours pour la recherche et l'enseignement et je lui souhaite de nombreuses éditions pour qu'il puisse, à l'aide d'une critique constructive, continuer à se développer dans le même esprit.

IRENÄUS EIBL-EIBESFELDT

Acknowledgements

This *Ethological Dictionary* would never have been accomplished without the help of the following colleagues, which read and commented upon the manuscript or parts of it; I wish to thank them for the long period of cooperation and for their constructive criticism:

Prof. Dr. FRANÇOIS BOURLIERE, Unité de Recherches Gérontologiques, Paris
Prof. Dr. RÉMY CHAUVIN, Laboratoire de Sociologie Animale, Rambouillet-Mittainville
Prof. Dr. EBERHARD CURIO, Arbeitsgruppe für Verhaltensforschung, Ruhr-Universität Bochum
GERARD DUBOST, Maître Assistant, Laboratoire d' Ecologie Générale, Brunoy
Prof. Dr. ERICH KLINGHAMMER, Dept. of Psychology, Laboratory of Ethology, Purdue University, West-Lafayette, Indiana
Dr. JAMES T. MARTIN, Faculty of Natural Sciences and Mathematics, Stockton State College, Pomona, New Jersey
Prof. Dr. JEAN MEDIONI, Laboratoire de Psychophysiologie, Faculté des Sciences, Toulouse
Prof. Dr. GEORGES THINES, Centre de Psychologie Expérimentale et Comparée, Université de Louvain, Pellenberg
PETER WIRTZ, M.Sc., Max-Planck-Institut für Verhaltensphysiologie, Seewiesen.

For their helpful comments as non-ethologist readers, especially on the French part of this work, I am very grateful to my friends JEAN-MARIE BETSCH, Maître Assistant, and Dr. GUY VANNIER, Maître de Recherches au C. N. R. S.

The most ungrateful task, the typing of the successive versions of the manuscript, was performed by my wife LENI, who also helped to read the proofs, for which I wish to express my warmest thanks.

I am indebted to many friends and colleagues. Unfortunately I cannot mention them all here, but I wish to thank them sincerely. Those who helped with illustrations are cited in the illustration reference list on page 221 to 222. The drawings have been elaborated by KATHARINA MUÑOZ-CUEVAS, Brunoy, and some of them by HERMANN KACHER, Seewiesen. Furthermore, I am grateful to the Verlag PAUL PAREY and especially to Dipl.-Ing. agr. CHRISTIAN GEORGI who took the great risk to edit this Dictionary, and also to PETER A. EMMERICH who was indefatigable and careful in the technical elaboration of this work.

Brunoy, Autumn 1976 ARMIN HEYMER

Contents

Einleitung

Es gibt wohl kaum Tierarten, bei denen die intraspezifische Verständigung zu einem so schier unüberwindbaren Hindernis geworden ist wie bei der Spezies Mensch. Die rasche Entwicklung zu Pseudoarten *sensu* ERIKSON (1966), trotz biologischer Monospezifität, hat Kommunikationsbarrieren entstehen lassen, die zum großen Teil nur noch über das nicht-verbale Ausdrucksverhalten überwunden werden können.

In der Vergangenheit sind verschiedentlich Versuche unternommen worden, ethologische Fachausdrücke von einer Sprache in eine andere zu übersetzen. In der Regel geschah dies in Form von Listen, in denen man einfach die entsprechenden Bezeichnungen beider Sprachen gegenüberstellte, hauptsächlich zwischen Englisch und Deutsch (SCHÄFER 1960, KOEHLER 1964, EISENBERG 1967). Es bestand also das Bedürfnis der Verständigung über unsere Kultursprachen hinweg. Die sich immer mehr entwickelnde ethologische Forschung im französischen Sprachraum ließ den Wunsch und die Notwendigkeit aufkommen, eine solche Fachausdrucksliste aufs Französische zu erweitern, doch sehr schnell mußte eingesehen werden, daß in Anbetracht der Komplexität und Mannigfaltigkeit der Ausdrucksweise in der Ethologie ein solches Unterfangen ohne erklärenden Text zu keiner rechten Hilfe führen würde. Es hätte dies eher zur Vergrößerung bereits bestehender Mißverständnisse in der Begriffsbildung geführt. Es genügt eben nicht zu wissen, daß → Erbkoordination (LORENZ 1939) im Englischen *fixed action pattern* und im Französischen *coordination héréditaire* heißt. Da die meisten Ethologen der Erde zumindest eine der drei Sprachen – Deutsch, Englisch, Französisch – beherrschen, entstand das Projekt eines *Dreisprachigen Ethologischen Wörterbuches,* um ihnen eine Arbeits- und Diskussionsbasis zu schaffen.

Die Ethologie, und das wurde schon oft gesagt, ist eine junge Wissenschaft, und folgerichtig bedient sie sich einer neuen, einem ständigen Wandel unterworfenen Terminologie. Viele Ausdrücke stammen aus der Umgangssprache, welche in der Verhaltensforschung eine ganz andere Bedeutung erlangt haben, andere wiederum wurden anderen Wissenschaftsbereichen, vor allem aus dem der Psychologie, entlehnt und haben ebenfalls einen Bedeutungswandel erfahren. Außerdem gibt es Verständigungsschwierigkeiten zwischen verschiedenen Schulen – LORENZ 1965, SKINNER 1965, CHAUVIN 1968, RICHELLE 1970 –, um nur die ganz rezenten Stellungnahmen aufzuführen.

Die gerechte Auswahl der Termini war schwierig; von insgesamt 4000 ursprünglich gesammelten Stichwörtern sind nur rund 1000 übriggeblieben. Die relativ große Anzahl von Ausdrücken aus dem Bereich von Aktionssystemen einzelner Arten (Honigbiene – zusammenfassend VON FRISCH 1965) oder Artengruppen (Entenvögel – HEINROTH 1911, LORENZ 1941) erscheint uns insofern gerechtfertigt, als sie in den Anfängen der Ethologie und in ihrer historischen Entwicklung eine bedeutende Rolle gespielt haben; außerdem gehören sie mit zu den schwierigsten Ausdrücken, deren Übersetzung und Erklärung auf jeden Fall für notwendig erachtet wird.

Die Ethologie – Biologie des Verhaltens – ist keineswegs nur die Verhaltensforschung am Tier, wie das vielerorts immer wieder zum Ausdruck kommt, sondern schließt ganz selbstverständlich auch die Erforschung des menschlichen Verhaltens mit ein, entfaltet sich doch auch das Verhalten des Menschen nach genetisch vorprogrammierten Gesetzmäßigkeiten. Um dieser Tatsache gerecht zu werden, sind aus dem Bereich der → Humanethologie (EIBL-EIBESFELDT und HASS 1966, EIBL-EIBESFELDT 1971) zahlreiche Beispiele aufgenommen worden.

Das vorliegende *Ethologische Wörterbuch* stellt einen Versuch dar, über Sprachgrenzen hinweg ein Minimum ethologischer Begriffe und Fachausdrücke zu *erklären;* – keineswegs aber soll dieser Versuch als eine endgültige Definition verstanden werden. Mit jedem neuen Beitrag in der ethologischen Forschung wird auch die Auseinandersetzung um die Begriffe und deren Aussagewert anhalten, und dafür möchte das vorliegende Wörterbuch als dreisprachige Diskussionsbasis ständig offenbleiben.

English Introduction

There are hardly any animal species in which intra-specific communication has become such an insurmountable obstacle as in the human species. The rapid development into pseudo-species *sensu* ERIKSON (1966), in spite of biological mono-specificity, has been the cause of barriers to communication which in part can only be overcome by non-verbal expressive behaviour.

In the past, various attempts have been made to translate ethological terms from one language into another. Generally this was done in the form of lists in which comparable terms were shown side-by-side. This was done primarily in English and German (SCHÄFER 1960, KOEHLER 1964, EISENBERG 1967). The need for better communication between German, English and French became quite apparent. The increasing development of ethological research done by French-speaking researchers sparked the wish as well as the necessity to expand such a list to include French. However, it soon became obvious that such an undertaking would have to include explanatory text material, in view of the complexity and diversity of expressions in ethological research and terminology. Without it an increase rather than decrease of clarity regarding ethological terminology might have resulted. It is not enough to know that → Erbkoordination (LORENZ 1939) is referred to as *fixed action pattern* in English and *coordination héréditaire* in French. Since most ethologists master at least one of the three languages, – German, English and French –, this project naturally grew into a *Tri-Lingual Ethological Dictionary* to provide them with a basis for discussion and further work.

Ethology is a young science and of necessity employs a new terminology that is subject to continuous modification and expansion. Many terms are derived from everyday language which have acquired a more specific meaning. Other terms were taken from other fields, especially from psychology and may have acquired new meaning in the process. Furthermore, there are differences between various schools of thought – LORENZ 1965, SKINNER 1965, CHAUVIN 1968, RICHELLE 1970 –, to name only the most recent sources.

The appropriate selection of terms was difficult. Out of an initial 4000 terms that were collected only about 1000 remained. The rather large number of expressions from the behaviour repertoires of individual species (honey bee – summary VON FRISCH 1965) or species' groups (geese and ducks – HEINROTH 1911, LORENZ 1941) seems to be justified in as much as they have played an important role in the beginnings of ethology and in its historical development. In addition, they contain the most difficult expressions whose translation and explanation must be considered as essential.

Ethology – the biology of behaviour – is not merely a study of behaviour of animals as many seem to think; it quite naturally includes the study of human behaviour as well since the behaviour of man also obeys to genetically pre-programmed laws. To reflect this state of affairs, a large number of terms stemming from → human ethology (EIBL-EIBESFELDT and HASS 1966, EIBL-EIBESFELDT 1971) have been included.

This *Ethological Dictionary* thus constitutes an effort to *explain* a minimum of ethological terms across the three languages; however, this attempt should not be taken as a final definition of these terms. With each new contribution from ethological research discussion of terms and their conceptual value will continue. This dictionary is intended as a tri-lingual basis for further discussion.

Introduction française

Il n'y a probablement pas d'espèce animale chez laquelle la communication intraspécifique soit devenue un obstacle aussi difficilement surmontable que chez l'Homme. Le développement rapide vers des pseudo-espèces *sensu* ERIKSON (1966), malgré la monospécificité biologique, a fait apparaître des barrières de communication qui, pour la plupart, ne peuvent être franchies qu'à travers des comportements expressifs non-verbaux.

Dans le passé, des essais sporadiques ont été entrepris pour traduire la terminologie éthologique d'une langue en une autre. En règle générale, ceci se faisait sous forme de listes dans lesquelles les notions correspondantes étaient tout simplement indiquées sans explication aucune, et ceci surtout entre l'anglais et l'allemand (SCHÄFER 1960, KOEHLER 1964, EISENBERG 1967). Il existait évidemment le besoin d'une communication au-delà de nos langues culturelles. La recherche éthologique qui s'est étendue et développée ces dernières années dans les pays francophones a fait naître le désir et la nécessité d'élargir une telle liste de termes au français; mais rapidement il est apparu, compte tenu de la complexité et de la diversité des expressions en éthologie, qu'une telle entreprise sans texte explicatif des différentes notions ne serait en aucune manière une aide appréciable. Ceci aurait plutôt agrandi les malentendus déjà existants à propos des différentes notions. Il n'est évidemment pas suffisant de savoir par exemple que → Erbkoordination (LORENZ 1939) est appelé *fixed action pattern* en anglais ou encore *coordination héréditaire* en français. Etant donné que la plupart des éthologistes maîtrisent au moins une des trois langues, à savoir l'allemand, l'anglais ou le français, il naquit alors le projet d'un *Vocabulaire Ethologique Trilingue* pour leur apporter une base de travail et de discussion.

L'éthologie, – et ceci a souvent été exprimé, est une science jeune; par conséquent, elle se sert d'une terminologie neuve et soumise à une transformation permanente. De nombreuses expressions sont issues du langage courant, ayant une signification toute autre en éthologie. D'autres ont été empruntées à des domaines scientifiques tels que la psychologie, et ont subi également un changement de signification. Par ailleurs, il ne faut pas négliger les difficultés d'intercommunication des différentes écoles. Pour ne citer que quelques toutes récentes prises de position, qu'il nous soit permis de nous référer à LORENZ (1965), SKINNER (1965), CHAUVIN (1968) et RICHELLE (1970).

Un choix judicieux des termes a été difficile et sur 4000 expressions éthologiques initialement collectées, environ 1000 seulement ont été retenues. Le très grand nombre de termes issus des études des systèmes d'action ou éthogrammes de différentes espèces, comme par exemple de l'Abeille (résumé VON FRISCH 1965) ou de groupes d'espèces tels que les Anatidae (HEINROTH 1911, LORENZ 1941), nous semble d'autant plus justifié que les recherches sur ces groupes zoologiques ont joué un rôle important au début de l'éthologie et lors de son développement historique. D'autre part, ces expressions figurent parmi les plus difficiles et leur traduction ou au moins leur explication ont paru nécessaires.

L'éthologie – biologie du comportement – n'est pas seulement la recherche comportementale sur l'animal, comme on l'exprime souvent, mais inclut tout naturellement la recherche sur le comportement humain étant donné que le comportement de l'Homme s'exprime et se développe également selon des lois génétiques programmées. En tenant compte de cette évidence, de nombreux exemples du domaine de → l'éthologie humaine (EIBL-EIBESFELDT et HASS 1966, EIBL-EIBESFELDT 1971) ont été pris en considération.

Le présent *Vocabulaire Ethologique* constitue une tentative pour *expliquer*, au-delà des frontières linguistiques, un minimum de termes et de notions éthologiques – mais en aucun cas, cet essai ne doit être considéré comme un ouvrage achevé dans lequel chaque terme garderait sa définition à jamais fixée. En effet, ce *Vocabulaire trilingue* doit être considéré comme base de discussion et veut rester ouvert au dialogue; chaque contribution nouvelle dans le domaine de la recherche éthologique devrait l'enrichir.

Hinweise zur Benutzung

Nach reichlicher Überlegung und bedingt durch die Tatsache, daß in den Anfängen der objektiven Erforschung des Verhaltens die Neubildung von Fachausdrücken sich hauptsächlich auf Veröffentlichungen in deutscher Sprache stützte – *Zeitschrift für Tierpsychologie* seit 1937 –, haben wir uns dafür entschieden, die deutschen Stichwörter in alphabetischer Reihenfolge aufzuführen, was es dem nicht Deutsch sprechenden Benutzer ermöglicht, schnell an die Übersetzung des Begriffes und an eine ihm zugängliche Erklärung zu gelangen. Außerdem war es für den Autor leichter, von seiner Muttersprache aus zu agieren, was sicher jedem Leser verständlich sein wird. Der Aufbau eines dreisprachigen Nachschlagewerkes bedingt auch ein gewisses Maß an Ordnung, Ökonomie und Synoptik, vor allem sollten Wiederholungen vermieden werden, und so mußten zwangsläufig sprachliche Ungleichheiten entstehen. Die zentrale Stellung des Englischen als vermittelnde Wissenschaftssprache hat uns veranlaßt, ihr auch im Rahmen des vorliegenden *Ethologischen Wörterbuches* die Vermittlerrolle zwischen Deutsch und Französisch aufzuerlegen.
Zur schnellen Auffindung wurde das entsprechende deutsche Stichwort in einer Groteskschrift gedruckt. Danach erfolgt eine kurze Erklärung des Begriffes, und es sind, falls erforderlich, Beispiele angeführt. Darauf folgt der entsprechende *englische* und *französische* Begriff in einer Groteskschrift mit der dazugehörigen Erklärung; dies trägt erheblich zur Übersichtlichkeit bei. Für eine bessere Verständigung werden vielerorts didaktische Abbildungen beigefügt. Soweit möglich, wurde versucht, jene Arbeit zu zitieren, in welcher der Ausdruck das erste Mal benutzt wurde, jedoch schien es in vielen Fällen vorteilhafter, rezente und weiterführende Literatur anzuführen bzw. auf einschlägige Lehrbücher zu verweisen.
Nicht immer war es möglich, ein Stichwort gerecht zu übersetzen; in solchen Fällen haben wir vorerst das deutsche Wort stehenlassen und seine Bedeutung erklärt, wie z. B. bei → *Giftsterzeln*, → *Manteln*, → *Saugtrinken* und → *Zirkeln* im Englischen sowie bei → *Absprungalter*, → *Antrinken*, → *Aufsteilen*, → *Aufstoßen*, → *Balzkette*, → *Bekanntheit*, → *Herden* und → *Jauchzen* im Französischen. Wir folgen dabei durchaus einem bereits üblichen Verfahren. In diesem Sinne sei lediglich auf PIERON (1968) – *Vocabulaire de la Psychologie* – verwiesen. Wieder andere deutsche Ausdrücke haben sich seit langem im Französischen und gelegentlich auch im Englischen eingebürgert. Beispiele hierfür sind → *Flehmen* und → *Laufschlag*. Andererseits wird im Französischen für → *Prellsprung* das englische → *stotting* benutzt, und auch → *tobogganning* hat sich eingebürgert.
Die häufig durch einen Pfeil gegebenen Verweise auf andere Begriffe innerhalb des erklärenden Textes (Beispiel: → *Übersprungverhalten*) weisen darauf hin, daß unter diesem Stichwort weitere ergänzende Erläuterungen zu finden sind. Bei Synonymen oder zu einem Begriffskomplex gehörenden Stichwörtern wird direkt auf jenen Ausdruck verwiesen, unter welchem die ausführliche Erklärung zu finden ist.

Beispiele für Synonyme:
Ablenkungsmanöver → Verleiten
Aktionskatalog → Ethogramm
E. behavioural repertoire □ F. répertoire comportemental

Beispiel für Begriffskomplex:
Einander-Lausen → Körperpflegehandlungen
E. allo-grooming □ F. épouillage mutuel

Für Leser, die von einem englischen oder einem französischen Stichwort ausgehen, befindet sich von Seite 223 bis Seite 229 ein englischer und von Seite 230 bis Seite 237 ein französischer Index mit dem entsprechenden Seitenverweis zur Auffindung der dreisprachigen Stichwörter mit ihren Erklärungen.

Beispiel: E. displacement activity p. 183 □ F. activité substitutive p. 184

Auf den Seiten 208–220 befindet sich ein ausführliches Literaturverzeichnis mit den im Text zitierten Arbeiten und allgemeinen Werken und Lehrbüchern der Ethologie und der Soziobiologie.

How to Use this Dictionary

After giving it much thought and necessitated by the fact that in the early days of ethology new terms originated primarily in German publications – namely in the *Zeitschrift für Tierpsychologie* since 1937 – we have decided to list the German key-words in alphabetical order. This will enable the non-German reader to gain quick access to the translation of the terms and their explanations. Furthermore, it was easier for the author to compile the material in his mother language, as the reader may readily appreciate. The compilation of a tri-lingual dictionary requires furthermore a certain measure of order, economy and synopsis. Hence, repetitions should be avoided as much as possible. As a result a certain unevenness in representation in the several languages must be expected. The key position of English as a mediating scientific language induced us to give the same role to it in the present *Ethological Dictionary* between German and French.

To enable the reader to find a word more easily, the German key word has been printed in grotesque. It is then followed by a short explanation and by examples if deemed necessary. The appropriate *English* and *French* terms in sanserif letters then follow. To further clarify the text, appropriate illustrations have been provided. As much as possible we strove to provide the reference(s) of the paper(s) in which the term first appeared. In other instances it seemed of advantage to include more recent and additional references, or to refer to appropriate text-books.

It was not always possible to precisely translate the key word. In such instances we have left the German word and explained its meaning, e. g. in English → *Giftsterzeln,* → *Manteln,* → *Saugtrinken* and → *Zirkeln,* and in French → *Absprungalter,* → *Antrinken,* → *Aufsteilen,* → *Aufstoßen,* → *Balzkette,* → *Bekanntheit,* → *Herden* and → *Jauchzen.* We herewith followed a common usage; see PIERON (1968) – *Vocabulaire de la Psychologie.* Other German expressions have long since become a part of the French and English language, e. g. → *Flehmen* and → *Laufschlag.* On the other hand the English word → *stotting* is used in French for the German word → *Prellsprung,* and → *tobogganning* is also frequently employed.

Frequently the reader is referred to other terms within the explanatory text (e. g. → *displacement activity*). This indicates that additional explanation may be found under that term. With synonyms or key words that are part of a larger concept, reference is made directly to that expression following which the detailed explanation is to be found.

Examples for synonyms:
Ablenkungsmanöver → Verleiten
Aktionskatalog → Ethogramm
E. behavioural repertoire □ F. répertoire comportemental

Example for larger concepts:
Einander-Lausen → Körperpflegehandlungen
E. allo-grooming □ F. épouillage mutuel

Readers who begin with an English or French term are provided with an English Index from p. 223 to p. 229, and a French Index from p. 230 to p. 237 which give the appropriate page for the tri-lingual key words and their explanations.

Example:
E. displacement activity p. 183
F. activité substitutive p. 184

On the pages 208 to 220 of the dictionary there is a complete list of references to all papers and books that are cited in the text, along with more general text-books on ethology and sociobiology.

Conseils aux lecteurs

Après mûre réflexion et compte tenu du fait que le début des recherches objectives sur le comportement animal et humain et la création des néologismes d'une terminologie adéquate étaient basées quasi essentiellement sur les publications en langue allemande – *Zeitschrift für Tierpsychologie* depuis 1937 –, nous avons jugé utile de présenter les expressions allemandes par ordre alphabétique ce qui permet aux utilisateurs non allemands de retrouver rapidement la traduction recherchée d'un terme donné, y compris son explication. Par ailleurs, il a été évidemment plus facile pour l'auteur d'agir à partir de sa langue maternelle ce que chaque lecteur, je l'espère, comprendra. La position centrale de l'anglais comme langue médiatrice dans le domaine de la science nous a incité à lui donner, dans le cadre du présent *Vocabulaire Ethologique,* ce même rôle médiateur entre l'allemand et le français.

Pour permettre au lecteur de trouver rapidement les termes allemands, nous les avons imprimés en antique, suivis d'une courte explication de la notion et, si nécessaire, d'un exemple. Les expressions anglaises et françaises correspondantes sont imprimées en sanserif, chacune suivie d'une explication. Cette manière de procéder permet une amélioration de la synoptique. Pour une meilleure compréhension, une iconographie didactique et adéquate y figure. Si possible, nous avons essayé de citer le travail original dans lequel l'expression a été utilisée pour la première fois, mais dans la plupart des cas, il a été plus judicieux de citer une bibliographie plus récente ou encore de renvoyer à un traité compétent.

Il n'a pas toujours été possible de trouver une équivalence parfaite entre les expressions correspondantes des différentes langues. Dans ces cas, nous nous sommes permis de faire figurer l'expression germanique originale et d'expliquer sa signification, comme par exemple pour → *Giftsterzeln,* → *Manteln,* → *Saugtrinken* et → *Zirkeln* en anglais ou pour → *Absprungalter,* → *Antrinken,* → *Aufsteilen,* → *Aufstoßen,* → *Balzkette,* → *Bekanntheit,* → *Herden* et → *Jauchzen* en français. Nous suivons là un exemple tout à fait courant et nous voudrions tout simplement, dans ce contexte, faire référence à PIERON (1968) – *Vocabulaire de la Psychologie.* D'autres expressions allemandes sont depuis longtemps adoptées en français et parfois même en anglais. Des exemples parlants à ce sujet sont → *Flehmen* et → *Laufschlag.* Par ailleurs, pour le terme → *Prellsprung,* on utilise en français depuis longtemps l'expression anglaise → *stotting* et même → *tobogganning* se trouve accepté.

Les différents termes ou expressions imprimés en *italique* et précédés d'une flèche à l'intérieur du texte explicatif, comme par exemple → *activité substitutive,* indiquent au lecteur qu'ils sont expliqués par eux-mêmes dans ce vocabulaire. Dans les cas d'une terminologie synonyme ou d'un complexe de notions, il est renvoyé directement au terme correspondant sous lequel l'explication a été donnée.

Exemples de synonymes:
Ablenkungsmanöver → Verleiten
Aktionskatalog → Ethogramm
E. behavioural repertoire
F. répertoire comportemental

Exemple de complexe de notions:
Einander-Lausen → Körperpflegehandlungen
E. allo-grooming □ F. épouillage mutuel

Pour les utilisateurs qui souhaitent partir d'un terme anglais ou français, un index alphabétique anglais se trouve de p. 223 à p. 229, et un index alphabétique français de p. 230 à p. 237 avec indication de la page leur permettant de retrouver aisément les explications trilingues à l'intérieur du vocabulaire.

Exemple:
E. displacement activity p. 183
F. activité substitutive p. 184

De page 208 à 220 figure une bibliographie avec les travaux cités dans le texte et un certain nombre d'ouvrages et traités généraux sur l'éthologie et la sociobiologie.

A

AAM, angeborener Auslösemechanismus □ Der *AAM* ist ein → *AM,* den die Art in phylogenetischer Anpassung an bestimmte Reizsituationen erworben hat. Der Ausdruck *AAM* soll nur dann verwendet werden, wenn die Prüfung am unerfahrenen, d. h. von diese Reaktion auslösenden Reizen isolierten Individuum ergeben hat, daß die Verknüpfung zwischen einem bestimmten Reiz und einer bestimmten Reaktion angeboren ist, und nur für die Bezeichnung eines *AM* beim unerfahrenen Tier, es sei denn, er bleibt in der Ontogenese unverändert bestehen (SCHLEIDT 1962 – dort weiterführende Literatur).

IRM, innate releasing mechanism □ The *IRM* is a →*RM* which the species has acquired by phylogenetic adaptation to a certain environmental situation. The term *IRM* should be used only, if it has been proved by experiments on inexperienced individuals, i. e. individuals isolated from the stimuli which release this reaction, that the connection between a certain stimulus and a certain reaction is innate. The term *IRM* denotes a *RM* in the inexperienced animal only, except in the rare cases in which this *RM* remains unchanged by experience (SCHLEIDT 1962 – see this article for more detailed references).

MDI, mécanisme déclencheur inné; MID, mécanisme inné de déclenchement □ Le *MDI* est un mécanisme déclencheur (→ *MD*) que l'espèce a acquis au cours de son évolution par adaptation à des situations déterminées de l'environnement. Le terme de *MDI* ne doit être utilisé que dans les cas où l'examen sur un individu sans expérience, c'est-à-dire un individu isolé des stimuli déclenchant une telle réaction, a montré que le rapport entre un stimulus déterminé et une réaction déterminée est inné. Le *MDI* désigne uniquement un mécanisme déclencheur chez un animal sans expérience, sauf si ce mécanisme reste inchangé au cours de l'ontogenèse (SCHLEIDT 1962 – voir cet article pour une bibliographie plus approfondie).

Abaufbewegung □ Eine epigame Verhaltensweise der Enten (Abb. 1 – *Anas platyrhynchos*); der Erpel taucht blitzschnell den Schnabel ins Wasser (Abb. 1 A) und reißt im nächsten Augenblick den Kopf wieder hoch (Abb. 1 B), ohne dabei die tief ins Wasser gesenkte Brust zu heben (LORENZ 1941, 1966).

1 A 1 B

down-up-movement □ An epigamic behaviour pattern in ducks (Fig. 1 – *Anas platyrhynchos*) in which the drake dips his bill in the water for an instant (Fig. 1 A) followed immediately by thrusting the head high out of the water (Fig. 1 B) without lifting the breast which has sunk deep in the water (LORENZ 1941, 1966).

mouvement de bas en haut □ Comportement épigame chez les Canards (Fig. 1 – *Anas platyrhynchos);* le malard plonge rapidement le bec dans l'eau (Fig. 1 A) et un instant plus tard, relève la tête très haut (Fig. 1 B) en gardant la poitrine immergée (LORENZ 1941, 1966).

Abdomenschwingen □ Das »Abdomen-Auf- und Abbewegen« bei Odonaten, das bei den ursprünglichen Lestidae noch dem Flügelputzen dient, hat bei den Calopterygidae diese Bedeutung verloren (Abb. 2 – *Calopteryx haemorrhoidalis*). Aus den einfachen Auf- und Abbewegungen ist ein rhythmisches Schwingen geworden, das in Verbindung mit der leuchtend gefärbten Unterseite der letzten drei Abdominalsegmente bei den ♂♂ Signalfunktion besitzt und somit zu einem ritualisierten Verhalten geworden ist (HEYMER 1973 a + b) → *Ritualisierung.*

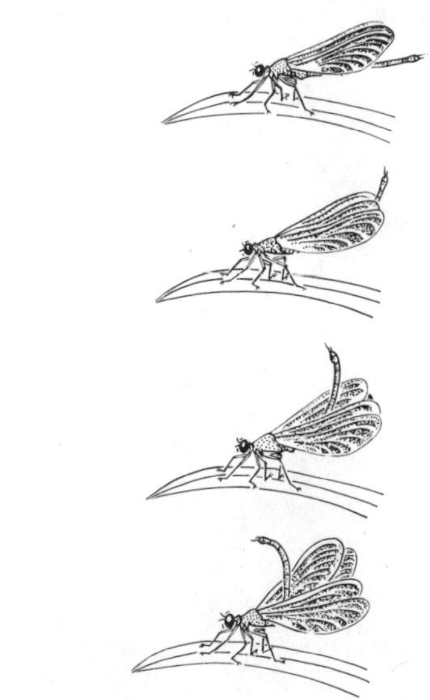

2

abdominal up-and-down movement □ In Odonata, abdominal up-and-down movement still serves wing-grooming in the primitive Lestidae, but has lost its original significance in the Calopterygidae (Fig. 2 – *Calopteryx haemorrhoidalis*). The simple up-and-down movement has developed into a rhythmic oscillation, which plays an integral part in the use of luminous colouration of the ventral aspect of the last three abdominal segments as a signal in the ♂♂. It has thus become a ritualized behaviour pattern (HEYMER 1973 a + b)→ *ritualization.*

mouvement pendulaire de l'abdomen □ Mouvement pendulaire vertical de l'abdomen chez les Odonates. Chez les Lestidae, groupe primitif, il sert au nettoyage des ailes tandis qu'il a perdu cette fonction chez les Calopterygidae (Fig. 2 – *Calopteryx haemorrhoidalis*); le mouvement pendulaire s'est transformé en un balancement rythmique qui, en liaison avec la coloration vive de la face inférieure des trois derniers segments abdominaux, a pris une fonction de signalisation et, de ce fait, est devenu un comportement ritualisé (HEYMER 1973 a + b) → *ritualisation.*

Ablageplatzzeigen → Eiablageplatzzeigen

Ablaichvorspiel → Balzkette

E. prespawning courtship behaviour □ F. comportement de parade nuptiale avant le frai

Ablenkung → Verleiten

E. distraction □ F. diversion

Ablenkungsmanöver → Verleiten

Ablösungszeremoniell □ Bei vielen Vogelarten die Ablösung von ♂ und ♀ am Nest. Ein besonders schönes Beispiel zeigt EIBL-EIBESFELDT (1965 und »Grundriß«) beim Galapagoskormoran *Nannopterum harrisi* (Abb. 3), der zum Nest und zum Partner zurückkehrend ein Geschenk in Form eines Tangbüschels mitbringt, um die Aggressivität des Partners auf die Gabe zu lenken.

3

relieving ceremony □ In many bird species, the relieving of the partner on the nest. A particularly good example is reported by EIBL-EIBESFELDT (1965 and »Ethology«) in the Galapagos Cormorant *Nannopterum harrisi* (Fig. 3) which returns to the nest with a gift in the form of a piece of seaweed upon which the partner can direct its aggression.

cérémonial de relève □ Désigne chez de nombreux Oiseaux la relève au nid entre les deux partenaires. Un exemple est rapporté par EIBL-EIBESFELDT (1965 et »Ethologie«) chez le Cormoran des Galapagos, *Nannopterum harrisi* (Fig. 3); l'Oiseau qui se dirige vers le nid apporte une touffe de varech et apaise ainsi le comportement agressif de son partenaire en détournant son attention sur la touffe.

Abreaktion □ Ein Verhalten zum Ableiten gestauter → *Aggression;* hierzu gehören Handlungen am → *Ersatzobjekt,* wie z. B. Türen zuschlagen, sowie → *Ventilsitten* zur »friedlichen« Abreaktion von Kampfenergie, etwa Gesangsduelle, Sportwettkämpfe usw. (EIBL-EIBESFELDT 1970).

re-directed behaviour □ Resulting from pent-up → *aggression;* behaviour directed at → *substitute objects,* such as slamming of a door, as well as harmless venting of aggression (→ *safety valve customs*) in the form of singing contests or competitive sports, etc. (EIBL-EIBESFELDT 1970).

»défoulement« □ Comportement ayant pour but de »décharger« une agressivité (→ *agression*) accumulée; il peut s'agir d'actions dirigées sur un → *objet de remplacement,* comme le claquement de portes, ou d'activités non agressives (→ *soupape de sûreté*) comme des duos chantés, manifestations sportives, etc. (EIBL-EIBESFELDT 1970).

Abschütteln → Totschütteln

Absprungalter □ Bei Lummen jenes Alter bzw. jener Zeitpunkt, an dem die Jungtiere ihre Brutfelsen verlassen und ins freie Wasser springen (TSCHANZ 1968).

jumping-off age □ The age or time period at which Guillemot hatchlings leave their cliff-side nests and leap into the water below (TSCHANZ 1968).

Absprungalter □ Ce terme allemand désigne l'âge ou le moment où les jeunes Guillemots quittent leur falaise d'éclosion en sautant directement dans l'eau libre (TSCHANZ 1968).

Absturzscheu □ Verschiedene Nestflüchter zeigen eine angeborene Scheu vor einem Abgrund, den sie vor Erfahrung mit Abstürzen am Seheindruck erkennen. Küken, Zicklein und Lämmer, aber ebenso auch 4 Wochen alte Katzen, die nie irgendwo abgestürzt waren, halten vor einem Abgrund, der mit einer Glasplatte bedeckt ist, inne (Abb. 4), während die weniger optisch orientierte Wanderratte ohne weiteres über den Abgrund hinausläuft (EIBL-EIBESFELDT »Grundriß«).

fear of falling □ Various nidifugous animals show an innate fear of a precipice which they recognize from the visual impression alone without themselves having experienced falling. Chicks, kids and lambs even as well as four week old kittens that have never fallen stop before a precipice that is covered with glass (visual cliff, Fig. 4) whereas the less optically oriented Norway rat moves without hesitation onto the glass plate (EIBL-EIBESFELDT »Ethology«).

peur du vide □ Plusieurs animaux nidifuges montrent une peur innée du vide avant même d'avoir fait toute expérience de chute. Les Poussins, les

4

Chevreaux et les Agneaux, mais aussi les Chats âgés de 4 semaines qui n'ont jamais fait de chute, s'immobilisent devant un espace vide recouvert d'une vitre (Fig. 4) tandis que le Rat gris dont l'orientation optique est moins importante, continue sa marche sans hésitation par-dessus le vide (EIBL-EIBESFELDT »Ethologie«).

Abwehrauge □ Magisches Symbol im menschlichen Kulturbereich unterschiedlicher Darstellungsform, dessen Ausgangs- oder Zielpunkt des Ritualisierungsweges stets das natürliche Auge ist. Das Grundschema entspricht in allen Fällen einer einwandfrei als Auge oder Augenpaar erkennbaren Darstellung. Der Anbringungsort und die Bedeutung gehören in die Funktionsbereiche des Beaufsichtigens, Abwehrens, Ablenkens und Schützens, aber auch des Wahrnehmens und des Blickfanges, wie z. B. in der Werbung. Abwehraugen befinden sich häufig an allen möglichen Gegenständen des täglichen Gebrauchs wie des Kults, an Hauseingängen und am Bug von Schiffen (Abb. 5 – Odysseus in einem Schiff mit Abwehraugen; griechisches Vasenbild). Noch heute finden wir Abwehraugen an jugoslawischen Frachtseglern. Als Blickabwehrsymbol hat das Abwehrauge besonders im osteuropäischen Raum und in der Türkei aktuelle Bedeutung (SELIGMANN 1922, 1927, KOENIG 1970 und vor allem 1975). Für Beispiele bei Tieren → *Augenflecke.*

protective eye □ Magic eye symbol in human cultures with different forms of presentation. The starting point or the goal of the process of ritualization is always the natural eye. The basic schema corresponds in all cases of representation to one or two clearly recognizable eyes. The localization and signification serves the function of guarding, defending, distraction and protection, but also of perception and eye-catching as in advertising. Protective eyes are frequently found on a variety of every-day objects as well as those connected with cults, at the entrance to houses and on the bows of ships (Fig. 5 – Odysseus in his ship with protective eyes; picture on a Greek vase). Even today protective eyes are to be found on Yugoslavian freight sailing vessels. As a visual distraction symbol protective eyes have meaning in East European countries and in Turkey (SELIGMANN 1922, 1927, KOENIG 1970 and especially 1975). For examples in animals see → *eye spots.*

oeil de défense □ Symbole magique qu'on trouve sous différentes formes de représentation dans les cultures humaines. L'œil naturel est le point de départ, mais également le but du processus de ritualisation. Le schéma de base correspond dans tous les cas

5

à la représentation d'un ou de deux yeux reconnaissables en tant que tels. Le lieu de représentation et la signification entrent dans les domaines fonctionnels de la surveillance, de la défense, de la distraction et de la protection, mais également de la perception et de l'attraction, comme par exemple dans la publicité. Les yeux de défense se trouvent souvent sur différents objets d'usage quotidien ou de culte, au-dessus des entrées de maisons et sur les proues des bateaux (Fig. 5 – Ulysse dans un bateau muni d'yeux de défense; peinture sur un vase grec). Aujourd'hui encore, on les trouve sur les cargos à voile yougoslaves. En tant que symbole de signification apotropéïque, l'oeil de défense possède une signification actuelle, en particulier dans les pays est-européens et en Turquie (SELIGMANN 1922, 1927, KOENIG 1970 et surtout 1975). Pour les exemples chez les animaux → *taches ocellaires.*

Abwehrspritzen → Harnspritzen
E. defensive urine spraying □ F. miction de défense

Achtmonatsangst → Fremdenfurcht
E. eight-months-fear □ F. peur des huit mois

Achtungserweisung □ Demonstrationsaufwand als ein Spiegel der Wertschätzung der Geladenen ist ein Muster der Achtungserweisung, vor allem, daß man dem Gast eine hohe Rangstellung zuweist. Wenn Schimpansen an einem Ranghöheren vorbeigehen, legen sie die etwas gekrümmte, mit dem Handteller nach unten gehaltene Hand zum Gruß an ihren Kopf; dies dürfte wohl ebenfalls als Achtungserweisung gelten (EIBL-EIBESFELDT 1970, KORTLANDT 1958).

showing »respect« □ Gesture which serves as reflection of the respect accorded to the object or recipient; in particular a display in which one accords a guest a high rank. If a chimpanzee passes another of higher rank, he lays the slightly folded palm of the hand on his head as a greeting; this may be viewed as a gesture of showing »respect« (EIBL-EIBESFELDT 1970, KORTLANDT 1958).

témoignage de respect □ Comportement et gestes d'un individu envers un autre de rang plus élevé. Chez les Chimpanzés, quand un animal passe devant un congénère de rang supérieur, il porte sa main, la paume dirigée vers le bas, à sa propre tête pour le saluer. Ce comportement peut être considéré comme une marque de respect envers son compagnon (EIBL-EIBESFELDT 1970, KORTLANDT 1958).

Adaptation → Anpassung

Adelphophagie = Kainismus → Verwandtenfresserei

Affektambivalenz □ Gleichzeitiges Auftreten miteinander unverträglicher Affekte, z. B. Liebe und Haß.

emotional ambivalence □ Simultaneous occurrence of incompatible emotions, e. g. love and hate.

ambivalence affective □ Apparition simultanée d'émotions incompatibles, par exemple amour et haine.

Affekthandlung □ Meist unbeherrschte, beschleunigte Handlungsabläufe ohne Einsicht, unter Kontrolle des autonomen Nervensystems.

emotional behaviour □ Mostly impulsive and rapid behaviour discharged without insight, under the control of the autonomic nervous system.

comportement émotif □ En général, déroulement d'un comportement impulsif et sans délibération, sous contrôle du système nerveux autonome.

Affektkrämpfe □ Krampfartig übersteigerte Affektausdrucksbewegungen, z. B. Lach-, Wein- und Schreikrämpfe.

convulsions during emotional outbursts □ Convulsively exaggerated emotional expressive movements, e. g. convulsive laughing, sobbing or screaming.

contractures émotives □ Mouvements expressifs d'origine émotive convulsivement exagérés; rire, pleurs ou cris convulsifs.

afferente Drosselung □ Hemmung im → *AAM,* derzufolge das Tier nicht mehr auf spezifische Reize von bestimmter Intensität anspricht. Beim Puter z. B. kann man die Lautäußerung des Kollerns durch Reize einer bestimmten Frequenz auslösen. Zuletzt reagiert das Tier nicht mehr, erweist sich aber als voll reaktionsbereit, wenn man mit einer anderen Frequenz reizt. SCHLEIDT (1954) spricht hier von afferenter Drosselung.

afferent throttling □ Inhibition in the → *IRM,* resulting in the termination of an animal's response to specific stimuli. In the turkey one can repeatedly elicit the gobbling sound by sound stimulation of a given frequency. Finally the turkey fails to respond further; however, the full responsiveness may still be elicited if one switches to a different sound frequency. SCHLEIDT (1954) refers to this as *afferente Drosselung.*

inhibition afférente □ Inhibition dans le → *MDI* empêchant un animal de réagir à des stimuli spécifiques. Chez le Dindon, par exemple, on peut déclencher le glouglou par des stimuli sonores d'une fréquence déterminée. Après plusieurs répétitions, l'Oiseau ne réagit plus, mais il peut répondre parfaitement si on le stimule avec une nouvelle fréquence. SCHLEIDT (1954) appelle ce phénomène *afferente Drosselung.*

Afferenz □ Erregung, Impuls oder Information, die über die afferenten Nervenfasern von der Peripherie zum Zentralnervensystem geführt werden. → *Reafferenzprinzip.*

afference □ Excitation, impulse or information moving from the periphery over afferent nerve fibres into the central nervous system. → *reafference principle.*

afférence □ Excitation, impulsion ou information qui sont conduites par les fibres nerveuses afférentes depuis la périphérie jusqu'au système nerveux central. → *principe de réafférence.*

Aggregation → anonymer Verband

Aggression □ Ein physischer Akt oder eine Drohhandlung durch ein Individuum, welches die Freiheit und die genetische Lebensfähigkeit eines anderen reduziert oder einschränkt. – Das allgemeine Interesse und die öffentliche Kontroverse über die Frage der Entstehung aggressiven Verhaltens veranlassen uns, einige der bisher entwickelten Modelle zur Erklärung der Aggressivität anzuführen (LORENZ 1963, HACKER 1971, PLACK 1973, HOLLOWEY 1974).

Während das *Frustrations-Aggressions-Modell* (DOLLARD *et al.*, 1939) und das *lernpsychologische Modell* von BANDURA und WALTERS (1963) meist dem *Lorenz-Freudschen Triebmodell* (LORENZ 1950, 1963) gegenübergestellt wurden, als gäbe es nur diese oder jene Möglichkeit der Erklärung aggressiven Verhaltens, hat MOYER (1968) in einer neueren Untersuchung aggressives Verhalten eindeutiger aufgegliedert, da die früheren Aufstellungen nicht vollständig oder nicht ausführlich genug waren. Nachfolgend sollen einige der bekanntesten Aggressionsformen aufgeführt werden:

1. *Aggression zum Beuteerwerb* – wird durch wenige Reize ausgelöst. Sie wird nur in geringem Maße durch Gonadenhormone beeinflußt. Das Beutefangverhalten wird besonders durch den seitlichen Hypothalamus gesteuert.

2. *Aggression zwischen ♂♂* – wird durch ein anderes ♂ der gleichen Art ausgelöst und (in vielen Fällen) durch das spezifische Demutsverhalten dieses ♂ gehemmt. Ihre Entwicklung ist hauptsächlich vom männlichen Hormon abhängig. Über die neurale Grundlage dieser Art von Aggression ist nur wenig bekannt; es wird jedoch vermutet, daß die septale Region daran beteiligt ist. – *Aggression zwischen ♀♀*, von MOYER nicht aufgeführt, kommt bei einer ganzen Anzahl von Tierarten vor.

3. *Angst-Aggression* – kommt nur in einigen Fällen vor, in denen keine Flucht mehr möglich ist. Sie wird durch Reizung des vorderen Hypothalamus bis zum ventro-medianen Nukleus verstärkt. Über die beteiligten endokrinen Faktoren ist wenig bekannt. Die *kritische Reaktion* (HEDIGER 1950), bei der ein in die Enge getriebenes Tier einen Feind oder einen Artgenossen angreift, dürfte ebenfalls zu dieser Kategorie gehören.

4. *Irritative Aggression* – kann durch Frustration, Entbehrung und Schmerz verstärkt werden. Diese Art von Aggression kann durch eine Vielzahl von Reizen ausgelöst werden. Der ventro-mediane Hypothalamus ist besonders daran beteiligt, und paradoxerweise wird sie sowohl durch Reizung als durch Ablation dieser Region verstärkt. Sie verringert sich durch Amygdalektomie, Kastration und Stimulation des caudalen Nukleus und des Septums.

5. *Territoriale Aggression – maternelle Aggression* – beide dürften spezifische Reaktionen auf eine bestimmte Situation sein, und die wenigen bisher vorliegenden Ergebnisse deuten auf eine Abhängigkeit von den Fortpflanzungshormonen hin. Über die neurale Grundlage dieser Aggressionsform ist so gut wie nichts bekannt.

6. *Instrumentale Aggression* – kann auf jeder der oben aufgeführten Arten der Aggression beruhen. Sie besteht in der Tendenz eines Lebewesens, sich aggressiv zu verhalten, wenn dieses Verhalten in der Vergangenheit gefördert und verstärkt wurde.

7. *Innergruppenaggression* – Aggressivität innerhalb einer Gruppe, ein sich z. Z. ausbreitendes Phänomen innerhalb unserer anonymen Gesellschaft, wohl eine Zivilisationserscheinung.

8. *Zwischengruppenaggression* – Kämpfe zwischen verschiedenen Gruppen, Populationen oder Klans innerhalb einer Art, wie das vor allem bei Hyänen sehr schön nachgewiesen werden konnte (KRUUG 1966, VAN LAWICK und VAN LAWICK-GOODALL 1970, 1972). Im ethologischen Sinne sind auch Auseinandersetzungen zwischen Völkern oder Nationen, im Ernstfall Kriege (destruktive Aggressionen), nichts anderes als Zwischengruppenaggressionen, bedingt durch die Entwicklung zur kulturellen → *Pseudospeziation* (EIBL-EIBESFELDT 1975). Diese Ansicht verdient unterstrichen zu werden in Anbetracht des eindeutig monospezifischen Status des Menschen.

Der Ausdruck *agonistisches Verhalten* sollte als generelle Bezeichnung vorgezogen werden, wenn nicht bekannt ist, um welche spezifische Art der Aggression es sich handelt. Dieser Ausdruck ist neutral und bezeichnet auf treffende Weise das gemeinte Verhalten.

aggression □ A physical act or threat of action by one individual that reduces the freedom or genetic fitness of another. – The current interest and public controversy over the question of causation of aggressive behaviour prompts us to give some previously postulated explanatory models (LORENZ 1963, HACKER 1971, PLACK 1973, HOLLOWEY 1974).

While the *frustration-aggression model* (DOLLARD *et al.* 1939) and the *learning theory model* of BANDURA and WALTERS (1963) have been contrasted with the *Lorenz-Freudian drive model* (LORENZ 1950, 1963) as if either one or the other were the way to explain aggressive behaviour, more recent work by MOYER (1968) has classified aggressive behaviour in more detail, indicating that the earlier classifications were either not complete or not detailed enough. Some of the best known forms of aggression are cited below:

1. *Predatory aggression* – elicited by a narrow range of stimuli. It is little influenced by either the presence or absence of gonadal hormones. The lateral hypothalamus is particularly involved in predation.

2. *Inter-male aggression* – elicited by a strange male of the same species and is inhibited (in many cases) by the particular submissive behaviour of that male. It is particularly dependent on the male hormone for

its development. Little is known about the neural basis of this type of aggression, except that the septal area is probably involved. – *Inter-female aggression,* not discussed by MOYER, is found in some other species, too.

3. *Fear-induced aggression* – occurs only in cases where escape has been attempted, but is not possible. It is increased by stimulation of the hypothalamus anterior to the ventromedial nucleus. Little is known about the endocrinology involved. The *critical reaction* (HEDIGER 1950) where a cornered animal attacks a predator or conspecific seems to fall into this category as well.

4. *Irritable aggression* – may be increased by frustration, deprivation, and pain. A wide range of stimuli will elicit irritable aggression. The ventro-medial hypothalamus is particularly involved and paradoxically, it is increased by either stimulation or ablation of that area. It is reduced by amygdalectomy, castration, and stimulation of the caudate nucleus and the septum.

5. *Territorial aggression – Maternal aggression* – both appear to be specific to the situation involved and what little evidence there is indicates that the reproductive hormones are involved. Essentially nothing is known about the neural basis of these kinds of aggression.

6. *Instrumental aggression* – may be based on any of the above classes. It consists of an increase in the tendency for an organism to engage in aggressive behaviour when that behaviour has been reinforced in the past.

7. *Intragroup aggression* – a phenomenon which is found in present-day anonymous civilized society in contrast to intergroup aggression.

8. *Intergroup aggression* – fighting between various groups, populations or clans within the same species; intergroup aggression has been carefully documented in hyenas (KRUUG 1966, VAN LAWICK and VAN LAWICK-GOODALL 1970, 1972). From the ethological point of view, strives between peoples and nations, and in an extreme case our wars (destructive aggression), can be considered as intergroup aggression, due to the development of cultural → *pseudospeciation* (EIBL-EIBESFELDT 1975). It is important to underline this point of view taking into account the monospecifity of man.

The term *agonistic behaviour* may be preferable as a general term when the specific kind of aggression involved is not known. It is free from excess meaning, yet denotes the behaviour in question.

agression □ Acte physique ou action de menace d'un individu envers un autre réduisant par là même sa liberté et sa potentialité génétique. – L'intérêt général et la controverse ouverte en ce qui concerne le développement d'un comportement agressif nous obligent à donner quelques explications détaillées sur les différents modèles qui ont été élaborés pour éclaircir les problèmes de l'agressivité (LORENZ 1963, HACKER 1971, PLACK 1973, HOLLOWEY 1974).

Dans le passé, le *modèle de l'agression-frustration* selon DOLLARD *et al.* (1939) et le *modèle de l'agression acquise* d'après BANDURA et WALTERS (1963) ont souvent été opposés au *modèle de l'agression pulsionnelle* tel que l'ont exposé LORENZ et FREUD (LORENZ 1950, 1963). Les débats se sont alors déroulés comme s'il n'existait que ces trois possibilités pour expliquer le comportement agressif. Cependant, MOYER (1968) a essayé d'expliquer les comportements agressifs avec plus de précision, les définitions antérieures n'ayant jamais été suffisamment complètes ou détaillées. Ci-après, nous citons quelques-unes des formes d'agression les plus connues:

1. *Agression prédatrice* – Elle est déclenchée par un nombre réduit de stimuli et influencée d'une manière minime par la présence ou l'absence d'hormones gonadotropiques. Le comportement de capture de proie est dirigé en particulier par l'hypothalamus latéral.

2. *Agression entre* ♂♂ – Le développement de cette agression est surtout dépendant des hormones mâles. Nous connaissons très peu les bases nerveuses de cette forme d'agression, mais il semble que la région septale y joue un rôle. Ce comportement intraspécifique est souvent inhibé par un comportement de soumission spécifique du ♂ ayant déclenché l'agression. – *L'agression entre* ♀♀ qui n'est pas citée par MOYER, existe chez un certain nombre d'espèces animales.

3. *Agression par peur* – Elle n'existe que dans très peu de cas, en particulier dans les situations où la fuite a été tentée, mais n'est pas possible. Cette agression est déclenchée par la stimulation de l'hypothalamus antérieur jusqu'au nucléus ventro-médian. Nous connaissons peu les facteurs endocrines impliqués. La *réaction critique* (HEDIGER 1950) par laquelle un animal (dans une situation sans issue) attaque un ennemi ou un congénère semble faire partie de cette catégorie d'agression.

4. *Agression par irritation* – Cette forme d'agression peut être déclenchée par de nombreuses stimulations différentes. Elle peut être renforcée par la frustration, la privation ou la douleur. L'hypothalamus ventro-médian est particulièrement impliqué et paradoxalement, cette agression est renforcée non seulement par stimulation, mais aussi par ablation de cette région. Par contre, cette forme d'agression diminue par amygdalectomie, par castration et par stimulation du nucléus caudal et du septum.

5. *Agression territoriale et agression maternelle* – Ces deux formes sont vraisemblablement des réactions spécifiques à des situations déterminées. Les quelques données actuellement connues indiquent une influence de l'hormone reproductrice. Nous ne connaissons pratiquement rien des bases neurales de cette agression.

6. *Agression instrumentale* – Cette forme d'agression

peut être basée sur pratiquement toutes les formes agressives citées ci-dessus. Elle consiste dans la tendance d'un animal à se comporter agressivement lorsque, dans le passé, ce comportement a été particulièrement renforcé ou soutenu.
7. *Agressivité intra-groupe* – Agressivité entre individus à l'intérieur d'un même groupe. Actuellement, phénomène répandu à l'intérieur de notre civilisation trop anonyme, probablement un phénomène culturel.
8. *Agression inter-groupes* – Combats et luttes entre différents groupes, populations ou clans à l'intérieur d'une même espèce; ce phénomène est particulièrement bien démontré chez les Hyènes (KRUUG 1966, VAN LAWICK et VAN LAWICK-GOODALL 1970, 1972). Au sens strict de l'éthologie, les disputes entre peuples ou nations, et dans le cas extrême les guerres (agression destructive), ne sont rien d'autre que des combats entre groupes, dus au développement vers une → *pseudo-spéciation* culturelle (EIBL-EIBESFELDT 1975). Ce point est à souligner en tenant compte de la monospécificité de l'Homme.
Lorsque la forme spécifique de l'agression n'est pas connue, il est proposé d'une manière générale d'utiliser l'expression *comportement agonistique*. Cette expression est neutre et désigne en tout cas d'une manière explicite le comportement en question.

Aggressionstrieb → Trieb-Hypothese
E. drive for aggression □ F. pulsion agressive

Aggressionsventil □ Verhaltensmechanismus zur Ableitung von aufgestauter Aggression, um eine tatsächliche aggressive Handlung zu verhindern → *Ventilsitten.*

aggressivity valve □ Behavioural mechanism for harmless discharge of pent-up aggression functioning to reduce actual aggressive acts → *safety valve customs.*

»soupape« de l'agressivité □ Mécanisme comportemental laissant échapper la pulsion agressive et empêchant ainsi l'agression → *soupape de sûreté.*

Aggressivdrohen □ Drohen mit Angriffscharakter, im Gegensatz zum → *Defensivdrohen.* Das angriffslustige Eichhörnchen z. B. legt die Ohren zurück (Abb. 6) und reibt die Nagezähne aneinander, beim Defensivdrohen richtet es die Ohren auf (EIBL-EIBESFELDT 1957 und »Grundriß«).

6

offensive threat □ Threat with intention to attack in contrast to → *defensive threat.* For example the squirrel showing offensive threat has the ears laid back (Fig. 6) and rubs the incisors together whereas in defensive threat the ears are erected (EIBL-EIBESFELDT 1957 and »Ethology«).

menace agressive □ Menace à caractère agressif, par opposition à la → *menace défensive.* L'Ecureuil agressif, par exemple, replie les oreilles (Fig. 6) et frotte les incisives les unes contre les autres tandis qu'il dresse les oreilles en menace défensive (EIBL-EIBESFELDT 1957 et »Ethologie«).

agonistisches Verhalten → Aggression
E. agonistic behaviour □ F. comportement agonistique

Aha-Erlebnis □ Das plötzliche, einfallsartige Verstehen eines gesuchten, aber vorher unbekannten Sinnzusammenhangs. Diese unvermittelte, nicht aufgrund einer schrittweisen Ableitung gewonnene → *Einsicht* ist mit dem befreienden Erlebnis *»aha – so ist es«* verbunden (BÜHLER 1922 a + b).

»aha«-experience □ The sudden insightful understanding of a previously unknown contingency. This sudden → *insight* which is not accomplished in a stepwise manner is combined with the expression *»aha – so that's it«* (BÜHLER 1922 a + b).

L'eurêka d'Archimede □ → *Compréhension soudaine* d'un contexte recherché, mais inconnu jusqu'alors. Une telle intuition qui se produit subitement et qui n'est pas due à une déduction progressive, est liée à l'expérience libératrice *»J'ai trouvé!«* (BÜHLER 1922 a + b).

Akinesis □ Durch Reflex bedingte Bewegungslosigkeit, Erstarrung als Folge einer Dauerkontraktion der Bewegungsmuskulatur; bekannt ist das *Sichtotstellen (Thanatose)* mancher Tiere, z. B. vieler Insekten, bei Gefahr.

akinesis □ Immobility determined by reflexes, torpidity resulting from continued contraction of the locomotory muscles; e. g. »freezing« of some animals, esp. insects, in danger.

acinèse □ Immobilisation déterminée par voie réflexe, suite d'une contraction permanente de la musculature locomotrice. Ce phénomène est connu en particulier sous forme de *thanatose* chez de nombreux Insectes en cas de danger.

Akkomodation □ 1. Adaptive, reversible Modifikation des Auges, speziell der Linse, welche es ermöglicht, ein ständig scharfes Bild wahrzunehmen, ganz gleich über welche Entfernung.
2. Bei Kindern die Fähigkeit zur Umbildung eines ursprünglichen Schemas, welche die Anpassung an neue Situationen ermöglicht.

accomodation □ 1. Adaptive reversible modification of the eye, specifically of the lens, which permits perception of a continuously sharp image regardless of the distance to the object.
2. In children the ability to transform an original model or perception such that it permits adaptation to new situations.

accomodation □ 1. Modification adaptative et réversible de l'œil, spécialement du cristallin, permettant d'assurer la netteté des images pour différentes distances de vision.
2. Faculté de l'enfant à transformer un schème initial pour s'adapter à une situation nouvelle.

Aktionsbereitschaft → Handlungsbereitschaft

Aktionsgemeinschaft □ Mitglieder einer Aktionsgemeinschaft führen zahlreiche Handlungen gemeinsam durch, z. B. Nahrungssuche, Gesang, Baden, Staubbaden, Gesellschaftsbalz. Beim Menschen Demonstrationen, Häuser besetzen, Spiel.

allomimetic group □ Members of an allomimetic group perform various behaviours together, for example, food gathering, singing, bathing, dust bathing, social courtship. Examples in humans are demonstrations or sport matches.

société à activités communes □ Les membres d'une telle société ont de nombreuses activités en commun, par exemple recherche de la nourriture, chant, bain, parade nuptiale collective. Chez les Hommes, nous connaissons comme exemples d'activités en commun les démonstrations, les activités sportives etc.

Aktionskatalog → Ethogramm
E. behavioural repertoire □ F. répertoire comportemental

Aktionspotential □ Durch Zellreizung verursachte Spannungsänderungen an den Membranen lebender Zellen, die zu einem Aktionsstrom führen. → *SAP.*

action potential □ Changes in electric potential on the membranes of living cells, resulting from cell stimulation, leading to an action current. → *SAP.*

potentiel d'action □ Variation transitoire du potentiel de membrane des cellules vivantes, provoquée par une stimulation de ces cellules et produisant un courant d'action. → *PAS.*

Aktionsraum □ Das gesamte Gebiet, das ein Individuum oder eine ständig organisierte Gruppe während der Zeit ihres Lebens bzw. ihres Bestehens betritt. Hierin sind sowohl verschiedene Saisonterritorien und Aufenthaltsgebiete wie auch die Wanderwege zwischen diesen eingeschlossen. EIBL-EIBESFELDT (»Grundriß«) sieht im Aktionsraum im Gegensatz zum → *Revier* eine neutrale Zone, also einen Raum, der nicht verteidigt wird. Reviere sind im Aktionsraum zu finden, falls sie bestehen. So verteidigen die Seelöwen-♂♂ (*Zalophus wollebaeki*) wohl einen bestimmten Uferstreifen an Land und im Wasser als Revier, nicht aber die Fischgründe im Meer, die selbstverständlich zum Aktionsraum gehören.

home range □ The entire area which is occupied by an individual animal or a stable group throughout the life of the animal(s). This includes the various seasonal territories and ranges as well as the migration routes between them. EIBL-EIBESFELDT (»Ethology«) considers the home range in contrast to the → *territory* as a neutral zone which is not defended. Territories would be within the home range if they occur. The male sea lion *Zalophus wollebaeki*, e. g., defends a given strip of beach both on shore and on the water, but does not defend the fishing grounds below which, however, are included in the home range.

domaine vital □ Espace qu'un individu ou un groupe organisé parcourt tout au long de son existence. Y sont inclus les différents territoires saisonniers et les régions dans lesquelles les animaux séjournent temporairement, y compris les chemins de migration et de déplacement. EIBL-EIBESFELDT (»Ethologie«) voit dans le domaine vital en opposition au → *territoire,* une zone neutre, donc un espace qui n'est pas défendu activement d'une manière générale. Les territoires s'ils existent, se trouvent de toute façon à l'intérieur du domaine vital. Par exemple, les Otaries ♂♂ *Zalophus wollebaeki* défendent une bande déterminée de la côte et de l'eau comme territoire alors que les zones de pêche en mer ne sont nullement défendues bien que ces dernières fassent partie du domaine vital.

Aktionsstrom → Aktionspotential
E. action current □ F. courant d'action

Aktionssubstanzen □ Von VON MURALT (1939) so bezeichnete Stoffe, die nach Reizung eines Nervs an den Nervenendigungen (mit Synapsen) freiwerden und (durch Membranänderungen) der Reizübertragung dienen.

neural transmitters □ Substances which are released upon stimulation of a neuron at the nerve endings (with synapses) and which aid in the transmission of stimuli by means of changes in the permeability of the membrane (VON MURALT 1939).

neuromédiateurs □ Substances qui, après stimulation d'un nerf, sont libérées au niveau des synapses et qui, par modification de la perméabilité des membranes, servent à la transmission des excitations des messages (VON MURALT 1939).

Aktionssystem → Ethogramm
E. system of actions □ F. répertoire comportemental

Aktivitätsanalyse □ Analyse eines optimalen, rationalen Verhaltens oder Prozesses im Rahmen eines geschlossenen Systems, in dem alle Verhaltens- oder Prozeßbedingungen berücksichtigt werden. → *Motivationsanalyse.*

activity analysis □ Analysis of an optimal, rational behaviour or process within the framework of a closed system, whereby all conditions of a process or behaviour are taken into consideration. → *motivational analysis.*

analyse d'activité □ Analyse d'un comportement ou d'un processus optimal et rationnel dans le cadre d'un système fermé en tenant compte de toutes les conditions du comportement ou du processus analysé. → *analyse motivationnelle.*

Alarmverhalten → Warnverhalten

Alleinaufziehen → Kaspar-Hauser-Versuch
E. rearing in isolation □ F. élevage dans l'isolement

Alles-oder-Nichts-Gesetz □ Besagt, daß Reize nach Überschreitung eines bestimmten Schwellenwertes (Reizschwelle) unabhängig von einer weiteren Steigerung der Reizstärke an den Einzelelementen erregbarer Gebilde der Organismen immer nur die gleiche fortleitbare Erregung (→ *Aktionspotential*) als maximale Reizantwort (»alles«) auslösen. Unterhalb dieses Wertes findet jedoch keine Erregungsausbreitung (»nichts«) statt. Der unterschwellige Reiz führt lediglich an der Reizstelle selbst zu einer in ihrer Größe der Reizstärke proportionalen lokalen, nicht fortleitbaren Reizantwort (BOWDITCH 1871).

all-or-none-law □ States that once a specific stimulus threshold has been reached, any stimulus can release no more than a certain transmittable excitation (→ *action potential*) which represents the maximum response value (»all«). This value cannot be exceeded even if the strength of a stimulus influencing the individual elements of excitable structures in the organism increases. Below this value, no transmission of excitation occurs (»none«). A stimulus below the threshold simply leads to a response at the stimulus location itself, proportional to the strength of the stimulus (BOWDITCH 1871).

loi du »tout ou rien« □ Elle dit que les stimuli, après avoir dépassé une certaine valeur limite (seuil absolu), ne peuvent, indépendamment d'une augmentation de leur force d'excitation au niveau des éléments excitables des organismes, déclencher qu'une même excitation propagée (→ *potentiel d'action*) comme réponse d'excitation maximale (»tout«). En dessous de cette valeur, il n'y a plus d'excitation propagée (»rien«). Un stimulus n'atteignant pas le seuil déterminé ne provoque, à l'endroit même de l'excitation, qu'une réponse locale (non propagée) et proportionnelle au logarithme de l'intensité du stimulus (BOWDITCH 1871).

allochthone Handlung □ Eine durch triebfremde Erregung gespeiste → *Übersprungbewegung*.

allochthonous behaviour pattern □ A behaviour motivated by a drive other than that of the underlying functional system with which the behaviour is usually associated → *displacement activity*.

comportement allochtone □ Comportement extrinsèque provoqué par une activation étrangère à la pulsion propre → *activité substitutive*.

Allogrooming → Körperpflegehandlungen

allomimetisches Verhalten → Stimmungsübertragung

E. allomimetic behaviour □ F. comportement allomimétique

Alternativbewegung □ Für gewisse Arten von Übersprungbewegungen schlägt LEYHAUSEN (1952) den in seinem Sinne hypothesenfreien Terminus Alternativbewegung vor. → *Übersprungbewegung*.

alternative movement □ Suggested by LEYHAUSEN (1952) as an unbiassed term for certain forms of → *displacement activity*.

mouvement alternatif □ Pour une certaine catégorie de mouvements de substitution, LEYHAUSEN (1952) propose ce terme qui, à son sens, est dépourvu de toute hypothèse sous-jacente. → *activité substitutive*.

Altruismus → Beistandsverhalten

altruistisches Verhalten → Beistandsverhalten

AM, Auslösemechanismus □ Der *AM* einer Reaktion umfaßt alle angenommenen Strukturen des Organismus, die an der selektiven Auslösung dieser Reaktion wesentlich beteiligt sind, nicht aber ihre motorischen Instanzen. Der Ausdruck *AM* ist immer dann zu verwenden, wenn nicht bekannt ist, ob die Verknüpfung zwischen Reiz(en) und Reaktion(en) durch phylogenetische oder ontogenetische Anpassung entstanden ist (SCHLEIDT 1962).

RM, releasing mechanism □ The term *RM* of a reaction includes all those hypothetical structures of a postulated CN mechanism of an organism which have an important part in releasing this reaction, but does not include their motor system. The term *RM* should be used whenever it is not known whether the connection between stimulus and reaction is originated by phylogenetic or ontogenetic adaptation (SCHLEIDT 1962).

MD, mécanisme déclencheur □ Le *MD* d'une réaction inclut toutes les structures hypothétiques d'un organisme qui participent au déclenchement sélectif d'une telle réaction, à l'exclusion de leur système moteur. Le terme de *MD* doit être employé dans tous les cas où l'on ignore si la liaison entre stimulus et réaction s'est développée par une adaptation phylogénétique ou ontogénétique (SCHLEIDT 1962).

ambivalentes Verhalten □ Verhalten, das sich aus zwei unterschiedlichen, neben- oder nacheinander auftretenden, meist unvollständigen Bewegungen zusammensetzt, wenn das gleiche Objekt verschiedene Tendenzen auslöst. Z. B. sind in einer bestimmten Stellung (Drohen) der Silbermöwen-♂♂ Anteile des Angriffs- und des Fluchtverhaltens kombiniert. → *Konfliktverhalten*.

ambivalent behaviour □ Behaviour which consists of two different, often incomplete reactions occuring simultaneously or in short, rapid succession. Can be observed when different sign stimuli releasing these reactions are present simultaneously. Ambivalence is displayed e. g. by ♂ herring gulls in their threat posture which combines attack and flight behaviour. → *conflict behaviour*.

comportement ambivalent □ Comportement qui consiste en deux réactions différentes, généralement incomplètes, qui peuvent apparaître en même temps ou en consécution rapide. Il se manifeste lorsque des stimuli-clés différents apparaissent en même temps. Une certaine posture de menace chez les ♂♂ du Goéland argenté, par exemple, traduit un compromis entre les tendances à l'attaque et à la fuite. → *comportement conflictuel*.

Ameisengäste → Ethoparasit

E. ant hosts □ F. myrmécophiles

Analogie □ Morphologische Strukturen, Körpermerkmale und Verhaltensweisen, die einander ähnlich sind und auf konvergenter Evolution beruhen, aber nicht auf gemeinsame Vorfahren schließen lassen. → *Konvergenz.* Im Gegensatz hierzu → *Homologie.*

analogy □ Referring to structures, physical processes or behaviour patterns that are similar owing to convergent evolution as opposed to common ancestry; hence, display analogy. → *convergence.* In contrast to → *homology.*

analogie □ Structures morphologiques, caractères corporels et comportements qui sont similaires en raison d'une évolution convergente, mais qui ne proviennent pas d'ancêtres communs. → *convergence.* En opposition à → *homologie.*

Anamnese = Erinnerung → Gedächtnis
E. memory □ F. mémoire

Anemotaxis → Taxis

»Anfangsreibung« □ Eine Bezeichnung aus der Technik dafür, daß ein Verhalten die vollständigste oder stärkstmögliche Ausprägung erst nach Durchlaufen entsprechend schwächerer Stufen während eines Handlungsablaufes erreicht (LORENZ 1939, VON HOLST 1950, LEYHAUSEN 1952).

initial friction □ A term expropriated from the physical sciences which means that the strongest possible and complete expression of a behaviour pattern is reached only after having passed through lower stages (LORENZ 1939, VON HOLST 1950, LEYHAUSEN 1952).

inertie réactive □ Le mot allemand *»Anfangsreibung«* est un terme du domaine de la technique qui, en éthologie, signifie qu'un comportement n'atteint son expresssion la plus forte qu'après être passé par des stades plus faibles, c'est-à-dire par des stades de démarrage (LORENZ 1939, VON HOLST 1950, LEYHAUSEN 1952).

angeboren □ Bezieht sich auf das Verhalten, das nicht erlernt bzw. nicht erworben, zum Zeitpunkt der Geburt vorhanden, aber nicht unbedingt auch aktionsfähig ist, sondern erst im Laufe der ontogenetischen Reifung in Erscheinung tritt.

innate □ Refers to behaviour not learned or acquired during ontogeny; present at birth, but not necessarily functional, may appear only in the course of ontogenetic maturation.

inné □ Concerne les comportements non appris ou non acquis; potentiellement ou réellement présents à la naissance, mais non pas obligatoirement fonctionnels, ne se manifestant qu'au cours de la maturation ontogénétique.

angeborener Auslösemechanismus → AAM

angeborener gestaltbildender Mechanismus, AGM □ Angenommenes System im ZNS, das unter den Gesetzen der Gestaltbildung bestimmte Informationsmuster, die die Sinnesorgane liefern, artspezifischen Handlungsformen über ein obligatorisches Lernen zuordnet; so können Vögel über Jahre hinweg zum selben Brutplatz zurückkehren (TEMBROCK 1968).

innate Gestalt-producing mechanism □ Hypothetical system in the CNS which according to the Gestalt laws assigns certain informational patterns, supplied by sense organs, to species-specific behaviour patterns by means of a learning process; in this way, for example, birds can find their way back to the same breeding area year after year (TEMBROCK 1968).

mécanisme inné de structuration □ Système hypothétique du système nerveux central qui, selon les lois de la formation de la »Gestalt«, établit une liaison entre certains éléments d'information émis par les organes sensoriels et des comportements specifiques par la voie d'un apprentissage obligatoire; ainsi, les Oiseaux peuvent retourner chaque année dans la même aire de reproduction (TEMBROCK 1968).

angeborenes Aktionssystem → Erbkoordination
E. innate action system □ F. système d'action inné

Angleichungstendenz □ HEDIGER (1964) bezeichnete damit die subjektive Angleichung eines artfremden Wesens an die eigene soziale Umwelt. Im Falle von Tier-Mensch-Beziehungen sind Zoomorphismen bzw. Anthropomorphismen die Folge.

assimilation tendency □ HEDIGER (1964) uses this term to refer to the assimilation of a non-conspecific into the social environment of a population. In the relationship between humans and animals, this leads to zoomorphisms and anthropomorphisms.

tendance à l'assimilation □ HEDIGER (1964) désigne par ce terme l'assimilation d'éléments étrangers à l'espèce, c'est-à-dire leur intégration dans l'environnement social. Dans les relations entre les animaux et l'Homme, ceci conduit à des zoomorphismes ou anthropomorphismes.

Angst □ In ihrer Grundbedeutung heißt Angst Enge bzw. Beklemmung; aus dem lat. *angustus = eng.* Im allgemeinen mit Beklemmung und quälender Verzweiflung einhergehender Gefühlszustand oder Affekt. Angst kann je nach Anlage, Charakter und Temperament des Menschen mehr oder weniger unterschiedlich durchlebt werden, insbesondere in bezug auf ihre Intensität – in extremer Form bis hin zum Gefühl der Todesangst. Wie jeder starke Affekt ist die Angst von auffallenden körperlichen Symptomen begleitet, wie z. B. erhöhte Pulsfrequenz, Atemnot, Schweißausbruch, Zittern sowie gesteigerte Blasen- und Darmtätigkeit = *Angstmiktion* und *Angstdefäkation.* Nach TEMBROCK (1961) wäre es zweckmäßig, in der Ethologie zwischen Angst und → *Furcht* zu unterscheiden. Nicht die Maus, die vor dem Verfolger flieht, hat »Angst«, sondern jene, die daran gehindert wird. Auch LEYHAUSEN (1967) hat sich als Ethologe mit dem Problem beschäftigt.

fear, »angst« □ Etymologically »angst« is derived

from the latin »*angustus*« = *closeness, confinement.* An affect or feeling associated with oppression or persisting despair. People experience fear differently according to their own predisposition, character and temperament especially with regard to its intensity including even the extreme form – fear of death. As in all strong emotions, physical symptoms accompany fear, as for example, increase in pulse rate, breathing, perspiration, trembling, as well as increased motility in the intestinal tract and urinary tract, e. g., *fear-induced defecation and urination.* According to TEMBROCK (1961) it is ethologically useful to distinguish between fear and → *fright.* The mouse which flees from its pursuer is not anxious, but the one which is prevented from escaping. LEYHAUSEN (1967) has also delt with this problem from an ethological view-point.

peur □ Dans sa signification fondamentale, peur veut dire serrement ou oppression; du latin *angustus* = *serré.* En général, état affectif accompagné d'angoisse et d'un désespoir torturant. La peur peut, selon les dispositions, le caractère et le tempérament de l'individu, être vécue d'une manière différente, en particulier en ce qui concerne son intensité, dans sa forme extrême jusqu'à la peur de mort. Comme tout état affectif violent, la peur est accompagnée de symptômes physiques apparents, comme par exemple augmentation de la fréquence du pouls, dyspnée, transpiration (sudation) spontanée, tremblements ainsi qu'une suractivation intestinale et vésiculaire (= *défécation et miction de peur*). Selon TEMBROCK (1961), il serait utile, en éthologie, de distinguer entre peur et → *crainte.* En effet, selon lui, ce n'est pas la Souris qui fuit devant son prédateur qui a peur, mais celle qui est empêchée de prendre la fuite. LEYHAUSEN (1967) a également traité les problèmes de la peur d'un point de vue éthologique.

Angst-Aggression → Aggression
E. fear-induced aggression □ F. agression par peur

Anhassen → Hassen
E. to mob □ F. houspiller

anonymer Verband □ Eine Gruppe von Tieren, in der sich die einzelnen Mitglieder nicht persönlich kennen. Neben bloßen *Aggregationen,* bei denen zwischen den Tieren keinerlei Attraktion besteht, kennen wir vor allem zwei Arten anonymer Verbände.

Im *offenen anonymen Verband* kennen die Tiere einander nicht individuell, es gibt jedoch eine soziale Attraktion, und ein abgesprengtes Individuum wird immer wieder den Verband aufsuchen. Artgenossen und auch Artfremde (gemischter Verband) werden akzeptiert.

Im *geschlossenen anonymen Verband* erkennen die Tiere das Gruppenmitglied nicht individuell, wohl aber am Geruch (→ *Gruppenduft*). Gruppenfremde werden nicht akzeptiert.

anonymous group □ A group of animals in which the members do not recognize each other. Besides simple *aggregations* in which there is no attraction, we recognize two types of anonymous groups.

In the *open anonymous group* the animals do not know each other as individuals; however, a social attraction exists such that an individual which strays from the group will attempt to find his way back. All conspecifics as well as members of other species (mixed-species groups) are accepted.

In the *closed anonymous group* the animals do not recognize each other as individuals; however, they do distinguish individuals as members of the group by their smell (→ *group scent*). Non-members are rejected.

groupement anonyme □ Il s'agit de groupements d'animaux dans lesquels les membres ne se reconnaissent pas individuellement. En dehors des simples *agrégations* dans lesquelles il n'y a aucune inter-attraction entre les individus, nous connaissons deux types différents de groupements anonymes.

Dans le *groupement anonyme ouvert,* les individus ne se reconnaissent pas individuellement, mais il existe une interattraction sociale, et un individu isolé va toujours rechercher un groupement. Les congénères et les individus d'autres espèces sont acceptés et le groupement est donc souvent plurispécifique, comme chez de nombreux Poissons par exemple.

Dans les *groupements anonymes fermés,* les membres ne se connaissent pas non plus individuellement, mais se reconnaissent par leur → *odeur sociale.* Les individus étrangers au groupe, même congénères, ne sont pas acceptés.

Anpassung □ In der Biologie ganz allgemein die Einstellung des Organismus auf die jeweiligen Umweltbedingungen. Insbesondere in der Ökologie und Evolutionslehre versteht man darunter die natürliche Auslese (Selektion) im Verlaufe der stammesgeschichtlichen Entwicklung (phyletische Anpassung, → *Mimikry,* → *Schutzanpassung*).

In der Sinnesphysiologie bezeichnet man als Anpassung die Abnahme der Erregbarkeit eines Sinnesorgans als Folge fortgesetzter Reizung.

In der Ethologie im engeren Sinne die Änderung der Auslöseschwelle einer Reaktion. Die Eintrittsschwelle kann absinken, wenn z. B. eine Verhaltensweise über längere Zeit nicht ausgelöst wurde, und kann sich erhöhen, wenn der auslösende Reiz in rascher Folge wiederholt dargeboten wird. Es handelt sich hierbei um eine Schwellenwertänderung (→ *Reizschwelle*). Weitere Beispiele → *Feindanpassung,* → *Fluchtdistanz.*

adaptation □ The accomodation of the organism to particular environmental conditions. In ecology and evolutionary theory refers to natural selection in the course of evolution (phylogenetic adaptation, → *mimicry,* → *protective adaptation*).

In sensory physiology it refers to the reduction of excitability of a sense organ following continuous stimulation.

In the ethologically narrow sense refers to the change in the releasing threshold for a reaction. The primary threshold may drop, for example, if the behaviour has not occurred for a long while and conversely may rise if the releasing stimuli are presented in rapid succession (→ *stimulus threshold*). For further examples see → *antipredator adaptation,* → *flight distance.*

adaptation □ En biologie, d'une manière générale, accomodation de l'organisme aux conditions de son environnement (*Umwelt sensu* UEXKÜLL). En particulier en écologie et en évolution, on entend par cette notion la sélection naturelle au cours du développement phylogénétique (adaptation phylétique, → *mimétisme,* → *adaptation protectrice*).

En physiologie sensorielle, on comprend par adaptation une diminution de l'excitabilité d'un organe sensoriel à la suite d'une stimulation continue ou répétée.

En éthologie, au sens plus restrictif, changement du seuil déclencheur d'une réaction. Le seuil d'entrée peut diminuer lorsque, par exemple, un comportement donné n'a pas été déclenché pendant une longue période, et peut augmenter, par contre, lorsque le stimulus déclencheur est présenté en successions rapides. Il s'agit, dans ce contexte, d'une modification de la valeur du seuil (→ *seuil absolu de réponse*). Pour d'autres exemples, voir → *adaptation au prédateur,* → *distance de fuite.*

ansteckendes Verhalten → Stimmungsübertragung

E. contagious behaviour □ F. comportement contagieux

»Ansteckung« → Stimmungsübertragung

E. »contagion« □ F. »contagion«

Antrinken □ Eine epigame Verhaltensweise bei vielen Schwimmenten. Ursprünglich ist das Antrinken (Abb. 7 – *Aix galericulata*) eine Ausdrucksbewegung rein sozialer Bedeutung und in der Familie der Anatidae weit verbreitet. Als männliche Balzbewegung tritt es gekoppelt mit anderen epigamen Handlungen auf, oft zusammen mit dem → *Scheinputzen* (LORENZ 1941).

7

display drinking □ Epigamic behaviour pattern in many Anatid ducks. Originally this behaviour pattern (Fig. 7 – *Aix galericulata*) subserved simple social functions and occurred throughout the family Anatidae. A connection with male courtship behaviour has developed such that it now occurs in combination with other epigamic patterns, in particular with the → *»pseudo-preening«* (LORENZ 1941).

»Antrinken« □ Comportement épigame chez de nombreux Canards de la famille des Anatidae. Initialement, le Antrinken (Fig. 7 - *Aix galericulata*) est un mouvement expressif de signification purement sociale et très répandu à l'intérieur de la famille des Anatidae. Dans le comportement de parade nuptiale du ♂, ce mouvement apparaît couplé avec d'autres comportements épigames, souvent avec le → *»pseudo-nettoyage«* (LORENZ 1941).

Antwortfunktion □ Verlauf der Ausgangsgröße eines Signalüberträgers bei definierten Änderungen des Eingangssignals. Die Antwortfunktion hängt ab vom Übertragungsverhalten des betrachteten Systems.

output function □ Variation in the output level in a signal transmitter after a defined change in the input. The output function depends on the transmission characteristics of the system under consideration.

fonction de transfert □ Variation de la grandeur de sortie (*output*) d'un transmetteur de signaux en fonction de changements définis du signal d'entrée (*input*). La fonction de transfert dépend des propriétés de transmission du système en question.

Anwesenheitssignale → Duftmarkierung

E. olfactory markings □ F. signaux de présence

aposematisches Verhalten → Warnverhalten

E. aposematic behaviour □ F. comportement aposématique

Appetenz → Appetenzverhalten

E. appetence □ F. appétence

Appetenzverhalten □ In der Verhaltensphysiologie: spezifisches Suchverhalten nach einer auslösenden Reizsituation, das über den *Auslösemechanismus* (→ *AM*) zur erstrebten → *Endhandlung* führt; erster Ausdruck einer spezifischen inneren Handlungsbereitschaft oder Stimmung. Ihr liegt ein gewöhnlich als Drang oder Trieb bezeichneter Mechanismus zugrunde. Das Appetenzverhalten hört auf, wenn die Endhandlung beginnt. – In der Verhaltenspsychologie: für unmittelbar oder mittelbar auf Bedürfnisbefriedigung zielende Handlung (CRAIG 1918, MEYER-HOLZAPFEL 1940).

appetitive behaviour □ Specific seeking behaviour for a releasing stimulus situation which leads to the desired → *consummatory act* by means of the *releasing mechanism* (→ *RM*); first sign of a specific internal action readiness or state, based on a mechanism commonly described as a drive. Appetitive behaviour terminates with initiation of the consummatory act. – In behavioural psychology: actions carried out to directly or indirectly satisfy a need (CRAIG 1918, MEYER-HOLZAPFEL 1940).

comportement appétitif, comportement d'appétence □ En éthologie: comportement de recherche spécifique à l'égard d'une situation stimulatrice déclenchante, comportement qui, à l'aide d'un *mécanisme déclencheur* (→ *MD*), conduit à → *l'acte consommatoire* recherché. Il s'agit de l'expression première d'une disponibilité spécifique interne ou d'une motivation. Le comportement d'appétence est basé sur un mécanisme qu'on désigne en général par le terme de pulsion ou de tendance. Il débouche sur l'acte consommatoire. – En psychologie: action dirigée d'une manière directe ou indirecte vers une satisfaction de besoins (CRAIG 1918, MEYER-HOLZAPFEL 1940).

Arbeitsteilung □ Bei allen sozialen Insekten eine nach Kasten und in der Zeit bedingte Teilung der verschiedenen Aktivitäten, die auch Zusammenarbeit beinhalten kann. Abb. 8 zeigt ein Beispiel von Arbeitsteilung und Zusammenarbeit bei der Weberameise (*Oecophylla longinoda*): Während eine Gruppe von Arbeitern die Blattränder aneinanderzieht und hält, vernähen andere Arbeiter die Blattränder, indem sie die Spinndrüsen ihrer Larven gegen die Blattränder drücken und so Spinnfäden hin und her weben (LINDAUER 1952, EIBL-EIBESFELDT »Grundriß«).

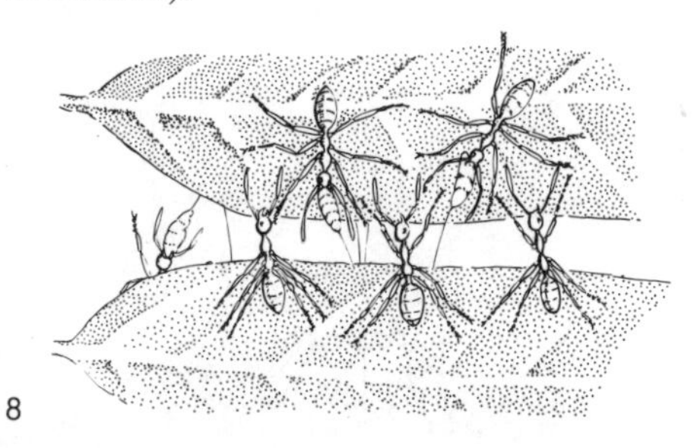

8

work division □ In all social insects division of the different activities among the castes and in the time which may include co-operation. Fig. 8 shows an example of *division of labour* in the Weaver Ant (*Oecophylla longinoda*): While a group of workers holds the leaf edges, another group cements the edges by pressing the silk gland of their larva against the leaf rim in such a manner that silk threads are woven back und forth (LINDAUER 1952, EIBL-EIBESFELDT »Ethology«).

division du travail □ Chez tous les Insectes sociaux, les activités sont réparties selon les différentes castes et dans le temps selon les différentes âges et peuvent également impliquer une coopération interindividuelle. Fig. 8 montre un exemple de division du travail et de coopération chez la Fourmi tisseuse (*Oecophylla longinoda*): Pendant qu'un groupe d'ouvrières tient les feuilles et rapproche leurs bords, un autre groupe d'ouvrières les relie en pressant les glandes tisseuses de leurs larves contre les bords des feuilles, tissant ainsi des fils d'un bord à l'autre (LINDAUER 1952, EIBL-EIBESFELDT »Ethologie«).

Artkumpan → Kumpan

E. conspecific companion □ F. compagnon conspécifique

Attrappe □ Im allgemeinen eine möglichst naturgetreue Nachbildung eines Gegenstandes, z. B. als Schaufensterauslage oder zur Täuschung des Gegners im Kriege oder zur Anlockung von Wild bei der Jagd. In der Ethologie bezeichnet man als Attrappe alle möglichen Nachbildungen, die zur Prüfung von Verhaltensreaktionen eines Tieres verwendet werden, um die für das betreffende Verhalten notwendigen → *Schlüsselreiz(e)* und → *Auslöser* zu ermitteln (siehe auch → *übernormaler Auslöser*). Solche Nachbildungen sind oft sehr substantiell und beschränken sich oft nur auf einen bestimmten Körperteil (Schnabel, Kopf, oder nur der Körper usw.). Selbst völlig unnatürliche Gegenstände, wie z. B. Kugeln und Würfel (als Eier), finden Verwendung. Um bei menschlichen Säuglingen in den ersten Monaten → *Lächeln* auszulösen, genügt eine einfache runde Scheibe mit zwei schwarzen Punkten als *Augenattrappe,* auch als *Punktattrappe* bezeichnet. Viele schöne Beispiele finden wir bei TINBERGEN »Instinktlehre«.

model □ In general a simulation of some natural object, e. g., maniquins in a show window, military camouflage, decoys for attracting wildlife. In ethology model refers to all objects used in behavioural tests for determining the → *key stimuli* and → *releaser(s)* of a particular behavioural pattern (see also → *supernormal releaser*). Such models are frequently quite abstract and may represent only certain body parts (beak, head, only the body, etc.). Even unnatural objects such as balls and cubes (egg imitations) have been used. For elicitation of → *smiling* in the infant during the first few months of his life a simple round disc with two black spots representing eyes (*eye spot model*) suffices. Many good examples of model experiments are found in TINBERGEN »The Study of Instinct«.

leurre □ En général, imitation la plus fidèle possible d'un objet, comme par exemple les articles factices des vitrines ou ceux destinés à tromper l'ennemi pendant la guerre ou encore à l'attraction du gibier pendant la chasse. En éthologie, on appelle leurre toutes sortes d'imitations permettant de tester les réactions comportementales d'un animal, en particulier pour élucider les → *stimuli-clé* ou → *déclencheurs* d'un comportement déterminé (voir aussi → *déclencheur supranormal*). De telles imitations sont souvent très schématisées et se résument, dans la plupart des cas, en une partie du corps seulement (bec, tête ou corps seul). On utilise souvent des objets tout à fait artificiels, comme par exemple des billes ou des cubes pour représenter des œufs. Pour déclencher le → *sourire* du nourrisson humain dans les premiers mois de sa vie, il suffit de lui présenter un simple disque rond avec deux points noirs représentant les yeux, objet également appelé *leurre à taches oculaires*. De nombreux exemples didactiques au sujet des expériences à leurre sont donnés par TINBERGEN »Etude de l'Instinct«.

Aufrafftrieb → Verlegen

E. gathering drive □ F. pulsion à ramasser

Aufreißen □ Paarungsnachspiel des Erpels bei einigen Schwimmenten. Unmittelbar nach dem Treten reißt der Erpel plötzlich Kopf und Hals, ohne sich hochzurecken, weit auf den Rücken (Abb. 9 A – *Virago castanea* ♂, B – *Nettion flavirostre* ♂). Dann, als wäre diese Rückwärtsbewegung nur das Ausholen zu einem übertriebenen Kopfnicken gewesen, schießt der Erpel mit der typischen Bewegungsweise des → *Nickschwimmens* davon und im Kreis um das ♀ herum. Wir finden dieses Paarungsnachspiel bei allen *Anas*-Arten (LORENZ 1941).

9 A 9 B

bridling □ Post-copulatory behaviour of the male duck in some species of ducks. Immediately after copulation the ♂ suddenly pulls back his head and neck, without stretching it upward, far onto his back (Fig. 9 A – *Virago castanea* ♂, B – *Nettion flavirostre* ♂). Then the ♂ dashes forward with → *nod-swimming* and circles the ♀, giving the appearance as if the pulling back of the head was merely an exaggerated nodding movement. This behaviour is found in all *Anas* species (LORENZ 1941).

rengorgement □ Comportement postcopulatoire du ♂ chez un certain nombre de Canards Anatidae. Juste après l'accouplement, le ♂ se rengorge sans s'étirer (Fig. 9 A – *Virago castanea* ♂, B – *Nettion flavirostre* ♂). Immédiatement après, comme si ce mouvement en arrière n'avait été qu'un mouvement intentionnel, il se précipite en avant et contourne à la nage la ♀ en hochant fortement la tête (→ *nage de coquetterie*). Ce jeu postcopulatoire se trouve chez toutes les espèces du genre *Anas* (LORENZ 1941).

Aufreitdrohung □ Bei vielen Säugern ist das männliche Aufreiten eine → *Rangdemonstration*, die gelegentlich auch von ♀♀ rangniederen Artgenossen gegenüber gezeigt wird. Bei Raufereien Jugendlicher beobachtet man Aufreiten von hinten, Umklammern und gelegentlich kräftige Lendenstöße. Auch beim Menschen ist Aufreiten eine Dominanz- und Drohgeste. → *Wutkopulation*.

mounting threat □ In many mammals mounting or riding up serves as a demonstration of higher rank and is also occasionally seen in the ♀ directed against other lower ranking ♀♀. During brawls, youths often show mounting from the rear, clutching and occasionally kicking at the groin. Hence in humans mounting may also act as a dominance or threat gesture. → *rage copulation*.

chevauchement de menace □ Chez de nombreux Mammifères, le chevauchement du ♂ est une démonstration de dominance qui peut se manifester également chez les ♀♀ envers des congénères d'un rang social inférieur. Au cours des batailles chez les enfants, on observe aussi un chevauchement par derrière, accompagné d'un accrochage par les bras et parfois de quelques coups de reins. Chez l'Homme, le chevauchement constitue donc également un geste de dominance et de menace. → *copulation de colère*.

Aufreitversuch □ Isoliert aufgezogene männliche Rhesus-Affen geraten durch ♀♀ im Östrus in Erregung und versuchen zu kopulieren, doch steigen sie nicht in der richtigen Weise auf, es bleibt beim Aufreitversuch (Abb. 10). Sie lernen es auch später nicht mehr.

mounting attempt □ Male Rhesus monkeys that are raised in isolation are excited by ♀♀ in oestrus and attempt to copulate. However, they do not mount in the proper fashion and copulation does not ensue (Fig. 10). Furthermore, they subsequently fail to learn the behaviour.

10

tentative de monte □ Les ♂♂ de *Macaca rhesus* élevés dans l'isolement sont excités par la présence de ♀♀ en œstrus et essaient de s'accoupler, mais la monte ne s'effectue pas correctement et ne dépasse pas le stade préliminaire (Fig. 10). Ils sont d'ailleurs incapables d'apprendre par la suite de monter correctement des ♀♀.

Aufspießen □ Würger der Familie Laniidae speichern gefangene Beutetiere auf, um bei Bedarf darauf zurückgreifen zu können. Zur Anlegung solcher Vorräte dienen vor allem Dornen und Astgabeln, und die Beute wird aufgespießt (Abb. 11 – *Lanius collurio)* oder eingeklemmt (WATSON 1910). Das Aufspießen ist eine angeborene Verhaltensweise. Die Orientierung der Spießbewegung braucht nicht gelernt zu werden, doch ist Lernen und Erfahrung zur Vergrößerung der Zielgenauigkeit notwendig. Nach LORENZ und ST. PAUL (1968) handelt es sich beim Aufspießen um eine »Verschränkung« einer angeborenen Bewegungskoordination mit einer erlernten Orientierung nach einer Spieß- und Klemmgelegenheit. Das Einklemmen in Astgabeln, vor allem größerer Beute, wie z. B. Mäuse und kleine Vögel, dient auch dem Zweck, diese besser mit dem Schnabel zerreißen zu können (BROSSET mündl. nach Beobachtungen an afrikanischen Arten).

11

impaling □ Shrikes of the family Laniidae store captured prey for later consumption. They utilize thorns to impale the prey and to wedge it into forks of tree branches (WATSON 1910). Impaling is an innate behaviour pattern (Fig. 11 – *Lanius collurio*). The orientation of the impaling movement does not have to be learned. However, improvement of the aim is a function of learning and experience. According to LORENZ and ST. PAUL (1968) impaling is an »interlocking« of an inborn behaviour pattern with a learned orientation towards an object suitable for impaling or wedging. The wedging into forks of tree branches of relatively big preys such as mice or small birds also helps the shrike to tear them up with its bill (BROSSET *pers. comm.,* observations on some African species).

embrochage □ Les Pies-Grièches de la famille des Laniidae stockent leurs proies pour avoir des réserves en cas de nécessité. Ces proies sont embrochées sur des épines d'arbres (Fig. 11 – *Lanius collurio*) ou coincées dans les bifurcations des branches (WATSON 1910). L'embrochage est un comportement inné, mais l'apprentissage et l'expérience sont nécessaires pour augmenter la précision du mouvement. Selon LORENZ et ST. PAUL (1968), l'embrochage résulterait de l'interaction d'une coordination héréditaire et d'une orientation acquise en vue d'un embrochage. Le coinçage dans des bifurcations de branches, surtout des proies relativement grandes telles que Souris et petits Oiseaux, sert aussi à pouvoir mieux les déchirer avec le bec (BROSSET *comm. pers.* selon ses observations chez quelques espèces africaines).

Aufsteilen □ Ein fast senkrechtes Strecken des Kopfes (mit Nase nach oben) und Halses, das dem Kopf-hoch-Recken (mit Nase nach vorn) nahesteht und bei verschiedenen Horntieren, Hirschen und Giraffen, auch Okapi, als Droh- oder Imponiergeste vorkommt. Bei einigen Arten tritt es auch in den Paarungszeremoniellen als Werbehaltung der ♂♂ auf; z. B. bei *Antilope cervicapra* (Abb. 12) ist es die wichtigste Imponier- und Werbegeste des ♂, das der Geiß seine weiße Kehlregion weist (WALTHER 1966, DUBOST *mündl.).*

upward stretch, »nose-up« posture □ An almost vertical stretching of head (nose upward) and neck which is related to the erect posture (nose forward), and is found as a threat or dominance display in several bovids, cervids, and giraffids. In certain species, it also occurs in the mating rituals as a male courtship display. For example, in *Antilope cervicapra* (Fig. 12), the ♂ stretching head and neck upward, presents his white throat in the same manner against a male rival in an agonistic encounter as toward a ♀ in courtship (WALTHER 1966, DUBOST *pers. comm.*).

»Aufsteilen« □ Etirement presque vertical du cou et de la tête qui existe chez certains Bovidae, Cervidae et Giraffidae, y compris l'Okapi, et qui correspond à un comportement de menace ou de parade, parfois présent lors du comportement précopulatoire. Chez *Antilope cervicapra,* par exemple (Fig. 12), c'est un des gestes principaux de la parade nuptiale du ♂ qui montre ainsi à sa ♀ la région blanche de sa gorge (WALTHER 1966, DUBOST *comm. pers.).*

12

Aufstoßen □ Ein epigames Verhaltenselement bei Enten, welches nur bei den Erpeln vorkommt (Abb. 13 A – *Anas acuta,* B – *Aix galericulata*) und gelegentlich, aber nicht immer, auf das vorangegangene → *Kurzhochwerden* folgt. Beim Mandarinenerpel ist es besonders auffallend (LORENZ 1941).

13 A 13 B

»burping« □ An epigamic display in the duck which only occurs in the drake (Fig. 13 A – *Anas acuta,* B – *Aix galericulata*) and which frequently, but not always follows the → *»head-up-tail-up«*. This display is particularly striking in the Mandarin Duck (LORENZ 1941).

»Aufstoßen« □ Comportement épigame chez les Canards qu'on observe uniquement chez les ♂♂ (Fig. 13 A – *Anas acuta,* B – *Aix galericulata*) et qui suit parfois, mais pas toujours le → »*Kurzhochwerden*« (tête-haute-queue-haute). Ce comportement est particulièrement apparent chez le ♂ d'*Aix galericulata* (LORENZ 1941).

Aufzucht unter Erfahrungsentzug → Kaspar-Hauser-Versuch
E. rearing in isolation □ F. élevage en isolement

Augenflecke □ Bei vielen Nachtfaltern und einigen Homopteren befinden sich auf den in Ruhelage verdeckten Hinterflügeln Augenflecke, die stark an Säugeraugen erinnern; oft haben sie sogar einen Augenspiegel. Bei Störung oder Gefahr werden die Flügel schnell gespreizt, und die Augenflecke werden sichtbar (Abb. 14 – *Automeris acutissima* ♂, Saturniidae). Wie Versuche ergeben haben, werden Freßfeinde damit abgeschreckt und vom Angriff abgehalten. Ähnliches gilt für Tagfalter wie z. B. *Vanessa jo,* der allerdings, wenn er die Flügel aufklappt, vier Augenflecke zeigt, vor denen Vögel nachweislich zurückschrecken.

eye spots □ In many nocturnal Lepidopterans and some Homopterans »eye spots« are present on the rear wings. These spots which are normally covered when the animal is resting, strongly resemble mammalian eyes. If disturbed, the animal quickly spreads his wings and the eye spots become visible (Fig. 14 – *Automeris acutissima* ♂, Saturniidae). Experiments have shown that potential predators are frightened by the eye spots and are less likely to attack. In this manner the butterfly, *Vanessa jo,* which has four eye spots on its extended wings is able to frighten away birds.

taches ocellaires □ Chez de nombreux Lépidoptères nocturnes et quelques Homoptères, on trouve sur les ailes postérieures, cachées en position de repos, des taches ocellaires qui ressemblent beaucoup aux yeux des Mammifères; parfois, ces taches ocellaires possèdent même un spéculum. S'ils sont dérangés ou en danger, ces Insectes écartent alors spontanément leurs ailes et les taches ocellaires apparaissent (Fig. 14 – *Automeris acutissima* ♂, Saturniidae). Les expériences effectuées ont démontré que les prédateurs potentiels sont ainsi effrayés et empêchés de capturer l'animal. Le même phénomène se retrouve chez quelques Lépidoptères diurnes, comme par exemple *Vanessa jo;* mais lorsque ce dernier ouvre ses ailes, il montre quatre taches ocellaires devant lesquelles, comme il a été prouvé, les Oiseaux prédateurs reculent.

14

Augengruß □ Im menschlichen Grußverhalten optische Kontaktaufnahme mit ruckartigem Anheben der Augenbrauen, meist gefolgt von Lächeln (Abb. 15 – Woitapmin aus Neuguinea). Das Anheben der Augenbrauen hat aber auch noch viele andere Bedeutungen. → *Brauenheben.*

eye-brow flash □ In human greeting gestures – the

15

raising of the eye-brows followed often by a smile (Fig. 15 – Woitapmin from New Guinea). A quick raising and lowering of the eyebrow may have many other meanings as well. → *eyebrow-raising.*

salut des sourcils □ Chez l'Homme, lors du salut, prise de contact visuel, accompagnée d'un haussement de sourcils et fréquemment d'un sourire (Fig. 15 – Woitapmin de la Nouvelle-Guinée). Le haussement des sourcils peut avoir bien d'autres significations. → *haussement des sourcils.*

Augentarnung □ Bei sehr vielen Korallenfischen sind die Augen durch eine schwarze Augenbinde getarnt (Abb. 16 A – *Chaetodon auriga*). Überdies entwickelten einige einen nachweislich Angriffe ablenkenden Augenfleck am Körperende (Abb. 16 B – *Forcipiger flavissimus*).

16 A

16 B

eye-camouflage □ The eyes of several species of coral fishes are obscured by a black eye bar (Fig. 16 A – *Chaetodon auriga*). Moreover, among these species some have developed an eye spot on or near the tail which appears to distract the attacker (Fig. 16 B – *Forcipiger flavissimus*).

camouflage des yeux □ De nombreux Poissons coralliens ont leurs yeux camouflés par une bande noire (Fig. 16 A – *Chaetodon auriga*). Par ailleurs, quelques espèces ont développé sur le corps une tache ocellaire qui détourne les attaques des yeux véritables (Fig. 16 B – *Forcipiger flavissimus*).

Ausdrucksverhalten □ Für einen Partner derselben oder einer anderen Art bestimmte Ausdrucksbewegung bei Mensch und Tier; besonders stark entwickelt zur innerartlichen und sozialen Kontaktpflege; oft schematisiert und ritualisiert, z. B. → *Demutsverhalten,* → *Drohmimik,* → *Lächeln,* → *Lachen,* → *Zähnefletschen* usw.

expressive behaviour □ Expressive movement directed at a partner of the same or a different species; observed in animals and man; especially well developed as a means of initiation and maintaining intraspecific social contact; often schematized and ritualized, e. g. → *submissive behaviour,* → *threat face,* → *smiling,* → *laughing,* → *bared teeth display,* etc.

comportement expressif □ Mouvement expressif destiné à un congénère ou un partenaire d'une autre espèce chez les animaux et l'Homme. Il est particulièrement bien développé pour le contact intraspécifique et social et est souvent schématisé ou ritualisé; par exemple → *comportement de soumission,* → *mimique de menace,* → *sourire,* → *rire,* → *montrer les dents* etc.

Auslösemechanismus → AM

Auslöser □ In der Verhaltensforschung Körpermerkmale und Verhaltensweisen, die ihre raum-zeitliche Gestalt der Anpassung an die Übermittlung von Informationen verdanken. Ihre Aufgabe besteht darin, beim Empfänger eine bestimmte Reaktion auszulösen. Zu den Auslösern gehören Farb- und Formmerkmale, Lautäußerungen, Düfte sowie verschiedene Bewegungsweisen, z. B. beim Balz- und Drohverhalten. Für eine optimale Informationsübertragung ist Auffälligkeit und Eindeutigkeit wichtig. Auslöser beim Menschen sind u. a. Uniformen, Rangabzeichen und aus dem Bereich des Verhaltens insbesondere die → *Universalien* Lächeln und Weinen. Die meisten Auslöser kommen im innerartlichen Bereich vor, doch gibt es auch Auslösebeziehungen zwischen verschiedenen Arten, wie z. B. bei der Symbiose, die eine gegenseitige Verständigung erforderlich macht; man spricht dann von interspezifischen Auslösern (Lorenz 1965, 1966, Burkhardt *et al.* 1966, Koenig 1970).

releaser □ In ethology, body markings and postures or behaviour patterns which owe their temporal-spatial gestalt to an adaptation through natural selection for the transmission of information. Their function is to release in the receiver a specific reaction. Releasers may be colour or form cues, vocalizations, odours as well as movement patterns, e. g. during courtship and fighting behaviour. For optimal transmission of information it is important that these signals be conspicuous and unambiguous. Releasers in humans are uniforms, signs of rank, ranging all the way to universal behaviours (→ *universals*) such as smiling and crying. Most releasers are found in intra-specific communication, however, they are also found between species, e. g. in symbiotic relationships which necessitate mutual understanding of intentions. The latter are called inter-specific releasers (Lorenz 1965, 1966, Burkhardt *et al.* 1966, Koenig 1970).

déclencheur □ En Ethologie, caractères morphologiques et modes de comportement qui, dans leur développement dans l'espace et dans le temps, ont acquis une adaptation à transmettre des informations. Leur fonction est de déclencher une réaction déterminée chez le récepteur. Les couleurs, les différents mouvements et modes de locomotion peuvent être des déclencheurs, comme par exemple pendant le comportement de parade nuptiale et le comportement de menace. Pour une transmission d'information optimale, il est important que ces déclencheurs soient apparents et sans équivoque. Chez l'Homme, nous connaissons comme déclencheurs culturels les uniformes et les insignes de rang et dans le domaine du comportement en particulier, les → *universaux* tels que sourire, pleurs etc. La plupart des déclencheurs sont intraspécifiques, mais il existe aussi des relations de déclenchement entre différentes espèces, comme par exemple dans le domaine de la symbiose où une compréhension réciproque est nécessaire. Dans ces cas précis, nous parlons de déclencheurs interspécifiques (LORENZ 1965, 1966, BURKHARDT *et al.* 1966, KOENIG 1970).

Ausmulden □ Scharrbewegungen, vor allem bei am Boden brütenden Vögeln, die meist nur eine Mulde im Sand scharren und dort die Eier hineinlegen (Abb. 17 – *Charadrius alexandrinus*). Ausmuldebewegungen beobachtet man aber auch bei anderen Arten während des Nestbaues.

17

hollowing out of nest □ Occurs mainly in ground breeding birds that dig out a depression in the substrate in which to lay the eggs (Fig. 17 – *Charadrius alexandrinus*). The motor patterns involved are also frequently found in other species during nest building.

excavation □ Mouvements effectués en particulier par les Oiseaux nichant au sol pour creuser une cuvette dans laquelle ils déposent leurs oeufs (Fig. 17 – *Charadrius alexandrinus*). Des mouvements semblables s'observent aussi chez d'autres espèces pendant la construction du nid.

Austragungsmodus → Kommentkampf
E. method of execution □ F. mode d'exécution

autochthone Handlung □ Benutzt für eine durch triebeigene Erregung gespeiste Handlung (KORTLANDT 1940). → *Appetenzverhalten,* → *Triebhandlung.*

autochthonous behaviour pattern □ Applied to a behaviour motivated by a drive with which the behaviour is normally observed (KORTLANDT 1940). → *appetitive behaviour,* → *instinctive behaviour pattern.*

comportement autochtone □ Comportement provoqué par une activation intrinsèque par rapport à la pulsion propre (KORTLANDT 1940). → *comportement appétitif,* → *activité instinctive.*

Automimikry □ Nachahmung eines Geschlechtes oder Lebensstadiums innerhalb ein und derselben Art im Dienste der Verständigung mit dem anderen Geschlecht oder einem anderen Lebensstadium. Ein gutes Beispiel ist die Nachahmung der weiblichen Geschlechtssignale bei einigen Primaten-Arten durch das ♂, welche vorwiegend in Beschwichtigungsgebärden auftreten. → *Bedeutungserweiterung,* → *Mimikry.*

automimicry □ The imitation by one sex or life stage for communication with another sex or life stage of the same species. An example is the imitation by the male of some monkey species of the female sexual signals, which they appear to employ in appeasement rituals. → *expansion in meaning,* → *mimicry.*

automimétisme □ Imitation d'un sexe ou d'un stade ontogénétique par l'autre sexe ou un autre stade de développement à l'intérieur d'une même espèce en vue d'une intercommunication. Un bon exemple d'imitation est connu des ♂♂ d'un certain nombre d'espèces de Primates qui imitent les signaux sexuels des ♀♀ pour les utiliser par la suite lors des gestes d'apaisement. → *changement de signification,* → *mimétisme.*

Autotomie → Schutzanpassung
E. autotomy □ F. autotomie

B

Badeverhalten → Komfortverhalten
E. bathing behaviour □ F. comportement de bain

Balz □ Paarungszeit der Vögel und Fische, im erweiterten Sinne auch der Reptilien und Amphibien. Als gleichsinnige Bezeichnung für Säugetiere → *Brunst.* Die Balzzeit ist hormonell bedingt; in gemäßigten Breiten bewirkt die Zunahme der Tageslänge einen Mechanismus im Hypothalamus und in der Hypophyse, der zu verstärktem Gonadenwachstum führt (vermehrte Produktion von Testosteron bzw. Östrogen). → *Gruppenbalz,* → *Imponierverhalten,* → *Werbeverhalten.*

courtship □ The German term *Balz* applies to the courtship period and pair formation only in birds and fish, and in a larger sense also to reptiles and amphibians. For mammals the German term is → *Brunst* or *Brunft*. The courtship period is stimulated by hormones; in the temperate zones the prolongation of daylight releases a mechanism in the hypothalamus and the hypophysis which leads to the growth of the gonads (increased production of testosteron or oestrogene). → *communal courtship*, → *display behaviour*, → *courtship behaviour*.

parade nuptiale □ Le terme allemand *Balz* se réfère à la période d'appariement et d'accouplement des Oiseaux (pariade) et des Poissons, et dans un sens plus large aux Reptiles et aux Amphibiens. La période de parade nuptiale nécessite une activation hormonale: l'augmentation de la photopériode déclenche au niveau de l'hypothalamus et de l'hypophyse un mécanisme qui induit la croissance des gonades (augmentation de la production de testostérone ou d'œstrogène). Pour les Mammifères, on parle de → *Brunst* ou *Brunft*. → *parade nuptiale collective*, → *comportement de parade*, → *comportement de parade nuptiale*.

Balzfarben □ Farben und Färbungsmuster, die nur im Zusammenhang mit → *Balz, Balzzeit* und dem entsprechenden → *Werbeverhalten* auftreten.

courtship colours □ Colours and colour patterns occurring solely in connection with courtship, the mating season, and behaviour related to → *courtship*. → *courtship behaviour*.

couleurs de parade nuptiale □ Couleurs et dessins qui apparaissent exclusivement en relation avec la → *parade nuptiale*, la période et le comportement correspondant. → *comportement de parade nuptiale*.

Balzflug → Schauflug
E. courtship flight □ F. vol de parade nuptiale

Balzfüttern □ Eine epigame Verhaltensweise, die sich historisch aus dem → *Jungefüttern* ableitet und zur Beschwichtigung des umworbenen Partners dient. Bei Seeschwalben z. B. überreicht das werbende ♂ dem ♀ einen Fisch. WICKLER (1969) postuliert, daß man vielleicht besser → *Begrüßungsfüttern* sagen sollte, da bei verschiedenen Vögeln die Futterübergabe während oder gar erst nach der Kopula erfolgt, – und dann kann man nicht mehr von Balz sprechen.

courtship feeding □ An epigamic behaviour pattern derived from → *feeding of young* which serves to appease the courted partner. For example, the male tern may present the ♀ with a fish. WICKLER (1969) suggests that it is perhaps better to call it → *greeting feeding* since various birds present the food only during or after copulation, and hence one can no longer speak of courtship.

nourrissage de parade nuptiale □ Comportement épigame historiquement dérivé du → *nourrissage des jeunes* et qui sert à l'apaisement du partenaire sexuel. Chez les Sternes, par exemple, le ♂ présente un Poisson à la ♀ au cours de la parade nuptiale. WICKLER (1969) suggère qu'on devrait appeler ce comportement plutôt → *nourrissage de salut* étant donné que chez certains Oiseaux la nourriture destinée au partenaire n'est parfois donnée que pendant l'accouplement ou même après, et dans ces cas, on ne peut plus parler de parade nuptiale.

Balzgesang → Werbeverhalten
E. courtship song □ F. chant de parade nuptiale

Balzhandlung → Werbeverhalten
E. courtship activity □ F. activité de parade nuptiale

Balzkette □ Eine Handlungskette oder Reaktionskette im Bereich der Balz und des Fortpflanzungsverhaltens. Es handelt sich um eine programmierte Handlungsfolge, bei welcher die verschiedenen Verhaltensabläufe aus einer festgelegten Kette aufeinanderfolgender Einzelhandlungen bestehen, die stets in weitgehend gleicher Weise ablaufen. Eine der vollständigsten Analysen solch einer Handlungskette kennen wir vom Stichling *Gasterosteus aculeatus* (TER PELKWIJK und TINBERGEN 1939, TINBERGEN 1942). Jede Reaktion des ♂ löst die nächstfolgende des ♀ aus (Abb. 18). Das Erscheinen des dickbäuchigen ♀ löst den Zickzacktanz des ♂ aus (2), worauf das ♀ seinen dicken silbrigen Bauch zeigt (3) und auf das ♂ zuschwimmt. Darauf macht das ♂ kehrt und schwimmt zum Nest (4), das ♀ folgt ihm (5), dann zeigt das ♂ den Nesteingang (6), welches wiederum das Einschlüpfen ins Nest auslöst (7). Der → *Schnauzentriller* des ♂ an der Schwanzwurzel des ♀ (8) löst schließlich die Laichabgabe aus. Die meisten Glieder dieser Handlungskette hängen von optischen Signalreizen (→ *Schlüsselreiz*) ab; die Eiablage erfordert mechanische Reize, die Besamung schließlich chemische und wahrscheinlich zugleich taktile. → *Reaktionskette*, → *Werbeverhalten*.

courtship chain □ A reaction chain in courtship

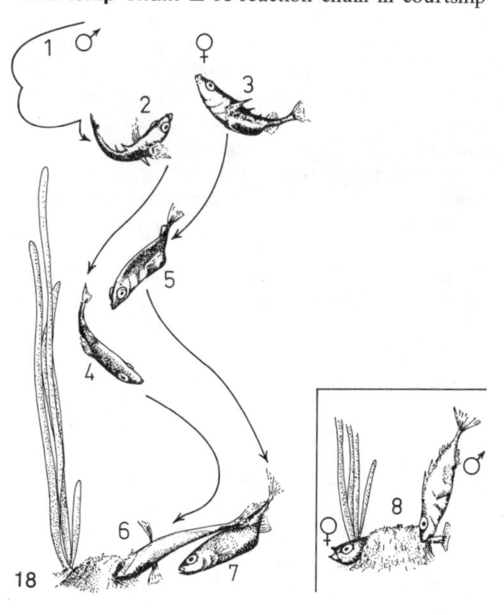

and reproductive behaviour. A sequence of pre-programmed activities in which the different behavioural elements always follow one another in the same manner. One of the best studies of such a reaction chain concerns the courtship behaviour of the three-spined stickleback, *Gasterosteus aculeatus* (TER PELKWIJK and TINBERGEN 1939, TINBERGEN 1942). Each behaviour pattern of the ♂ releases the following reaction of the ♀ (Fig. 18). The appearance of the ♀ with her swollen belly releases the zig-zag-dance of the ♂ (2), then the ♀ shows her swollen silvery belly (3) and swims towards the ♂. The ♂ turns away and swims to the nest (4), the ♀ follows him (5), the ♂ shows the nest entrance (6) which induces the ♀ to glide into the nest (7). The → *quivering* of the ♂ at the base of the ♀ tail (8) elicits spawning by the ♀. Most of the behavioural components of this courtship chain depend on visual sign stimuli *(→ key-stimulus);* the spawning requires mechanical stimulation, and the fertilization by the ♂ requires chemical and probably also tactile stimulation. → *reaction chain,* → *courtship behaviour.*

»Balzkette« □ Enchaînement des éléments comportementaux lors de la parade nuptiale. Il s'agit d'une suite de mouvements programmés dont les différents éléments comportementaux se suivent l'un après l'autre et toujours de la même manière (réaction en chaîne). Un des meilleurs exemples est la parade nuptiale de l'Epinoche, *Gasterosteus aculeatus* (TER PELKWIJK et TINBERGEN 1939, TINBERGEN 1942). Chaque réaction du ♂ déclenche la suivante de la ♀ (Fig. 18). L'apparition de la ♀ au ventre enflé déclenche chez le ♂ la danse zigzagante (2), ensuite la ♀ montre son ventre enflé et argenté (3) et se rapproche du ♂. Le ♂, de son côté, fait demi-tour et se dirige vers le nid (4) et la ♀ le suit (5). Le ♂ montre à la ♀ l'entrée du nid (6) ce qui incite la ♀ d'entrer dans le nid (7). Le → *frémissement du museau* du ♂ effectué à la base de la queue de la ♀ (8) déclenche alors la ponte. La plupart des composantes de cette parade nuptiale dépendent de stimuli-signaux optiques *(→ stimulus-clé);* la ponte exige des stimulations mécaniques et la fertilisation enfin nécessite des stimulations chimiques et probablement tactiles. → *réactions en chaîne,* → *comportement de parade nuptiale.*

Balzverhalten → Werbeverhalten

Band zwischen Partnern □ Verhaltensweisen eines objektiv feststellbaren Zusammenhaltes (persönliche Bindung) zweier Paar-Partner. → *Partnerbindung.*

E.□ Behaviour patterns used for objective determination of cohesion in the interaction between two individuals of a pair. → *partner bonding.*

F. □ Comportement indiquant d'une manière objective une liaison et des rapports précis entre les deux partenaires d'un couple. → *lien entre partenaires.*

Bedeutungserweiterung □ Die sexuellen Präsentiergebärden weiblicher Mantelpaviane, *Papio hamadryas,* haben zugleich beschwichtigende Wirkung und werden mit dieser Funktion auch von den ♂♂ benutzt, die in weiterer Angleichung an die ♀♀ auch deren rote Schwellkörper um die Ano-Genitalregion nachahmten. Für weitere interessante Fälle von Bedeutungserweiterung siehe WICKLER (1965 a + b) und EIBL-EIBESFELDT »Grundriß«. → *Automimikry.*

expansion in meaning, change in signal function □ The sexual presentation gestures of the female hamadryas baboon, *Papio hamadryas,* have an appeasing effect, and these gestures with the swollen red anogenital region are imitated by the males for the same function. For further interesting examples of change in meaning see WICKLER (1965 a + b) and EIBL-EIBESFELDT »Ethology«. → *automimicry.*

changement de signification □ Les gestes de présentation sexuelle des femelles Babouins, *Papio hamadryas,* ont en même temps un effet apaisant et sont utilisés, pour cette fonction, par les ♂♂ qui, par une ressemblance encore plus poussée, imitent également les callosités rouges et enflées de la région anogénitale des ♀♀. Pour d'autres exemples de changement de signification, voir WICKLER (1965 a + b) et EIBL-EIBESFELDT »Ethologie«. → *automimétisme.*

Bedeutungsträger □ Nach VON UEXKÜLL (1921, 1937) haben nur solche Objekte Merkmale, die für das Leben eines Tieres von Bedeutung sind. Sie werden dadurch zu Bedeutungsträgern des Subjektes.

cue bearer □ According to VON UEXKÜLL (1921, 1937) only those aspects of the environment serve as cues which are of significance in the life of the animal; such aspects are therefore called cue bearers.

objets significatifs □ Selon VON UEXKÜLL (1921, 1937), seuls les objets qui ont une importance pour la vie d'un animal constituent pour lui des stimuli-signaux. Ils sont ainsi devenus des *porteurs de signification.*

Bedeutungswechsel → Bedeutungserweiterung

bedingter Reflex → Reflex

E. conditioned reflex □ F. réflexe conditionné

Bedürfnisbefriedigung → Appetenzverhalten

E. need satisfaction □ F. satisfaction de besoins

Befriedungsverhalten → Beschwichtigungsverhalten

Begattung → Kopulation

E. mating □ F. accouplement

Begleitfische → Lotsenfische

Begrüßungsfüttern □ Bei Vögeln die Futterübergabe zwischen zwei Paar-Partnern zur Begrüßung (Abb. 19 A – *Corvus corax,* B – *Sterna hirundo*). Man sprach bisher von → *Balzfüttern.* Stattdessen sollte man aber eher von Begrüßungsfüttern sprechen, da Balz bzw. Kopulationseinleitung nicht immer obligatorisch sind oder die Futterübergabe erst während bzw. nach der Begattung erfolgt (WICKLER 1969). Bei vielen Arten wird gar nicht mehr gefüttert, sondern die Begrüßung beschränkt sich aufs → *Schnäbeln.*

»greeting feeding« □ In birds the feeding ritual between a mated pair upon greeting (Fig. 19 A – *Corvus corax,* B – *Sterna hirundo*). One previously referred to this as → *courtship feeding,* however, »greeting feeding« is more precise since courtship or copulation induction are either not always necessary or the ritual may occur subsequent to mating (WICKLER 1969). In many species, actual feeding no longer occurs; rather, the greeting occurs as → *billing.*

nourrissage de salut □ Chez les Oiseaux, scène de nourrissage entre les deux partenaires d'un couple qui se saluent (Fig. 19 A – *Corvus corax,* B – *Sterna hirundo*). Ce comportement a été appelé jusqu'à présent → *nourrissage de parade nuptiale.* Il est cependant plus logique de l'appeler nourrissage de salut, étant donné que la parade nuptiale et les préliminaires à l'accouplement ne sont pas obligatoirement liés à ce comportement et d'autre part que la nourriture destinée au partenaire n'est parfois donnée que pendant l'accouplement ou même après (WICKLER 1969). Chez de nombreuses espèces, il n'y a plus → *offrande de nourriture* et le salut se limite au → *becquetage.*

19 A

19 B

Begrüßungsklappern □ Begrüßungszeremonie des Storches, wobei die Waffe, der spitze Schnabel, weggewendet wird; der Kopf wird dabei klappernd zurückgeworfen (Abb. 20 – *Ciconia ciconia*). Auch als Klapperzeremonie bezeichnet.

stork greeting ceremony □ Greeting display in which the stork turns the pointed »weapon-like« beak away by throwing the head backwards in a flapping manner (Fig. 20 – *Ciconia ciconia*).

craquètement de salut □ Cérémonie de salut et de retrouvailles chez les Cigognes pendant laquelle le bec pointu qui pourrait être considéré comme une arme est détourné du partenaire (Fig. 20 – *Ciconia ciconia*). Pendant le craquètement, la tête est rejetée en arrière.

20

Begrüßungslächeln → Lächeln
E. greeting smile □ F. sourire de salut

Begrüßungsriten → Grußformeln
E. greeting rituals □ F. rituels de salutation

Beharrungstendenz □ Es gibt neben Anpassungsphänomenen auch ausgesprochene Beharrungstendenzen, d. h. einmal Erreichtes wird beibehalten, und dabei kommt es oft vor, daß neue Spezialisierungen alte Differenzierungen überlagern.

persistence determination □ Besides adaptive changes we can observe at times a marked persistence in the maintenance of a particular characteristic, i. e. a characteristic once achieved is maintained, and thus it often happens that new specializations superimpose previous differentiations.

persistance phylogénique □ Dans l'évolution des traits éthologiques, il se manifeste, en plus des phénomènes d'adaptation, une tendence à la persistance, c'est-à-dire que des structures une fois développées sont maintenues, et il se peut alors que de nouvelles spécialisations se trouvent »recouvertes« par des différentiations antérieures.

Beinstellen □ Eine bei Kindern häufige Form der Herausforderung, welche darin besteht, daß man einen Vorbeilaufenden über das schnell vorgeschobene Bein stolpern läßt. Eine Tätlichkeitshandlung, wohl bei allen Menschen.

tripping □ A kind of provocation often observed in children. The subject suddenly holds out his leg, deliberately causing another to stumble over it. An act of violence, probably general in all humans.

croc-en-jambe, croche – pied □ Une provocation commune chez les enfants qui consiste à faire trébucher un passant sur la jambe rapidement avancée. Il s'agit d'un acte de violence qui se trouve probablement chez tous les Hommes.

Beißintention → Scheinbeißen
E. intention to bite, bite intention □ F. morsure intentionnelle

Beißkuß □ Der Beißkuß kommt als rhythmisch wiederholtes Beknabbern oder als gehemmtes Zubeißen vor. Plötzliches, bisweilen recht festes Zubeißen beobachtet man als Neckerei im Liebesvorspiel beim Menschen, oft von der Frau als Antwort auf eine leichte Herausforderung und als Abwehr (Abb. 21 – Liebespaar in Paris; Eibl-Eibesfeldt 1970).

21

bite-kiss □ Occurs as rhythmically repeated nibbling or as inhibited bite attempts. Sudden rather intense bites sometimes occur during sexual fore-play in humans in which the female may respond defensively to the male's advances (Fig. 21 – couple of lovers in Paris; Eibl-Eibesfeldt 1970).

baiser mordant □ Le baiser mordant existe en tant que mordillement rythmique ou comme morsure inhibée. Une morsure soudaine, parfois assez forte, s'observe en tant que taquinerie dans le jeu prénuptial chez l'Homme, souvent utilisée par la femme comme réaction à une légère provocation ou comme défense (Fig. 21 – couple d'amoureux à Paris; Eibl-Eibesfeldt 1970).

Beißordnung → Hackordnung → Rangordnung

E. bite order □ F. ordre hiérarchique de morsure

Beistandsverhalten □ Im tierischen und menschlichen Verhalten das uneigennützige Zuhilfeeilen und Helfen von gefährdeten Artgenossen und Freunden. Von Lorenz (1931) wohl zuerst beim Kolkraben und bei Dohlen beschrieben. Altruistisches Verhalten gibt es in verschiedenen Formen auch bei Primaten. Es wird auch von Delphinen (Abb. 22; Siebenaler u. Caldwell 1956, Pilleri u. Knuckey 1969) und kürzlich vom Mara (Dubost u. Genest 1974) beschrieben. Ein schönes Beispiel von Beistandsverhalten beim Menschen konnte Eibl-Eibesfeldt (1972) in einer Buschmanngesellschaft beobachten und unbemerkt filmen.

altruistic behaviour □ The protection and assistance of endangered companions occurring in both animals and man. This social defense behaviour was described in detail by Lorenz (1931) for the crow and the jack daw. Helping behaviour can also be observed in diverse forms among primates. It has been described in dolphins (Fig. 22; Siebenaler and Caldwell 1956, Pilleri and Knuckey 1969) and in maras (Dubost and Genest 1974). A fine example of supportive behaviour in humans was caught on film, unnoticed, by Eibl-Eibesfeldt (1972) among a group of Bushman children.

22

comportement altruiste □ Dans le comportement animal et humain, assistance aux compagnons en danger. Ce comportement a été décrit d'une manière très précise par Lorenz (1931) chez le Grand Corbeau et chez les Choucas. Il existe sous différentes formes chez les Primates, mais aussi chez les Dauphins (Fig. 22; Siebenaler et Caldwell 1956, Pilleri et Knuckey 1969) et a été récemment décrit chez les Maras (Dubost et Genest 1974). Eibl-Eibesfeldt (1972) a pu observer et filmer un bel exemple de comportement d'entraide dans une société de Boschimans.

Bekanntheit □ Durch wiederholtes Wahrnehmen erzeugte Eindrucksqualität, Gleiches mit Gleichem und Ähnlichem zu verknüpfen. Das aus Erfahrung resultierende Wiedererkennen kann als Leistung des vitalen Gedächtnisses unbewußt vollzogen werden.

familiarity □ The quality of impression resulting from repeated perceptions such that similarity or sameness is recognized. The experience resulting from repeated recognition may unconsciously serve active memory function.

Bekanntheit □ La faculté, résultant d'une perception répétée, d'établir un rapport entre impressions identiques ou similaires. La reconnaissance résultant de l'expérience peut se produire inconsciemment en tant que performance de la mémoire vitale.

Beruhigungssaugen □ Weint ein Kleinkind, so wird ihm sogleich die Brust angeboten (Abb. 23 A). Ist die eigene Mutter nicht anwesend, so kann auch eine andere so trösten. Eibl-Eibesfeldt (1972) beschreibt es von !ko-Buschleuten; ich selbst sah es bei Pygmäen (Heymer 1974). Das Kind saugt an der Brust, nicht weil es hungrig ist, sondern um sich zu beruhigen (Abb. 23 B – Bayaka-Pygmäen aus Zentralafrika). In unserer Gesellschaft gibt man in der gleichen Situation den Schnuller.

comfort sucking □ Often, a child starting to cry is immediately offered the breast (Fig. 23 A). In the absence of the mother, this act can be performed by another woman to comfort the child. Eibl-Eibesfeldt (1972) reports this behaviour occurring among the !ko Bushmen, and I have observed it among Pygmies (Heymer 1974). The child sucks at the breast, not because it is hungry, but to pacify itself (Fig. 23 B – Bayaka Pygmies from Central Africa). In our society the infant is comforted in the same situation with a rubber nipple.

tétée de consolation □ Quand un enfant pleure,

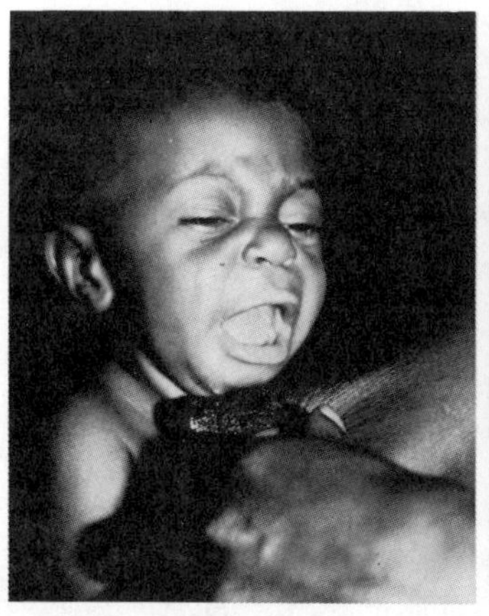
23 A

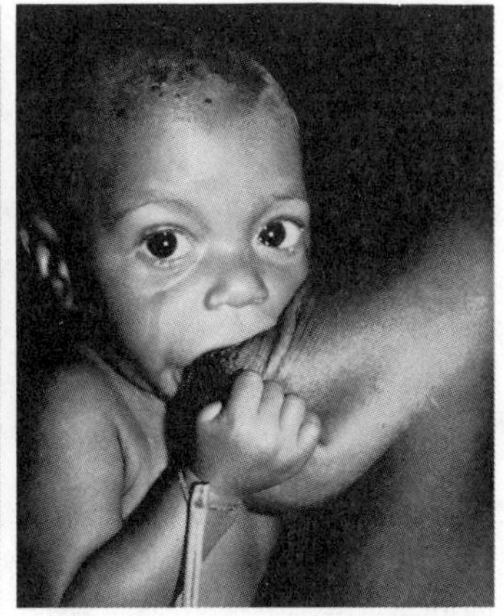
23 B

on lui offre immédiatement la mamelle (Fig. 23 A). Si la mère de l'enfant n'est pas présente, une autre femme peut la remplacer. EIBL-EIBESFELDT (1972) a décrit ce comportement chez les !ko Boschimans; moi-même, je l'ai observé chez les Pygmées (HEYMER 1974). L'enfant tète, non pas parce qu'il a faim, mais pour se consoler (Fig. 23 B – Pygmées Bayaka d'Afrique Centrale). Dans notre société, on donne dans la même situation la tétine.

Beschädigungskampf □ Im Gegensatz zum → *Kommentkampf* können sich die kämpfenden Tiere schwer verletzen. Allerdings gegen Kampfsituationen, die lebensgefährlich werden können, sind oft Signale und Auslösesysteme entwickelt, die den Angriff im kritischen Augenblick absolut hemmen. Sie bestehen meist im Darbieten gefährdeter Körperstellen, die zudem ein bestimmtes Zeichnungsmuster zur Schau stellen können. Soweit er die intraspezifische Aggression betrifft, ist der Beschädigungskampf eine seltene Erscheinung, da in der Regel infolge der eingebauten Hemmechanismen antagonistische Auseinandersetzungen nur in Form eines Kommentkampfes ausgetragen werden. Bei Ratten (Abb. 24 – *Rattus norvegicus*), Mäusen, Löwen gibt es intraspezifische Ernstkampfhandlungen nur gegenüber gruppenfremden Individuen.

24

damaging fight □ Fighting animals sometimes damage each other quite severely. This is not the case in → *ritualized fight*. But even for damaging fights, should the risk of fatality prevail, signals and releasers often have been evolved to completely inhibit attack at the crucial moment. Such mechanisms usually involve exposing vulnerable parts of the body. These body parts may even sport a particular visual pattern design. As far as intraspecific aggression in animals is concerned, damaging fight is an exceptionally rare occurrence since usually due to the built-in inhibitory mechanisms antagonistic encounters are settled by a ritualized fight. In rats (Fig. 24 – *Rattus norvegicus*), mice and lions intraspecific true fights occur only with non-group individuals.

combat sanglant □ Contrairement à ce qui se passe au cours d'un → *combat rituel,* les animaux peuvent ici se blesser grièvement. Cependant, dans les situations de combat qui pourraient entraîner la mort d'un partenaire, il existe souvent des signaux et des systèmes déclencheurs qui arrêtent la lutte d'une manière absolue au moment critique. Il s'agit généralement de l'exposition des points corporels vulnérables et en même temps, de la mise en valeur d'une coloration particulière. Chez les animaux, le combat intraspécifique authentique est un phénomène très rare, car grâce à l'existence de mécanismes inhibiteurs, les luttes sont, en général, menées selon certaines règles empêchant des blessures graves ou la mort de l'adversaire. Chez les Rats (Fig. 24 – *Rattus norvegicus*), les Souris et les Lions, on n'observe des actions agressives authentiques qu'envers des individus étrangers au groupe.

Beschwichtigungsgebärde → Beschwichtigungsverhalten

E. appeasement gesture □ F. geste d'apaisement

Beschwichtigungslaute → Beschwichtigungsverhalten

E. appeasement vocalization □ F. vocalisation d'apaisement

Beschwichtigungsverhalten □ Verhaltensweisen, die den Aggressionstrieb von Artgenossen neutralisieren oder ablenken und eine Umstimmung bewirken. Während Drohgebärden in erster Linie auf einige Entfernung hin der Distanzierung dienen, ist es Aufgabe von Demuts- und Beschwichtigungsgebärden, artschädigenden Kampf bei naher Distanz zu verhindern. Als Ausgangssituation zu ihrer Ritualisierung sieht TINBERGEN (1959) einen Konflikt zwischen → *Angst* (wohl besser: → *Furcht*) mit Tendenz zur Flucht und Bleibenwollen. Mit der Beschwichtigungsgebärde zeigt ein Tier, daß es auf Aggression verzichtet. In diesem Sinne werden Drohstrukturen, etwa die schwarze Maske der Lachmöwen, vom Artgenossen weggewendet (→ *Hinterkopfzudrehen*). Weiterhin können auch Lautformen wie Beschwichtigungslaute die soziale Unterordnung anzeigen oder den arteigenen Angreifer hemmen. In der Paarbindung dienen Beschwichtigungsgebärden oft der gegenseitigen Duldung der Partner. In diesem Sinne kann auch das → *Lächeln* des Menschen als Beschwichtigungsgebärde betrachtet werden.

appeasement behaviour □ Behaviour patterns which neutralize or redirect the aggressive drive of conspecifics and serve to effect a change of mood. While threat displays function to increase the interindividual distance before a confrontation occurs, submissive gestures and appeasement displays serve at close range to prevent actual injury from intraspecific fights. The origin and ritualization of such displays occurs according to TINBERGEN (1959) in the conflict between → *fear* with flight tendency and tendency to stand one's ground. With the appeasement display the animal can show its nonaggressive intentions. Hence, structures subserving threat such as the black mask of the laughing gull are turned away (→ *head flagging*). Moreover, appeasement vocalizations may indicate social submission and/or aim at inhibiting aggression by a conspecific. Within the pair-bond such appeasement display frequently promotes tolerance of the partner. In this respect, one can view → *smiling* as an appeasement display.

comportement d'apaisement □ Comportement qui neutralise ou détourne la pulsion agressive des congénères et entraîne un changement de motivation. Alors que les gestes de menace servent en premier lieu à garder la distance, les gestes d'apaisement doivent surtout éviter des luttes dangereuses. TINBERGEN (1959) voit comme situation initiale de leur ritualisation un conflit entre la → *peur* (mieux: → *crainte*) avec une tendance à la fuite et la volonté de rester. Par un geste d'apaisement, l'animal montre qu'il renonce à un comportement agressif. Il soustrait ses armes à la vue du congénère; ainsi, par exemple, la Mouette rieuse cache son masque noir par un détournement de la tête (→ *présentation de l'arrière de la tête*). D'autre part, des vocalisations d'apaisement peuvent également indiquer une subordination sociale ou servir à inhiber l'attaque d'un congénère. Dans le couple, les gestes d'apaisement servent souvent à la tolérance réciproque des partenaires. Dans ce sens, on peut considérer le → *sourire* chez l'Homme comme un geste d'apaisement.

Beschwörungsrituale □ Gemeint sind hier Verhaltensweisen beim Menschen zur Beschwörung von Geistern und Feinden, also einen vermeintlichen Schutz verleihendes Verhalten.

rituals of entreaty, rituals of implorement, rituals of adjuration □ Behaviour displayed by humans to implore or adjure spirits and enemies; thus, designed to afford protection.

rituels d'exorcisation □ Modes de comportement chez l'Homme ayant pour but d'exorciser des esprits et des ennemis; il s'agit donc d'un comportement conférant une protection.

Beutefanghandlung □ Unter Beutefanghandlung oder Beutehandlung versteht man schlechthin jene Handlungen der im Beutefangverhalten enthaltenen Einzelelemente, die ein Tier seine Beute ergreifen lassen. → *Endhandlung.*

prey-catching behaviour □ Defined as simply those actions of predatory behaviour which definitely and finally lead the animal towards the actual prey-capture, i. e. consummatory movement (→ *consummatory act*) of seizing the prey.

prédation □ La prédation comprend tous les mouvements et éléments comportementaux conduisant un animal vers sa proie et aboutissant à la saisie de celle-ci. → *acte consommatoire.*

Beutefangstimmung □ Die Aktivierung jener Triebe, die das Tier zum Beutefang stimulieren bzw. antreiben und auch bis zur Endhandlung erhalten bleiben. Beim Beutespiel aber muß die Beutefangstimmung nicht aktiviert sein. Das ist sogar unwahrscheinlich, da es sonst schwerlich möglich wäre, das Ausbleiben der echten Beutehandlungen zu erklären.

predatory motivation □ In the case of such motivation, those drives are activated which stimulate or impel the animal to hunt prey; these drives remain in operation until the consummatory act is complete. The motivation for predation, however, needs not to be activated in the case of predatory play. In fact, such drive activation is improbable since there is an absence of real predatory activity.

motivation de prédation □ Par motivation à la prédation, il faut entendre une activation des pulsions qui stimulent l'animal à la capture de proie et qui persistent jusqu'au moment où l'acte consommatoire est exécuté. Pendant le jeu avec la proie, cependant, cette motivation ne semble pas être activée, sinon il serait difficile d'expliquer l'absence de véritables mouvements de capture.

Beutehandlung → Beutefanghandlung

Beuteschema □ RÄBER (1949) hat durch seine Untersuchungen an Eulen die Kenntnis der beim Beuteerwerb beteiligten Faktoren wesentlich erweitert. Dabei gibt es ein Beuteschema »Maus«, für das u. a. folgende Reizkombination gilt: »fester Körper mit bewegten Gliedern oder bewegten Muskelpartien« als optimal sowie »kleiner als Meerschweinchen von 250 g« als Größenbegrenzung. Das Beuteschema »Vogel« wird durch die Reizkombination »eiförmiger Körper mit anliegendem Gefieder und deutlich sichtbarer, in der verlängerten Körperachse liegender Schwanz« dargestellt. Die Bewegung ist hierbei unwichtig.

prey schema □ An important contribution to the understanding of factors involved in the owl's prey recognition and capture was made by RÄBER (1949). The basic »mouse« prey pattern includes the following stimulus combination: »solid body with moving extremities or muscle units« (optimal); also »the figure must be smaller than a guinea pig of 250 g«. The basic prey pattern for »bird« is represented by the stimulus combination of »egg-shaped body with folded feathers, strongly noticeable tail lying on the extended axis of the body«. In the latter case, movement is unimportant.

schème perceptif de la proie □ RÄBER (1949), par ses recherches sur les Chouettes et les Hiboux, a considérablement élargi nos connaissances sur les différents facteurs contribuant à la prédation. Il existe, par exemple, un schème perceptif de la proie »souris« qui est constitué, entre autres, de la combinaison des stimuli suivants: »corps compact avec extrémités ou parties de la musculature en mouvement« comme stimulus optimal ainsi que »plus petit qu'un Cobaye de 250 g« comme limite de taille. Le schème perceptif de la proie »oiseau« est constitué par la combinaison stimulatrice »corps ovale avec plumage serré, queue bien visible dans le prolongement du grand axe du corps«. Le mouvement, dans ce contexte, est sans importance.

Beuteschmarotzer □ Tiere, die anderen Arten die Beute abjagen, auch *Kleptobiose* genannt. Beispiel: Stercoraridae.

prey parasite □ Animal that steals the prey caught by another, also called *cleptobiosis*. Example: Stercoraridae.

prédateur parasite □ Animal qui dérobe la proie capturée par une autre espèce, également appelé *cleptobiose*. Exemple: Stercoraridae.

Beutespießen → Aufspießen

Beutesprung → Mäuseln

Beutespucken □ Es handelt sich bei Fischen um ein gezieltes Spucken aus dem Wasser heraus nach Beuteobjekten im Luftraum. Im Hungerzustand konnte bei *Colisa lalia* gesteigerte Spuckbereitschaft beobachtet werden. Jede Spuckhandlung besteht aus einer Serie von 1 bis 10 (meist 5) Einzelspuckern, die im Abstand von etwa $^1/_3$ bis $^1/_4$ sec zur Beute hochgehen. Das Spuckverhalten war bislang insbesondere vom Schützenfisch *Toxotes jaculatrix* (Abb. 25) bekannt (LÜLING 1969, 1973, VIERKE 1973).

water spitting □ Squirting of water drops by fish to catch prey above the water. The fish *Colisa lalia* displays an increased tendency to spit at insects above the surface when deprived of food. Each instance consists of squirting a series of from 1 to 10 (mostly 5) single drops every $^1/_3$ to $^1/_4$ second at the food

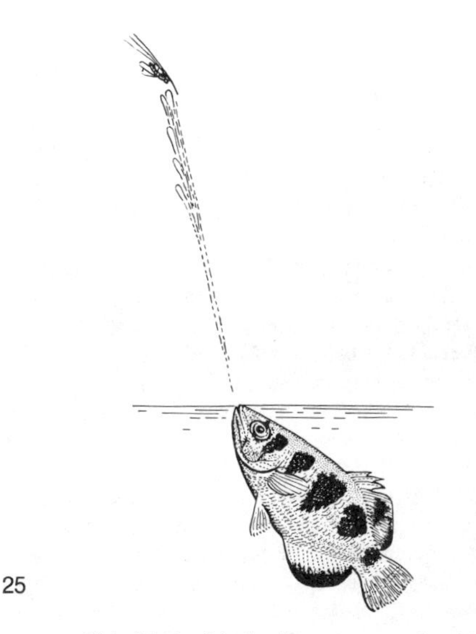

25

source. This kind of behaviour occurs also in the Archer Fish, *Toxotes jaculatrix;* Fig. 25 (LÜLING 1969, 1973, VIERKE 1973).

crachement d'eau □ Il s'agit, chez certains Poissons, d'un jet d'eau orienté de la surface de l'eau vers des proies situées au-dessus de celle-ci. La disposition à cracher augmente chez *Colisa lalia* au fur et à mesure que l'animal est plus affamé. Chaque émission se compose d'une série de 1 à 10 jets indépendants (5 en géneral) effectués à des intervalles de $^1/_3$ à $^1/_4$ sec. Ce comportement de capture de proie a été décrit surtout chez *Toxotes jaculatrix;* Fig. 25 (LÜLING 1969, 1973, VIERKE 1973).

Beutestimmung → Beutefangstimmung

Beutetöten → Totschütteln

E. prey-killing □ F. mise à mort de la proie

Bewegungsablauf → Erbkoordination

E. locomotory sequence □ F. processus de locomotion

Bewegungsbestandteil □ Das Einzelelement eines Bewegungsablaufes bzw. die verschiedenen Phasen der Fortbewegung. → *Gangart,* → *Erbkoordination.*

motor component act □ The individual elements of an action pattern, the various phases of a movement. → *locomotory pattern,* → *fixed action pattern.*

composante motrice □ Plusieurs composantes motrices constituent le processus de locomotion ou les différentes phases d'une locomotion ou d'autres mouvements coordonnés complexes. → *schème locomoteur,* → *coordination héréditaire.*

Bewegungshemmung → Käfigverblödung

E. motor inhibition F. inhibition motrice

Bewegungsmuster → Gangart

E. locomotor pattern □ F. type de locomotion

Bewegungsspiele → Spielverhalten

E. motoric play □ F. jeux de mouvement

Beziehungsmerkmal □ Von einem Beziehungsmerkmal kann man sprechen, wenn zwei bestimmte Merkmale zueinander in einem bestimmten Verhältnis stehen. Man kann in einem solchen Fall auch von einem konfigurativen oder figuralen Merkmal reden. Ein solches ist z. B. der »rote Fleck an der Schnabelspitze«, der die Futterbettelreaktion bei Silbermöwen auslöst (TINBERGEN u. PERDECK 1950).

relational characteristic, relational trait □ A relational characteristic can be defined when two specific traits stand in a particular relationship to one another. Here we could also speak of a configurational or figural characteristic. One example is the »red spot on the beak« of herring gulls, which releases food begging responses in the young (TINBERGEN and PERDECK 1950).

caractère relationnel □ On peut parler de caractère relationnel lorsque deux caractères déterminés présentent entre eux un rapport déterminé. Dans ce cas, on peut également parler d'un caractère configurationnel. Un tel caractère est, par exemple, la »tache rouge à la pointe du bec« chez le Goéland argenté qui déclenche chez les Poussins le comportement de quête de nourriture (TINBERGEN et PERDECK 1950).

Beziehungsschema → Beziehungsmerkmal
E. relational schema □ F. schéma relationnel

Bezugsperson □ Unter Bezugsperson versteht man eine Person, die vom Subjekt, z. B. einem Kind, trotz dessen zahlreichen Kontakten mit anderen Leuten immer wieder bevorzugt wird, z. B. die Mutter. → *Mutter-Kind-Beziehung*.

reference person □ A reference person is an individual for whom another, e. g. a child, shows consistent preference despite opportunities offered for frequent contact with other individuals. The example most commonly named is a child's mother. → *mother-child relationship*.

personne de référence □ Il s'agit d'une personne pour laquelle une autre, malgré de multiples contacts avec des personnes étrangères, montre une préférence marquée; par exemple, l'enfant pour sa mère. → *relation mère-jeune*.

Bienentänze □ Zusammenfassender Begriff für die verschiedenen Orientierungstänze der Bienen. Die drei wichtigsten sind: → *Rundtanz*, → *Schwänzeltanz* und → *Sicheltanz*.

bee dances □ Collective term referring to the various orientation dances of the honeybee. The three most important dances are: the → *round dance*, → *waggle dance* and → *sickle dance*.

danses des abeilles □ Cette expression couvre les différentes danses d'orientation chez les Abeilles. Les trois plus importantes sont: la → *danse circulaire*, la → *danse frétillante* et la → *danse en croissant*.

Bindegespräch → Wechselgespräch

Bindetrieb □ Der Zusammenschluß einer Gruppe von Tieren zum Verband setzt nicht nur die Existenz bandstiftender und aggressionsbeschwichtigender Verhaltensweisen voraus, sondern auch den Antrieb, diese Mittel einzusetzen. Diesen Antrieb einer Appetenz nach der Nähe des Artgenossen oder zum Elter nennen wir Bindetrieb (EIBL-EIBESFELDT 1970, 1972). FISCHER (1965) und WÜRDINGER (1970) verwenden den Ausdruck »Bindungstrieb«.

bonding drive □ The formation of a group of animals into a social unit presupposes not only the existence of appeasement behaviour and bond-maintenance behaviour, but also the drive to utilize these behaviours. This appetence to remain near the parent or the conspecific is called bonding drive (EIBL-EIBESFELDT 1972, 1973). FISCHER (1965) and WÜRDINGER (1970) use the expression *Bindungstrieb*.

pulsion de cohésion sociale □ Le groupement de plusieurs animaux en une association dépend non seulement de l'existence de modes de comportement apaisant l'agression ou créant des liens, mais aussi d'une pulsion tendant à l'utilisation de ces moyens. Une telle pulsion entraîne une appétence à rester à proximité du congénère ou du parent; elle est désignée par le terme *Bindetrieb* (EIBL-EIBESFELDT 1970, 1972, 1973). FISCHER (1965) et WÜRDINGER (1970) utilisent le mot *Bindungstrieb*.

Bindungskopulation → Kopulation
E. bond-oriented copulation □ F. accouplement de cohésion.

Bindungsgespräch → Wechselgespräch

Bindungstrieb → Bindetrieb

Bioakustik □ Die Erforschung von Lautäußerungen und Lauterzeugungen bei Tieren. Die Aufnahmen der Laute erfolgen mit Tonbandgeräten, und die so gewonnenen Aufzeichnungen dienen ihrerseits zur Herstellung von → *Klangspektrogramm(en)*. In den Bereich der Bioakustik gehört auch die Erforschung der lauterzeugenden Organe sowie der Aufnahme- und Gehörorgane (TEMBROCK 1959, BUSNEL 1964).

bioacoustics □ The study of vocalizations and sound production in animals. The sounds are recorded on tape and the recording is used to make a sound spectrogram (→ *sonogram*). Bioacoustics also include the study of the sound production mechanisms, the auditory and sound reception organs (TEMBROCK 1959, BUSNEL 1964).

bioacoustique □ Etude des vocalisations et productions sonores chez les animaux. L'enregistrement des vocalisations et productions sonores se fait à l'aide de magnétophones et les documents enregistrés servent par la suite à l'élaboration de → *sonogramme (s)*. L'étude des organes producteurs (émetteurs) et récepteurs de sons fait également partie du domaine de la bioacoustique (TEMBROCK 1959, BUSNEL 1964).

Bittbewegung → Händepatschen
E. begging movement □ F. mouvement de sollicitation

Blickkontakt □ Ein Verhaltenselement im menschlichen Kontaktverhalten, z. B. beim auf Distanz Flirten. Nach Aufnahme des Blickkontaktes folgt meist → *Lächeln* und → *Augengruß,* danach aber Kopfsenken mit Lidschluß als ritualisierte Form einer Ausweichreaktion bzw. des »Flüchtens« (Abb. 26 – Turkana-Frau aus Kenia). Kurz danach wird der Blickkontakt meist erneut aufgenommen.

eye contact □ A behavioural element in human social interaction, e. g., during flirting at a distance. After eye contact is established, → *smiling* or → *eyebrow flash* generally follows. Next, the head is lowered followed by closing the eyelids, representing a ritualized form of avoidance or escape (Fig. 26 – Turkana woman from Kenya). Shortly thereafter, the eye contact is generally resumed.

26

contact visuel □ Un élément dans les interactions sociales chez l'Homme, en particulier pour établir un contact, comme par exemple lors du flirt à distance. L'établissement du contact visuel est généralement suivi d'un → *sourire* ou d'un → *haussement des sourcils,* mais ensuite, la tête est abaissée, avec fermeture des paupières en tant que forme ritualisée d'évitement ou de fuite (Fig. 26 – femme Turkana du Kenya). Peu de temps après, le contact visuel est généralement renouvelé.

Blockierungssituation □ Es handelt sich um eine Situation, deren Reizmuster beim Tier nicht ohne weiteres anspricht, sondern z. T. widerstrebende Verhaltenstendenzen in Gang setzt. Möglicherweise sind solche Blockierungs- und Konfliktsituationen der phylogenetische Ausgangspunkt zur Ausbildung semantischer Reaktionen (BLUME 1967).

conflict blocking □ In a situation of this kind, the animal does not necessarily respond to the stimulus pattern involved. Instead, opposing behaviour tendencies result. It is possible that such obstructional and conflict situations form the starting point for the development of semantic reactions (BLUME 1967).

»bloquage« comportemental □ Il s'agit d'une situation où l'animal réagit avec hésitation à une stimulation. Parfois même, des comportements contradictoires sont activés. Il est probable que de telles situations conflictuelles sont le point de départ du développement phylogénétique de réactions sémantiques (BLUME 1967).

Bodenfeindverhalten □ Das Bodenfeindverhalten kann Angriff und Flucht einschließen, wie wir es an folgendem Beispiel sehen können: Der gereizten Henne wird als Bezugsobjekt ein ausgestopfter Iltis präsentiert (Abb. 27). Dauert die Reizung lange genug, so erscheinen nacheinander: Drohen (A) – Angriff (B) – Flucht (C) (VON HOLST u. VON ST. PAUL 1960).

27

behaviour towards terrestrial ground predators □ Behaviour towards ground predators includes both attack und flight components as may be seen in the following example: A hen stimulated by a stuffed polecat will respectively threaten (Fig. 27 A), attack (B) and flee (C) if the stimulation lasts long enough (VON HOLST and VON ST. PAUL 1960).

comportement envers des prédateurs terrestres □ Ce comportement comprend attaque et fuite, comme le montre l'exemple suivant: Une Poule est stimulée par la présentation d'un Putois empaillé (Fig. 27). Si la stimulation est suffisamment prolongée, on observe les comportements suivants de la part de la Poule: menace (A), puis attaque (B) et enfin fuite (C) (VON HOLST et VON ST. PAUL 1960).

Bodenforkeln □ Bei Boviden und Cerviden das Wühlen (*Bodenwühlen*) und Forkeln mit dem Gehörn oder Geweih im Boden, nach DUBOST (mündl.) vor allem während der Fortpflanzungszeit vom α-Tier vor jüngeren männlichen Tieren als Imponiergehaben ausgeführt (Abb. 28 – *Cervus nippon*).

28

ground-rutting □ Digging and rooting into the ground by Bovidae and Cervidae with their horns or antlers which according to DUBOST (*pers. comm.*) occurs primarily during the breeding season and is exhibited by the *alpha*-animal during displays directed at younger males (Fig. 28 – *Cervus nippon*).

râclement du sol □ Chez les Bovidae et Cervidae, le fouissement du sol avec les cornes et les bois; selon DUBOST (*comm. pers.*) il se manifeste particulièrement pendant la période de reproduction par l'animal *alpha* en tant que comportement d'intimidation envers des congénères mâles plus jeunes (Fig. 28 – *Cervus nippon*).

Bodenpicken □ Bodenpicken kommt bei Hühnervögeln in mehreren Situationen vor, und zwar lockt die Glucke damit ihre Küken (Abb. 29 A –→ *Futterlocken*); auf gleiche Weise lockt damit der Hahn seine Hennen herbei (Abb. 29 B). In Anwesenheit eines Gegners aber tritt Bodenpicken auf, das von Faktoren kontrolliert wird, die mit agonistischem Verhalten zusammenhängen (SCHENKEL 1956, 1958, WICKLER 1969, FEEKES 1972).

29 A 29 B

ground pecking □ Ground pecking occurs in gallinacious birds in different situations; the hen entices her chicks to follow with ground pecking (Fig. 29 A –→ *feeding enticement*) whereas the cock may use the same behaviour in holding his hens together (Fig. 29 B). However, when ground pecking occurs during an encounter with an opponent, it is under control of factors underlying agonistic behaviour (SCHENKEL 1956, 1958, WICKLER 1969, FEEKES 1972).

picorage au sol □ Ce comportement se manifeste chez les Gallinacés dans plusieurs situations. La Poule, par exemple, attire ainsi ses Poussins (Fig. 29 A –→ *attraction alimentaire*), mais par le même comportement, le Coq attire ses Poules (Fig. 29 B). Par contre, en présence d'un adversaire, ce comportement est contrôlé par des facteurs liés à un comportement agonistique (SCHENKEL 1956, 1958, WICKLER 1969, FEEKES 1972).

Bodenwühlen → Bodenforkeln

Brauenheben □ Das *schnelle Brauenheben* beim Menschen wurde zuerst in der Flirt- und Grußsituation beobachtet und deshalb von EIBL-EIBESFELDT

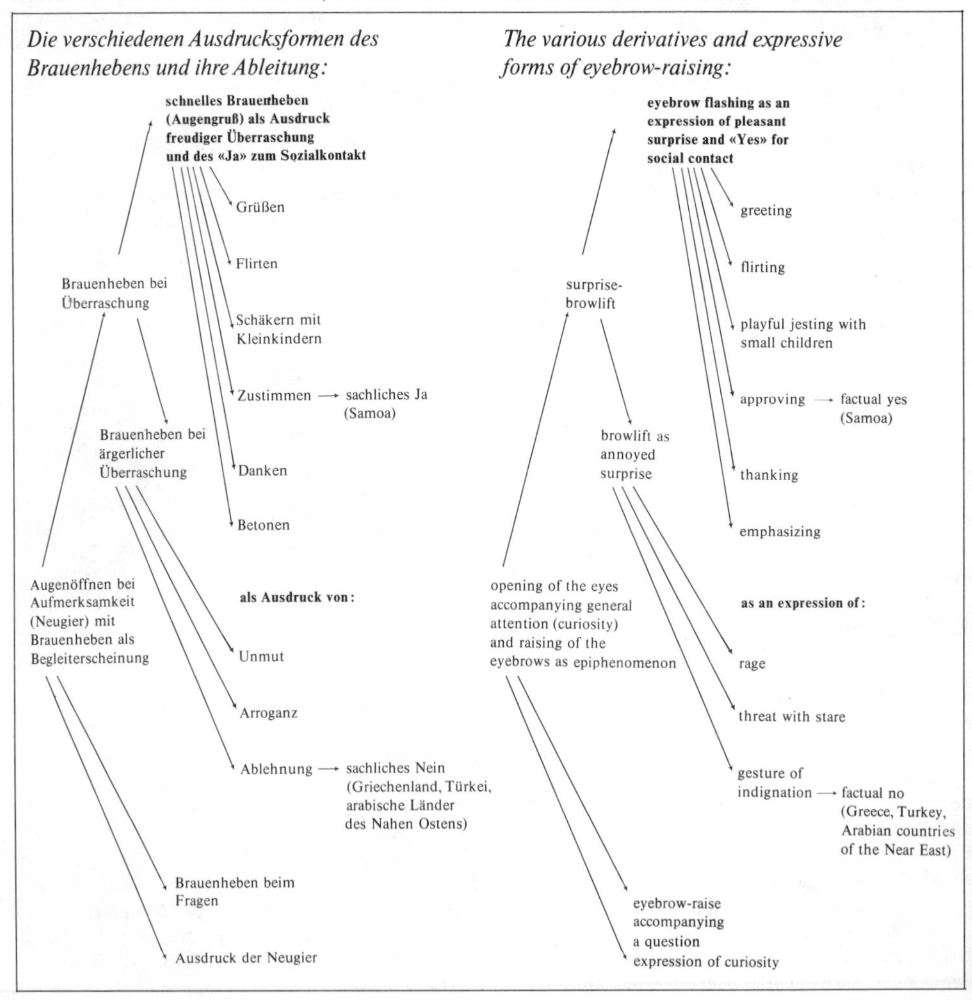

(1968) → *Augengruß* genannt. Später wurden dann weitere Bedeutungen bekannt, und als allgemeinen und weniger restriktiven Begriff hat der gleiche Autor (1972 a + b) »*schnelles Brauenheben*« vorgeschlagen.

eyebrow-raising □ Rapid raising of the eyebrow in humans has first been observed in connection with flirtation and greeting and was hence called → *eyebrow flash* by EIBL-EIBESFELDT (1968). Subsequently, additional implications were found, and hence the same author (1972 a + b) proposed the less restrictive, more general term *eyebrow-raising*.

haussement des sourcils □ Chez l'Homme, ce mouvement rapidement esquissé a d'abord été observé lors des saluts et du flirt; c'est pourquoi EIBL-EIBESFELDT (1968) l'a dénommé → *salut des sourcils*. Plus tard, d'autres significations ont été trouvées et le même auteur (1972 a + b) a proposé d'appeler ce comportement *haussement des sourcils* ce qui est plus général et moins restrictif.

Les différentes formes expressives du haussement des sourcils et leurs dérivés:

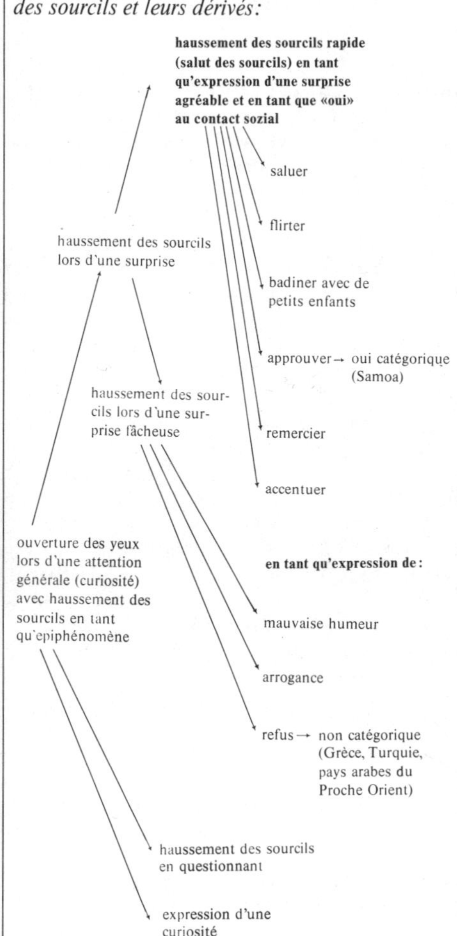

Brunft → Brunst

Brunftfeige → Brunstfeige

Brunftharem → Herden

E. heat harem in Cervids and Bovids □ F. harem de rut chez les Cervidae et Bovidae

Brunst □ Bei Säugetieren ein durch Sexualhormone gesteuerter, periodisch auftretender Zustand geschlechtlicher Erregbarkeit und der Paarungsbereitschaft während der sog. Brunstzeit. Die Brunst (Östrus nur für ♀♀) tritt bei manchen Arten nur einmal jährlich (monöstrisch) oder mehrmals jährlich in bestimmten Abständen (polyöstrisch) auf. Die Brunst ist u. a. von der Reifung der Geschlechtszellen abhängig und äußert sich in besonderen Ausprägungen der sekundären Geschlechtsmerkmale (BERMANT und DAVIDSON 1974). Bei vielen Säugern, z. B. bei Insektivoren, Fledermäusen und Nagern, verlagern sich die Hoden nur während der Fortpflanzungszeit aus der Leibeshöhle in den Hodensack (Skrotum).

Beim Fuchs und beim Dachs spricht man von *Ranz,* bei Hasen und Kaninchen von *Rammelzeit,* beim Schwarzwild von *Rauschzeit.* Den Östruszustand bei Katzen nennt man *Rolligkeit* und bei Pferden *Rossigkeit.* Die Brunst wird häufig von tiefgreifenden Veränderungen im Verhalten begleitet. → *Werbeverhalten.*

heat, rut □ In mammals a periodically occurring state of sexual excitability and motivation for mating controlled by the sexual hormones. In some species *oestrus* (only for ♀♀) occurs only once a year (monoestric species), in other species several times at determinate intervals (polyoestric species). Rut and oestrus depend on the maturation of the sexual cells and appear in a particular manifestation of the secondary sexual characters (BERMANT and DAVIDSON 1974). In many mammals, e. g., in insectivores, chiropterans and rodents, the testes move into the scrotum only during the reproduction period.

In foxes and badgers the reproduction period is termed *Ranz,* a term stemming from hunting jargon in German; in hares and rabbits *Rammelzeit* (shag time). The German term is derived from *Rammler,* the ♂ of hare and rabbit, whereas shag refers to sexually motivated chasing. In the wild boar the reproduction period is called *Rauschzeit.* The oestrus state in felids is called *Rolligkeit* and in horses *Rossigkeit.* Rut and oestrus are frequently accompanied by profound changes in behaviour. → *courtship behaviour.*

rut □ Chez les Mammifères, état d'excitabilité sexuelle périodique et de disposition à l'accouplement (période de rut), sous le contrôle des hormones sexuels. Chez un certain nombre d'espèces, *l'oestrus* (seulement pour les ♀♀) n'apparaît qu'une fois par an (espèces monoestriennes), mais chez d'autres peut se manifester plusieurs fois par an à intervalles réguliers (espèces polyoestriennes). Le rut ou l'oestrus sont entre autres dépendants de la maturation des cellules sexuelles et s'expriment souvent par

des manifestations particulières des caractères sexuels secondaires (BERMANT et DAVIDSON 1974). Chez de nombreux Mammifères, comme par exemple chez les Insectivores, les Chauves-Souris et les Rongeurs, les testicules ne descendent dans le scrotum que lors de la période de reproduction.
Chez les Lapins et les Lièvres, *Rammelzeit* désigne la période de la parade nuptiale et d'accouplement; le terme allemand est vraisemblablement dérivé du mot *Rammler,* désignant le ♂ de ces animaux. Chez le Renard et le Blaireau, on parle de *Ranz,* terme allemand emprunté au langage des chasseurs. La période de reproduction et d'accouplement chez le Sanglier est appelé *Rauschzeit.* L'état d'oestrus chez les Felidae est appelé *Rolligkeit* et chez les Chevaux *Rossigkeit.* Le rut et l'oestrus sont souvent accompagnés de modifications profondes dans le comportement. → *comportement de parade nuptiale.*

Brunstfeige □ Bei Gemsen unmittelbar hinter den Hörnern gelegene paarige Hautdrüse, die sich zur Brunst stark entwickelt, aber auch sonst funktionsfähig ist. Sie dient u. a. zur → *Reviermarkierung.*

postcornual skin gland □ Paired skin gland located directly behind the horns of chamois; most prominently developed during the rutting season, but functional at other times, as well. It serves, e. g., to → *territory marking.*

glande post-cornuale □ Chez les Chamois et les Isards, cette glande subit un développement considérable lors du rut, mais elle est fonctionnelle d'une manière permanente et sert, entre autres, au → *marquage du territoire.*

Brunstkampf □ Die Brunstkämpfe sind ausschließlich durch den Konkurrenten bestimmt, und zahlreiche Brunstkämpfe müssen nicht territorial gebunden sein. → *Geweihkampf.*

heat fight, fight between sexual rivals □ Such fights, determined exclusively by the rival, are not necessarily territorial. → *locking antlers.*

combat de rut □ Ces combats sont exclusivement déclenchés par la présence d'un rival et ne sont pas forcément liés au territoire. → *lutte avec les bois.*

Brunstschwielen □ An der Innenseite des Unterarms befindliche Schwielen bei männlichen Fröschen und Kröten während der Fortpflanzungszeit, die zum Festhalten des ♀ bei der Besamung des Laiches dienen. → *Klammergriff.*

claspers □ Structures on the inner side of the male frog's or toad's lower legs for clasping the ♀ during fertilization of spawn. → *clasping hold.*

callosités tibiales □ Structures présentes en période reproductrice chez les ♂♂ des Amphibiens Anoures à qui elles permettent de s'accrocher à la ♀ pendant la fécondation de la ponte. → *étreinte.*

Brustsuchen → Suchautomatismus
E. breast seeking □ F. recherche de la poitrine

Brutablösung → Ablösungszeremoniell

Brutpflegefüttern → Jungefüttern

Buckelstellung □ Drohstellung bei vielen Katzenarten, auch *Drohbuckeln* genannt. Die extreme Buckelstellung ist breitseits zum Gegner orientiert (*Breitseitsdrohen,* Abb. 30 – Hauskatze; der Gegner befindet sich schräg unterhalb). Wird der Angriff in Buckelstellung ausgeführt, so galoppiert die Katze quer zu ihrer Längsachse auf den Gegner zu (LORENZ 1951, LEYHAUSEN 1973).

30

humped posture □ Threat posture in many cat species; in the extreme form the animal presents laterally to the opponent (*lateral threat,* Fig. 30 – domestic cat; the opponent ist diagonally below). When attacking the cat gallops toward opponent in a sidestepping fashion with the body axis diagonal to the opponent (LORENZ 1951, LEYHAUSEN 1973).

posture du »dos rond« □ Posture de menace chez de nombreux Felidae qui, dans le cas extrême, est orientée latéralement par rapport à l'antagoniste (*menace latérale,* Fig. 30 – Chat domestique; l'adversaire se trouve en contre-bas). Lorsqu'une attaque est exécutée à partir de cette posture, le chat se jette sur l'adversaire en effectuant un galop de biais (LORENZ 1951, LEYHAUSEN 1973).

C

Caecotrophie → Coecotrophie

Chemorezeptoren □ Sinneszellen oder Sinnesorgane, die der Wahrnehmung chemischer Reize dienen, vor allem die Geruchs- und Geschmackssinneszellen bzw. -organe.

chemoreceptors □ Sensory cells or sense organs for the perception of chemical stimuli; primarily the cells or organs of smell and taste.

chémorécepteurs □ Cellules et organes sensoriels permettant la perception des stimulations chimiques, en particulier cellules ou organes olfactifs et gustatifs.

Chemotaxis → Taxis

circadianer Rhythmus → Tagesrhythmik

Coecotrophie □ Die orale Aufnahme von Blinddarm-(Coecum)-Inhalt (nicht von Blinddarmkot, wie ALTMANN [1969] schreibt). Wir bezeichnen diese aus dem Coecum stammenden Materialien mit HARDER (1949), der die Herkunft aus dem Blinddarm einwandfrei nachweisen konnte, als »Coecotrophe« und das Verhalten der oralen Wiederaufnahme als Coecotrophie. Ein solches Verhalten ist von allen Lagomorpha und Rodentia bekannt (Abb. 31 – *Dolichotis patagonum*). Die Coecotrophe zeichnet sich im Vergleich zum normalen Kot vor allem durch einen höheren Gehalt an Rohprotein aus. Der Entzug der Coecotrophe setzt, wie Versuche ergeben haben, die Widerstandskraft der Tiere herab, die gleichzeitig starke Anfälligkeit gegenüber Coccidiose zeigen und an der Seuche eingehen können (HARDER 1949, FRANK, HADELER und HARDER 1951). Coecotrophie wurde kürzlich auch bei einem phyllophagen Prosimier, der Gattung *Lepilemur*, nachgewiesen (HLADIK *et al.* 1971, CHARLES-DOMINIQUE und HLADIK 1971).

caecotrophy □ The oral intake of caecum contents (but not of caecum feces as reported ALTMANN, 1969). The material ingested and called »caecotroph« arises from the caecum as clearly demonstrated by HARDER (1949). This behaviour occurs in all lagomorphs and rodents (Fig. 31 – *Dolichotis patagonum*). The caecotroph is distinguished from normal feces by a much higher content of raw protein. Take away this source of food and the animal's resistance drops; he becomes highly susceptible to coccidiosis and may die from it (HARDER 1949, FRANK, HADELER and HARDER 1951). Caecotrophy has also recently been observed in a prosimian of the genus *Lepilemur* (HLADIK *et al.* 1971, CHARLES-DOMINIQUE and HLADIK 1971).

31

caecotrophie □ Ingestion orale du contenu du caecum, comportement connu chez tous les Lagomorphes et Rongeurs (Fig. 31 – *Dolichotis patagonum*). La provenance de ces matières caecotrophes a été clairement démontrée par HARDER (1949). Elles se distinguent des matières fécales normales surtout par une plus grande teneur en protéine brute. Les animaux privés de matières caecotrophes, comme l'ont démontré des expériences, ont une résistance diminuée et sont alors très sensibles à la coccidiose et peuvent mourir de cette maladie (HARDER 1949, FRANK, HADELER et HARDER 1951). La caecotrophie a très récemment été découverte chez un Prosimien phyllophage du genre *Lepilemur* (HLADIK *et al.* 1971, CHARLES-DOMINIQUE et HLADIK 1971).

D

Daueraktivität □ Die sog. »accessorischen« Zellen des Riechkolbens befinden sich in Daueraktivität; sie sind ständig ansprechbar. Durch den Nachweis eines »Bestandsstromes« im unbelichteten Wirbeltierauge ist auch für die Netzhaut die Daueraktivität sehr wahrscheinlich gemacht (ADRIAN 1950, LEYHAUSEN 1954).

spontaneous activity, resting level activity □ The so-called »accessory« cells of the olfactory bulb are continuously firing without any apparent external stimulation. Similarly a continuous afferent discharge occurs in the vertebrate eye in the absence of light and hence spontaneous activity is most likely a property of the retina, as well (ADRIAN 1950, LEYHAUSEN 1954).

activité spontanée □ On a pu démontrer que les cellules »accessoires« du bulbe olfactif se trouvent en activité permanente, même en l'absence de toute stimulation. Dans l'œil non éclairé des Vertébrés, une »activité permanente« a aussi été mise en évidence; de ce fait, il est très probable qu'une décharge spontanée existe également dans la rétine (ADRIAN 1950, LEYHAUSEN 1954).

Dauerparasitismus → Inquilinismus
E. permanent parasitism □ F. parasitisme permanent

Dauertanz □ Im Bienenstock spontan auftretender Tanz nach dem Sonnenstand ohne erneute Sonnensicht. LINDAUER (1954, 1955) entdeckte, daß Spurbienen *(→ Kundschafterinnen),* die für ein schwärmendes Volk – oft schon vor dem Schwärmakt – auf Wohnungssuche gehen, die Lage einer neuen Nistgelegenheit oft stundenlang ohne neuerlichen Ausflug durch wiederholte Tänze den Stockgenossen bekanntgeben. Nach VON FRISCH (1965) hielt eine solche *Dauertänzerin,* ohne wieder auszufliegen und ohne sich am Himmel über den Sonnenstand orientieren zu können, da das Flugloch verschlossen war, an der Kompaßrichtung eines 7,5 km entfernten Zieles noch am Tage darauf fest. Sie zeigte also die Flugrichtung zum neuen Nestplatz zu jeder Tageszeit immer richtig an. → *Verrechnungsmechanismus.*

persistent dance □ Spontaneous sun-oriented dance in the hive performed by → *scouting bees* without new view of the sun. LINDAUER (1954, 1955) discovered that bees, which seek out new locations for the colony before swarming occurs, may report the locations of these nest sites to the other bees by a persistent dance. According to VON FRISCH (1965) such a *persistent dancer-bee* continued to portray the correct position relative to the sun of a source 7.5 km away even though the entrance to the hive had been closed experimentally the day before. Consequently the flight direction to the new nest site can be given correctly at all different times of the day. → *compensatory mechanism.*

danse persistante □ Danse spontanée à l'intérieur de la ruche selon la position solaire et sans nouvelle vision du soleil. LINDAUER (1954, 1955) a démontré que des → *Abeilles exploratrices* à la recherche d'une nouvelle habitation, bien avant l'essaimage, effectuent des danses persistantes pendant des heures pour indiquer à leurs congénères le site du nouveau nid sans quitter la ruche. Selon VON FRISCH (1965), une telle *danseuse persistante,* sans quitter la ruche et sans voir le soleil pour s'orienter – car la sortie de la ruche avait été fermée expérimentalement –, a encore maintenu son indication directionnelle le jour suivant et indiqué d'une manière exacte la direction de vol à n'importe quelle heure de la journée. → *mécanisme de compensation.*

Dauertänzerin → Dauertanz
E. persistent dancer-bee □ F. danseuse persistante

Defäkation → Koten

Defäkationsverhalten → Kotzeremoniell
E. defecation behaviour □ F. comportement de défécation

Defensivdrohen □ Im Gegensatz zum → *Aggressivdrohen* ein Drohen, das während des Zurückweichens und bei der Verteidigung auftritt. Verteidigt sich ein in die Enge getriebenes Eichhörnchen, so droht es durch Aufrichten der durch Haarbüschel vergrößerten Ohren (Abb. 32 – *Sciurus vulgaris),* während es diese beim Aggressivdrohen anlegt (EIBL-EIBESFELDT 1957).

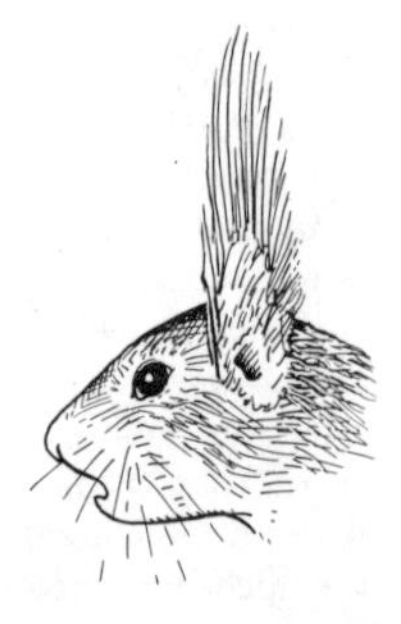

32

defensive threat □ In contrast to → *offensive threat,* this type of threat occurs during withdrawal. A squirrel driven into a corner threatens by erecting the hair tufts on its ears (Fig. 32 – *Sciurus vulgaris*), whereas during offensive threat the ears are laid flat (EIBL-EIBESFELDT 1957).

menace défensive □ Par opposition à la → *menace agressive,* c'est une forme de menace particulière qui apparaît souvent pendant le retrait de l'animal et au cours du comportement de défense. Un Ecureuil incapable de fuir menace alors en dressant les oreilles (Fig. 32 – *Sciurus vulgaris*) alors qu'elles sont repliées en arrière au cours de la menace agressive (EIBL-EIBESFELDT 1957).

Defensivverhalten → Demutsverhalten
E. defensive behaviour □ F. comportement défensif

Demonstrationsbewegung □ Eine Bewegung mit demonstrativem Charakter, um die Aufmerksamkeit anderer Individuen auf sich zu lenken, insbesondere im Kontext des Werbeverhaltens und des Imponierens, auch gegenüber Gegnern. Viele Erpel z. B. berühren bei der → *Balz* oft flüchtig ihren Flügel mit dem Schnabel, um dabei auf ihren prächtigen Spiegel zu weisen. Beim Mandarinenerpel wurde diese Handlung wirklich zu einer Demonstrationsbewegung, da er besonders auffallende Schmuckfedern hat, die er beim → *Scheinputzen* zeigt.

demonstrative movement □ A movement with expressive character which attracts attention of other individuals and which occurs particularly in → *courtship* and in display against rivals. For example, many male ducks touch the wing feathers during courtship display as if they wished to point out their colourful speculum. In the mandarin drake this behaviour has actually become a demonstrative movement. He has particularly striking feathers which are exposed during → *pseudo-preening.*

mouvement démonstratif □ Comportement à caractère démonstratif pour attirer l'attention d'autres individus, surtout pendant le comportement de parade nuptiale envers les ♀♀, mais aussi envers les rivaux. Pendant la → *parade nuptiale,* de nombreux Canards ♂♂, par exemple, touchent un instant l'aile avec leur bec comme s'ils voulaient montrer leur miroir alaire souvent très coloré. Chez le Mandarin ♂, ce mouvement est devenu un véritable mouvement démonstratif. Ce Canard possède, en effet, des plumes décoratives apparentes qui sont exposées pendant le → *pseudo-nettoyage.*

Demonstrationscharakter □ Gewisse Körpermerkmale, die bei → *Demonstrationsbewegung(en)* besonders gezeigt und präsentiert werden und somit Demonstrationscharakter bekommen haben. Auch Verhaltensweisen können Demonstrationscharakter haben, z. B. gewisse → *Gangart(en)* und das → *Verleiten.*

demonstrative character □ Certain anatomical characters that are presented during the → *demonstrative movement* and which have hence become demonstrative characters. Also behaviour patterns as for example some characteristic → *locomotory pattern(s)* and → *distraction display* can become demonstrative characters.

caractère démonstratif □ Un certain nombre de caractères morphologiques sont mis en valeur au cours de → *mouvement(s) démonstratif(s)* et reçoivent ainsi un caractère démonstratif. Certains modes de comportement peuvent également avoir un caractére démonstratif, par example certains modes de locomotion (→*schème locomoteur*) ou la→ *manoeuvre de diversion.*

Demutsgebärde → Demutsverhalten
E. submissive gesture □ F. geste de soumission

Demutshaltung → Demutsverhalten
E. submissive posture □ F. attitude de soumission

Demutsstellung → Demutsverhalten
E. submissive posture □ F. posture de soumission

Demutsverhalten □ Verhaltensweisen der Unterwerfung, um den Angriff zu verhindern; sich defensiv zeigen. Von MERTENS (1946) auch *katasematisches Verhalten* genannt. Die in diesem Sinne entwickelten Demutsstellungen sind Körperhaltungen, die ein Tier annimmt, wenn es sich, z.B. im innerartlichen Rivalenkampf, geschlagen gibt. Sie verhindern die ernsthafte Beschädigung oder gar Tötung von Artgenossen und sind daher arterhaltend. Sie sind angeboren und enthalten oft Elemente kindlichen oder sexuellen Verhaltens. Verwundbare Körperstellen werden ungeschützt dargeboten, Waffen wie Zähne, Geweih, Hörner usw. demonstrativ weggedreht. Die Demutsstellung wurde bereits von DARWIN (1872) beim Hund beschrieben. Wölfe bzw. unsere Hunde legen sich wie Jungtiere auf den Rükken und liegen still. Affen zeigen als Beschwichtigungsgebärde u. a. die weibliche Kopulationsstellung. – Auch in menschlichen Verhaltensweisen zeigt sich die Demutsstellung, hier aber ritualisiert in Form von bestimmten Begrüßungsformen: Verbeugung, Knicks, Ziehen des Hutes, Niederknien, Niederwerfen.

submissive behaviour □ Behaviour of subordination in order to prevent attack; defensiveness. Referred to by MERTENS (1946) as *catasematic behaviour*. Submissive body postures developed at this effect are adopted by an animal, e. g., when it is beaten in an intraspecific fight. Such postures inhibit serious injury or even killing by conspecifics, and, therefore, serve to maintain the species. They are innate and often consist of juvenile or sexual behaviour elements. Generally, particularly vulnerable parts of the body are exposed to the attacker; weapons such as teeth, antlers, horns, etc. are turned away in a demonstrative manner. The submissive posture of the dog was already described by DARWIN (1872). Dogs and wolves both roll like puppies on their back and remain motionless. Apes assume the female's copulation stance during appeasement. Also submissive postures are present in human behaviour in a ritualized form, such as in certain forms of greeting, bowing, tipping the hat, falling to the knees or falling to the ground.

comportement de soumission □ Modes de comportement ayant pour fonction d'éviter une attaque; se tenir sur la défensive. MERTENS (1946) parle aussi de *comportement catasématique.* Dans ce sens, des postures de soumission se sont développées qu'un animal adopte lorsqu'il est vaincu au cours d'un combat intraspécifique. Elles évitent des blessures graves ou même la mort du congénère et possèdent de ce fait un caractère adaptatif au service du maintien de l'espèce. Elles sont innées et comportent souvent des éléments d'un comportement enfantin ou sexuel. Les endroits du corps particulièrement vulnérables sont souvent exposés; les armes telles que les dents, les bois, les cornes etc. sont détournées d'une manière démonstrative. L'attitude de soumission a déjà été décrite par DARWIN (1872) chez le Chien. Les Loups par exemple et aussi nos Chiens se couchent sur le dos comme le font les jeunes au cours du jeu et restent immobiles. Chez les Singes, on voit souvent comme comportement de soumission ou d'apaisement la posture adoptée par la ♀ pendant l'accouplement. – Nous connaissons également des attitudes de soumission dans le comportement humain, mais sous une forme ritualisée dans certains comportements de salut: révérence, coup de chapeau, agenouillement, prosternation.

Desafferentierung □ Die Durchtrennung aller afferenten Nerven, z. B. die hinteren Wurzeln der Spinalnerven, die Trennung des Gehirns vom Rükkenmark; Versuche zur experimentellen Prüfung einer eventuellen endogenen Erregungsproduktion (VON HOLST 1969). OZORIO DE ALMEIDA und PIERON (1924) haben in ihren Experimenten beim Frosch sämtliche Hautnervenafferenzen ausgeschaltet und festgestellt, daß dies einen vollkommenen Zusammenbruch des gesamten Muskeltonus zur Folge hat, verstärkt noch durch die optische Desafferentierung. Diese Versuche sprechen gegen eine vollständige Autonomie der Nervenzentren, da es genügt, dem Frosch 1 cm^2 innervierter Haut zu lassen, um ihm einen fast normalen Muskeltonus zu erhalten.

desafferentation □ The severing of all afferent nerves, for example the dorsal roots of the spinal nerves; separation of the brain from the spinal cord. Concerns experimentation on a potential endogenous central energy of activation (VON HOLST 1969). OZORIO DE ALMEIDA and PIERON (1924) cut the total afferent innervation from the skin in an experiment with frogs and determined that the entire muscle tone was lost. This impairment was accentuated if the optic nerve was also severed. These experiments suggest that the central nervous system is nearly, but not completely, autonomous since one sq. cm. of intact innervated skin suffices to permit almost normal muscle tone.

désafférentiation □ Phonétiquement préférable à

déafférentiation. – Section de tous les nerfs afférents, par exemple les racines postérieures des nerfs spinaux, séparation du cerveau de la moelle épinière, pour tester expérimentalement l'existence éventuelle d'une production d'excitation endogène (VON HOLST 1969). OZORIO DE ALMEIDA et PIERON (1924) ont effectué, chez la Grenouille, la suppression complète des afférences sensorielles cutanées et ont constaté un effondrement général du tonus musculaire, encore aggravé par la désafférentiation visuelle. Ces expériences s'opposent évidemment à une autonomie complète du fonctionnement des centres nerveux. En effet, il suffit de laisser à la Grenouille 1 cm^2 de peau innervée pour lui conserver un tonus musculaire à peu près normal.

Desorientierung □ Vorübergehende oder dauernde Unfähigkeit eines Individuums, seine Umwelt oder seinen eigenen Standpunkt darin sinnvoll zu erfassen. Es kann sich dabei um eine räumliche oder auch zeitliche Desorientierung handeln.

disorientation □ Temporary or permanent inability of an individual to maintain a point of reference with respect to his surroundings. Spatial disorientation or temporal disorientation.

désorientation □ L'impossibilité temporaire ou permanente pour un individu de percevoir son environnement ou sa propre position à l'intérieur de celui-ci. Désorientation spatiale, désorientation temporelle.

»Dichten« □ Halblauter Motivgesang der Jungvögel.

subsong □ Fledglings' vocalizations which lack the structure of an adult's.

chant mineur □ Emission vocale semi-structurée de jeunes Oiseaux encore impubères.

Dissoziation □ Trennung zweier ursprünglich verbundener Komponenten, z. B. der Liebe und des Triebes zur Begattung, auch bei Gänsen (LORENZ). – In der Psychologie Bezeichnung für den Prozeß der Auflösung bzw. des Zerfalls von assoziativen Denk-, Vorstellungs- und Verhaltensverbindungen durch Vergessen bzw. Verlernen. Krankhafte Formen der Dissoziation äußern sich z. B. in unkoordinierten Bewegungen, Erinnerungsstörungen u. ä.

dissociation □ Dissociation between affection or love and the sex drive within a pair-bond in geese (LORENZ). – In psychology the process that dissolves connections and associations responsible for integrated coordinated mental function; forgetting. Pathological forms of dissociation occur in memory disturbances, poorly coordinated movements, etc.

dissociation □ Dissociation entre les liens affectifs et la pulsion à l'accouplement, décrite par LORENZ chez les Canards et les Oies. – En psychologie, notion qui désigne le processus de la dissolution ou décomposition des liaisons associatives de la pensée, de la représentation et du comportement par l'oubli et le désapprentissage. Les formes pathologiques de la dissociation sont visibles par exemple dans les troubles de la coordination motrice, les défaillances de mémoire etc.

Distanzgruß □ Als Distanzgruß gilt bei vielen Völkern das Heben der rechten Hand, wobei dem Grußpartner die offene Handfläche zugewendet wird, wohl um zu zeigen, daß man keine Waffe trägt. Abb. 33 zeigt eine so grüßende Karamojo-Frau aus Uganda (EIBL-EIBESFELDT 1972, 1973).

33

distance greeting □ Occurs in many cultures by raising the right hand with an open palm which would show that no weapon is concealed. Fig. 33 shows a Karamojo woman from Uganda greeting in this manner (EIBL-EIBESFELDT 1972, 1973).

salut à distance □ Nous connaissons chez de nombreux peuples et cultures le salut à distance qui se fait en levant la main droite. La paume de la main ouverte est alors dirigée vers le partenaire à saluer, probablement pour lui montrer que la main ne porte pas d'arme. Fig. 33 montre une femme Karamojo de l'Ouganda saluant de cette manière (EIBL-EIBESFELDT 1972, 1973,).

Distanzierungsverhalten □ Verhaltensweisen, die dazu dienen, die für Einzelindividuen oder Gruppenverbände notwendige Distanz untereinander zu halten. Hierzu gehört u. a. das → *Revierverhalten* und die → *Individualdistanz.*

spacing behaviour □ A behaviour pattern that serves to maintain the appropriate distance between individuals or groups; includes → *territorial behaviour* and → *individual distance.*

comportement d'espacement □ Mode de comportement qui sert aux différents groupements ou individus à maintenir entre eux la distance appropriée. A ce comportement s'attachent, entre autres, le → *comportement territorial* et la → *distance interindividuelle.*

Distanztiere → Individualdistanz
E. distance type animals □ F. animaux du type distant

Dominanzordnung → Rangordnung

Dominanz-Unterlegenheits-Verhältnis → Dominanzverhalten

E. dominance-submission-relation □ F. relation de dominance et de subordination

Dominanzverhalten □ Demonstration der Überlegenheit zur Erlangung oder Erhaltung einer dominierenden α-Stellung innerhalb einer sozialen Gemeinschaft. Die Ablösung eines α-Tieres durch ein bisher rangniederes Gruppenmitglied bezeichnet man als Dominanzwechsel. Bei Schimpansen z. B. lockert ein dominantes ♂ gegenüber einem brünstigen ♀ oft den Beziehungstyp, so daß es unter den ♂♂ zu einem solchen Dominanzwechsel kommen kann. Die sexuell reifen ♀♀ ihrerseits sind den unreifen überlegen und zeigen untereinander ein stabiles *Dominanz-Unterlegenheits-Verhältnis* (YERKES 1940).

dominance behaviour □ Demonstration of superiority in pursuing or maintaining the dominant α-position within a social group. The removal of an α-animal by a previously lower ranking group member is called change of leadership. In chimpanzees, e. g., the dominant ♂ while pursuing a ♀ in heat may often relax his role to the point that a change of leadership among the ♂♂ may occur. The sexually mature ♀♀ have higher rank than the immature ♀♀ and show a stable *dominance-submission-relation* within their own group (YERKES 1940).

comportement de dominance □ Démonstration de la supériorité pour l'obtention ou le maintien d'une position dominante à l'intérieur d'un groupement social. Le relais d'un animal α par un membre du groupe jusqu'alors d'un rang inférieur est appelé changement de dominance. Chez les Chimpanzés, par exemple, un ♂ dominant assouplit son mode de relation sociale avec une ♀ en oestrus, de sorte qu'il se produit souvent un changement de dominance parmi les ♂♂. Les ♀♀ sexuellement matures, de leur côté, dominent les ♀♀ immatures et montrent entre elles des *relations de dominance et de subordination* stables (YERKES 1940).

Doppelhalstauchen □ Eine epigame Verhaltensweise, die bei Schwänen meist kurz vor der Begattung mehrmals nacheinander ausgeführt wird (Abb. 34 – *Cygnus atratus*; PETZOLD 1964).

34

neck-over-neck diving □ An epigamous behaviour pattern of swans just before mating in which the neck is dipped into the water several times in succession (Fig. 34 – *Cygnus atratus*; PETZOLD 1964).

plongée croisée des têtes □ Un comportement épigame chez les Cygnes qu'on observe souvent juste avant l'accouplement et qui peut être exécuté plusieurs fois de suite par les deux partenaires simultanément (Fig. 34 – *Cygnus atratus*; PETZOLD 1964).

Doppelverdauung → Coecotrophie

E. double digestion □ F. digestion double

Drang □ Bereitschaft zu einem bestimmten Verhalten; Instinkthandlung. *Drang* stellt den aktivierten → *Trieb* dar, welcher seinerseits als latenter Zustand zu verstehen ist. Der Drang zu einem artspezifischen Handeln beruht auf Erregungsvorgängen im Zentralnervensystem, die sich auch experimentell reproduzieren lassen.

urge □ Readiness for a particular behaviour (instinctive behaviour). *Urge* (*Drang*) represents the activated *drive* (*Trieb*) which is considered as the basic underlying factor. The urge to perform a species-specific behaviour results from excitatory processes in the CNS which are experimentally reproducible.

pulsion □ Disposition à accomplir un comportement déterminé (comportement instinctif). La *pulsion* (*Drang*) peut être considérée comme la forme actualisée ou activée de la *tendance* (*Trieb*) qui, elle, constitue la forme latente. La pulsion d'une action spécifique est basée sur des processus d'excitation dans le système central nerveux et qui peuvent être reproduits expérimentalement.

Dreiecksverhältnis → Rangordnung

E. triangular relationship □ F. hiérarchie triangulaire

Dressur □ Die Bildung von Assoziationen oder von bedingten Reflexen. Durch Dressur wird der unbedingte Reiz (Originalreiz) durch einen bedingten (Signalreiz) ersetzt, der nun das dem unbedingten zugeordnete Verhalten auslöst. Originalreiz und Signalreiz können denselben oder verschiedenen Reizmodalitäten angehören (RICHELLE 1966, BUCHHOLTZ 1973, RENSCH 1973, MACKINTOSH 1974, ANGERMEIER 1976).

conditioning □ The construction of associations or of conditioned reflexes. During conditioning the unconditioned stimulus is replaced by the conditioned stimulus as the factor that releases the given response. The unconditioned stimulus and conditioned stimulus may belong to the same or a different stimulus modality (RICHELLE 1966, BUCHHOLTZ 1973, RENSCH 1973, MACKINTOSH 1974, ANGERMEIER 1976).

conditionnement □ La constitution d'associations ou de réflexes conditionnés. A l'aide du conditionnement, le stimulus inconditionné (stimulus original) est remplacé par un stimulus conditionné (stimulus signal) qui alors déclenche la réponse normalement liée au stimulus inconditionné. Stimulus original et stimulus signal peuvent appartenir aux mêmes modalités de stimulation, ou à des modalités différentes (RICHELLE 1966, BUCHHOLTZ 1973, RENSCH 1973, MACKINTOSH 1974, ANGERMEIER 1976).

Drohbuckeln → Buckelstellung

Drohgähnen □ Eine besondere, bei Primaten, wohl auch beim Nilpferd, vorkommende Ausdrucksbewegung, wobei der Mund wie beim Gähnen weit aufgerissen wird. Jedoch wird dieses Verhalten nicht durch endogene physiologische Ursachen ausgelöst (Abb. 35 – ungerichtetes Drohgähnen bei einem erwachsenen ♂ von *Miopithecus talapoin*).

threat yawning □ A threat posture occurring particularly in Primates, but also in the hippopotamus, in which the mouth is opened widely as in yawning; this behaviour, however, is not released by endogenous physiological factors (Fig. 35 – undirected threat yawning in an adult ♂ of *Miopithecus talapoin*).

bâillement de menace □ Un geste de menace particulier existant par exemple chez les Primates et les Hippopotames au cours duquel la bouche est ouverte à l'extrême comme pour un bâillement; cependant, ce bâillement n'est pas déclenché par des stimuli physiologiques endogènes (Fig. 35 – bâillement de menace non dirigé chez un ♂ adulte de *Miopithecus talapoin*).

35

Drohgruß □ Grußgebärden enthalten oft Elemente des Drohverhaltens. Das Begrüßungsgesicht des Pferdes enthält zwei deutliche aggressive Gebärden, die jedoch durch die aufgerichteten Ohren gemildert werden. Beim wirklichen Drohen sind die Ohren angelegt. Zwei Drohgebärden, die Aufrechthaltung und die Vorwärtshaltung, sind Teil der Begrüßungszeremonie bei der Lachmöwe. Drohgruß-Verhalten gibt es in verschiedenen Varianten auch bei Menschen (Abb. 36 – Krieger aus Uganda). Es wird von EIBL-EIBESFELDT (1970) ausführlich beschrieben.

threat greeting □ Greeting gestures frequently contain elements of threat behaviour. The facial expression of greeting in the horse contains two clearly aggressive elements which are canceled by the erected ears. In true threat the ears are laid flat. Two threat gestures of the laughing gull, the erect posture and the forward posture, are also parts of the greeting ceremony. Threat-greeting behaviour also occurs in man in various forms (Fig. 36 – warrior from Uganda). It is described in detail by EIBL-EIBESFELDT (1972).

36

salut à composantes de menace □ Les gestes de salut contiennent souvent des éléments du comportement de menace. L'expression faciale de salut chez le Cheval contient deux expressions typiquement agressives qui sont cependant apaisées par la position verticale des oreilles; pendant la véritable menace, les oreilles sont couchées. La cérémonie de salut chez la Mouette rieuse contient également deux gestes de menace, la position redressée et la position étirée en avant. Le comportement de salut à composantes de menace existe aussi sous diverses formes chez l'Homme (Fig. 36 – guerrier d'Ouganda); EIBL-EIBESFELDT (1970, 1972) le rapporte en détail.

Drohimponieren → Drohverhalten
E. intimidation display □ F. manoeuvre d'intimidation

Drohmiene → Drohmimik

Drohmimik □ Gesichtsausdruck des Drohens bei Wut und Zorn. Der Mandrill zieht dabei die Lippenwinkel herab und zeigt die Eckzähne (Abb. 37 A). Wir Menschen tun das ebenfalls, obwohl unsere Eckzähne keineswegs mehr so lang sind (Abb. 37 B – wütendes vierjähriges Mädchen aus Deutschland, C – Wut mimender Schauspieler aus Japan).

threat face □ Facial expression of threat during anger and rage. The mandrill exposes the canine teeth by lifting the lips (Fig. 37 A). Humans do the same although their canine teeth are no longer enlarged (Fig. 37 B – expression of rage in a 4 years old girl from Germany, C – Japanese actor showing rage).

mimique de menace □ Expression faciale de menace pendant la rage et la colère. Le Mandrill abaisse le coin de ses lèvres et met ainsi en évidence ses canines puissantes (Fig. 37 A). Les Hommes font de même bien que leurs canines ne soient que peu apparentes (Fig. 37 B – expression de colère d'une

fillette de 4 ans d'Allemagne, C – acteur japonais jouant la colère).

Drohstellung → Drohverhalten
E. threat posture □ F. posture de menace

Drohverhalten □ Abweisendes Verhalten mit aggressiver Motivation, z. B. Drohimponieren zur Einschüchterung des Gegners. Das Drohverhalten ist angeboren, also arttypisch, und enthält stets Komponenten des Angriffs- und Fluchtverhaltens. Es ist auch ein Ausdrucksverhalten, wodurch das Tier seine Kampfbereitschaft anzeigt. Viele Drohstellungen, charakteristische Körperhaltungen bei Tieren und Menschen mit Angriffskomponenten, haben sich ganz offenbar aus Ansprungbewegungen entwickelt. Das Drohbuckeln (→ *Buckelstellung*) der Katze ist eine typische Drohstellung.

threat behaviour □ Avoidance behaviour with aggressive motivation, for example intimidation display with elements of threat. Threat behaviour is innate and therefore species-specific, and always contains components of attack and fleeing behaviour. It is also an expressive behaviour by which the animal exhibits his readiness to fight. Many threat postures, characteristic postures in animals and man, often with elements of attack, have apparently developed from lunging movements. The threat posture of the cat (→ *humped posture*) is a typical example.

comportement de menace □ Comportement de dissuasion à motivation agressive, par exemple une manœuvre d'intimidation contenant des éléments de menace. Ce comportement de menace est inné, donc propre à l'espèce, et contient toujours des composantes d'attaque et de fuite. Il est également un comportement expressif par lequel l'animal indique sa disposition au combat. De nombreuses postures de menace, attitudes caractéristiques chez les animaux et l'Homme avec des composantes d'attaque, se sont probablement développées à partir des mouvements intentionnels de bond. Le → *dos rond* chez les Chats est une posture de menace typique.

Drosselung → afferente Drosselung
E. throttling □ F. inhibition

Duettgesang □ Ein Zeremoniell im Sinne der Paarbildung und Paarbindung, das der Zusammenführung der Partner dient. Solche Ruf- und Gesangsduette, in denen die Partner ihre Anteile rhythmisch und motivlich aufeinander abstimmen, sind bei Vögeln verschiedenster Ordnungen bekannt (THORPE 1963, DIAMOND und TERBORGH 1968, WICKLER 1972 – dort weitere Literatur). Typisch sind Duettgesänge für dauerehige Vögel, bei denen die Geschlechter äußerlich nicht zu unterscheiden sind (Abb. 38 – Afrikanischer Schmuckbartvogel, *Trachyphonus d'arnaudii;* Capitonidae). WICKLER (1972) hat Duettieren im Freiland auch zwischen

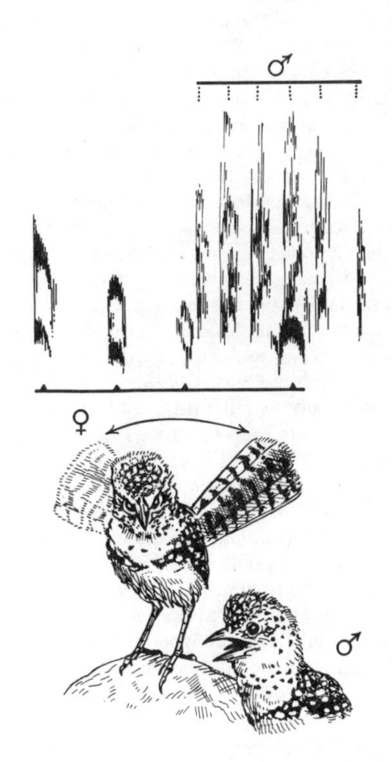

38

artverschiedenen Vögeln festgestellt. Duettgesang gibt es ferner unter den Säugern beim Siamang, *Symphalangus syndactylus* (LAMPRECHT 1970), einem dauermonogam lebenden Gibbon, und beim Großen Mara, *Dolichotis patagonum,* einem in Dauerehe lebenden Nager (DUBOST und GENEST 1973).

antiphonal song □ A behavioural ceremony serving pair formation and pair-bonding which permits the partners to keep in touch with each other. Such antiphonal songs in which the partners co-ordinate the rhythm and the types of call of their respective song parts are known in several orders of birds (THORPE 1963, DIAMOND and TERBORGH 1968, WICKLER 1972 – see for further references). Duetting is typical of birds living in permanent monogamy and in which the two sexes show no differences in their external appearance (Fig. 38 – African Barber, *Trachyphonus d'arnaudii;* Capitonidae). WICKLER (1972) described interspecific duetting in free living birds. Duetting also occurs in the siamang, *Symphalangus syndactylus* (LAMPRECHT 1970), a permanently monogamous gibbon, and in the great mara, *Dolichotis patagonum,* a monogamous rodent (DUBOST and GENEST 1973).

chant antiphonique □ Comportement au service de la formation du couple et des liens entre les partenaires. De tels duos dans lesquels les partenaires accordent et synchronisent leurs contributions vocales dans leur rythme et dans leurs motifs, sont connus chez des Oiseaux appartenant à des ordres différents (THORPE 1963, DIAMOND et TERBORGH 1968, WICKLER 1972 – voir pour références plus détaillées). Les chants antiphoniques sont typiques pour les Oiseaux vivant en monogamie permanente et chez lesquels les sexes ne sont pas reconnaissables au plumage (Fig. 38 – Barbier africain, *Trachyphonus d'arnaudii;* Capitonidae). WICKLER (1972) a également observé dans la nature des duos entre Oiseaux d'espèces différentes. Nous connaissons par ailleurs des duos chez les Mammifères, comme par exemple chez le Siamang, *Symphalangus syndactylus* (LAMPRECHT 1970), un Gibbon vivant en monogamie permanente, et chez le Grand Mara, *Dolichotis patagonum,* un Rongeur monogamique d'Amérique australe (DUBOST et GENEST 1973).

Duftmarken → Duftmarkierung
E. scent marks □ F. marques odorantes

Duftmarkierung □ Das Anbringen von Anwesenheitssignalen oder »chemischen Grenzsteinen«, die bei verschiedenen Tieren der innerartlichen Verständigung dienen. Die Abgabe von *Duftstoffen* erfolgt mit Hilfe spezieller *Duftorgane,* wie z. B. Duftschuppen (Androkonien), Dufthaare, Duftpinsel und Drüsen. Diese *Duftsubstanzen,* auch *Riechstoffe* genannt, werden in unwägbaren Mengen, bei Säugern auch Exkrementen beigemischt, an Gegenstände, an die Luft oder ins Wasser abgegeben und von Individuen der gleichen oder anderer Arten mit Hilfe des Geruchssinnes wahrgenommen (→ *Pheromone*). Bienen und Ameisen legen *Duftstraßen* zur Wegmarkierung an, um Nestgenossen das Auffinden von Futterstellen zu ermöglichen. Eine besondere Art des Duftmarkierens ist das *Markierungstrommeln* bei Mäusen und Hamstern, wobei die Hinterfüße zunächst an den Flankendrüsen entlangreiben und dann auf den Boden trommeln (FRANK 1956). Weitere Beispiele für Duftmarkierung: → *Harnmarkierung,* → *Harnspritzen,* → *Reviermarkierung.*

scent marking □ The deposition of scent marks or »chemical boundaries« that in various animal groups functions as a kind of species-specific communication. The *odorous substances* are excreted by special scent organs such as scent scales (androconia), scent hairs, scent brushes and glands. These odorous substances are deposited in extremely small quantities on objects, into air or into water; in mammals, they may also be added to the excrements. They are perceived by individuals of the same or other species via the olfactory sense (→ *pheromones*). Bees and ants lay out *olfactory trails,* assisting other colony members in finding food sites. An interesting type of olfactory marking is the *marking drumming* in mice and hamsters, in which the hind feet first scratch the lateral scent glands and then drum on the ground (FRANK 1956). For many other examples of olfactory marking: → *urine marking,* → *urine spraying,* → *territory marking.*

marquage odorant □ Dépôt de signaux de présence ou »bornes chimiques« par différents animaux au service de la communication intraspécifique. Les *substances odorantes* sont sécrétées par des organes odorants spéciaux tels qu'écailles odorantes (androconies), poils (pinceaux odorants) et glandes. De telles substances sont libérées en quantités infinitésimales dans l'air ou dans l'eau ou déposées sur le substrat pour être perçues par les congénères ou par des individus d'autres espèces (→ *phéromones*). Les Mammifères additionnent ces substances à leurs excréments, urine et fèces. Les Abeilles et les Fourmis établissent de véritables *traces olfactives* pour marquer ainsi leur chemin, permettant à leurs congénères du même nid de trouver les emplacements de nourriture et le chemin de retour au gîte. Le *tambourinage de marquage* constitue une autre forme intéressante de marquage odorant chez les Souris et Hamsters qui effectuent d'abord avec leurs pattes postérieures des mouvements de grattage de leurs glandes des flancs pour effectuer ensuite une sorte de *tambourinage sur le sol* (FRANK 1956). Autres exemples de marquage odorant: → *marquage d'urine,* → *aspersion d'urine,* → *marquage du territoire.*

Duftorgane → Duftmarkierung
E. scent organs □ F. organes olfactifs

Duftsignale → Duftmarkierung
E. olfactory signals □ F. signaux odorants

Duftspur = Duftstraße → Duftmarkierung

Duftstoffe → Duftmarkierung
E. odorous substances □ F. substances odorantes
Duftstraße → Duftmarkierung
E. olfactory trail □ F. trace odorante
Dulosis → Sklavenhalterei
Durchseihen → Seihen

E

EAAM – durch Erfahrung ergänzter angeborener Auslösemechanismus □ Der *EAAM* ist ein → *AAM*, den ein Individuum in ontogenetischer Anpassung an bestimmte Situationen durch Gewöhnung oder Lernen ergänzt hat. Der Ausdruck *EAAM* soll nur dann verwendet werden, wenn experimentell festgestellt wurde, daß in dem untersuchten ontogenetischen Stadium des Individuums ein *AAM*, über den die Reaktion auch weiterhin ausgelöst werden kann, als Grundgerüst noch funktionsfähig ist (SCHLEIDT 1962).
IRME – innate releasing mechanism modified by experience □ The *IRME* is an → *IRM* which in an individual has been modified, in ontogenetic adaptation, to a certain situation, either by habituation or sensitization and/or learning. The term *IRME* should be used only if it has been proved experimentally that, at the examined ontogenetic stage of the individual, an *IRM* is still functioning as a skeleton which is able, by itself, to release the reaction (SCHLEIDT 1962).
MDIE – mécanisme déclencheur inné complété par l'expérience □ Le *MDIE* est un → *MDI* qu'un individu, au cours d'une adaptation ontogénétique à des situations déterminées, a complété par habituation ou apprentissage. Le terme de *MDIE* ne doit être utilisé que dans le cas où il a été prouvé expérimentalement que dans le stade ontogénétique considéré de l'individu persiste un *MDI* qui en est la base fonctionnelle et par lequel la réaction reste déclenchable (SCHLEIDT 1962).
EAM – erworbener Auslösemechanismus □ Der *EAM* ist ein → *AM*, den ein Individuum in ontogenetischer Anpassung an bestimmte Situationen erworben hat. Der Ausdruck *EAM* soll nur dann verwendet werden, wenn experimentell festgestellt wurde, daß in dem untersuchten ontogenetischen Stadium des Individuums die Reaktion nicht über einen → *AAM* ausgelöst werden kann (SCHLEIDT 1962).
ARM – acquired releasing mechanism □ The *ARM* is an → *RM* which an individual has acquired in adaptation to a certain situation in its ontogeny. The term *ARM* should be used only if it has been proved experimentally that the reaction cannot be released by an → *IRM* (SCHLEIDT 1962).
MDA – mécanisme déclencheur acquis □ Le *MDA* est un → *MD* qu'un individu a acquis au cours d'une adaptation ontogénétique à des situations déterminées. Le terme de *MDA* ne doit être utilisé que dans le cas où il a été prouvé expérimentalement que dans le stade ontogénétique considéré de l'individu, la réaction ne peut être déclenchée par un → *MDI* (SCHLEIDT 1962).
Echolotung → Echo-Orientierung
Echo-Orientierung □ Dieses Phänomen wurde erstmals von SPALLANZANI (1793) an Fledermäusen entdeckt; sie stoßen während des Fluges Ultraschall-Orientierungslaute aus, um anhand des Echos Gegenstände, wie Hindernisse und Beute, zu lokalisieren. Gewisse Nachtfalter können diese »Feindlaute« wahrnehmen und fliehen. Auch bei Zahnwalen (Odontoceti) wurde Echo-Orientierung nachgewiesen, die bei Delphinen bisher am besten untersucht ist.
echolocation □ This phenomenon has been recorded for the first time by SPALLANZANI (1793) in bats; they produce ultra-sonic emissions during flight that are used to locate by means of the echo obstacles and prey. Certain moths are capable of detecting these sounds which then release escape manoeuvres. Echolocation occurs also in whales (Odontoceti) and has most thoroughly been studied in dolphins.
écholocalisation □ Ce phénomène a été découvert pour la première fois par SPALLANZANI (1793) chez les Chauves-Souris; celles-ci émettent au cours de leur vol des ultrasons pour localiser à l'aide de l'écho des objets tels que des obstacles et leur proie. Un certain nombre de Papillons nocturnes sont capables de percevoir ces émissions sonores ce qui leur permet de fuir. L'orientation par écholocalisation a également été mise en évidence chez les Baleines Odontoceti et a été particulièrement bien étudiée chez les Dauphins.
Echo-Ortung → Echo-Orientierung
Echopeilung → Echo-Orientierung
Eckzahndrohen □ Drohverhalten mit Ausdrucksmimik bei einigen Hirscharten. Die Mimik entsteht durch ein Hochziehen der Oberlippe entlang des ganzen Mundspaltes, wodurch nahe den Mundwinkeln die rudimentären Eckzähne des Oberkiefers sichtbar werden. Die Augen zeigen infolge der Blickrichtung das Weiß der Augäpfel, da der drohende Hirsch schräg frontal zum bedrohten Rivalen steht. Nacken- und Rückenlinienhaare werden dabei aufgestellt. Diese Verhaltensweise ist an ganz bestimmte soziale Auseinandersetzungen gebunden. Sie spielt eine wichtige Rolle gegenüber Rivalen beim Vorgang der Rangumkehr nach dem Geweihabwurf und beim → *Herden* der ♀♀ (BÜTZLER 1974). Dieses Entblößen und Präsentieren der rudimentären Eckzähne ist von jenem Drohverhalten stammesgeschichtlich abgeleitet, das wir vor allem von primitiven Wiederkäuern kennen, welche noch gut entwickelte Eckzähne besitzen; – *Tragulidae*, sowie *Moschus*, *Muntiacus* und *Elaphodus* (DUBOST 1971, 1975).
canine-tooth threat □ A facial expression repre-

senting threat in several deer species. The upper lip is drawn back along the entire upper jaw exposing the rudimentary canine teeth in the corner of the mouth. The eyes at the same time appear white to the rival due to the diagonal stance of the animal; only the side of the eyeball is exposed to the opponent. The hair on the neck and back is also erected. Canine-tooth threat is associated with specific types of confrontations, i. e., it is important in the change of rank which often follows antler shedding and in → *herding* of the ♀♀ (BÜTZLER 1974). The exposing and presenting of the vestigial teeth can be regarded as a behaviour pattern phylogenetically derived from the threat gesture of the most primitive ruminants, like the *Tragulidae, Moschus, Muntiacus* and *Elaphodus,* which have at present very developed canine teeth (DUBOST 1971, 1975).

menace en découvrant les canines □ Comportement de menace très expressif chez un certain nombre de Cervidae ♂♂. Cette mimique se produit par un retroussement de la lèvre supérieure qui découvre ainsi les canines rudimentaires de la mâchoire. L'animal qui menace se tient en position oblique par rapport à son rival; le regard est dirigé latéralement vers celui-ci, ainsi le blanc des yeux devient très apparent. Les poils de la nuque et de la ligne dorsale sont alors érigés. Cette menace est effectuée également par un ♂ qui veut éloigner un rival qu'envers les ♀♀ lors du → *Herden* (BÜTZLER 1974). Elle semble cependant être davantage utilisée après la chute des bois où elle joue un rôle important lors de l'inversion de la hiérarchie sociale. Elle dérive phylogénétiquement de la menace qu'effectuent des Ruminants primitifs encore pourvus de canines bien développées, tels que *Tragulidae, Moschus, Muntiacus* et *Elaphodus* (DUBOST 1971, 1975).

Efferenz □ Erregungsleitung vom Zentralnervensystem in die Körperperipherie im Gegensatz zur → *Afferenz.* Über die efferenten Nerven wird die Funktion der Skelettmuskulatur, der inneren Organe und aller anderen Effektoren gesteuert. → *Reafferenzprinzip.*

efference □ Nerve impulses running from the central nervous system to the periphery in contrast to → *afference.* The efferent nerves control the function of the skeletal musculature as well as certain inner organs and effectors. → *reafference principle.*

efférence □ Transmission des influx depuis le système nerveux central vers la périphérie du corps, par opposition à l'→ *afférence.* Les nerfs efférents dirigent le fonctionnement de la musculature squelettique, des organes internes et de tous les autres effecteurs. → *principe de réafférence.*

Eiablageaufforderungsverhalten □ Bei einigen revierbesitzenden Libellen, insbesondere bei *Orthetrum,* bei welchen es kein → *Eiablageplatzzeigen* gibt, wird das ♀ erst *nach* der Begattung vom ♂ durch auffälliges Flugverhalten aufgefordert, ihm zu folgen, und so zum Eiablageplatz geführt (Abb. 39 –

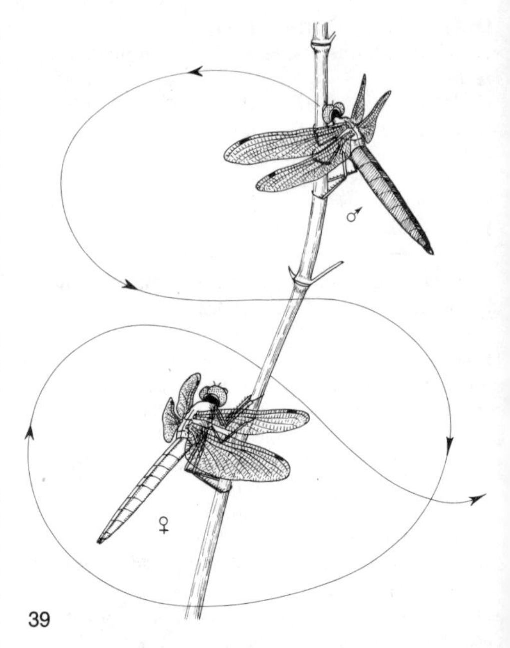

39

Orthetrum brunneum; HEYMER 1969).

egg-laying eliciting behaviour □ In some territorial dragonflies, in particular *Orthetrum,* which lack a → *display showing of the egg-laying site,* the ♂ may entice the ♀ *after* copulation towards the egg-laying site by a conspicuous flight pattern (Fig. 39 – *Orthetrum brunneum;* HEYMER 1969).

invitation à la ponte □ Chez un certain nombre d'Odonates territoriaux, en particulier chez le genre *Orthetrum,* la présentation du lieu de ponte n'a lieu qu'*après* l'accouplement. A l'aide d'un vol particulier, le ♂ invite la ♀ à le suivre et la conduit ainsi vers le lieu de ponte (Fig. 39 – *Orthetrum brunneum;* HEYMER 1969).

Eiablageplatzzeigen □ Bei einigen revierbesitzenden Libellen, insbesondere bei den Calopterygidae, eine epigame Verhaltensweise. Das ♂ führt dabei sein ♀ *vor* der Begattung zum Eiablageort und zeigt durch auffällige Verhaltensweisen dem ♀ an, daß es hier seine Eier abzulegen habe (Abb. 40 – *Calopteryx haemorrhoidalis*). Das ♀ findet diesen Ort nach der Begattung allein wieder (HEYMER 1973).

display showing of the egg-laying site □ An epigamic behaviour pattern of some territorial dragonflies, in particular in the Calopterygidae, in which the ♂, *before* mating the ♀, leads her to a site where she may lay her eggs. He indicates the spot by a conspicuous sequence of movements (Fig. 40 – *Calopteryx haemorrhoidalis*), and the ♀ subsequently recognizes the site and can return to it alone (HEYMER 1973).

présentation du lieu de ponte □ Comportement épigame chez un certain nombre de Libellules terri-

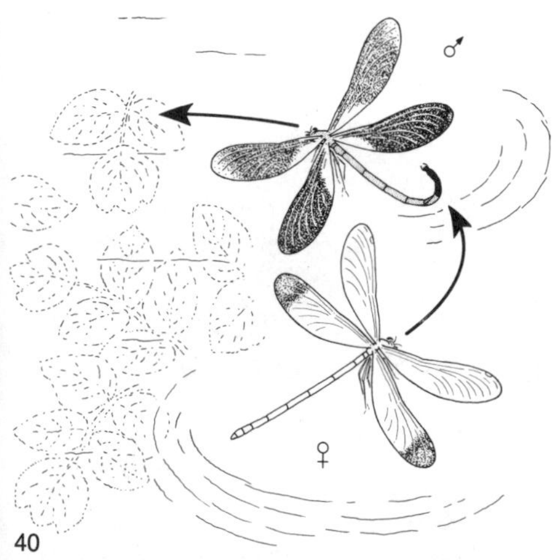

40

toriales, en particulier chez les Calopterygidae. Le ♂ conduit *avant* l'accouplement la ♀ vers le lieu de ponte et lui montre, par un comportement particulier, que c'est là qu'elle doit déposer ses œufs (Fig. 40 – *Calopteryx haemorrhoidalis*). Après l'accouplement, la ♀ se dirige seule vers le lieu de ponte (HEYMER 1973).

Eiattrappen → Eiflecke

E. imitation eggs □ F. imitation d'oeufs

Eiflecke □ Eiattrappen an der Afterflosse verschiedener afrikanischer maulbrütender Cichliden (Abb. 41 A – *Haplochromis wingatii,* B – *H. burtoni,* C – *H. schubotzi,* D – *Schubotzia eduardiana*). Es handelt sich nicht um → *Augenflecke.* Das ♀ nimmt seine gelegten unbesamten Eier sehr schnell mit dem Maul auf; danach besamt das ♂ den leeren Boden, wobei das ♀ versucht, die Eiflecke wie die richtigen Eier aufzunehmen: dabei bekommt es die Spermien ins Maul. Die Eiflecke haben also Signalfunktion (WICKLER 1962).

egg spots □ Imitations of eggs on the anal fin of various African mouth-breeding cichlids (Fig. 41 A – *Haplochromis wingatii,* B – *H. burtoni,* C – *H. schubotzi,* D – *Schubotzia eduardiana*). They are different from → *eye spots.* The ♀ lays the eggs and takes them into her mouth. While the ♀ attempts to pick up the egg-like spots on the ♂, the ♂ spawns and in this manner the eggs are fertilized. Hence egg spots function as a signal (WICKLER 1962).

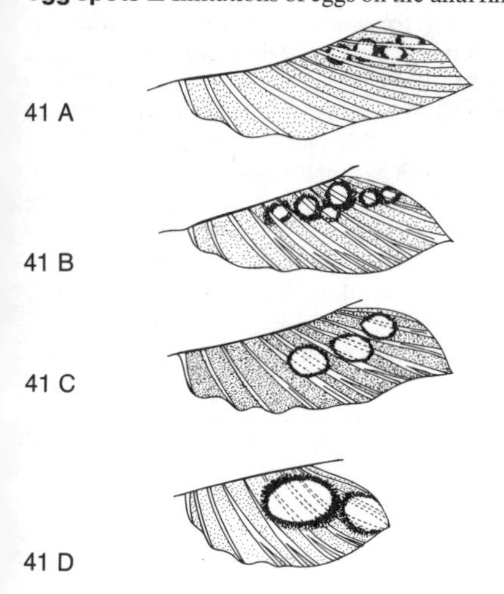

41 A

41 B

41 C

41 D

taches imitant des oeufs □ Taches se trouvant sur la nageoire anale d'un certain nombre de Cichlidés africains, incubateurs buccaux (Fig. 41 A – *Haplochromis wingatii,* B – *H. burtoni,* C – *H. schubotzi,* D – *Schubotzia eduardiana*). Il ne s'agit pas de → *taches ocellaires.* Chez ces espèces, la ♀ qui vient de pondre reprend très vite ses propres œufs non fécondés dans la bouche, puis le ♂ éjacule sur le sol. Comme la ♀ suit le ♂ de très près et essaie de saisir également ces imitations d'œufs sur la nageoire anale du ♂, elle prend ainsi le sperme dans sa bouche. Ces taches assument donc une fonction signalétique (WICKLER 1962).

Eigenerregbarkeit → Erregung

E. autochthonous excitation □ F. excitation autochtone

Einander-Lausen → Körperpflegehandlungen

E. allo-grooming □ F. épouillage mutuel

Einemsen □ Eine angeborene Verhaltensweise bei manchen Vögeln, die darin besteht, das giftige Sekret der Ameisen ins Gefieder zu bringen (STRESEMANN 1935, LÖHRL 1965, QUERENGÄSSER 1973). Einige Arten setzen sich dazu direkt in die Ameisenhaufen, andere nehmen einzelne Ameisen in den Schnabel und streichen sich damit über das Gefieder (Abb. 42 – *Sturnus vulgaris*).

42

anting □ An innate behaviour in several bird species that serves to spread the poisonous secretions of ants over the plumage (STRESEMANN 1935, LÖHRL 1965, QUERENGÄSSER 1973). Some species sit down directly onto the ant hill whereas others take single ants into their beak and rub them over the plumage (Fig. 42 – *Sturnus vulgaris*).

bain de fourmis □ Comportement inné chez un certain nombre d'Oiseaux qui cherchent à se badigeonner le plumage avec les sécrétions venimeuses des Fourmis (STRESEMANN 1935, LÖHRL 1965, QUERENGÄSSER 1973). Dans ce but, quelques espèces se posent directement sur la fourmilière tandis que d'autres prennent les Fourmis une par une dans leur bec pour enduire leur plumage (Fig. 42 – *Sturnus vulgaris*).

Einsicht □ Man spricht in der Verhaltensforschung von Einsicht, wenn ein Lebewesen eine ihm neue Situation oder Aufgabe meistert, ohne daß dazu vorgebildete angeborene Verhaltensweisen (→ *AAM*) oder Instinktbewegungen zur Verfügung stehen und ohne daß es durch vorhergehendes Probieren die Lösung findet; BÜHLER (1922) spricht vom → *Aha-Erlebnis* als Kriterium der Einsicht. Sie besteht im Erkennen der Zusammenhänge, Ursachen und Wirkungen eines Sachverhaltes oder Geschehens, wahrscheinlich basierend auf einer gewissen Vertrautheit mit verschiedenen Anteilen, die zur Lösung des Problems beitragen (KÖHLER 1963, MÜLLER 1965). → *Versuchs-Erfolgs-Verhalten.*

insight □ Ethologists speak of insight when an individual is able to master a new situation or task without the help of pre-programed innate responses (→ *IRM*) and without previously having attempted the solution; BÜHLER (1922) calls the → *»aha«-experience* a criterion of insight. It consists of recognition of the relationships, causes and effects of a particular contingency or occurrence which is apparently based on familiarity with separate parts necessary for the solution (KÖHLER 1963, MÜLLER 1965). → *trial and success behaviour.*

compréhension soudaine □ En éthologie, on parle de »Einsicht« (compréhension subite) lorsqu'un animal résout une situation nouvelle ou un problème nouveau sans qu'il y soit génétiquement prédisposé (→ *MDI*) ou qu'il dispose de mouvements instinctifs appropriés et sans qu'il ait pu préalablement procéder à des essais; selon BÜHLER (1922), le → *Aha-Erlebnis* ou *l'eurêka d'Archimède* est le critère déterminant de la compréhension subite. Celle-ci est basée sur la reconnaissance des relations, des causes et des effets d'une situation, probablement dépendante d'une certaine familiarité avec les différentes composantes nécessaires à la solution du problème (KÖHLER 1963, MÜLLER 1965). → *comportement par essai et réussite.*

Eirollbewegung □ Eine bei vielen bodenbrütenden Vögeln vorkommende Bewegungsfolge (→ *Erbkoordination*), mit der Eier, die außerhalb des Nestes liegen, in die Nestmulde zurückgerollt werden (zusammenfassend bei WICKLER 1961). Legt man z. B. einer Graugans ein Ei außerhalb ihres Nestes hin, dann greift sie mit dem Schnabel über das Ei hinweg und rollt es, mit der Schnabelunterseite ziehend und sorgfältig balancierend, ins Nest zurück (Abb. 43 – *Anser anser*). Nimmt man der Gans das

43

Ei weg, wenn sie schon zum Eirollen angesetzt hat, dann läuft die Eirollbewegung im Leerlauf weiter, jedoch ohne die orientierende Bewegung (LORENZ und TINBERGEN 1938).

egg-rolling □ Motor pattern (→ *fixed action pattern*) frequently observed in ground breeding birds by means of which eggs that have rolled outside the nest are retrieved (summarized in WICKLER 1961). If one places the egg, e. g. of a graylag goose outside of its nest, it will retrieve the egg with the bill by gently rolling it and balancing it with the underside of the bill (Fig. 43 – *Anser anser*). If the egg is removed during the procedure, the goose completes the instinctive egg-rolling anyway, but without the oriented movement (LORENZ and TINBERGEN 1938).

roulage des oeufs vers le nid □ → *Coordination héréditaire* fréquente chez de nombreux Oiseaux nidificateurs au sol, par laquelle les œufs se trouvant en dehors du nid sont ramenés dans celui-ci en les roulant avec le bec (résumé chez WICKLER 1961). Lorsqu'on pose, par exemple chez l'Oie cendrée, un œuf en dehors du nid, elle place son bec au-devant de l'œuf et roule celui-ci vers le nid en le tirant et en le balançant avec précaution à l'aide de la face inférieure du bec (Fig. 43 – *Anser anser*). Si on enlève l'œuf alors que l'Oie a déjà commencé à le tirer, le mouvement de roulage se poursuit et s'achève à vide, mais sans le mouvement d'orientation (LORENZ et TINBERGEN 1938).

Ekelreaktion □ Kontraktion der Rachenmuskulatur, Speichelsekretion und Körpergefühle, die denen des Erbrechens ähnlich sind. Auch bei Tieren, insbesondere experimentell, auslösbar (VON HOLST und VON ST. PAUL 1960).

disgust reaction □ Contraction of the pharyngeal musculature, salivation and feelings resembling those accompanying vomiting. Also experimentally elicitable in animals (VON HOLST and VON ST. PAUL 1960).

réaction de dégoût □ Contraction de la musculature du pharynx, sécrétion de salive et sensations corporelles similaires à celles du vomissement. Cette réaction peut être déclenchée expérimentalement chez les animaux (VON HOLST et VON ST. PAUL (1960).

Ektohormone → Pheromone

Elternhocker → Tragling
E. »parent clinger« □ F. »agrippeur«
Elternkumpan → Kumpan
E. parent companion □ F. compagnon parental
Endhandlung □ Auf das → *Appetenzverhalten* folgt als letzte Phase die Endhandlung. Sie beendet den jeweils aktivierten Verhaltensablauf und führt unter natürlichen Bedingungen zu einer Triebbefriedigung (CRAIG 1918). Bei Handlungen aus vielen Gliedern haben spätere mehr Endhandlungscharakter als ihnen vorausgehende; ein Glied kann zugleich »Appetenzverhalten« für das folgende und »Endhandlung« für das vorausgehende sein.
consummatory act □ The last phase of an → *appetitive behaviour* sequence is the consummatory act. It ends the elicited series of actions and leads under normal conditions to drive reduction (CRAIG 1918). In behaviour sequences that consist of many components, the last ones have rather more the quality of a consummatory act than the preceding ones; a given component may simultaneously represent an »appetitive behaviour« for the following one and a »consummatory act« for the preceding one.
acte consommatoire □ Dans l'activité instinctive, l'acte consommatoire vient à la suite du → *comportement d'appétence.* Il termine la séquence du comportement activé et conduit dans des conditions naturelles à la disparition de la pulsion (CRAIG 1918). Lors d'actions composées de plusieurs actes consécutifs, les derniers ont davantage un caractère d'acte consommatoire que ceux qui précèdent; ainsi un acte peut être en même temps »comportement d'appétence« pour celui qui suit et »acte consommatoire« pour celui qui précède.
Entbehrungserlebnis → Frustration
Entfernungsweisung □ Bienentänzerinnen können beim → *Schwänzeltanz* ihre Nestgenossen über die Entfernung der Futterquelle unterrichten. Je größer die Entfernung ist, desto langsamer werden die Umläufe (zusammenfassend bei v. FRISCH 1965) → *Richtungsweisung.* Feldsperlinge zeigen durch verschiedene Laute an, wie weit sie wegfliegen wollen (DECKERT 1968).
distance indication □ Bee dancers can inform their hive-mates about the distance of a food source with the → *waggle dance.* The greater the distance the slower the bee turns (summarized in v. FRISCH 1965) → *indication of direction.* Field sparrows indicate by different calls how far they will fly away (DECKERT 1968).
indication de distance □ Chez les Abeilles, les danseuses peuvent indiquer à leurs congénères à l'aide d'une → *danse frétillante* la distance à laquelle se trouve une source de nourriture. Plus la distance est grande, plus la vitesse de la danse est faible (résumé chez v. FRISCH 1965) → *indication de direction.* Les Moineaux friquets indiquent par des cris différents la distance qu'ils vont parcourir (DECKERT 1968).
Enthemmungshypothese □ Erklärung für Übersprungverhalten; nach EIBL-EIBESFELDT (Grundriß) können prinzipiell alle Übersprungbewegungen mit der Enthemmungshypothese erklärt werden. Die TINBERGEN'sche Hypothese des zentralen Übersprungs ist bis heute weder widerlegt noch bewiesen. → *Übersprungbewegung.*
disinhibition hypothesis □ An explanation for displacement behaviour; EIBL-EIBESFELDT (Ethology) suggests all displacement behaviour can be explained in this way. The hypothesis of central displacement set up by TINBERGEN has so far neither been proven nor disproven. → *displacement activity.*
hypothèse de la désinhibition □ Une explication pour le comportement de substitution; selon EIBL-EIBESFELDT (Ethologie), tous les mouvements de substitution peuvent, en principe, être expliqués par cette hypothèse. L'hypothèse de TINBERGEN de la substitution centrale n'est, jusqu'à ce jour, ni réfutée ni prouvée. → *activité substitutive.*
epigame Verhaltensweisen □ Ausdrucksbewegungen, die mit dem Fortpflanzungsverhalten unmittelbar zusammenhängen, werden als epigam bezeichnet. Entsprechend sind dann nichtepigame Verhaltensweisen solche, die im Normalfall außerhalb des Fortpflanzungsverhaltens auftreten.
epigamic behaviour □ Expressive movements which are directly related to reproduction. Conversely, non-epigamic behaviours refer to those patterns which normally occur apart from reproduction.
comportement épigame □ Mouvements expressifs en relation directe avec le comportement reproducteur. Par conséquent, les comportements non épigames sont ceux qui normalement apparaissent en dehors du contexte du comportement reproducteur.
Erbkoordination □ Im Verhaltensrepertoire eines Tieres trifft man auf wiedererkennbare, mithin »formkonstante« Bewegungen, die vom Tier nicht erst gelernt werden müssen und die, so wie körperliche Merkmale, Kennzeichen der Art sind. Es handelt sich gewissermaßen um ein genetisch fixiertes *angeborenes Können.* Man nennt solche angeborenen Bewegungsweisen Erbkoordinationen oder Instinktbewegungen, wobei der Name Erbkoordination bereits ausdrückt, daß das Angeborensein das entscheidende Kriterium dieser Bewegungsabläufe ist (LORENZ 1939, SCHLEIDT 1974).
fixed action pattern □ German term translated means: *inherited movement coordination.* In the behavioural repertoire of an animal one repeatedly observes certain movements which have a constant form and which need not to be learned, and those movements like morphological traits are characteristic of the species. It refers to a kind of genetically fixed *inborn ability.* One calls such innate movements fixed action patterns or innate behaviour patterns. Here the »fixed« does not mean rigid and invariable. *Fixed action pattern* has become established in Eng-

lish usage, and hence the meaning is sometimes misunderstood (LORENZ 1939, SCHLEIDT 1974).

coordination héréditaire □ Dans le répertoire comportemental d'un animal, on rencontre des mouvements identifiables, car constants dans leur forme, que l'animal n'a pas besoin d'apprendre et qui sont des caractères de l'espèce tout comme les traits morphologiques. Il s'agit en quelque sorte d'un *pouvoir (savoir) inné* génétiquement fixé. On appelle ces mouvements innés des coordinations héréditaires ou des mouvements instinctifs. Le terme de coordination héréditaire exprime clairement que l'inné est le critère décisif de ces processus locomoteurs (LORENZ 1939, SCHLEIDT 1974).

Erdmagnetfeld → Magnetfeldorientierung
E. earth magnetic field □ F. champ magnétique terrestre

Erfahrung □ Entsteht durch Speichern von Erlebtem oder Erlerntem. Nach heutigem Wissensstand setzt Erfahrungsbildung, z. B. echtes Lernen, ein einigermaßen zentralisiertes Nervensystem voraus. Beispiele für tierische Erfahrungsinhalte sind Ortserinnerung, z. B. an das eigene Revier, und erlernte Handlungsabläufe.

experience □ Arises through storage of learned contingencies. According to our present state of knowledge, true retention of experience presumes the presence of a somewhat centralized nervous system. Examples of experience formation (retention) in animals are site orientation memory, such as of the territory, and learned behaviour sequences as well as stimulus-response relationships.

expérience □ Elle se forme par le stockage du vécu et de l'appris. D'après nos connaissances actuelles, la formation d'une expérience, comme par exemple l'apprentissage véritable, présuppose l'existence d'un système nerveux central organisé. Des exemples pour la formation d'une expérience chez les animaux sont la mémorisation des lieux, la capacité de retour au gîte et les séquences comportementales apprises.

Erfahrungsentzug → Kaspar-Hauser-Versuch

Erinnerung → Gedächtnis
E. memory □ F. mémoire

Erkundungsverhalten □ Verhalten zur Herstellung der arttypischen Raum-Zeit-Beziehung, dem obligatorische Lernvorgänge zugeordnet sind. Durch das Erkundungsverhalten lernt ein Tier seinen Nestort und alle Raumstrukturen kennen, die der Orientierung im Lebensraum dienen.

exploratory behaviour □ Behaviour which produces the species typical orientation in time and space necessary for effective learning. Through exploratory behaviour the animal learns to know its nest site and its surroundings.

comportement exploratoire □ Comportement permettant l'établissement d'un système de relations spatio-temporelles propres à une espèce donnée et faisant appel à des processus d'apprentissage obligatoires. A l'aide du comportement explorateur, l'animal apprend la localisation du gîte et toutes les relations spatiales nécessaires à son orientation dans son domaine vital.

Erlebnisenttäuschung → Frustration

Ernstkampf → Beschädigungskampf

Ernstverhalten □ Im Gegensatz zum → *Spielverhalten* und dem → *Kommentkampf* sind alle Hemmungen abgebaut, und es kommt zu unkontrollierten Kämpfen. → *Beschädigungskampf.*

serious behaviour, behaviour with true intent □ In contrast to → *play behaviour* and → *ritualized fight* all inhibitions are removed and real uncontrolled fighting occurs. → *damaging fight.*

comportement de combat authentique □ Contrairement au → *comportement ludique* et au → *combat rituel,* tous les mécanismes inhibiteurs sont abolis et il se produit alors des luttes incontrôlées et dommageables. → *combat sanglant.*

Erregbarkeit → Erregung
E. excitability □ F. excitabilité

Erregung □ Die unmittelbare, charakteristische, zeitlich begrenzte nervöse Antwort, die in einem lebenden Objekt durch die Einwirkung eines äußeren Reizes oder durch innere Ursachen ausgelöst wird. Die Erregbarkeit ist eine der Grundeigenschaften der Lebewesen, die darauf beruht, daß eine als Reiz wirkende Milieu- oder Zustandsänderung einen nach eigenen Gesetzen ablaufenden Erregungsprozeß auslöst. Weiter schreibt LEYHAUSEN (1967), daß »Angstinstinkte eine besonders hohe Eigenerregbarkeit haben und daher mit Wahrscheinlichkeit meist aktionsbereiter sind als alle anderen Instinkte«.

Die Versuche von VON HOLST (1934, 1935) mit nicht völlig desafferenzierten Fischen haben gezeigt, daß eine offenbar endogene Erregungsproduktion des Zentralnervensystems vorliegt, ein Motor der Bewegung, der nicht von außen angestoßen zu werden braucht, und zweitens, daß diese zentralen Impulse auch zentral koordiniert werden. Eine solche zentrale Automatie dürfte auch der Atembewegung – Kiemendeckelbewegung bei Fischen – zugrunde liegen (LISSMANN 1946).

excitation □ The behavioural response of an organism to an exterior stimulus or internal change in milieu; this nervous response is immediate, characteristic and limited in time. Excitability is a basic property of living organisms in which changes in the milieu or organismic state elicit an arousal response with inherently defined characteristics. LEYHAUSEN (1967) suggests that »fear instincts have a particular high autochthonous excitation and hence are probably more easily elicited than all other instincts«.

The experiments of VON HOLST (1934, 1935) show that an endogenous production of excitation of the CNS occurs even in incompletely deafferentiated fish implying both potential independance from external stimuli and central coordination of motor function. Such a spontaneous excitation mechanism probably also controls breathing – gill cover movements in fish (LISSMANN 1946).

excitation □ Réponse comportementale ou réaction nerveuse limitée dans le temps, immédiate et caractéristique, déclenchée chez un être vivant par l'effet d'une stimulation externe ou de causes endogènes. L'excitabilité est une propriété fondamentale de tout être vivant basée sur l'effet stimulateur d'un changement dans le milieu externe ou interne qui déclenche un processus d'excitation se déroulant selon ses lois propres. D'autre part, LEYHAUSEN (1967) dit que »les instincts de peur possèdent une excitation autochtone particulièrement élevée et sont, de ce fait, probablement plus prédisposés à l'action que tous les autres instincts«.
Les expériences de VON HOLST (1934, 1935) sur des Poissons incomplètement désafférentiés ont montré qu'il existe probablement une production d'un potentiel d'excitation endogène dans le système nerveux central, un entretien automatique de l'activité motrice sans la nécessité d'une stimulation extérieure. Ces impulsions endogènes obéiraient à une coordination centrale. Les mouvements respiratoires – mouvements des opercules branchiaux chez les Poissons – sont sans doute basés aussi sur un tel automatisme central (LISSMANN 1946).

Erregungsübertragung → Stimmungsübertragung
E. contagious excitation □ F. transmission d'excitation

Ersatzbefriedigung □ LEYHAUSEN (1967) bezeichnet das Betreiben gefährlichen oder gefährlich scheinenden Sports als eine Art Ersatzbefriedigung zur Abreaktion von Angstinstinkten. Andere wiederum finden Ersatzbefriedigung in einem Kriminalroman oder Schauerfilm. Ersatzbefriedigung findet man bei Menschen und Tieren, aber auch noch auf vielen anderen Ebenen.

substitute gratification □ LEYHAUSEN (1967) suggests that dangerous or dangerous-looking sports serve as a releasing valve for fear instincts and, hence function as substitute gratification. Others find substitute gratification in murder stories or horror films. Substitute gratification occurs in both man and animals on many different levels.

satisfaction substitutive □ LEYHAUSEN (1967) considère la pratique de sports dangereux ou paraissant tels comme une sorte de satisfaction de remplacement pour éliminer les instincts de peur. D'autres individus trouvent une satisfaction substitutive dans la lecture de romans policiers ou dans les films d'horreur. Les phénomènes de satisfaction substitutive chez l'Homme et chez les animaux sont présents également à de nombreux autres niveaux.

Ersatzhandlung → Ersatzobjekt
E. redirected activity □ F. activité de redirection

Ersatzobjekt □ Bezeichnung für einen Gegenstand, der bei langem Ausbleiben adäquat auslösender Situationen die Endhandlung auslösen kann, ohne daß sich ihr biologischer Sinn erfüllt. Die Ersatzhandlung, eine am »falschen Objekt« ausgeführte Triebhandlung, stillt den zugrundeliegenden → *Trieb* nicht, kann ihn aber beschwichtigen.

substitute object □ Term used for an object that can elicit a consummatory act in the absence of appropriate stimulation and without necessarily fulfilling the biological need underlying the behaviour. The redirected activity, an act directed towards a substitute object or organism, does not satisfy the basic → *drive* but can hold it at abeyance.

objet de remplacement, substitut □ Terme désignant un objet qui peut déclencher l'acte consommatoire en cas d'absence prolongée de situations adéquates, mais sans que la signification biologique de celui-ci soit remplie. L'activité de redirection sur un objet inadéquat lorsque l'objet préalablement visé ne peut être atteint, ne fait pas disparaître la → *pulsion* sous-jacente, mais l'apaise.

Erwerbkoordination, erlernte Bewegungsfolgen □ In England lernten z. B. Meisen, Milchflaschen zu öffnen. Durch neue Kombinationen ihrer angeborenen Verhaltensweisen des Nahrungserwerbs entwickelten sie verschiedene Bewältigungsmethoden. Die Gewohnheit des Flaschenöffnens breitete sich geographisch aus, was für → *Tradition(sbildung)* spricht (FISHER und HINDE 1949). Solche erlernten Bewegungsfolgen kann man als Erwerbkoordination bezeichnen, obgleich in ihnen → *Erbkoordination(en)* als Elemente enthalten sind.

acquired behaviour pattern □ In England tits, e. g., have learned to open milk bottles; such strategies for food acquisition have developed from new combinations of innate behaviour sequences. The habit of opening milk bottles spread geographically suggesting the formation of a → *tradition* (FISHER and HINDE 1949). Such learned behaviour patterns are called an acquired behaviour pattern though components of → *fixed action pattern(s)* are embodied in them.

coordination acquise – performances comportementales apprises □ En Angleterre, par exemple, des Mésanges ont appris à décapsuler les bouteilles de lait. A l'aide de nouvelles combinaisons des éléments de leur comportement alimentaire inné, elles ont développé plusieurs méthodes. L'habitude de décapsuler les bouteilles s'est propagée géographiquement ce qui implique une formation de → *tradition* (FISHER et HINDE 1949). De telles capacités apprises peuvent être considérées comme une coordination acquise bien que des → *coordination(s) héréditaire(s)* élémentaires y soient impliquées.

Ethogenese □ Die Entwicklung und → *Reifung* von angeborenen Verhaltensweisen im Laufe der postembryonalen bzw. Jugendentwicklung. Bislang sprach man von *Ontogenese des Verhaltens* oder von *Verhaltensontogenese*. Da der von ERNST HAECKEL

(1866) geprägte Ausdruck aber seit langem in der Embryologie benutzt wird, was auch seiner Etymologie eher entspricht, schlägt JAISSON (1974) die Einführung von Ethogenese vor.

ethogenesis □ The development and → *maturation* of innate behaviour patterns during postembryonic and early-childhood development. Until recently, the term *ontogenesis of behaviour* or *behavioural ontogenesis* has been used. However, since the term ontogenesis coined by ERNST HAECKEL (1866) is commonly employed in reference to embryology, which is also more in conformity with its etymology, JAISSON (1974) suggested the introduction of the term ethogenesis.

éthogenèse □ Développement et → *maturation* des comportements innés au cours du développement postembryonnaire ou juvénile. Jusqu'alors, on parlait de *l'ontogenèse du comportement.* Comme le terme ontogenèse, depuis sa création par HAECKEL (1866), est surtout utilisé en embryologie, ce qui est d'ailleurs plus conforme à son étymologie, JAISSON (1974) propose l'introduction du terme éthogenèse.

Ethogramm □ Grundlage jeder ethologischen Untersuchung ist das Ethogramm, der genaue Katalog und die genaue Beschreibung aller einer Tierart eigenen Verhaltensweisen, einschließlich aller vorhandenen Lautäußerungen (Lautinventar). Die Forderung, jede Verhaltensuntersuchung mit der Aufstellung eines Verhaltensinventars zu beginnen, erhob bereits JENNINGS (1906), der von Aktionssystem sprach. Diese Forderung wurde seitdem von KONRAD LORENZ immer wieder erhoben. Ein brauchbares Ethogramm des Menschen fehlt bis heute.

ethogram □ The basis of every ethological study is the ethogram, an exact catalogue of all behaviour patterns occurring in an animal species, including all vocal patterns (vocal inventory). The necessity of compiling an inventory of specific behaviour before beginning experimental studies was already proposed by JENNINGS (1906) who called such an inventory an action system. Since that time KONRAD LORENZ raised this claim again and again. A satisfactory ethogram of man has not yet been established.

éthogramme □ La base de toute analyse comportementale est l'éthogramme, c'est-à-dire l'inventaire complet et la description exacte de tous les types de comportement d'une espèce animale, y compris tous les signaux vocaux (vocabulaire sonore). Cette exigence de commencer chaque analyse éthologique par l'établissement d'un inventaire comportemental a déjà été formulée par JENNINGS (1906); il appelait cela un système d'action. Une exigence qui depuis lors a toujours été réitérée par KONRAD LORENZ. Cependant, un éthogramme convenable de l'Homme n'a pas encore été établi jusqu'à présent.

Ethologie – Biologie des Verhaltens □ Die objektive Erforschung von Tier und Mensch aus biologischer Sicht und unter besonderer Hervorhebung des artspezifischen Verhaltens, seiner Angepaßtheit (Funktion) und seiner Evolution. Die Ethologie bedient sich der biologischen Methoden, die in der Systematik, Morphologie und Embryologie bereits mit Erfolg zur Klärung von Vorgängen und Zusammenhängen eingesetzt wurden, und wendet diese auf die Untersuchung des Verhaltens an. Sie bemüht sich sowohl um eine ganzheitliche Erfassung normalen Verhaltens als auch um eingehende Analyse der einzelnen Phasen im Verhalten eines Organismus. Der Begriff Ethologie wurde von GEOFFROY SAINT-HILAIRE (1854) eingeführt, jedoch meinte er damit genau das, was HAECKEL (1866) *Ökologie* nannte. Einen eigenen Wissenschaftszweig entsprechend unserer heutigen Biologie des Verhaltens begründete DOLLO (1895) und *expressis verbis* (1909), der auch den Ausdruck Ethologie übernahm. Zwei inzwischen entstandene spezielle Arbeitsrichtungen der vergleichenden Verhaltensforschung sind die → *Humanethologie* und die → *Kulturethologie.* Einen historischen Überblick über das erste Jahrhundert Tierethologie gibt KLOPFER (1974).

Ethology – Biology of Behaviour □ The objective study of animals and man from a biological point of view with emphasis on species-typical behaviour, its adaptiveness (function) and evolution. Ethology applies the same methods for the study of behaviour which systematics, morphology and embryology have used so successfully in unravelling processes and relationships. Furthermore, it constitutes a holistic approach to the study of normal behaviour while it embodies a detailed analysis of all phases of an organism's behaviour. The term ethology was introduced by GEOFFROY SAINT-HILAIRE (1854), but it had exactly the same meaning as HAECKEL's (1866) *ecology.* An independant scientific branch corresponding to the actual biology of behaviour was founded by DOLLO (1895) and *expressis verbis* (1909) who also adopted the term ethology. Two specialized fields of research, → *human ethology* and → *cultural ethology,* have developed within comparative ethology. An historical survey of the first century of animal ethology is given by KLOPFER (1974).

éthologie – Biologie du comportement □ Etude objective de l'animal et de l'Homme d'un point de vue biologique, avec un accent particulier sur le comportement spécifique, son adaptation (fonction) et son évolution. L'éthologie se sert des méthodes biologiques ayant fait leur preuve en systématique, en morphologie et en embryologie où elles ont contribué à éclaircir les différents processus biologiques, et elle les applique à l'étude du comportement. L'éthologie cherche autant à établir l'approche holistique du comportement normal qu'à entreprendre des analyses des diverses phases comportementales d'un organisme. Le terme d'éthologie a été créé par GEOFFROY SAINT-HILAIRE (1854) qui désignait par ce mot exactement ce que HAECKEL (1866) appelait alors *écologie.* Une science indépendante équivalente à ce que nous entendons aujourd'hui par »biologie du comportement« a été revendiquée par DOLLO (1895) et consacrée *expressis verbis* (1909), tout en utilisant le terme d'éthologie. Depuis lors, deux directions de travail spécialisées de l'étude comparée du comportement se sont développées sous les appellations de → *éthologie humaine* et de → *éthologie culturelle.* Une vue d'ensemble sur le premier siècle d'éthologie animale est donnée par KLOPFER (1974).

Ethometrie □ Von JANDER (1968) geprägter Ausdruck zur Bezeichnung von exakten ethologischen Meßverfahren zur Funktionsanalyse von Auslösemechanismen und Schlüsselreizen.

ethometry □ Term coined by JANDER (1968) to denote the quantitative and functional analysis of releasing mechanisms and key stimuli.

éthométrie □ Notion forgée par JANDER (1968) pour désigner les méthodes de métrologie éthologique, servant à l'analyse fonctionnelle des mécanismes déclencheurs et des stimuli-clés.

Ethomimikry → Mimikry
E. ethomimicry □ F. éthomimétisme

Ethoparasit □ Verschiedenen Coleopteren ist es gelungen, sich bei Ameisen einzunisten; sie werden nicht nur von den Ameisen gefüttert, sondern überlassen ihnen auch die Ei- und Larvenpflege. Eine Anzahl von Arten haben es erreicht, durch Angebot der entscheidenden Schlüsselreize die Fütterung auszulösen. Diese Ameisengäste sind somit zu Ethoparasiten geworden. → *Trophallaxie.*

ethoparasite □ Various Coleopterans have become social parasites of ants in that they live in the ant colony and are fed by the ants. A number of forms have succeeded, by offering the appropriate key stimuli, in eliciting feeding behaviour from ants. Such social parasites are called ethoparasites. → *trophallaxis.*

éthoparasite □ Quelques Coléoptères ont réussi à s'introduire dans les fourmilières; ils sont non seulement nourris par les Fourmis, mais laissent à ces dernières également le soin de leurs œufs et de leurs larves. Un certain nombre d'espèces ont réussi, en fournissant les stimuli-clés déterminants, à déclencher le nourrissage. Ces hôtes de Fourmis sont, de ce fait, devenus des éthoparasites (Myrmécophiles, Termitophiles). → *trophallaxie.*

Ethospezies → Ethosystematik
E. ethospecies □ F. espèce éthologique

Ethosystematik □ Anhand ethologischer Merkmale gewonnene Aussagen über die Phylogenie. WHITMAN (1899, 1919) und HEINROTH (1911) entdeckten unabhängig voneinander an Wirbeltieren, daß bestimmte Verhaltensweisen ebenso konstante und kennzeichnende Merkmale von Arten, Gattungen und noch größeren Einheiten des zoologischen Systems sein können wie etwa morphologische und meristische Merkmale. Die ethologische Forschung hat immer deutlicher zeigen können, daß der Bauplan von Verhaltensweisen ganz wie der des Körperbaues in der Erbmasse der Art verankert ist. Es gibt viele Tierarten, die einander sehr ähnlich sind und die man morphologisch kaum trennen kann. Man unterscheidet sie aber leicht an ihrer Fortpflanzungsweise, an ihren Futterpflanzen oder ganz allgemein an ihrem unterschiedlichen Verhalten. Solche Arten nennt man *Ethospezies* (EMERSON 1956). Individuen einer Population, die sich in bestimmter Weise verschieden verhalten, bezeichnet man als *Ethotypen* (CURIO 1975). Dieser Ausdruck wurde bisher auf angeborene, d. h. genetische Unterschiede angewendet.

ethosystematics □ Statements about phylogeny based on ethological (behavioural) characteristics. WHITMAN (1899, 1919) and HEINROTH (1911) discovered independently of each other in vertebrates that certain behaviour patterns are constant and characteristic traits of animal species, genera and even greater systematic categories in the same way as morphological and meristic characters. Ethological investigations have clearly demonstrated that the structure of behaviour patterns is fixed in the inheritance of a species just as much as its morphological structure. There are many species that are very similar morphologically, but which can be distinguished by their patterns of reproduction, food preferences or by their behaviour in general. Such species are called *ethospecies* (EMERSON 1956). Individuals of a population showing a different behaviour than their conspecifics are called *ethotypes* (CURIO 1975). Previously, this term was applied to innate (inborn) differences.

systématique éthologique □ Conclusions phylogénétiques obtenues à l'aide de caractères éthologiques (comportementaux). WHITMAN (1899, 1919) et HEINROTH (1911), indépendamment l'un de l'autre, ont découvert chez les Vertébrés que certains modes comportementaux sont des caractères aussi constants et aussi significatifs des espèces, des genres ou même d'unités supérieures du système zoologique que le sont par exemple les caractères morphologiques et méristiques. La recherche éthologique a clairement démontré que l'organisation du comportement est aussi profondément ancrée dans le patrimoine héréditaire de l'espèce que les caractères morphologiques. De nombreuses espèces sont très semblables entre elles et difficiles à séparer morphologiquement. Par contre, il est facile de les distinguer à l'aide de leurs différents modes de reproduction et de nutrition ou, d'une manière plus générale, de leur comportement très différencié. De telles espèces sont appelées des *espèces éthologiques* (EMERSON 1956). Les individus d'une même population montrant dans des cas particuliers un comportement différent de leurs congénères sont appelés *éthotypes* (CURIO 1975) alors que cette expression ne désignait jusqu'à présent que des différences innées, c'est-à-dire génétiques.

Ethotypen → Ethosystematik
E. ethotypes □ F. éthotypes, types éthologiques
Explorationsverhalten → Erkundungsverhalten
Exterozeptoren □ Sinnesorgane, welche Außenreize wahrnehmen: Auge, Ohr, Getast, Wärme- und Kältesinn, Strömungs- und elektrischer Sinn. Im Gegensatz zu den auf Innenreize ansprechenden → *Propriozeptoren.*
exteroceptors □ Sense organs which perceive stimuli emanating from outside the body: eyes, ears, sense of touch, warmth and cold receptors, current and electrical receptors; in contrast to receptors that respond to internal information. → *proprioceptors.*
extérocepteurs □ Organes sensoriels percevant des stimulations extérieures: yeux, oreilles, sens tactile, perception de la chaleur et du froid, du courant et du sens électrique; en opposition aux → *propriocepteurs* servant à la perception des stimulations endogènes.

F

Fächelbiene → Flügelfächeln
E. fanning bee □ F. abeille ventileuse
Fächeln □ Das Befächeln der Gelege durch das brutpflegende ♂ bei verschiedenen Fischen (Blennioidei, Pomacentridae, Gasterosteidae). Nachdem z. B. das Stichlings-♂ 3 bis 4 Gelege besamt hat, flaut sein Sexualtrieb ab, und es beginnt, mit fächelnden Bewegungen der Brustflossen zu ventilieren: Es kommt in Fächelstimmung (VAN IERSEL 1953). Eine andere Art von Fächeln finden wir bei Bienen. → *Flügelfächeln.*
fanning □ The fanning of the eggs by the ♂ in various fish species (Blennioidei, Pomacentridae, Gasterosteidae). After the ♂ stickleback, e. g., has fertilized 3 to 4 clutches of eggs, his sex drive wanes and he begins to move the pectoral fins rhythmically indicating onset of fanning mood (VAN IERSEL 1953). Another kind of fanning occurs in honey-bees. → *wing fanning.*
ventilation □ Activité de ventilation par le ♂, pour aérer la ponte, chez un certain nombre de Poissons (Blennioidei, Pomacentridae, Gasterosteidae). Chez l'Epinoche ♂, par exemple, après fécondation de 3 ou 4 pontes, la pulsion sexuelle baisse; le ♂ commence alors à effectuer des mouvements de ventilation avec ses nageoires pectorales près de ses œufs. La motivation à ventiler est de plus en plus forte (VAN IERSEL 1953). Une ventilation différente existe chez les Abeilles. → *ventilation alaire.*
Fächerversuch □ In einer bestimmten Entfernung vom Stock werden Duftplatten in Winkelabständen von 15° fächerförmig aufgestellt, um den Streukegel ausfliegender Neulinge von Sammelbienen zu prüfen (zusammenfassend bei VON FRISCH 1965).
fanning out experiment □ Plates of odoriferous attractant are placed at a certain distance from the hive at angles of 15 degrees in a fanlike pattern to determine the flight dissemination pattern of new juvenile forager bees (summarized in VON FRISCH 1965).
expérience de l'éventail □ Il s'agit d'une expérience où, à une distance déterminée de la ruche, un certain nombre de plaques odorantes sont disposées en éventail et à 15° d'angle l'une par rapport à l'autre pour contrôler la dispersion des jeunes butineuses au moment de leur première sortie (résumé chez VON FRISCH 1965).
Fanghandlung → Beutefanghandlung
Feedback → Rückkoppelung
Fegen □ Ausdruck aus der Jägersprache; das Reiben, Schlagen und Wetzen des ausgebildeten Geweihs an Stämmchen und Sträuchern, um den Bast abzuscheuern (Abb. 44 – *Cervus canadensis*). Bei einigen Arten fegen die Hirsche nicht nur deshalb, sondern sie tun dies auch besonders heftig, wenn ein Rivale dabei zusehen kann. Es handelt sich dabei um eine ritualisierte Verhaltensweise, da das Bastabscheuern gegenüber dem Imponieren zurücktritt. Es ist so zum *Imponierfegen* geworden und beeindruckt entsprechend (GRAF 1956, BÜTZLER 1974).

44

antler-rubbing □ The rubbing and scratching of the antlers on branches and limbs which scrapes off the cuticle (Fig. 44 – *Cervus canadensis*). In some varieties of deer, this behaviour is especially pronounced in the presence of a rival. In this case, antler-rubbing can be considered as a ritualized behaviour pattern since scraping of the cuticle has lost its primary significance and has been modified to *display-polishing* (GRAF 1956, BÜTZLER 1974).
râclement des bois □ Chez les Cervidae, le fait de frotter et d'aiguiser les bois complètement développés contre des troncs d'arbres et des buissons pour les débarasser de leur velours (Fig. 44 – *Cervus canadensis*). Chez quelques espèces, les ♂♂ ne frottent pas uniquement leurs bois à cet effet, mais ils le font aussi d'une façon particulièrement marquée lorsqu'un rival est à proximité. Il s'agit alors d'un comportement ritualisé étant donné que le fait de débarasser les bois de leur velours devient secondaire et

que le comportement est principalement destiné à intimider les rivaux. Il s'agit donc plutôt d'un *râclement d'intimidation* (GRAF 1956, BÜTZLER 1974).

Fehlprägung → Objektprägung → Prägung
E. inappropriate imprinting □ F. fausse empreinte

Feindablenkung → Verleiten
E. predator diversion tactics □ F. diversion de l'ennemi

Feindanpassung □ Verhaltensweisen und Strukturen, die so dem Verhalten des Feindes angepaßt sind, daß sie vor ihm schützen. An einen gut geschützten, tarnfarbigen Zackenbarsch kann man sehr nahe herankommen, ebenso an einen sich drükkenden Feldhasen, während bunte Fische schneller flüchten und sich verstecken. Es gibt nach HEDIGER (1943) eine artspezifische → *Fluchtdistanz,* die durch individuelle Erfahrung variiert werden kann; das Tier paßt sich an den Feind an.

anti-predator adaptation □ Concerns behaviours or structures such as protective colouration which have evolved as defense mechanisms against predators. One can approach a well-camouflaged fish or a crouched hare quite closely whereas a bright coloured fish flees from the intruder sooner. According to HEDIGER (1943), each species has a specific → *flight distance* which may vary according to individual experience; the animal adapts itself to its predator.

adaptation au prédateur □ Comportement étroitement adapté au comportement du prédateur, mais qui est également en corrélation avec la coloration protectrice de l'animal. Il est facile d'approcher un Poisson bien camouflé ou un Lièvre tapi dans un creux tandis que des Poissons aux couleurs vives prennent relativement vite la fuite. Selon HEDIGER (1943), il existe une → *distance de fuite* spécifique qui peut varier selon l'expérience individuelle d'un animal; l'animal s'adapte donc étroitement au comportement du prédateur.

Feindattrappe □ Ein Objekt, welches den potentiellen Feind einer Tierart repräsentiert und zu Untersuchungen der für das Feinderkennen spezifischen → *Schlüsselreiz(e)* dient. Für Enten und Gänse z. B. löst eine Greifvogel-Flugbildattrappe (Abb. 45), nach links bewegt, Flucht- bzw. Alarmverhalten aus; das Feindschema ist hauptsächlich durch dunkle Silhouette, kurzer Hals, langer Schwanz gekennzeichnet. Bewegt man die → *Attrappe* dagegen nach rechts (Kormoran- oder Entenflugbild), so reagieren die Tiere nicht darauf (TINBERGEN 1948, CURIO 1963). Versuche von SCHLEIDT (1961 a+ b) stellen allerdings ein *angeborenes Feindschema* »Raubvogel« in Frage; siehe hierzu auch MÜLLER (1961).

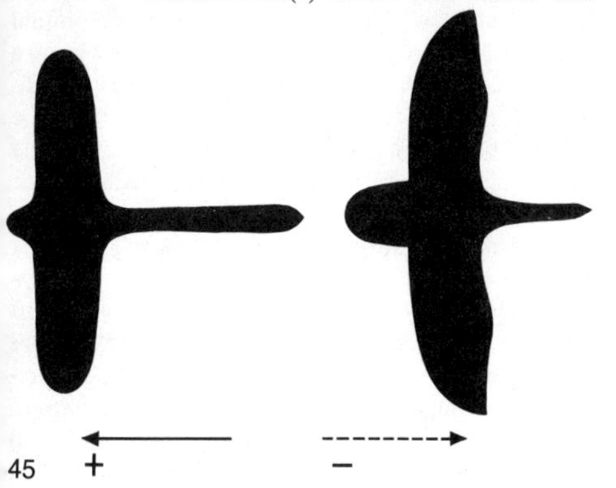

45

predator model □ An object representing the potential predator of an animal species and serving to study the specific → *key stimulus* in enemy recognition. For ducks and geese, for example, a silhouette of a bird of prey (Fig. 45) if moved to the left, releases fright or flight behaviour; the predatory configuration consists of a dark silhouette, short neck and long tail. If the same → *model* is moved to the right (cormorant or duck silhouette), they do not react (TINBERGEN 1948, CURIO 1963). The experiments made by SCHLEIDT (1961a + b), however, refute the existence of an *innate predator releaser;* see also MÜLLER (1961).

leurre représentant le prédateur □ Objet représentant l'ennemi potentiel d'une espèce animale et servant à l'étude des déclencheurs pour la reconnaissance du prédateur. Chez les Canards et les Oies, par exemple, un → *leurre* figurant l'image d'un Rapace en vol (Fig. 45) déclenche, s'il avance vers la gauche, un comportement d'alarme ou de fuite; ce schème est caractérisé par une silhouette sombre, un cou court et une queue longue. Lorsque ce même leurre est déplacé vers la droite (silhouette d'un Cormoran ou d'un Canard), les animaux ne réagissent pas (TINBERGEN 1948, CURIO 1963). Les expériences de SCHLEIDT (1961 a + b), cependant, mettent en doute l'existence d'un *schème inné du prédateur;* voir aussi MÜLLER (1961).

Feindreaktion □ Ein vom Feind ausgehender Reiz (Feindreiz), der ein bestimmtes Verhalten gegenüber diesem auslöst: Flucht oder → *Verleiten,* → *Warnverhalten.*

reaction to predators □ Response to a predator stimulus which releases a certain behaviour: flight or → *distraction display,* → *warning behaviour.*

réaction aux prédateurs □ Comportement déclenché par un stimulus émanant d'un prédateur: fuite ou → *manoeuvre de diversion,* → *comportement d'avertissement.*

Feindreiz → Feindreaktion

Feindschema → Feindattrappe
E. predator releaser □ F. schème perceptif ou inné du prédateur

Feindschutz → Schutzanpassung
E. anti-predator behaviour □ F. comportement anti-prédateur

Fellknabbern → Körperpflegehandlungen
E. fur nibbling □ F. mordiller la fourrure

Fellpflege → Körperpflegehandlungen
E. grooming the fur □ F. soins au pelage
Fellsträuben → Haarsträuben
Fernorientierung → Orientierung
E. distant orientation □ F. orientation lointaine

Fixierreaktion des Säuglings □ Ein ruhiges Fixieren der Schallquelle mit den Augen bei einem blindgeborenen Säugling, wenn die Mutter zu ihm spricht. Dann hört auch der Blindennystagmus auf. Fixierreaktionen kommen im übrigen bei allen neugeborenen sehenden Säuglingen, insbesondere beim Stillen, vor und dienen offenbar dazu, sich das Gesicht der Mutter »einzuprägen«.

fixation reaction behaviour in infants □ The quiet orientation toward a sound source with the eyes in the congenitally blind infant, e. g. when the mother speaks to it. Then the nystagmus ceases. Fixation reactions also occur in all seeing new-born infants, especially during nursing, and probably serve to memorize the features of the mother's face.

fixation oculaire chez les nourrissons □ Ce comportement s'observe chez les nourrissons aveugles-nés qui fixent calmement une source auditive avec leurs yeux, en particulier lorsque la mère leur parle. En même temps, les mouvements oculaires erratiques (nystagmus des aveugles) s'arrêtent. Du reste, ces mouvements de fixation oculaire s'observent chez tous les nouveaux-nés voyants, en particulier durant la tétée; ils ont pour fonction probable de mémoriser les traits de la mère.

Flehmen □ Eine bei Säugetieren weit verbreitete Verhaltensweise, bestehend aus Mundöffnen und Entblößen der Zähne durch Aufstülpen oder Hochschlagen der Oberlippe, Schließen der Nasenöffnungen bei mehr oder minder starkem Anheben des Kopfes (Abb. 46 – *Syncerus caffer* ♂). Es tritt be-

46

sonders während der Fortpflanzungszeit auf, wenn männliche Tiere Harn der Weibchen mit den Lippen aufgenommen haben, sowie nach Beriechen der Vulva. Erstmals ausführlich beschrieben von KARL MAX SCHNEIDER (1930–1933). Im Englischen (auch *lip-curl*, SCHALLER 1973) und im Französischen hat sich das deutsche »Flehmen« allgemein durchgesetzt (LEUTHOLD 1973, DUBOST 1975).

Flehmen □ A widely distributed behaviour pattern among mammals in which the mouth is opened, the teeth exposed by retracting the upper lip (*lip-curl;* SCHALLER 1973), the nasal openings closed, the head jerked back (Fig. 46 – *Syncerus caffer* ♂). It occurs particularly in the reproductive season after the male has licked the female's urine or vulva. First described by KARL MAX SCHNEIDER (1930–1933), the German term *Flehmen* is now in general usage in both English and French (LEUTHOLD 1973, DUBOST 1975).

Flehmen □ Comportement particulièrement répandu chez les Mammifères: la bouche est légèrement ouverte, les dents rendues visibles par retroussement de la lèvre supérieure, les narines fermées et la tête plus ou moins redressée (Fig. 46 – *Syncerus caffer* ♂). Ce comportement est particulièrement fréquent pendant la période de reproduction lorsqu'un mâle a prélevé l'urine de la femelle avec ses lèvres ou a flairé la vulve de cette dernière. Décrit pour la première fois d'une manière détaillée par KARL MAX SCHNEIDER (1930–1933). Faute d'une équivalence valable en français, l'expression allemande *Flehmen* est adoptée d'une manière générale (DUBOST 1975).

Fluchtbereitschaft □ Ein Zustand, der sich insbesondere durch erhöhte Spannung und Aufmerksamkeit mit sofortiger »Bereitschaft zur Flucht« auszeichnet.

readiness to flee □ A state characterized by increased tension and vigilance leading to an immediate »readiness to flee«.

disposition à la fuite □ Un état d'extrême vigilance et de grande tension chez l'animal avec une disposition particulière à la fuite immédiate.

Fluchtdistanz □ Derjenige Abstand, bei dessen Unterschreiten ein Tier vor einem Raubfeind oder einem Gegner (Rivalen) die Flucht ergreift. Die Fluchtdistanz kann von Art zu Art, aber auch innerhalb der Art von Individuum zu Individuum verschieden sein.

flight distance □ The distance at which an animal will flee from a predator or rival. Flight distance varies from species to species and from individual to individual.

distance de fuite □ Distance déterminée à la limite de laquelle un animal fuit devant son prédateur potentiel ou son rival. La distance de fuite peut varier d'une espèce à l'autre, mais également d'individu à individu à l'intérieur d'une même espèce.

Flügelfächeln □ Eine der sozialen Verständigung dienende Verhaltensweise bei Bienen. Die vor dem Flugloch sitzenden Arbeiterinnen haben, den Kopf zum Flugloch gerichtet, den Hinterleib schräg nach oben gehalten, das Duftorgan ausgestülpt und fächeln dabei heftig mit den Flügeln, um den Ankömmlingen den Geruch der Duftdrüse als Orientierungshilfe entgegenzuwerfen. VON FRISCH (1965) bezeichnet diese Fächelbienen als »sterzelnde« Bienen, was nicht mit dem → *Giftsterzeln* zu verwechseln ist.

wing-fanning □ A social form of communication in bees. Workers sitting before the entrance to the hive orient their heads towards the entrance, raising their abdomen upwards at an angle, extending their scent organ and fanning their wings. In this manner they help other hive members to find the entrance by fanning the odors emanating from their scent glands into their direction. VON FRISCH (1965) named these bees *sterzelnde* bees, which is not to be confused with → *Giftsterzeln.*

ventilation alaire □ Comportement servant à la communication sociale chez les Abeilles. Les ouvrières, postées devant la ruche, ont la tête dirigée vers l'entrée et l'abdomen redressé. L'organe odorant évaginé, elles ventilent avec leurs ailes pour propager le produit de leur glande odorante en direction des arrivants. VON FRISCH (1965) a utilisé pour ce comportement le terme de *Sterzeln* qui ne doit pas être confondu avec la → *ventilation d'alarme* (Giftsterzeln) et pour cette raison, nous préférons, d'une manière générale, de parler de ventilation alaire.

Flugkumpan → Kumpan
E. flight companion □ F. compagnon de vol

Fluglaufen □ Schwäne, Gänse, Enten, Taucher, Bleßhühner, Teichhühner usw. beim Auffliegen über dem Wasser.

skimming □ Swans, geese, ducks, divers, moore hens, coots, etc. above water.

course d'essor □ Chez les Cygnes, Oies, Canards, Plongeons, Poules d'eau, Foulques etc., course sur l'eau avant l'envol.

Folgehandlungen □ Nach der Entleerung erfolgen bei den meisten Säugern obligatorisch spezifische Handlungen, z. B. olfaktorische Kontrolle; olfaktorische Kontrolle und → *Flehmen;* olfaktorische Kontrolle mit Scharren (ALTMANN 1969).

obligatory concomitant behaviour □ After urination or defecation in mammals certain obligatory behaviours occur, e. g. olfactory control; olfactory control plus → *Flehmen;* olfactory control plus scratching (ALTMANN 1969).

actions post-excrétoires □ Après la défécation ou la miction, il y a chez la plupart des Mammifères des actions spécifiques obligatoires, par exemple un contrôle olfactif accompagné de → *Flehmen* ou de mouvements de grattage ou sol (ALTMANN 1969).

Folgeverhalten → Nachfolgereaktion → Prägung
E. following behaviour □ F. comportement de suivre

Forkeln → Geweihkampf

Fremdeln → Fremdenfurcht

Fremdenfurcht □ Eine Reaktion des Kleinkindes, mit der unbekannte Menschen abgelehnt werden. Diese Entwicklung kann etwa im 8. Lebensmonat ihren Höhepunkt erreichen; man spricht auch von der *Achtmonatsangst* und man sagt dann von ihm: »Es fremdelt«. Oft äußert sich dieses Verhalten durch Kopfabwenden und Vermeidung des Blickkontaktes. Werden solche Kinder dann angefaßt, kommt es zu Protestaktionen und zum Protestgeschrei (Abb. 47 – Bayaka-Pygmäenkind aus Äquatorialafrika). Dieses Verhalten mildert sich langsam mit dem Heranwachsen (EIBL-EIBESFELDT 1972, 1973, HASSENSTEIN 1972, 1973, HEYMER 1974).

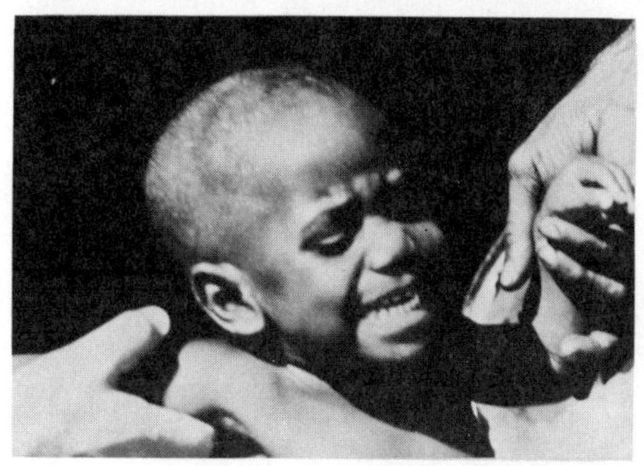

47

fear of strangers □ The negative reaction of infants to unfamiliar people which reaches its peak at the age of eight months; also termed *eight-months fear.* This shyness behaviour is often expressed by turning away the head and avoidance of eye contact. Attemps to touch such children result in protest actions and protest cries (Fig. 47 – Bayaka Pygmy child from Equatorial Africa). This behaviour gradually wanes as the child grows older (EIBL-EIBESFELDT 1972, 1973, HASSENSTEIN 1972, 1973, HEYMER 1974).

peur des étrangers □ Chez les petits enfants, une réaction de refus de contact vis-à-vis de personnes inconnues. Ce phénomène peut atteindre son point culminant vers le 8ème mois, fait dont dérive l'expression allemande *Achtmonatsangst* (= *peur des huit mois*). Le plus fréquemment, les enfants détournent la tête et évitent le contact visuel. S'ils sont touchés par des personnes étrangères, ils réagissent par des gestes et des cris de protestation (Fig. 47 – enfant des Pygmées Bayaka de l'Afrique Equatoriale). Ce phénomène s'atténue au fur et à mesure du développement de l'enfant (EIBL-EIBESFELDT 1972, 1973, HASSENSTEIN 1972, 1973, HEYMER 1974).

Fremdputzen → Körperpflegehandlungen
E. allogrooming □ F. toilettage réciproque

Freßgemeinschaft □ Eine lose Ansammlung von verschiedenen Tierarten beim gemeinsamen Mahle (Fraße), z. B. Aasfresser an einem Kadaver. DEEGENER (1918) stellte hierfür den Begriff *Symphagium* auf, um den Unterschied zu → *Kommensalismus* zu verdeutlichen.

feeding aggregation □ A loose association of various animal species during feeding, e. g. scavengers feeding on a cadaver. DEEGENER (1918) proposed the term *symphagium* to prevent confusion with → *commensalism.*

association alimentaire □ Il s'agit d'une association temporaire et lâche de différentes espèces animales à un même point d'alimentation, par exemple de différents charognards autour d'un cadavre. DEEGENER (1918) a forgé pour une telle association le terme de *symphagium* pour marquer la différence avec le → *commensalisme.*

Freßschutz → Mimikry

E. aposematic mimicry □ F. mimétisme aposématique

frühontogenetische Anpassung □ Verhaltensweisen und angeborenes Können neugeborener Lebewesen, z. B. die Bettelbewegungen von Jungvögeln, die *Sperreaktion (→ Sperren)* oder die Fähigkeit des nur 2 cm großen neugeborenen Riesenkänguruhs, mit Hilfe seiner Vorderbeine bis in den Brutbeutel der Mutter zu krabbeln. Auch die von den Imagines abweichenden Lebensweisen vieler Insektenlarven sind frühontogenetische Anpassungen oder *Kainogenesen,* die allerdings nach der Verwandlung verloren gehen. Viele weitere Beispiele bei EIBL-EIBESFELDT (Grundriß).

early ontogenetic adaptation □ Behaviour and innate ability in the neonate, e. g. the begging of young birds (→ *gaping*) or the struggling of the 2 cm long newborn kangaroo which crawls with its front legs to the broodpouch of its mother. Also, insect larvae which deviate from the imago form frequently constitute early ontogenetic adaptations or *cainogenesis* which disappear after metamorphosis. EIBL-EIBESFELDT (Ethology) lists many other examples.

adaptation ontogénétique précoce □ Modes de comportement et capacités innés chez les nouveaux-nés. Les mouvements de sollicitation alimentaire chez les Oisillons nidicoles ou leur réaction de → *bâillement de quête* ou encore la faculté du Kangourou nouveau-né, ne mesurant que 2 cm de long, à avancer uniquement à l'aide de ses pattes antérieures jusqu'à la poche marsupiale de sa mère, sont des adaptations ontogénétiques précoces ou *cainogenèses.* Les modes de vie des larves d'Insectes extrêmement différents de ceux des imagos en sont d'autres exemples bien qu'ils disparaissent au moment de la métamorphose. De nombreux autres exemples sont cités par EIBL-EIBESFELDT (Ethologie).

Frustration □ Eine Erlebnisenttäuschung durch Ausbleiben eines erwarteten und/oder geplanten Handlungserfolges, von dem die Befriedigung primärer oder sekundärer Bedürfnisse abhängt. Nach Beobachtungen einiger Psychologen tritt unter bestimmten Umständen im Anschluß an eine Frustration regelmäßig aggressives Verhalten auf, dessen Stärke direkt proportional zur Stärke einer Frustration sein soll. Dies führte zur Aufstellung der umstrittenen *Aggressions-Frustrations-Hypothese* (→ *Aggression*). Statt Aggression kann umgekehrt auch Depression die Folge von Frustration sein.

frustration □ The experiencing of disappointment resulting from failure of a planned or expected event to occur which itself would satisfy some primary or secondary need. According to the observations of some psychologists, under certain conditions, aggressive behaviour will regularly follow frustration, and its strength is directly related to the degree of frustration. This led them to propose the controversial *Aggression-Frustration Hypothesis (→ aggression).* Depression rather than aggression, however, may also be the result of frustration.

frustration □ Expérience d'une déception déclenchée par la non-réalisation d'un évènement attendu et/ou voulu dont dépend la satisfaction de besoins primaires ou secondaires. Selon les observations de quelques psychologues, une frustration entraîne régulièrement, sous certaines conditions, un comportement agressif dont la puissance serait proportionnelle à la force de la frustration. Ces observations ont conduit à l'établissement de *l'hypothèse agression-frustration,* d'ailleurs vivement contestée (→ *agression*). Parfois, la conséquence d'une frustration peut aussi être une dépression.

Frustrations-Aggressions-Modell → Aggression

E. frustration-aggression model □ F. modèle de l'agression-frustration

Führungslaut □ Lautäußerungen und Rufe, die z. B. bei Hühnern, Enten und Gänsen das Nachfolgen der Küken auslösen, können als Führungslaute bezeichnet werden.

call note □ The vocalizations and calls, e. g. in hens, ducks and geese, which release following in the chicks.

cri de pilotage □ Les vocalisations et cris, par exemple ceux qui, chez les Poules, Canards et Oies, déclenchent la réaction de suivre chez les jeunes, peuvent être considérés comme des cris de pilotage.

Führungsschwimmen □ Verschiedene Schwimmweisen und Lockbewegungen, die bei Buntbarschen das Folgen der Jungen auslösen; bei Pomacentriden nach dem → *Signalsprung* das Schwimmen des ♂ vor dem paarungswilligen ♀, um es zum Laichplatz zu führen. Man spricht auch von Führungsschwimmen, wenn Wasservögel führend vor ihren Jungen herschwimmen.

lead swimming □ Various movements or swim patterns in adult cichlids which release following in the young; swimming pattern of the male sunfish (→ *signal jump*) in front of a receptive female prior to leading her to the spawning site. In waterfowl may refer to the leading of the young.

nage de pilotage □ Différents modes de nage et mouvements d'attraction déclenchant chez les jeunes Cichlidae le comportement de suivre leurs parents. Chez les Pomacentridae, c'est une nage particulière du ♂ pour guider la ♀ réceptive vers le lieu de frai après le → *saut-signal* de celui-ci. On parle aussi de nage de pilotage chez les Oiseaux aquatiques précédant et guidant leurs jeunes.

Führungswechsel = Dominanzwechsel → Dominanzverhalten

Funktionserweiterung □ Innerhalb einer Stammesreihe zu beobachtende Erscheinung, daß ein Organ neben seiner Hauptfunktion eine oder mehrere Nebenfunktionen übernimmt. So können z. B. Fischflossen neben ihrer Funktion als Fortbewegungsorgane auch als Stütz- und Kopulationsorgane dienen.

expansion of function □ Proliferation in function of an organ observable during evolution. Fish fins, e. g., aside from their main locomotor function, have also developed functions as supporting or copulatory organs.

expansion fonctionnelle □ A l'intérieur d'une lignée évolutive, on peut observer qu'un organe déterminé peut, à côté de sa fonction principale, assumer une ou plusieurs fonctions supplémentaires. Les nageoires des Poissons, par exemple, sont primairement des organes locomoteurs, mais ont parfois acquis les fonctions d'organes copulateurs ou d'organes d'appui.

Funktionskreis □ Nach der Umweltlehre von VON UEXKÜLL (1921) Bezeichnung für die Zuordnung bestimmter Organe und Verhaltensweisen eines Lebewesens zu bestimmten Teilen seiner Umwelt. Diese Umwelt bildet einen wahrnehmbaren, von den Rezeptoren herausgefilterten Ausschnitt der Umgebung, und in ihm liegen diejenigen Eigenschaften (Merkmale), die für eine Lebensbewältigung notwendig sind; rückgekoppelt bestimmen sie als Wirkmale phylogenetisch vorprogrammierte Verhaltensweisen. Sobald ein Merkmal auftritt, wird es mit einer Wirkung beantwortet. Dies führt zur Tilgung des Wirkmales (Abb. 48; VON UEXKÜLL und KRISZAT 1934, VON UEXKÜLL 1937).

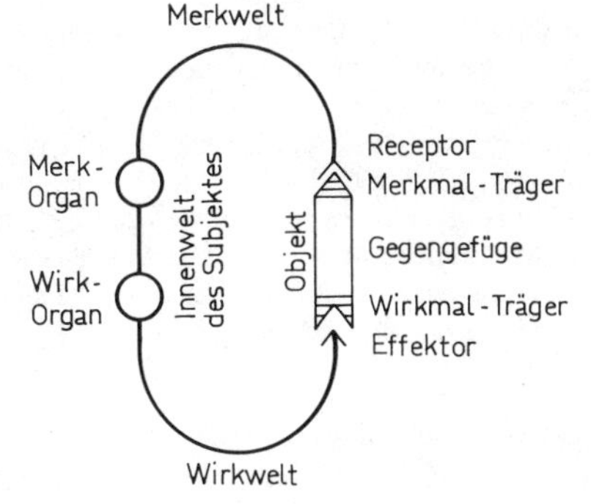

48

functional cycle □ Ecological term proposed by VON UEXKÜLL (1921) to describe the relation of an animal's organs and behaviour patterns to particular aspects of its environment. The animal's environment is essentially that part of its surroundings filtered out by its sensory receptors, containing certain objects whose recognition and perception is necessary for the animal's survival. These objects act as effector cues to control the animal's evolutionary preprogrammed behaviour. As soon as such an object occurs it is acted upon by the animal in some manner thereby eliminating the effector cue (Fig. 48; VON UEXKÜLL and KRISZAT 1934, VON UEXKÜLL 1937).

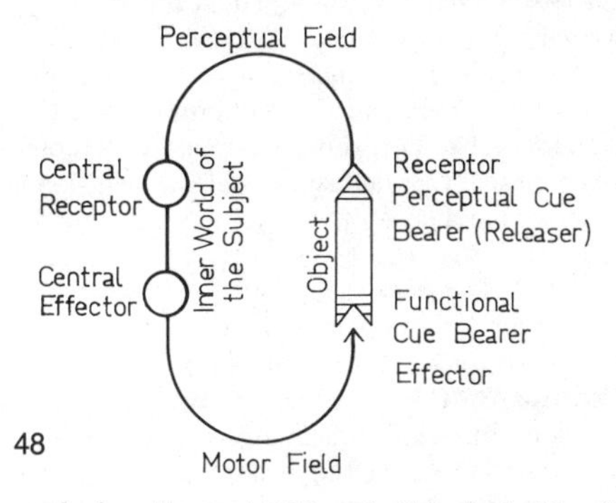

48

cycle fonctionnel □ Dans la théorie de VON UEXKÜLL (1921), cette notion désigne la relation entre certains organes et conduites d'un être vivant et certains compartiments de son *Umwelt,* c'est-à-dire de son monde phénoménal. *L'Umwelt* constitue en effet une portion de l'environnement, telle qu'elle est perçue et filtrée par les organes de réception sensorielle extéroceptive: elle se compose d'un ensemble de *Merkmale,* c'est-à-dire de propriétés signifiantes de l'environnement, généralement d'une importance critique pour la survie de l'organisme. En retour, ces propriétés signifiantes confèrent le statut de *Wirkmale,* c'est-à-dire de signaux opératoires, à des modalités comportementales génétiquement programmées. Sitôt qu'un *Merkmal* survient dans le champ perceptif du sujet, il entraîne une action en retour et la disparition momentanée du *Wirkmal* correspondant (Fig. 48; VON UEXKÜLL et KRISZAT 1934; VON UEXKÜLL 1937).

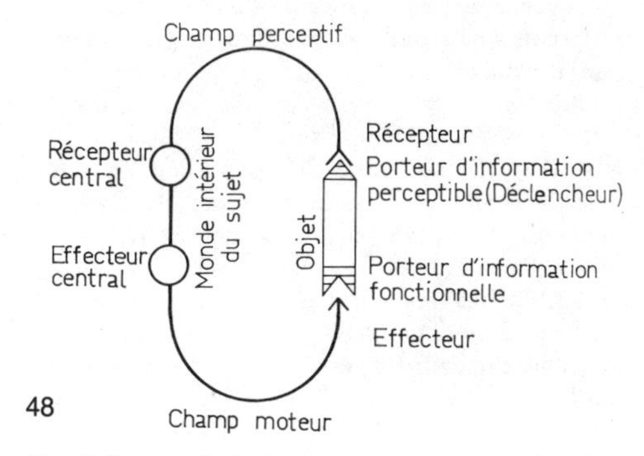

48

Funktionsspiele □ Kleinkindverhalten, – es sind Spiele mit dem eigenen Körper, wie Strampeln, Kriechen, Lallen, oder mit einfachen Gegenständen, wie Betasten und Greifen, im frühen Kindesalter. Das Kind folgt dabei dem Drang nach lustvoller Bewegung und übt seine Muskeln und Sinnesorgane und ihr Zusammenspiel. Auch im Jungtierverhalten (TEMBROCK 1961).

functional play □ Behaviour of small children, either undirected, e. g. kicking, crawling, jabbering, or directed toward simple objects, e. g. touching or grasping. In this manner, the child is able to exercise his muscles and sense organs and their coordination. Also occurs in young animals (TEMBROCK 1961).

jeux fonctionnels □ Comportement des jeunes enfants. Il s'agit de jeux, soit avec leur propre corps (gigoter, ramper, babiller) soit avec des objets simples (toucher, saisir). L'enfant suit en cela une pulsion à effectuer des mouvements plaisants et exerce ainsi ses muscles et ses organes sensoriels ainsi que leur coordination. Des jeux fonctionnels existent également chez de nombreux jeunes animaux, Mammifères en particulier (TEMBROCK 1961).

Funktionswechsel □ Innerhalb einer Stammesreihe beobachtbare Erscheinung, daß ein Organ nach Verlust seiner ursprünglichen Funktion neue, andere Funktionen übernehmen kann; auch Lautäußerungen und Verhaltensweisen können einem Funktionswechsel unterliegen. So wurde z. B. aus dem → *Futterlocken* bei verschiedenen Phasianiden eine Werbehandlung des Hahns zum Anlocken der ♀♀ (SCHENKEL 1956, 1958). → *Bodenpicken.*

change of function □ An evolutionary phenomenon whereby an organ which ceases its original function may take over a new function; vocalizations and behaviour patterns may be subject to alteration of function. Thus the → *food calling* of various phasianids became a courtship behaviour in males which serves to attract females (SCHENKEL 1956, 1958). → *ground pecking.*

changement de fonction □ Au cours de l'évolution, un organe, après avoir perdu sa fonction primitive, peut en assumer une nouvelle; les vocalisations et d'autres comportements peuvent également subir un changement de fonction. Ainsi par exemple, le comportement → *d'attraction alimentaire* chez les différents Phasianidae est devenu un élément fixe dans le comportement de parade nuptiale du Coq lui permettant d'attirer ainsi ses Poules (SCHENKEL 1956, 1958). → *picorage au sol.*

Furcht □ Gefühl des Bedrohtseins und Eintreten einer allgemeinen Verschärfung der Sinneswahrnehmung. Furcht ist im Unterschied zu Angst objektbezogen, d. h. sie tritt nur angesichts einer konkreten Gefahr auf, hat aber wohl keine physiologischen Reaktionen, wie Angstdefäkation und Angstmiktion, zur Folge, wie dies bei der → *Angst* der Fall ist. TEMBROCK (1961) hält es für zweckmäßig, in der Ethologie zwischen Angst und Furcht zu unterscheiden.

fright □ The feeling of being threatened with its accompanying alertness and acumen. Fright unlike »angst« or fear is object-oriented, i. e., it occurs only in response to tangible danger. Fright probably does not result in physiological arousal such as fear-induced defecation and urination, as does → *fear.* TEMBROCK (1961) suggested it would be useful to distinguish in ethology between fear and fright.

crainte □ Sentiment d'être menacé, accompagné d'un renforcement général de la perception sensorielle. La crainte serait, à la différence de la peur, dépendante d'un objet et par conséquent, n'apparaîtrait que devant une menace concrète. Cependant, elle ne déclencherait pas de réactions physiologiques telles que la défécation et la miction, comme ceci se produit généralement en cas de → *peur.* Selon TEMBROCK (1961), il serait souhaitable de distinguer en éthologie la peur de la crainte.

Futterbetteln □ Bettelbewegungen, um meist von den Eltern Nahrung zu erhalten. Beim Werbeverhalten kommen zwischen Partnern solche Bettelbewegungen als Infantilismen wieder vor, die man als ritualisiertes Futterbetteln bezeichnen kann. → *Balzfüttern,* → *Begrüßungsfüttern.*

food begging □ Begging by the young for food from the parents. In courtship, a ritualized form of begging may occur between partners. → *courtship feeding,* → *greeting feeding.*

sollicitation alimentaire □ Mouvements de quête pour obtenir de la nourriture, en général des parents. Pendant la parade nuptiale, de tels mouvements de sollicitation alimentaire existent entre partenaires. Dans ce contexte, ce sont des infantilismes devenus mouvements de sollicitation alimentaire ritualisés. → *nourrissage de parade nuptiale,* → *nourrissage de salut.*

Futterhorten → Futterverstecken
E. food hoarding □ F. amassement de nourriture

Futterlocken □ Einmal das Heranlocken der Küken durch die Henne, die mehrmals auf den Boden pickt und dazu »tuck–tuck« ruft. Mit denselben Verhaltensweisen kann der Hahn die Hennen zu sich rufen (SCHENKEL 1956, 1958). Das Futterlocken hat hier eine Ritualisierung erfahren. → *Bodenpicken.*

food-calling □ In one case, the enticing of the chicks by the hen through pecking at the ground and emitting a cluck-cluck sound (food calling). The cock may call the hens with the same behaviour pattern (SCHENKEL 1956, 1958), and hence this is a ritualized form of feeding enticement. → *ground pecking.*

attraction alimentaire □ Chez les Poules, les jeunes poussins sont attirés vers leur mère à l'aide de mouvements de picorage sur le sol, accompagnés de vocalisations »touc-touc«. Par le même comportement, le coq fait venir les poules auprès de lui (SCHENKEL 1956, 1958). Dans ce cas, le comportement a subi une ritualisation. → *picorage au sol.*

Futterspender □ In der *Skinner-Box* verwendetes Gerät, das auf bestimmte Leistungen eines Tieres, z. B. Bedienung eines Pedals, eine bestimmte Menge Futter als Belohnung spendet.

food dispenser □ Device used in the *Skinner box* which automatically dispenses a given amount of food following a certain performance by the animal.

distributeur de nourriture □ Dans une *boîte de Skinner,* dispositif qui, en réponse à des performances déterminées d'un animal, par exemple la manipulation d'une pédale, distribue une quantité donnée de nourriture comme récompense.

Futtertradition □ Bei vielen Säugern, aber auch bei Vögeln, lernen die Jungen von der Mutter oder von beiden Eltern, was man frißt, und es können sich auf diese Weise auf bestimmte Gruppen beschränkte, lokale Futtertraditionen entwickeln.

feeding tradition □ In certain mammals and bird groups, the young learn food preferences from the mother or the parents and in this manner certain regional diet peculiarities (feeding tradition) may develop.

tradition alimentaire □ Chez de nombreux Mammifères, mais aussi chez les Oiseaux, les jeunes apprennent de leur mère ou de leurs parents ce que l'on peut manger; il se développe ainsi une certaine tradition alimentaire, limitée à des groupes d'animaux déterminés.

Futterübergabe □ Ein häufig verbreitetes Verhalten im Dienste der Paarbildung und als Geste der Kontaktbereitschaft. Bei den räuberischen Tanzfliegen (Empididae) laufen die ♂♂ Gefahr, bei der Paarung gefressen zu werden, und um dieser zu entgehen, überreichen sie dem ♀ eine eingesponnene Beute. Bei der Art *Hilaria sartor* handelt es sich nur um eine symbolische Futterübergabe, denn das ♂ hat nur irgendeinen ungenießbaren Gegenstand eingesponnen (REUTER 1913, MEISENHEIMER 1921). Ritualisierte Futterübergabe finden wir auch bei Seeschwalben; die ♂♂ bieten ihren ♀♀ beim Werben einen Fisch an (→ *Begrüßungsfüttern*). Futterübergabe gibt es auch bei sozialen Insekten (→ *Trophallaxie*). Futterüberreichen als Zeichen der Kontaktbereitschaft ist bei Menschen, auch bei kleinen Kindern, sehr verbreitet. → *Mund-zu-Mund-Füttern.*

feeding ceremonies □ A widely distributed behaviour pattern serving pair-bond formation and facilitating close contact. In the predatory Empidid flies, the ♂ runs the risk of being eaten during pairing and he avoids this problem by giving the ♀ a silk encasement containing a prey. This is strictly symbolic in *Hilaria sartor* since the ♂ encloses only an unappetizing morsel (REUTER 1913, MEISENHEIMER 1921). A ritualized feeding ceremony occurs also among terns where the ♂ presents the ♀ with a fish (→ *greeting feeding*). Food exchange is also found in social insects (→ *trophallaxis*). In humans, presentation of food as a gesture of contact readiness is frequent in both small children and adults. → *mouth-to-mouth feeding.*

offrande de nourriture □ Comportement très répandu lors de la formation du couple et comme geste de contact. Chez les Diptères Empididae, les ♂♂ risquent d'être dévorés pendant l'accouplement et pour éviter cela, ils apportent à leur ♀ une proie enveloppée dans un cocon. Chez l'espèce *Hilaria sartor,* ce geste n'est que symbolique, car le ♂ n'a enveloppé dans un cocon qu'un objet quelconque et non un aliment (REUTER 1913, MEISENHEIMER 1921). Un nourrissage ritualisé existe également chez les Oiseaux Sternidae où les ♂♂ apportent à leur ♀ courtisée un Poisson (→ *nourrissage de salut*). Un transfert d'éléments nutritifs existe également chez les Insectes sociaux (→ *trophallaxie*). Chez l'Homme, y compris chez les petits enfants, l'offrande s'observe comme geste facilitant l'établissement d'un contact inter-individuel. → *nourrissage de bouche à bouche.*

Futterverstecken □ Bei vielen Vögeln und Säugetieren wird für den Winter Nahrung gesammelt (*Futterhorten*) und an bestimmten Stellen versteckt. Diese Verhaltensweisen des Versteckens von Nahrung für den Winter wurden von EIBL-EIBESFELDT (1951) beim Eichhörnchen und von GWINNER (1962) bei Kolkraben eingehend untersucht. Die Futterversteckhandlungen bestehen aus einer starren Kette angeborener Verhaltensweisen, die weitgehend erfahrungsunabhängig heranreifen. → *Reifung.*

food hiding □ Many birds and mammals gather food for the winter (*food hoarding*) and hide it at certain places. These behaviour patterns of food hiding have been studied in detail by EIBL-EIBESFELDT (1951) in squirrels and by GWINNER (1962) in ravens. The food stashing behaviour consists of a fixed chain of innate movements which mature independently of experience. → *maturation.*

dissimulation de nourriture □ Nous connaissons chez de nombreux Oiseaux et Mammifères le fait d'amasser de la nourriture pour l'hiver et de la cacher à des endroits précis. Cette dissimulation des réserves alimentaires pour l'hiver a été particulièrement bien analysée chez l'Ecureuil par EIBL-EIBESFELDT (1951) et par GWINNER (1962) chez le Grand Corbeau. Les composantes de ce comportement consistent en une chaîne rigide d'éléments comportementaux innés qui subissent une maturation progressive indépendante de l'expérience. → *maturation.*

G

Gähnen → Drohgähnen → Räkelsyndrom → Stimmungsübertragung

E. yawning □ F. bâillement

Gangart □ Die durch bestimmte rhythmische Bewegung der Extremitäten und ihrer Glieder gekennzeichneten Fortbewegungsweisen eines Tieres. Zu den allgemeinen Bewegungsformen der Säugetiere gehören das Schreiten, Traben und Galoppieren. Bei den Lagomorphen verlangsamt sich der Galopp zum Hoppeln, bei den Caniden kann sich der Trab zum »Schnüren« abwandeln. Der schnelle Lauf des Menschen schließlich ist ein modifizierter Trab. Es ist eine wesentliche Aufgabe ethologischer Analysen, derartige allgemeine Bewegungsformen zu kennzeichnen und zu beschreiben, denn sie liefern in mannigfachen Abwandlungen das »Substrat« für reaktionsspezifische Bewegungsabläufe. Zur tetrapoden Fortbewegung gehören auch der jeweils artspezifische → *Kreuzgang* und → *Paßgang*. Verschiedene spezielle Gangarten gibt es bei Arthropoden, wo wir myriapode, octopode und hexapode Fortbewegungsweisen kennen.

locomotory pattern □ The ambulatory patterns of an animal characterized by rhythmic movements of the extremities and limbs. Stepping, trotting and galloping are common ambulatory patterns of mammals. In lagomorphs one finds a slowed gallop called hopping, and in canids trotting may change into loping. In humans, the quick run is also a modified trot. One of the basic tasks of ethological analysis constitutes the determination and description of such general locomotor types. Thus, in a variety of ways, the »substrate« for reaction-specific locomotory sequences can be worked out. In tetrapods ambulation also encompasses the species-specific → *cross gait* and the → *pacing*. Furthermore, some special locomotory patterns occur in Arthropods such as myriapod, octopod and hexapod locomotion.

schème locomoteur □ Mode de locomotion rythmique et déterminé par les mouvements des extrémités d'un animal. Les principales formes de locomotion chez les Mammifères sont la marche, le trot et le galop. Chez les Lagomorphes, le galop peut se ralentir et devenir le sautiller; chez les Canidae, le trot se modifie en se ralentissant et devient un »filer«, la course rapide chez l'Homme n'est qu'un trot modifié. Dans les analyses éthologiques, il est important de caractériser et de décrire de telles formes locomotrices étant donné qu'elles fournissent, sous de multiples formes différenciées, le »substrat« nécessaire au déroulement des activités motrices spécialisées. Deux autres aspects fondamentaux de la locomotion tétrapode sont → *l'allure croisée* et → *l'amble*. Par ailleurs, il existe des schèmes locomoteurs spécifiques chez les différents Arthropodes tels que la démarche myriapode, octopode et hexapode.

Gattenfüttern → Balzfüttern → Begrüßungsfüttern → Futterübergabe
E. partner feeding □ F. nourrissage du partenaire

Gattenkumpan → Kumpan
E. sexual companion □ F. compagnon sexuel

Gebrauchshandlung → Werkzeughandlung

Gedächtnis □ Die Fähigkeit des Zentralnervensystems, Informationen abrufbar zu speichern (Erinnerung). Wie die Nervenzellen die Information aufnehmen, d. h. durch welche physiologischen Vorgänge Erregungen bzw. Spuren hinterlassen, ist noch weitgehend ungeklärt. Mit Ausnahme der Mesozoen und Porifera besitzen alle vielzelligen Tiere ein Gedächtnis. Ein sehr wichtiges Erinnerungsvermögen für das Tier ist die Fähigkeit, sein Nest bzw. sein Revier wiederzufinden (SEIFERT 1950). Tieren mit komplizierten Gehirnen sind *Wahlhandlungen* möglich und schließlich auch Handlungen aufgrund von Begriffsbildungen und letztlich einsichtige Handlungen (→ *Einsicht*). Die Erinnerungszeit an eine einmal gezeigte Futterstelle z. B., die beim zweiten Versuch sofort wieder aufgesucht wird, steigt von 5 min bei Prosimiern auf 15 min bei Gibbons, bis auf 2 h und mehr bei Schimpansen. Beim Menschen, der die meisten Informationen im Neocortex speichert, kann der Aufschub seines Handelns praktisch unbegrenzt sein; er besitzt die Fähigkeit, Erlebnisinhalte der Vergangenheit wieder bewußt werden zu lassen. Man hat herausgefunden, daß man zwischen einem *Kurzzeitgedächtnis* und einem *Langzeitgedächtnis* unterscheiden kann, die, wie neuere Versuche an Cephalopoden gezeigt haben, auch in verschiedenen Hirngebieten lokalisiert sind (BOYCOTT 1965). Eine erschöpfende Abhandlung des Problems Gedächtnis mit weiterführender Literatur finden wir bei FOPPA (1968), BUCHHOLTZ (1973), RENSCH (1973) und ANGERMEIER (1976).

memory □ The capability of the CNS to store information for subsequent retrieval. It is still not known how the nerve cells process the information, i. e., what physiological processes encode the excitations or engram. With exception of the Mesozoans and Poriferans, all multicellular organisms possess a memory. A very important memory capacity for an animal is to find back to its nest site or territory (SEIFERT 1950). In animals with increasingly advanced nervous systems *discrimination operations* are possible, then concept formation and lastly in the most advanced forms → *insight* operations. The duration of memory for a food cache location shown once to the animal increases from 5 min in prosimians to 15 min in gibbons and 2 hours or more in chimpanzees. In humans which store most information in the neocortex the postponement of such an action is practically unlimited. They have the ability to remind themselves of past experiences. Two forms of memory are distinguished – *short term* and *long term* which also are localized in different brain regions as shown by experiments on Cephalopods (BOYCOTT 1965). An exhaustive discussion of the subject with an introduction to the literature is offered by FOPPA (1968), BUCHHOLTZ (1973), RENSCH (1973) and ANGERMEIER (1976).

mémoire □ Capacité du système nerveux central de stocker des informations prêtes à l'appel. Jusqu'à présent, peu de données existent sur la manière dont les cellules nerveuses reçoivent les informations ou sur les mécanismes physiologiques par lesquels des excitations ou des traces sont mises en trame. A l'exception des Mésozoaires et des Porifères, tous les Pluricellulaires sont pourvus d'une mémoire. Une capacité de mémoire importante pour un animal consiste par exemple dans le fait de retrouver son nid, son gîte ou son territoire (SEIFERT 1950). Les animaux possédant des cerveaux complexes sont en mesure d'effectuer des *actions discriminatoires;* ils peuvent agir sur la base de notions préformées et finalement sont capables d'une → *compréhension soudaine.* La durée de mémorisation de l'emplacement d'une source alimentaire montrée auparavant et qui est immédiatement recherchée lors d'une deuxième expérience, est de 5 min chez les Prosimiens et s'élève à 15 min chez les Gibbons pour atteindre finalement 2 heures chez les Chimpanzés. Chez l'Homme qui stocke la plupart de ses informations dans le néocortex, le retardement de son action est pratiquement illimité. Il possède la faculté de retenir d'une manière permanente les expériences du passé et de les faire revivre. On a pu prouver qu'il est possible de distinguer entre une *mémoire immédiate* et une *mémoire à long terme* qui, selon les expériences effectuées chez les Céphalopodes, seraient localisées dans des régions cérébrales distinctes (BOYCOTT 1965). Le problème de la mémoire est traité d'une manière exhaustive par des auteurs récents tels que FOPPA (1968), BUCHHOLTZ (1973), RENSCH (1973) et ANGERMEIER (1976) où nous trouvons également une bibliographie plus détaillée.

Gedrängefaktor → Massensiedlungseffekt

Gefangenschaftserscheinungen → Käfigverblödung

E. captivity-related phenomena □ F. phénomènes de captivité

Gefiederkraulen → Körperpflegehandlungen

E. nibble-preening □ F. »mordillement« des plumes

Gefiedersträuben □ Bei Vögeln während aggressiver Auseinandersetzungen und beim Drohen ein Aufrichten der Federn, um das Erscheinungsbild eindrucksvoll zu verändern. Der Effekt ist ein scheinbar größerer Körper. Dem Gefiedersträuben kommt wohl ursprünglich eine wärmeregulatorische Funktion zu.

ruffling, shuffling □ Erection of the feathers during aggressive encounters or threat. This presents an impressive appearance; the body appears larger. Originally ruffling served a temperature-regulating function.

ébouriffement des plumes □ Composante comportementale permettant de modifier la silhouette d'une manière impressionnante lors de manifestations agonistiques chez de nombreux Oiseaux. Son effet principal est une apparence plus grande du corps. Primairement, l'ébouriffement des plumes a une fonction de régulation thermique.

Geländemarken → Landmarken

Gemeinschaftsbalz → Gruppenbalz

Genitalpräsentieren □ Ein bei Primaten, den Menschen eingeschlossen, vorkommendes Imponierverhalten gegenüber Artgenossen (PLOOG *et al.* 1963, WICKLER 1966, CHRISTEN 1974). Die einfachste Form ist wohl das bloße Zeigen besonders bunt gefärbter Genitalien, wie z. B. bei *Cercopithecus aethiops* (→ *Wachesitzen*). Beispiele mit erigiertem Penis, das man auch als *Phallusdrohen* bezeichnet, kennen wir von vielen Simiern (Abb. 49 A – *Nasalis larvatus*) und auch vom Menschen (Abb. 49 B – Der Riese von Cerne Abbas in Dorset, England). Von den Papuas kennen wir ein ritualisiertes Phalluspräsentieren in Form von Verwendung auffälliger Penishüllen (Abb. 49 C – Papua aus Kugome, Neu-

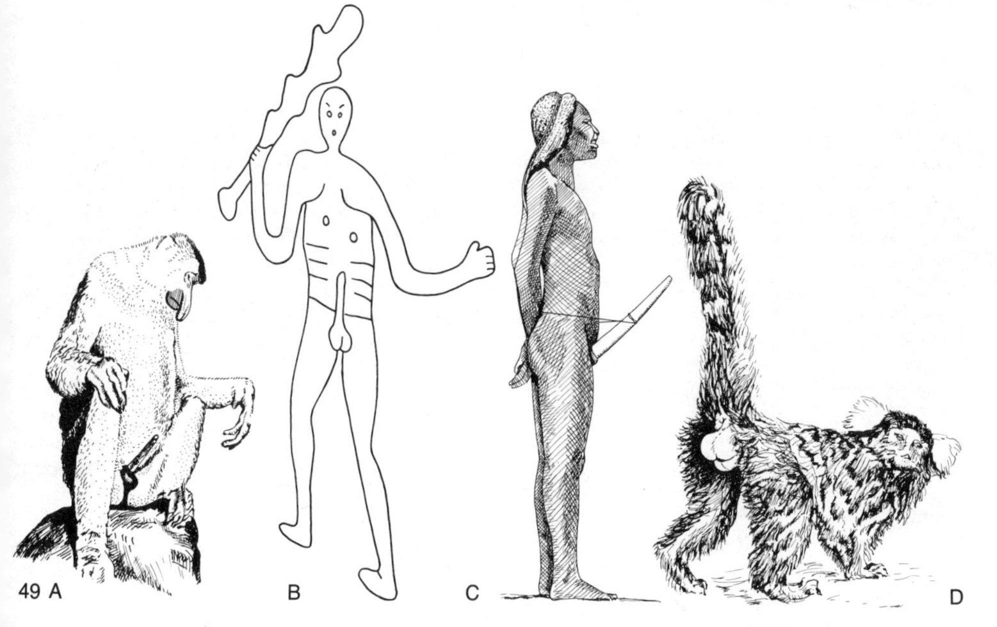

guinea). Niedere Primaten, wie z. B. *Hapale* und *Cebuella,* präsentieren mit hoch erhobenem Schwanz das angeschwollene und oft bunt gefärbte Skrotum nach rückwärts (Abb. 49 D – *Hapale jacchus*). Bei Buschmädchen gehört das Genitalpräsentieren in den Kontext des → *Spottverhaltens* (EIBL-EIBESFELDT 1972). → *Schampräsentieren,* → *Schamweisen.*

genital presentation □ A form of presenting behaviour found in primates, including humans, directed at conspecifics (PLOOG *et al.* 1963, WICKLER 1966, CHRISTEN 1974). The simplest form seems to be the mere presence of colored genitals, e. g. *Cercopithecus aethiops* (→ *sentinel behaviour*). Examples with erected penis, also called *phallic threat,* were found in many simians (Fig. 49 A – *Nasalis larvatus*) and also in humans (Fig. 49 B – The giant of Cerne Abbas in Dorset, England). Papuans show ritualized phallic displays in which they use conspicuous gourds which cover their penises (Fig. 49 C – Papuan from Kugome, New Guinea). Lower primates, e. g. *Hapale* and *Cebuella* present while holding their tails high, thus exposing the swollen and often colored scrotum to the rear (Fig. 49 D – *Hapale jacchus*). Girls of the Bushmen show genital presentation in the context of → *mocking behaviour* (EIBL-EIBESFELDT 1972). → *vulva presentation,* → *pubic presentation.*

présentation des organes génitaux □ Chez les Primates, l'Homme y compris, il s'agit d'un comportement d'intimidation envers les congénères (PLOOG *et al.* 1963, WICKLER 1966, CHRISTEN 1974). La forme la plus primitive est incontestablement la simple présentation des organes génitaux mâles d'une coloration particulièrement vive, comme par exemple chez *Cercopithecus aethiops* (→ *comportement de sentinelle*). Des exemples avec un pénis en érection, également appelé *menace phallique,* sont connus chez de nombreux Simiens (Fig. 49 A – *Nasalis larvatus*), mais également chez l'Homme (Fig. 49 B – Le géant de Cerne Abbas en Dorset, Angleterre). Chez les Papous, nous connaissons une forme de présentation phallique ritualisée avec utilisation d'étuis péniens apparents (Fig. 49 C – Papou de la région de Kugome, Nouvelle Guinée). Les Primates inférieurs, par exemple *Hapale* et *Cebuella,* présentent vers l'arrière, la queue relevée, leur scrotum enflé et souvent très coloré (Fig. 49 D – *Hapale jacchus*). Chez les jeunes filles Boshimans, la présentation des organes génitaux fait partie des attitudes de défi et du → *comportement de moquerie* (EIBL-EIBESFELDT 1972). → *présentation vulvaire,* → *présentation pubienne.*

Geotaxis → Taxis

Geruchsabzeichen → Gruppenduft

E. scent marks □ F. marques de reconnaissance olfactive

Gesangsduelle → Ventilsitten

E. song duels □ F. duels chantés

Gesangsprägung → Prägung

E. song imprinting □ F. empreinte du chant

Gesäßweisen □ In verschiedenen menschlichen Kulturen und bei verschiedenen Rassen gibt es Gesäßweisen als → *Spottverhalten* gegenüber anderen Personen. Es kommt bei beiden Geschlechtern vor und hat auch die Bedeutung einer aggressiven Drohung. Buschleute klemmen oft vorher Sand zwischen die Gesäßbacken, um ihn dann gegenüber der zu verspottenden Person bei einer tiefen Verbeugung zu entlassen; dabei werden auch oft Winde abgelassen. Sie symbolisieren gewissermaßen den Akt des Defäkierens. Bei europäischen Kindern beiderlei Geschlechts im Alter von zwei Jahren sah HEYMER Gesäßweisen gegenüber den Eltern als Protestaktion, ohne daß dieses Verhalten vom Kind vorher je gesehen wurde. Bei Erwachsenen im europäischen Kulturraum finden wir Gesäßweisen in ritualisierter Form im Tanzverhalten wieder (Abb. 50 A – altgriechische Vasenbemalung; B – Striptease-Tänzerin, Folies-Bergères).

50 A 50 B

buttocks display □ In various human cultures and races display of the buttocks represents scorn or mockery towards other persons. It occurs in both sexes and has the significance of an aggressive threat display. Bushmen may grasp sand between their pelvic cheeks and release it during a deep bow directed toward the person they are mocking; often they release gas at the same time. This is a symbolic display of the defecation act. European children of both sexes at two years of age may be observed to display the buttocks to the parents as an act of protest though they never had seen this behaviour before. In European adults, buttocks display occurs in a ritualized form in dancing behaviour (Fig. 50 A – painting on an ancient Greek vase; B – striptease dancer, Folies-Bergères). → *mocking behaviour.*

présentation du postérieur □ Dans différentes races et cultures humaines, il existe un comportement de présentation du postérieur comme geste de moquerie vis-à-vis d'autres personnes. Ce comportement existe dans les deux sexes et a également une signification de menace agressive. Chez les Boshimans, on observe qu'ils introduisent d'abord du sable entre les fesses pour ensuite le libérer lors d'une présentation du postérieur en position très inclinée devant la personne dont il s'agit de se moquer. Ce comportement est souvent accompagné de pets. Il s'agit là en quelque sorte d'une symbolisation de l'acte de défécation. HEYMER a observé chez les enfants européens à partir de l'âge de 2 ans et chez les deux sexes, une présentation du postérieur envers leurs parents en tant qu'action de protestation sans que ces enfants aient jamais pu voir ce comportement auparavant. Chez les Européens adultes, nous retrouvons la présentation du postérieur sous une forme ritualisée lors du comportement de danse dans différents contextes (Fig. 50 A – peinture sur vase de l'antiquité grecque; B – danseuse de striptease du cabaret des Folies-Bergères). → *comportement de moquerie.*

Geschenküberreichen → Grußformeln
E. present giving □ F. offre d'un cadeau

Geschlechtskumpan → Kumpan
E. sexual companion □ F. compagnon sexuel

Geschwisterkumpan → Kumpan → Kinderfamilie
E. sibling companion □ F. compagnon fraternel

Gesellschaftsbalz → Gruppenbalz

Gesichtverbergen → Verlegenheitsgebärde
E. hiding the face □ F. cacher le visage

Gespenstreaktion □ Attrappen von *Astatotilapia strigigena,* denen der lackartige Glanz, den ein klarer Schleimüberzug der gesunden Fischhaut verleiht, fehlte, lösten bei mit ihresgleichen aufgewachsenen Fischen wilde Panik aus, und SEITZ (1940) prägte dafür den Ausdruck Gespenstreaktion.

spook reaction □ Models of *Astatotilapia strigigena* which lack the sheen of a normal healthy fish's epithelium, cause panic reactions in the adult fish. From this, SEITZ (1940) coined the term spook reaction.

réaction au fantôme □ (traduction littérale du terme allemand). – Des leurres du Cichlide *Astatotilapia strigigena* auxquels manque la brillance qu'une couche de mucus claire confère normalement à la peau de ces Poissons, déclenchent chez les Poissons élevés entre congénères une réaction de panique. SEITZ (1940) a créé pour désigner ce phénomène le terme de réaction au fantôme.

Gestaltspsychologie □ Wahrnehmungspsychologie, die als Grundlage von Verhaltensantworten übersummative Ganzheiten annimmt, welche aus ihren Elementen allein als Antworten auf Beziehungs- oder konfigurale Reize nicht erklärbar sind. Bei der Honigbiene z. B. wurde im relativen Wahlversuch nachgewiesen, daß eine Dressur auf »stärker gegliedert« gegenüber »schwächer gegliedert« möglich ist (KÖHLER 1971).

gestalt psychology □ A branch of psychology that attributes behavioural responses to the recognition of the sum of all elements of a configuration (in contrast to simple cues). Honey bees for example, can be taught to distinguish between »more segmented« and »less segmented« stimuli (KÖHLER 1947).

psychologie de la »Gestalt« □ Psychologie de la perception supposant comme bases des activités perceptives des entités supra-sommatives qui ne pourraient être totalement réduites à leurs éléments constituants. Chez l'Abeille, il a pu être démontré, sur la base d'un dressage, qu'une discrimination visuelle est possible entre des formes plus ou moins structurées (KÖHLER 1964).

Gestaltsverfremdung → Konturverfremdung

Geweihfegen → Fegen

Geweihkampf □ Kampfverhalten bei Hirschen, die mit gesenkten Köpfen und verschränkten Geweihen einen *Schiebekampf* ausführen (Abb. 51 – *Cervus nippon*). Beim Forkeln kann der eine Partner den anderen auch seitlich treffen, was häufig das Verenden des Geforkelten zur Folge hat. Besonders gefährlich ist das Forkeln durch Spießerhirsche. Beim → *Bodenforkeln* wühlt nur das Einzeltier mit dem Geweih im Boden.

51

locking antlers □ Agonistic behaviour in male deer fighting with lowered heads and interlocked antlers (Fig. 51 – *Cervus nippon*). Horned animals can often fatally injure their opponents especially if the opponent is struck broadside. Antler combat in prickets is particularly dangerous. In contrast to → *ground-rutting,* in which single male animals root in ground with their antlers.

lutte avec les bois □ Comportement agonistique chez les Cerfs qui, tête baissée, luttent avec les bois emmêlés (Fig. 51 – *Cervus nippon*); à ne pas confondre avec le → *râclement du sol,* comportement où un animal seul fouit le sol avec ses bois. Lors de ces combats, il peut arriver que les rivaux se blessent latéralement ce qui entraîne souvent la mort de l'animal atteint. Les luttes sont particulièrement dangereuses avec les daguets.

Giftsterzeln □ Alarmverhalten bei der Honigbiene; die Wächterinnen alarmieren auf diese Weise (Abb. 52) die Nestgenossen zur Bekämpfung von in der Nähe des Nestes befindlichen Feinden und Eindringlingen. Der Stachel ist ausgeschoben, das Alarmpheromon wird aus dem Stachelrinnenpolster freigesetzt, und das Flügelschwirren beschleunigt seine Verteilung (MASCHWITZ 1964).

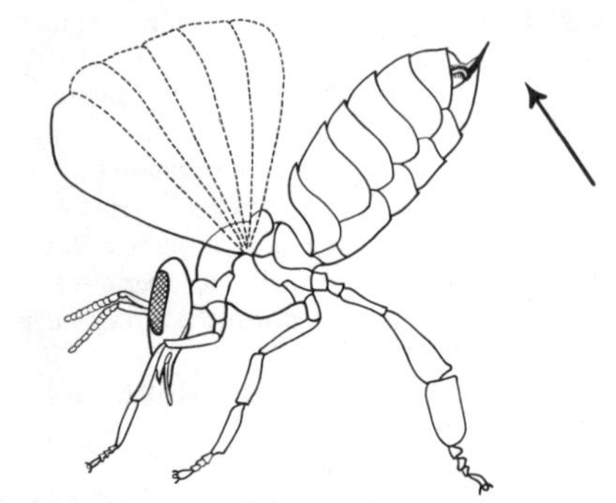

52

»Giftsterzeln« □ *Alarm substance ventilation behaviour* in honey bees; the guards arouse their nestmates (Fig. 52) to defend the nest from enemies or invaders by fanning an alarm substance into the nest. The stinger is exposed, the alarm pheromone is released and fanning with the wings aids in its dissemination (MASCHWITZ 1964).

ventilation d'alarme □ Comportement d'alerte chez les Abeilles; des surveillantes alertent ainsi (Fig. 52) leurs compagnes de ruche pour attaquer les intrus. Le dard est évaginé, le phéromone d'alarme est ainsi libéré par la canalisation du dard et le mouvement de ventilation des ailes assure et accélère sa répartition (MASCHWITZ 1964).

Gockern → Würgeln

Grasrupfen □ Treffen zwei Silbermöwen an ihren Reviergrenzen aufeinander, kommt es meist bei beiden Tieren zum Kampf, jedoch sind in einer solchen Situation meist zwei in sich unvereinbare Triebe aktiviert, nämlich Angriff und Flucht, und so kommt es fast regelmäßig zu einem Verhalten, welches dem Nistmaterialsammeln ähnelt. Die Tiere rupfen dabei heftig und sichtlich erregt an Grasbüscheln (Abb. 53 – *Larus argentatus*). Da die Vögel das Gras aber nicht aufsammeln, sondern nur daran zerren, ist es kein Übersprungnisten, sondern vom Kampfverhalten überlagert; daraus ergibt sich der Unterschied zwischen echtem Nistbau und Drohen. Dieses Grasrupfen einmal als Drohen in höchster Kampferregung gedeutet, ist aber wohl eher eine → *Übersprungbewegung,* bedingt durch den gleichzeitig aktivierten Drang von Flucht und Angriffsbereitschaft (GOETHE 1956, TINBERGEN 1958).

53

grass-pulling □ When two herring gulls happen to meet at their territorial boundaries, a confrontation often occurs. Here two mutually inconsistent drives are activated at once; flight and fight. The result is frequently a behaviour resembling gathering of nest material; the opponents frantically pull up tufts of grass (Fig. 53 – *Larus argentatus*). Since the birds do not collect the grass, but simply tear at it, it cannot be considered true nest building. Grass pulling, although interpretable as threat, is probably better considered as a → *displacement activity* arising from simultaneously activated fight and flight drives (GOETHE 1956, TINBERGEN 1958).

»desherbage« □ Lorsque deux Goélands argentés se rencontrent à la limite de leurs territoires, il s'engage généralement un combat, mais dans une telle situation, chacun étant »chez soi«, deux pulsions contradictoires (attaque et fuite) sont activées et il se produit alors un comportement qui ressemble au ramassage de matériaux de nidification. Les animaux, apparemment très excités, tirent avec leur bec sur des touffes d'herbes (Fig. 53 – *Larus argentatus*), mais comme l'herbe n'est pas ramassée, mais seulement arrachée, il ne s'agit pas non plus d'une nidification substitutive véritable, mais d'une interférence avec le comportement de combat. Cet arrachage d'herbe est parfois interprété comme une menace pendant une extrême excitation agressive, mais il semble qu'il s'agisse plutôt d'une → *activité substitutive* libérée par l'activation simultanée d'une tendance de fuite et d'une disposition agressive (GOETHE 1956, TINBERGEN 1958).

Grenzflächenappetenz □ Es ist ein abnormes Verhalten, welches insbesondere bei aus einem Schwarm entnommenen Fischen in einem Aquarium zu beobachten ist. Es beruht darauf, daß sich der Fisch an den Aquarienscheiben spiegelt (BRESTOWSKY 1972). Grenzflächenappetenz erweist sich als sozial motivierte Reaktion auf einen vermeintlichen Artgenossen. Nach SCHUETT (1933) haben Sauerstoffmessungen ergeben, daß der »Artgenosse als Spiegelbild« eine »beruhigende« Wirkung hat. Dies trifft aber nur für Schwarmfische zu, – bei solitär lebenden Fischen bewirkt das Spiegelbild genau das Gegenteil (WIRTZ und DAVENPORT 1976).

appetence for mirrors □ An abnormal behaviour observable in fish which have been removed from a group and placed alone. These fish remain near their own reflection in the glass (BRESTOWSKY 1972) indicating a socially motivated reaction to supposed conspecifics. According to SCHUETT (1933) oxygen measurements have shown that the »mirror image conspecific« has a calming effect. This applies only to swarm fishes. In solitary fishes a mirror image releases the opposite effect (WIRTZ and DAVENPORT 1976).

comportement »narcissique« □ Comportement anormal qui s'observe en particulier chez des Poissons prélevés dans un banc de congénères et isolés en aquarium. Ce Poisson est attiré par son propre reflet sur les parois vitrées de l'aquarium (BRESTOWSKY 1972). Ce comportement s'avère comme étant une réaction à motivation sociale envers un congénère présumé. SCHUETT (1933) a effectué des mesures de consommation d'oxygène et a constaté que le »congénère en tant qu'image reflétée« exerce un effet tranquilisant. Ceci n'est évidemment valable que pour les Poissons vivant en banc. Chez les Poissons solitaires, l'image reflétée d'un congénère produit un effet tout à fait contraire (WIRTZ et DAVENPORT 1976).

Grundkoordination □ Unter Grundkoordinationen des Verhaltens versteht man jene speziellen Haltungs- und Bewegungselemente, die immer zu gleicher Zeit in dem betreffenden Zusammenhang auftreten, z. B. spezifische, meist arttypische Verhaltensweisen und Körperstellungen, die bei den meisten Säugern mit der Entleerung gekoppelt auftreten.

basic movement pattern or coordination □ The special postural and locomotor elements that always occur together in given context, e. g. typical, usually species-specific behaviour patterns and positions in mammals accompanying evacuation.

coordination motrice de base □ Eléments posturaux et moteurs qui vont toujours de pair avec une même action, par exemple comportements et postures spécifiques accompagnant chez la plupart des Mammifères la miction et la défécation.

Grunzpfiff □ Eine epigame Ausdrucksbewegung mit Lauterzeugung vieler Anatiden-Erpel. Wie beim gewöhnlichen Sichschütteln wird der Schnabel zunächst gesenkt, die Aufrichtung des Körpers läuft dann im Gegensatz zum gewöhnlichen Sichschütteln der des Kopfes soweit voraus, daß beim Aufrichten des Körpers der Schnabel noch tief nach unten zeigt (Abb. 54 A – *Anas acuta;* B – *Nettion flavirostre*). Diese merkwürdige Zusammenkrümmung hat ihre mechanische Bedeutung offenbar in der Spannung der Luftröhre, denn gerade im Augenblick ihres Maximums erfolgt ein lauter scharfer Pfiff, gefolgt von einem tiefen Grunzton (LORENZ 1941).

54 A 54 B

grunt whistle □ An epigamic expression with accompanying vocalization in many Anatid drakes. As during normal shaking movements the bill is at first lowered, the raising out of the water of the entire body then precedes the raising of the head so that the bill still points downwards (Fig. 54 A – *Anas acuta;* B – *Nettion flavirostre*). This unusual bending of the neck results in a compression of air in the trachea, which when released at last causes a sharp whistling sound followed by a grunt (LORENZ 1941).

grognement sifflé □ Mouvement expressif accompagné d'un sifflement dans le comportement épigame chez les malards de nombreux Anatidae. Comme lorsque le ♂ s'ébroue, le bec est dirigé vers le bas, puis le corps se redresse immédiatement, le bec restant toujours dirigé vers le bas (Fig. 54 A – *Anas acuta;* B – *Nettion flavirostre*). Cette position incurvée a très probablement sa signification mécanique dans la tension de la trachée-artère, car c'est précisément au moment de la tension maximum que le malard produit un sifflement aigu qui est suivi d'un grognement sourd (LORENZ 1941).

Gruppenbalz □ Das gemeinsame Balzen vieler Vögel sowie Gruppenwerbung bei einigen Säugern; oft an einem dafür bestimmten Platz, wie es z. B. Kampfläufer, *Philomachus pugnax,* und Birkhähne, *Lyrurus tetrix,* sowie bestimmte Fledermäuse tun. Auch Austernfischer, *Haematopus ostralegus,* balzen in Gruppen (Abb. 55), haben aber keine bestimmten Plätze. Gruppenbalz finden wir auch bei Insekten, z. B. bei Mücken, wo sich mehrere oder viele ♂♂ zum gemeinsamen »Tanzen« zusammenfinden, offenbar um die ♀♀ nach dem Prinzip der → *Reizsummenregel* besonders wirkungsvoll anzulokken. → *Männerbalzplatz.*

55

communal courtship □ Communal courtship of many individuals in birds and mammals on a → *lek* (arena display), e. g. in ruffs, *Philomachus pugnax,* and in the black grouse, *Lyrurus tetrix,* as well as in some species of bats. Oyster catchers, *Haematopus ostralegus,* also exhibit courtship in groups (Fig. 55), but not on a traditional courtship ground. Collective courtship also occurs in insects, e. g. in mosquitos where a great number of ♂♂ flock together for a common »dance«, apparently in order to entice the ♀♀ more efficiently according to the → *law of heterogenous summation.*

parade nuptiale collective □ La parade nuptiale s'effectue en commun chez de nombreux Oiseaux et quelques Mammifères à un endroit déterminé (→ *arène*) comme le font le Chevalier combattant, *Philomachus pugnax,* et le Tétra-Lyre, *Lyrurus tetrix* ainsi que certaines Chauves-Souris. La parade nuptiale en commun chez l'Huîtrier-Pie, *Haematopus ostralegus* (Fig. 55) est bien connue, mais elle n'est pas liée à un endroit précis. Nous connaissons une parade nuptiale collective également chez les Insectes, comme par exemple chez les Moustiques où de nombreux ♂♂ se rassemblent pour une »danse« en commun, probablement pour mieux attirer les ♀♀ selon la → *règle de la sommation hétérogène des stimuli.*

Gruppenbewußtsein □ Bei Menschen das besondere Gefühl der Angehörigkeit oder Zugehörigkeit zu einer bestimmten Gruppe und/oder Organisation. Das Gruppenbewußtsein einer wohlgeordneten Gruppe steigert die Aggressivität gegen Gruppenfremde (SCHMIDT 1960, EIBL-EIBESFELDT 1975).

group consciousness □ In man, the feeling of belonging to a given group or organization. Group consciousness in a well integrated group increases aggressivity against outside groups (SCHMIDT 1960, EIBL-EIBESFELDT 1975).

conscience de groupe □ Chez les Hommes, le sentiment particulier d'appartenir à un groupe ou une organisation déterminée. Une telle conscience à l'intérieur d'un groupe bien structuré et hautement organisé peut augmenter l'agressivité envers des individus étrangers au groupe (SCHMIDT 1960, EIBL-EIBESFELDT 1975).

Gruppenbindung □ Die Gruppenbindung ist eine sehr wichtige aggressionshemmende Einrichtung, die durch verschiedene Mechanismen aufrecht erhalten wird. Einer von vielen ist z. B. die Bindung über das Kind, und zwar bei Menschen ebenso wie bei nicht-menschlichen Primaten (EIBL-EIBESFELDT 1972, 1973).

group cohesion □ Group cohesion is a means of limiting aggression and exists in several forms, e. g. the cohesion deriving from care of the child, which occurs in man as well as nonhuman primates (EIBL-EIBESFELDT 1972, 1973).

cohésion sociale □ La cohésion sociale est une propriété très importante inhibant l'agressivité et maintenue par différents mécanismes. Un de ces mécanismes parmi beaucoup d'autres est par exemple le lien créé par l'enfant chez l'Homme et les jeunes chez tous les autres Primates infra-humains (EIBL-EIBESFELDT 1972, 1973).

Gruppenduft □ Ein verbindendes Signal unter Mitgliedern großer Gruppen zum gegenseitigen »Sich-Erkennen«. Ratten und Mäuse aus Rudeln bzw. Großfamilien z. B. markieren einander geruchlich und erkennen sich an den *Geruchsabzeichen.* Auch bei vielen sozialen Hymenopteren erkennen die Individuen einander am *Volksduft.* Wahrscheinlich gibt es in diesen Fällen keine oder nur bedingte individuelle Erkennung.

group scent □ An olfactory signal among members of large groups for recognizing cohorts. Mice and rats of a troop for example mark each other and can recognize each other by these *scent marks.* A similar phenomenon exists among social Hymenopterans. In these cases, there is probably no or only conditional individual recognition.

odeur sociale □ Un signal de cohésion et de reconnaissance réciproque entre les membres de grands groupes. Les Rats et les Souris vivant en bande par exemple procèdent à des marquages olfactifs sociaux et se reconnaissent à l'aide de ces *signaux odorants.* Chez de nombreux Hyménoptères sociaux, les individus se reconnaissent également entre eux à l'aide de *l'odeur sociale.* Dans ces groupes, il n'y a probablement pas de reconnaissance individuelle ou uniquement sous certaines conditions.

Gruppeneffekt → Stimmungsübertragung
E. group effect □ F. effet de groupe

Gruppenjagd □ Das Jagen von lebender Beute durch Gruppen von untereinander zusammenarbeitenden Individuen bei Tieren bzw. Menschen. Diese Verhaltensweisen sind insbesondere bei Treiberameisen, Pelikanen und Wölfen bekannt, aber auch typisch für Pygmäenhorden.

group predation □ The hunting and retrieving of living prey by groups of cooperating animals and men. This behaviour pattern is developed, for example, in army and driver ants, pelicans, wolves, and is typical for Pygmies.

prédation en groupe □ La chasse des animaux de proie par des groupes composés d'individus coopérants chez les animaux et chez l'Homme. Ce comportement est particulièrement connu chez les Fourmis Magnans, les Pélicans et les Loups; il est aussi particulièrement typique chez les Pygmées.

Gruppenkämpfe = Zwischengruppenaggression → Aggression
E. inter-group fighting □ F. combats entre groupes

Gruppenrevier → Revier
E. group territory □ F. territoire de groupe

Gruppenzugehörigkeit → Gruppenbewußtsein
E. group membership □ F. appartenance à un groupe

Grußformeln □ Art und Weise des Sich-Begrüßens bei Menschen. Die bei vielen Völkern und Kulturen unterschiedlichen Grußriten unterliegen dennoch einheitlichen Grundprinzipien. Überall ist es die Aufgabe des Grußes, ein Band zu stiften oder zu erhalten und Aggressionen zu beschwichtigen. Meist enthalten Grußformeln Komponenten einer symbolischen Unterwerfung, wie z. B. viele Verbeugungen vor dem zu Begrüßenden oder den Hut lüften und damit das bare Haupt darbieten; außerdem gehören sich vor Herrschern Niederknien oder Auf-den-Boden-Werfen als Grußformeln in den Kontext der *symbolischen Unterwerfung.* Auch *Geschenküberreichen* gehört zu den Grußformeln und hat eine beschwichtigende Funktion, besonders bei Erstbesuchen in einem fremden Haus.

greeting etiquette □ Manner in which humans respond upon meeting. Greeting etiquette may differ from culture to culture, but the basic principle underlying it is the same. In any case, greeting serves to establish or to maintain a bond and to appease aggression. Often greeting etiquette contains components of symbolic submission, such as repeated bowing, lifting the hat and thus showing the bare head; falling to the knees or falling to the ground in front of rulers or royalty also enters as a form of greeting into the context of *symbolic submission. Present giving* also occurs in the greeting etiquette and has an appeasement function, as have all symbolic submissive gestures, e. g. at a first visit to someone's house.

formules de salut □ Différentes manières de se saluer chez les Hommes. Les rites de salut chez les différents peuples et cultures sont assez variés, mais reposent malgré tout sur des principes de base communs. Partout, le rôle du salut est de créer des liens ou de les conserver et d'apaiser des agressivités. Souvent, les formules de salut contiennent des composantes d'une soumission symbolique, comme par exemple les révérences devant la personne à saluer ou de tirer son chapeau et de se présenter ainsi à tête nue; d'autre part, se mettre à genoux devant les régnants ou même de se coucher par terre sont à considérer comme une *soumission symbolique* dans le contexte des formules de salut. *L'offre d'un cadeau* est également une formule de salut et possède une fonction apaisante, particulièrement au moment d'une première visite dans une maison étrangère.

Grußriten → Grußformeln

E. greeting ritual □ F. rituel de salut

Gynopaedium → Mutterfamilie

H

Haarereißen □ Bei ernsthaften Auseinandersetzungen von Kindern häufig angewandte aggressive Handlung, indem sie sich in den Haaren des Gegners verkrallen und daran heftig reißen. Nach bisherigen Beobachtungen tritt diese Verhaltensweise während Streitereien spontan auf.

hair-pulling □ A frequent tactic in serious quarrels among children in which the child grabs and pulls on the opponent's hair. Previous observations suggest that this behaviour pattern occurs spontaneously during arguments.

tirer les cheveux □ Geste agressif qui intervient souvent lors de disputes entre enfants qui s'agrippent mutuellement par les cheveux qu'ils tirent fortement. Selon les observations faites jusqu'à ce jour, ce comportement apparaît spontanément au cours des controverses.

Haarsträuben □ Bei vielen Säugern ein mit dem Drohen gekoppeltes Aufrichten des Pelzes (*Fellsträuben*) oder des Haarkleides, um das Erscheinungsbild eindrucksvoll zu verändern. Der Effekt ist ein scheinbar größerer Körper. Für Vögel → *Gefiedersträuben.*

piloerection □ In many mammals the erection of the fur or hair in connection with threat as a means of effecting a more impressive appearance. The result is an apparently greater body. In birds → *ruffling.*

horripilation □ Chez de nombreux Mammifères, le hérissement des poils est une composante du comportement de menace et modifie d'une manière impressionnante la silhouette: son effet principal est que le corps apparaît bien plus grand. Pour les Oiseaux, → *ébouriffement des plumes.*

Hackdistanz → Individualdistanz

E. peck distance □ F. distance de becquetage

Hackordnung □ An einem Futterplatz genießen einige Hennen gewisse Vorrechte. Sie dürfen zuerst zur Futterstelle und hacken andere, rangniedrigere Hennen. Der Ausdruck Hackordnung ist als Phänomen gleichbedeutend mit → *Rangordnung* und nur für Vögel verwendbar, während Rangordnung insofern ein Überbegriff ist, als er auf alle Hierarchiesysteme angewandt werden kann (SCHJELDERUP-EBBE 1922 a + b, MARLER 1955).

peck order □ Some hens enjoy certain privileges during feeding, i.e., they have initial access to the feeding site and peck at other lower ranking hens. The term peck order is synonymous with → *rank order,* but is only used for birds. The more general term rank order can be used for all hierarchical systems (SCHJELDERUP-EBBE 1922 a + b, MARLER 1955).

hiérarchie de becquetage □ Sur les lieux d'alimentation, un certain nombre de Poules possèdent des privilèges. Elles peuvent venir se nourrir les premières et donnent des coups de bec aux Poules de rang inférieur. Cette hiérarchie de becquetage correspond en tant que phénomène à la → *hiérarchie sociale,* mais ne peut s'appliquer qu'aux Oiseaux alors que la hiérarchie sociale est une notion surordonnée valable pour d'autres organisations sociales (SCHJELDERUP-EBBE 1922 a + b, MARLER 1955).

Hackrecht → Hackordnung

E. peck right □ F. droit de becquetage

Halspumpen → Pumpen

E. nodding □ F. mouvement de pompage

Hals-Schulter-Reaktion → Schulterreaktion

Hamstern □ Das Aufsammeln und Speichern von Nahrungsreserven, so wie es der Hamster und viele andere Nager tun. → *Futterverstecken.*

hoarding □ The gathering and storing of food reserves as occurs in the hamster and many other rodents. → *food concealment.*

amassement □ Accumulation de réserves de nourriture chez certains Rongeurs dont le Hamster, d'où dérive le mot allemand. → *dissimulation de nourriture.*

Händepatschen □ Bittbewegungen kommen bei menschlichen Säuglingen vom 9. Monat an vor, indem sie die Hände gegeneinander patschen. Dem Ursprung nach ist das möglicherweise ein Greifen nach der Mutter.

hand patting □ Begging movements can be observed in human babies, beginning at the age of 9 months. The hands are smacked or patted together. Possibly, the origin of this behaviour lies in the movement of reaching out for the mother.

battement des mains □ Chez le nourrisson humain apparaissent à partir du 9e mois des mouvements de sollicitation en battant des mains. L'origine de ce comportement pourrait être un mouvement pour saisir la mère.

Handgreifreflex □ Eine charakteristische Reaktion des neugeborenen Säuglings. Berührt man die Handfläche, so schließen sich die Finger fest um den berührenden Gegenstand, und zwar in einer geordneten Abfolge der Fingerbewegungen (Abb. 56). Dieser *Klammerreflex* ist ganz offensichtlich ein Verhaltensrudiment, denn so kräftig er bei manchem Neugeborenen ist, ermöglicht er ihm nicht, mangels eines Haarkleides am menschlichen Körper, sich an der Mutter festzuhalten (PRECHTL 1955, HASSENSTEIN 1973).

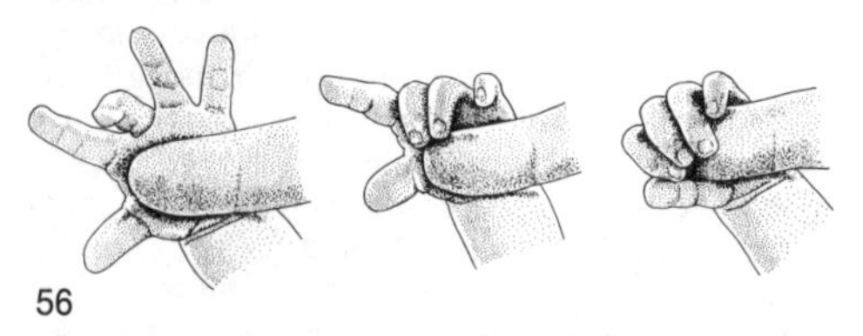

56

grasping reflex □ A characteristic reaction of the newborn infant. If one touches an infant's palm, then the fingers close around the object in an orderly sequence (Fig. 56). Despite the strength of this *clamping reflex* in some infants, it is apparently a behavioural rudiment. Since the human body lacks a hair covering, the infant cannot hold the mother by this means (PRECHTL 1955, HASSENSTEIN 1973).

réflexe de préhension □ Réaction caractéristique des nouveaux-nés humains. En réponse au contact de la paume de la main, les doigts se referment fortement autour de l'objet selon un ordre déterminé (Fig. 56). Ce réflexe de préhension est très probablement une relique comportementale, car malgré la puissance de cette préhension chez certains nouveaux-nés, il ne leur permet pas, à défaut de poils sur le corps humain, de s'accrocher après leur mère (PRECHTL 1955, HASSENSTEIN 1973).

Handheben → Distanzgruß
E. hand-raising □ F. salut de la main

Handlung am Ersatzobjekt → Ersatzobjekt
E. response to a substitute object □ F. activité sur un objet de remplacement

Handlungsbereitschaft □ Die Bereitschaft (Stimmung) zu einem bestimmten Verhalten ist einfach die festgestellte Häufigkeit oder auch Intensität, mit der dieses Verhalten auftritt; allgemeiner und weniger deterministisch: die Wahrscheinlichkeit seines Auftretens unter bestimmten Umständen. Die Bereitschaft hängt ab von der gegebenen (inneren) Situation, von Außenreizen usw. Der Ausdruck Handlungsbereitschaft wurde kürzlich (BECKER-CARUS, SCHÖNE *et al.* 1972) ausführlich diskutiert. Es wurde vorgeschlagen, in Zukunft Handlungsbereitschaft an Stelle von Motivation (und auch → *Trieb*) zu benutzen; da vor allem der Begriff Motivation in der Humanpsychologie und Ethologie sehr uneinheitlich gebraucht wird, sollte er in der Ethologie nicht weiter verwendet werden. Im Englischen und Französischen muß man wohl weiterhin »motivation« sagen.

motivation (lit.: *readiness to act*) □ The motivation for a particular behaviour is effectively the measured frequency or intensity with which the behaviour occurs, or generally speaking, the probability of its occurrence under certain circumstances. This readiness to act depends on certain given internal factors as well as external stimuli. The term has recently been reviewed by BECKER-CARUS, SCHÖNE *et al.* (1972); it was proposed that »Handlungsbereitschaft« rather than motivation (or → *drive*) be used in ethology since the term motivation has acquired diverse connotations in psychology as well as ethology.

motivation □ La disponibilité à une action correspond à la mesure de la fréquence ou de l'intensité avec laquelle ce comportement se manifeste; d'une manière plus générale et moins déterminée, c'est la probabilité de l'apparition de ce comportement sous certaines conditions. La disponibilité à une action est dépendante d'une situation donnée (interne), de stimuli externes, etc. La notion de la disponibilité à une action ou motivation a récemment été largement discutée (BECKER-CARUS, SCHÖNE *et al.* 1972); il a été proposé que le terme de *Handlungsbereitschaft* remplace à l'avenir la notion de motivation et aussi celle de *tendance* (→ *pulsion*) étant donné que la notion de motivation en particulier est utilisée avec des connotations très variées, tant en psychologie qu'en éthologie.

Handlungskette → Balzkette → Reaktionskette

Hänseln = Necken → Scherzpartnerbeziehung

Harem □ Weibchen-Gruppe oder Weibchen-Rudel, welches von einem ♂ betreut und zusammengehalten wird, welches außerdem andere ♂♂ daran hindert, mit diesen seinen ♀♀ zu kopulieren. → *Herden.*

harem □ A group of females guarded by a ♂ who prevents other ♂♂ from mating with them. → *herding.*

harem □ Groupe ou harde de ♀♀, gardé par un ♂ empêchant également d'autres ♂♂ de copuler avec ses ♀♀. → *Herden.*

Harnlecken □ Wird von männlichen und weiblichen Tieren ausgeführt; beim Wolf z. B. ist das Lekken des Harnes der ranghöchsten Fähen nur dem Leitwolf gestattet. Harnlecken kommt auch noch bei anderen Säugern vor. Die ♀♀ können dabei auch den Harn der Jungen lecken. Beim Hund lecken ♂♂ und ♀♀ die Jungen, die, wenn etwas älter, in Unterlegenheit auf dem Rücken liegen. Bei männlichen

Wiederkäuern folgt auf Harnlecken am ♀ oft das → *Flehmen.*

urine licking □ A behaviour pattern among mammals in ♂♂ and ♀♀; e. g. in wolves the licking of alpha ♀ urine is only done by the alpha ♂. The ♀♀ also lick the urine of their young. In dogs ♂♂ and ♀♀ lick the young which when a little older lie submissively on their backs. In ♂♂ ruminants urine licking on ♀♀ is often followed by → *Flehmen* (lip-curl).

léchage d'urine □ Comportement existant chez les ♂♂ et ♀♀ des Mammifères. Chez le Loup, par exemple, le léchage de l'urine de la ♀ dominante n'est autorisé qu'au ♂ alpha. Le léchage d'urine existe aussi chez de nombreux autres Mammifères. Les ♀♀ peuvent aussi lécher l'urine de leurs jeunes. Chez les Chiens, ♂♂ et ♀♀ lèchent les jeunes qui alors, à partir d'un certain âge, se couchent sur le sol en position de soumission. Chez les ♂♂ des Ruminants, le léchage d'urine chez la ♀ est souvent suivi du → *Flehmen.*

Harnmarkierung □ Eine bei vielen Säugern verbreitete Handlung der Markierung von Wegen und Grenzen des Reviers mit Harn (→ *Reviermarkierung*, HEDIGER 1957). Hundeartige heben dabei ein Hinterbein an und markieren nach der Seite in kurzen Stößen; Feliden spritzen im allgemeinen nach rückwärts (Abb. 57 A – *Lynx pardina*). Viele Hirsche spritzen während der Brunft Harn gegen ihre Bauchdecke und Vorderbeine, so daß sie an diesen Stellen fast schwarz aussehen (*Selbstmarkierung*). Der stark riechende Hirsch wirkt wahrscheinlich anziehend auf die Geschlechtspartner und drohend Rivalen gegenüber. Bei einigen anderen Arten, insbesondere bei Nagern, wird auch der Paarpartner mit Harn markiert, → *Harnspritzen* (ALTMANN 1969). Eine besondere Form des Duftmarkierens ist das *Harnwaschen* oder *Urinwaschen* bei einigen niederen Primaten. Die Tiere benetzen zuerst die Handflächen mit Urin und reiben diese danach in schnellen Bewegungen an den Fußsohlen (Abb. 57 B – *Microcebus murinus*), so daß dies beim Herumklettern Spuren hinterläßt (HILL 1938, EIBL-EIBESFELDT 1953). Neben einigen Prosimiern (*Galago, Loris, Nycticebus, Microcebus*) zeigen auch die Gattungen *Cebus, Saimiri* und *Aotus* ein offensichtlich homologes Verhalten (NOLTE 1958, ANDREW und KLOPMAN 1972).

57 A

57 B

urine marking □ A widespread behaviour pattern among mammals in which the paths and boundaries of the territory are marked with urine (→ *territorial marking*, HEDIGER 1957). Canids raise a hind leg and urinate to the side in short sweeps. Cats spray their urine to the rear (Fig. 57 A – *Lynx pardina*). During the breeding season many deer spray urine against their belly and forelegs resulting in almost black staining of these areas (*self-marking*). The strong smell produced probably acts to attract the sex partner and to threaten or to repulse rivals. Certain other groups, especially rodents, also spray the partner with urine, → *urine spraying* (ALTMANN 1969). A particular form of scent marking, the *urine washing,* occurs in several lower Primates. The animal moistens its palms with urine and then wipes it on his foot soles with rapid swipes (Fig. 57 B – *Microcebus murinus*) so that he leaves a trail as he moves about (HILL 1938, EIBL-EIBESFELDT 1953). In addition to several prosimiens (*Galago, Loris, Nycticebus, Microcebus*) a homologous behaviour occurs in the *Cebus, Saimiri* and *Aotus* groups (NOLTE 1958, ANDREW and KLOPMAN 1972).

marquage d'urine □ Chez de nombreux Mammifères, manoeuvre très répandue pour marquer des pistes et des limites territoriales avec de l'urine (→ *marquage du territoire*, HEDIGER 1957). Les Canidae procèdent au marquage en soulevant un membre postérieur et en effectuant des giclées d'urine répétées; de nombreux Felidae effectuent généralement un jet vers l'arrière (Fig. 57 A – *Lynx pardina*). Certaines espèces de Cervidae, lors du rut, arrosent d'urine leur ventre et leurs membres antérieures de sorte que ces endroits, imprégnés d'urine, paraissent presque noirs (*auto-marquage*). Le Cerf émet ainsi à proximité immédiate de son corps une forte odeur qui a très probablement un pouvoir attractif pour les ♀♀ et un pouvoir menaçant et répulsif vis-à-vis des rivaux. Chez quelques autres espèces, en particulier chez les Rongeurs, le partenaire sexuel est souvent marqué avec de l'urine, → *aspersion d'urine* (ALTMANN 1969). Une forme particulière de marquage d'urine, le *lavage à l'urine,* existe chez un certain nombre de Primates inférieurs: les animaux urinent d'abord sur la paume de leur main et effectuent, d'une manière simultanée, des frottements rapides sur la plante des pieds (Fig. 57 B – *Microcebus murinus*) ce qui leur permet de laisser des traces en marchant (HILL 1938, EIBL-EIBESFELDT 1953). Ce comportement est décrit chez un certain nombre de Prosimiens (*Galago, Loris, Nycticebus, Microcebus*). Il apparaît également et probablement d'une manière homologue chez les genres *Cebus, Saimiri* et *Aotus* (NOLTE 1958, ANDREW et KLOPMAN 1972).

Harnprüfen □ Ein bei vielen Säugern verbreitetes Verhalten, bei dem das ♂ die Begattungsbereitschaft des ♀ prüft. Das männliche Tier nähert sich einem harnenden ♀ und läßt sich den Harn über die Nase laufen; oft wird der Harn inhaliert und die Oberlippe schüsselförmig nach oben gebogen. → *Flehmen* kann darauf folgen.

urine sampling □ A widespread behaviour among mammals whereby the ♂ tests the readiness of the ♀ for mating. The ♂ approaches the urinating ♀ and allows her urine to run over his nose; the smell of the urine is inhaled and the upper lip turned upward to serve as a cup. This behaviour may be followed by → »*Flehmen*« (lip-curl).

examen de l'urine □ Chez de nombreux Mammifères, comportement très répandu par lequel le ♂ »teste« la disposition de la ♀ à l'accouplement. Il s'approche d'une ♀ en train d'uriner et laisse couler de l'urine sur son nez; souvent il l'aspire en retroussant légèrement la lèvre supérieure. Ce comportement peut être suivi de → »*Flehmen*«.

Harnriten □ Richtige Harnriten gibt es bei den Hottentotten; nach der Beschneidung werden die Knaben vom Beschneider beharnt, gleiches geschieht, wenn die Jünglinge in die Männergemeinschaft aufgenommen werden. Harnzeremonielle kommen auch bei der Verehelichung vor, wobei der Bräutigam und die Braut vom Zeremonienmeister beharnt werden; dann wendet er sich an die Umsitzenden. Beharnt wird außerdem noch der erfolgreiche Jäger. Möglicherweise handelt es sich bei dieser sehr altertümlichen Menschengruppe um ein analoges (nicht verwandtes) Verhalten, das im Laufe der Zeit zu Riten umgewandelt wurde, bei denen man Wasser benutzt; cf. Taufe (EIBL-EIBESFELDT 1970). – Viele Nager benässen Rudelmitglieder mit Harn und markieren sie auf diese Weise als zur Gruppe gehörig. → *Gruppenduft.*

urination ceremony □ Occurs among the Hottentots in several forms; the newly circumcised youth is urinated upon by the circumciser; a similar rite follows his acceptance into the adult male society; the marriage ceremony may involve the marriage official urinating first upon the bride and bridegroom and then upon the bystanders. Moreover, a successful hunter may be urinated upon when he returns. Possibly the urination ceremony which still persists in very primitive cultures may be analogous to certain rites in which water is used, e. g., baptism (EIBL-EIBESFELDT 1970). – Many rodents are known to urinate upon group members serving perhaps as a means of recognizing group members. → *group scent.*

cérémonial urinaire □ De véritables rites urinaires existent chez les Hottentots; après la circoncision, les garçons sont arrosés d'urine par le circonciseur; il en est de même lorsque les adolescents sont introduits dans la société des hommes adultes. De telles cérémonies existent aussi au moment d'un mariage où le jeune couple est arrosé d'urine par le sorcier. Le même rite sert à honorer le succès d'un chasseur. Il est probable que les cérémonies urinaires chez les Hommes dits paléolithiques constituent un comportement analogue, mais non apparenté, aux rites qui se sont modifiés au cours des temps et pour lesquels on utilise aujourd'hui de l'eau, comme par exemple le baptême (EIBL-EIBESFELDT 1970). – Chez de nombreux Rongeurs, les membres d'un même groupe sont également aspergés d'urine et ainsi marqués comme appartenant à un groupe déterminé. → *odeur sociale.*

Harnspritzen □ Verhaltensweisen unterschiedlicher Bedeutung bei Lagomorphen und Hystricomorphen, welche von KUNKEL und KUNKEL (1964) zusammenfassend abgehandelt wurden. Nach den bislang vorliegenden Beobachtungen ergeben sich folgende Situationen, in denen mit Harn gespritzt wird: Nicht paarungsbereite ♀♀ wehren ♂♂, die aufreiten wollen, ab, indem sie mit Harn nach rückwärts spritzen (Abwehrspritzen); bekannt von Hase, Kaninchen und Mara (DUBOST und GENEST 1974). Während der Werbung spritzen die ♂♂ nach den ♀♀; Hase, Kaninchen, Meerschweinchen, Urson (Baumstachler) und Mara (Abb. 58 – *Dolichotis patagonum*). Im Angriffs- und Verteidigungsverhalten

58

wird ebenfalls mit Harn gespritzt, sowie bei der Verteidigung des in die Enge getriebenen Tieres gegen den Menschen. Letzteres konnte aber auch bei einigen Feliden beobachtet werden (eig. Beob.).

urine spraying □ Behaviour patterns occurring in several contexts in lagomorphs and hystricomorph rodents discussed together by KUNKEL and KUNKEL (1964). Urine spraying is known to occur in the following contexts: Non-receptive ♀♀ may repulse a ♂ attempting to mount by spraying urine backwards; this is known from hares, rabbits and maras (DUBOST and GENEST 1974). ♂♂ may spray the ♀♀ during courtship, e. g., hares, rabbits, guinea pigs, porcupines and maras (Fig. 58 – *Dolichotis patagonum*). In attack and defensive behaviour urine spraying may occur; an animal cornered by man, e. g., several cat species (pers. obs.), may react in this way.

aspersion d'urine □ Comportement surtout connu chez les Lagomorphes et les Hystricomorphes et dont les différentes significations ont été récemment discutées par KUNKEL et KUNKEL (1964). Selon les observations faites jusqu'à ce jour, l'aspersion d'urine s'effectue dans les situations suivantes: les ♀♀ se défendent contre les ♂♂ qui s'approchent en tentative de monte en les aspergeant à l'aide d'un puissant jet d'urine (giclée de défense); ce comportement est surtout connu chez les Lièvres, les Lapins et les Maras (DUBOST et GENEST 1974). Au cours de la parade nuptiale, les ♂♂ aspergent les ♀♀ avec de l'urine; Lièvres, Lapins, Cobayes, Ursons et Maras (Fig. 58 – *Dolichotis patagonum*). L'aspersion d'urine s'observe également au cours de comportements d'attaque et de défense et a été également constatée envers l'Homme lorsqu'il est l'agresseur. Ce dernier phénomène a également été constaté chez un certain nombre de Felidae (observ. pers.).

Harnwaschen → Harnmarkierung
E. urine washing □ F. lavage à l'urine

Harnzeremoniell → Harnriten

Hassen □ Haßreaktion von Singvögeln, die einzeln oder im Schwarm einen Feind (Greifvögel) angreifen und mit besonderen Haßlauten und Scheinangriffen den Freßfeind von allen Seiten bedrängen bzw. verfolgen (Abb. 59 – Bachstelzen hassen auf einen Sperber).

59

mobbing □ Aversive response of song birds to birds of prey in which they swarm around and pursue their enemy while giving typical mobbing calls and performing sham attacks (Fig. 59 – wagtails mobbing at a sparrow-hawk).

houspillage □ Réaction collective et agressive de certains Passereaux qui attaquent en groupe des Rapaces en poussant des cris particuliers (de houspillage) et en portant des pseudo-attaques (Fig. 59 – Bergeronnettes poursuivant un Epervier).

Haßlaute → Hassen
E. mobbing calls □ F. cris de houspillage

Hecheln □ Bei vielen Enten, Raben- und Greifvögeln eine Hitzereaktion; ein schnelles Atmen mit weit geöffnetem Schnabel zur Abkühlung bei großer Hitze, auch Hitzehecheln genannt (Abb. 60 – hechelnde junge Rabenkrähe). Hundeartige hecheln auffällig nach sehr schnellem Lauf, sowie auch viele andere Arten (PRINZINGER 1976).

60

panting □ A response to overheating in many ducks, corvids, birds of prey, etc., characterized by fast breathing with the beak wide open (Fig. 60 – panting in a young carrion crow). Canids also pant after running, as do many other species (PRINZINGER 1976).

halètement □ Respiration très rapide, la bouche ou le bec largement ouvert, – l'état de la polypnée thermique. Chez de nombreux Corvidae, Canards et Rapaces, une réaction à une chaleur intense pour obtenir un refroidissement du corps (Fig. 60 – halètement chez une jeune Corneille). Chez les Canidae comme chez de nombreuses autres espèces, on observe un halètement après une course rapide (PRINZINGER 1976).

Heimbezirk → Aktionsraum

Heimfindevermögen → Orientierung
E. homing ability □ F. capacité de retour au gîte

Hemmung □ Die Unterdrückung eines Zustandes oder die Verhinderung bzw. Verlangsamung oder Unterbrechung eines Vorgangs. In der Ethologie spricht man von Hemmung, wenn eine Verhaltensweise durch bestimmte endogene oder allochthone Reize blockiert wird. Die gegenseitige Hemmung von Verhaltensweisen führt häufig zu Konflikten und zu → *Übersprungbewegung(en)*. Weiter können gewisse Verhaltensweisen eine Aggressions- oder gar Tötungshemmung hervorrufen; z. B. → *Demutsgebärde(n)*. In der Neurophysiologie versteht man unter Hemmung eine vorübergehende Aktivitätsminderung von Nervenzellen.

inhibition □ The suppression of a condition or state; the retardation, hindrance or interruption of a process. Ethologically inhibition refers to blockage of a behaviour pattern through certain endogenous or allochthonous stimuli. The reciprocal inhibition of behaviours frequently leads to conflict or → *displace-*

ment activities. Additionally, certain behaviour patterns can induce inhibition of aggression or killing, e. g. → *submissive gesture(s)*. In neurophysiology inhibition refers to a temporary reduction in activity of nerve cells.

inhibition □ Répression d'un état ou empêchement, ralentissement ou interruption d'un processus. En éthologie, nous parlons d'inhibition lorsqu'un comportement est bloqué par une certaine stimulation endogène ou allochtone. L'inhibition réciproque des comportements conduit souvent à un état conflictuel et à des → *activité(s) substitutive(s)*. D'autre part, un certain nombre de comportements peuvent produire une inhibition de l'agressivité ou une inhibition du geste mortel, comme par exemple les → *geste(s) de soumission*. En neurophysiologie, on entend par inhibition une diminution temporaire de l'activité des cellules nerveuses.

Herausforderungsritual □ Dieses Verhalten, beschrieben beim Buntbock, *Damaliscus dorcas,* dient der Behauptung in einem Revier. Es umfaßt mindestens 30 verschiedene Verhaltensweisen (DAVID 1973). Nur Revierbesitzer, die für die Fortpflanzung in der Population sorgen, halten ein vollständiges Herausforderungsritual durch. Junggesellen, die von einem Revierbesitzer herausgefordert werden, laufen davon.

challenge ritual □ This behaviour pattern, described for the Bontebok, *Damaliscus dorcas,* is a form of territorial maintenance involving at least 30 different behaviour elements (DAVID 1973). Only those individuals in the population actively concerned with reproduction exhibit a complete challenge ritual; in such cases unmated individuals challenged by the territory owner invariably retreat.

rituel de défi □ Ce comportement qui a été très bien décrit chez *Damaliscus dorcas* sert à la dominance dans un territoire. Il est composé d'une trentaine d'éléments comportementaux différents (DAVID 1973). Seuls les occupants d'un territoire assurant la reproduction à l'intérieur d'une population sont capables d'effectuer un tel rituel jusqu'au bout. Les jeunes ♂♂ défiés par un ♂ dominant occupant un territoire ne s'engagent pas dans un combat et prennent la fuite.

Herden □ Eine epigame Verhaltensweise bei brunftterritorialen männlichen Hirschen und Boviden in Zusammenhang mit dem Erobern und Verteidigen des Weibchenrudels bzw. Brunftharems (BUECHNER und SCHLOETH 1965 – *restraining display of the ♂♂*). Die Rothirsche haben dazu ein besonderes Droh- und Imponierverhalten entwickelt, welches DARLING (1937) erstmals ausführlich beschrieb. Der Platzhirsch verfolgt sich vom Rudel entfernende ♀♀, indem er diese seitlich aufholt und auf der dem Rudel abgewandten Seite im flachen Kreisbogen drohend imponierend umschreitet. Als weitere Komponenten treten → *Eckzahndrohen* als mimischer Ausdruck und Imponierschreiten in Körperbreitseitsstellung als dynamisches Ausdrucksverhalten auf. Das ♀ wendet sich daraufhin vom Hirsch weg, also dem Rudel hin, und wird dann vom Hirsch linear zum Rudel zurückgejagt (BÜTZLER 1974). In einer Arbeit über den Banteng spricht HALDER (1976) von *Hüten* oder *Hüteverhalten.*

herding □ An epigamic behaviour pattern in territorial male Cervids and Bovids connected with conquering and defending the harem during rut (BUECHNER and SCHLOETH 1965 – *restraining display of the ♂♂*). The red deer has developed a particular threat and intimidation display which was first described in detail by DARLING (1937). The red deer buck pursues a ♀ which leaves the herd overtaking her on the side. He moves in a broad arc on the side of the ♀ away from the herd threatening and intimidating. → *Canine-tooth threat* and intimidation display of presenting the body broadside may accompany herding. The ♀ turns away from the stag, and hence toward the herd, then he chases her in a direct fashion back to the harem (BÜTZLER 1974). HALDER (1976) in a similar description of the banteng referred to this as »guarding behaviour« (*Hüten* or *Hüteverhalten).*

Herden □ Comportement épigame chez les ♂♂ territoriaux d'un certain nombre de Cervidae et Bovidae grégaires en rapport avec la conquête et la défense de la harde des ♀♀ (BUECHNER et SCHLOETH 1965 – *restraining display of the ♂♂*). *Cervus elaphus* a, à ce titre, développé un comportement particulier de menace et d'intimidation décrit pour la première fois par DARLING (1937). Par une marche imposante, le Cerf poursuit chaque ♀ qui essaie de s'éloigner en la contournant latéralement du côté opposé à la harde pour tenter de la rabattre vers le troupeau. La ♀ se détourne alors du Cerf, c'est-à-dire se dirige vers la harde et à ce stade, elle est chassée en ligne droite vers les autres ♀♀. Le ♂, lors du comportement du Herden, peut effectuer également la → *menace en découvrant les canines* (BÜTZLER 1974).

Hesmosis → Volksteilung

Hetzen □ Die verbreitetste weibliche Werbehandlung bei Entenvögeln. Die Ente wendet sich dem Gatten oder dem umworbenen Zukünftigen zu, schwimmt oder geht hinter ihm her und droht gleichzeitig über die Schulter weg nach einem anderen artgleichen ♂ hin (Abb. 61 – *Tadorna tadorna*). Bei gesellig balzenden Arten dient dieses Verhalten dazu, ein ♂ aus der Gruppe auszusondern, um es gegen seinesgleichen angriffslustig zu stimmen. Es kommt auch bei bereits verpaarten Enten vor und dient offenbar der Partnerbindung (LORENZ 1941).

inciting □ The most widely distributed courtship behaviour in female ducks. The ♀ orients towards her partner or her prospective mate, follows him and threatens at the same time towards another conspecific ♂ (Fig. 61 – *Tadorna tadorna*). In communally courting species this behaviour serves to isolate a ♂ from the group and to stimulate him to become ag-

61

gressive against other conspecific ♂♂. The behaviour is also found in already existing pairs and apparently helps to maintain the pair bond (LORENZ 1941).

»provocation« □ Terme proposé, à défaut de mieux, pour *Hetzen*. Comportement de provocation qui fait partie intégrante, chez les Canards ♀♀, du répertoire comportemental de la parade nuptiale. La cane qui suit de près son futur partenaire ♂, menace en même temps un autre ♂ de la même espèce (Fig. 61 – *Tadorna tadorna*). Ce comportement sert aussi à choisir un conjoint parmi les ♂♂ en parade collective. La ♀ provoque un ♂ déterminé en l'incitant à attaquer ou à être agressif envers d'autres ♂♂. Le même comportement existe aussi chez les Canards déjà appariés et est probablement au service du maintien du couple (LORENZ 1941).

Hexenkreise → Paarungskreisen
E. lit.: witches circles □ F. »ronds de sorcière«

Hierarchie □ Der Ausdruck Hierarchie bezeichnet das Auftreten von Dingen nach einer bestimmten Ordnung, z. B. die Hierarchie von Verhaltensweisen bzw. ihr hierarchischer Aufbau; die Hierarchie von Zentren (Zentrenhierarchie); die Hierarchie von Stimmungen (Stimmungshierarchie) usw. Im Englischen und Französischen versteht man in der Regel darunter die → *Rangordnung*.

hierarchy □ The occurrence of things in a certain order, such as the hierarchical organization of behaviours, hierarchical organization of neural centers, mood hierarchy, etc. Particularly in English and French synonymous with → *rank order*.

hiérarchie □ Expression qui désigne l'apparition de certains phénomènes selon un ordre déterminé; citons la hiérarchie des différents comportements ou leur superposition hiérarchique, etc. De même, il existe aussi une hiérarchie des centres nerveux, une hiérarchie des motivations, etc. En anglais et en français, on entend en général par la → *hiérarchie sociale* une forme d'organisation fréquente chez les Vertébrés.

Hierarchiesystem → Rangordnung
E. hierarchical system □ F. système hiérarchique

Himmelskompaßorientierung → Orientierung
E. sky compass orientation □ F. orientation sur repères célestes

Hintenherumkratzen □ Bei dieser Kratzweise, im Gegensatz zum → *Vornherumkratzen*, senken die Vögel einen Flügel und schwingen das gleichseitige Bein über ihn hinweg, um sich am Kopf zu kratzen (Abb. 62 – *Motacilla alba*). Nur wenige Vögel kratzen sich so. Nach HEINROTH (1930) und WICKLER (1970) ist das Hintenherumkratzen als ein altes Wirbeltiererbe anzusehen, da auch Eidechsen und Säuger sich so kratzen, und sie meinen, das Kratzen am Kopf in dieser Form habe die Umwandlung der Vorderextremität überlebt.

62

scratching over the wing □ The scratch pattern in birds in which the wing is dropped and the bird raises the ipsilateral leg over and above it to scratch the head (Fig. 62 – *Motacilla alba*); in contrast to → *direct scratching*. Only a small number of birds scratch in this manner. According to HEINROTH (1930) and WICKLER (1970) scratching over the wing is a vestige of the original vertebrate line which has survived the reorganization of the forelimbs since lizards and mammals also scratch in this fashion.

grattage par-dessus l'aile □ Lors de ce comportement, en opposition au → *grattage direct de la tête*, les Oiseaux abaissent une aile et soulèvent la patte du même côté par-dessus cette aile pour se gratter la tête (Fig. 62 – *Motacilla alba*). Très peu d'espèces d'Oiseaux se grattent de cette manière. Selon HEINROTH (1930) et WICKLER (1970), ce grattage par-dessus l'aile est à considérer comme un vieil héritage des Vertébrés étant donné que les Reptiles et quelques Mammifères se grattent de la même manière, et les auteurs supposent que cette manière de se gratter la tête a survécu à la transformation du membre antérieur en aile.

Hinterkopfzudrehen □ Eine Beschwichtigungsgebärde, die bei Lachmöwen dazu dient, während der Balz bzw. während des Paarbildungszeremoniells die aggressionsauslösende schwarze Gesichtsmaske zu verbergen (Abb. 63 A – *Larus ridibundus*). Wahrscheinlich handelt es sich um einen vorübergehenden Kontaktabbruch mit Fluchtintention (MANLEY 1960, CHANCE 1962). Dieses Verhalten wird auch Wegsehen genannt. Bei einigen Schwimmenten gehört das Hinterkopfzudrehen in den Kontext der epigamen Verhaltensweisen. Bei der Werbung dreht der Erpel der Ente seinen Hinterkopf zu, um ihr das prächtig gefärbte Gefieder zu zeigen (Abb. 63 B – *Anas acuta*); oft wird das Bakkengefieder dabei gespreizt (LORENZ 1941).

head-flagging □ An appeasement gesture during the mating and pair-formation ceremony in the black-headed gull which consists of turning away the black facial mask normally associated with aggression (Fig. 63 A – *Larus ridibundus*). This behaviour constitutes most likely a temporary breaking of contact with intention to flee (MANLEY 1960, CHANCE 1962). The behaviour is also called »looking away«. In some species of ducks this behaviour is part of the epigamic behaviour patterns. In courtship the drake presents the back of his head towards the ♀, thus displaying the colourful feathers to her (Fig. 63 B – *Anas acuta*); frequently, the cheek-feathers are spread out (LORENZ 1941).

présentation de l'arrière de la tête □ Chez la Mouette rieuse, geste d'apaisement qui a pour fonction de dissimuler le masque noir déclenchant l'agression, au cours de la parade nuptiale et de la formation des couples (Fig. 63 A – *Larus ridibundus*). Il s'agit probablement d'une rupture de contact temporaire avec intention de fuite (MANLEY 1960, CHANCE 1962). Chez un certain nombre d'espèces de la famille des Anatidae, la présentation de l'arrière de la tête fait partie du comportement épigame. Au cours de la parade nuptiale, le malard présente l'arrière de sa tête à la cane pour lui montrer le beau plumage de cet endroit (Fig. 63 B – *Anas acuta*); souvent, il ébouriffe, de plus, son plumage jugal (LORENZ 1941).

Hitzehecheln → Hecheln

Hochzeitsflug □ Kopulationsflug der geflügelten Königinnen und der ♂♂ bei sozialen Insekten. → *Volksteilung*.

nuptial flight □ The mating flight of the winged queens and ♂♂ of an insect society. → *colony fission*.

vol nuptial □ Vol d'accouplement des reines et des ♂♂ ailés chez les Insectes sociaux. → *fission de colonie*.

Homologie □ Die Gleichartigkeit von Organen, Strukturen und Verhaltensweisen im Sinne ihrer stammesgeschichtlichen Herkunft. Der Nachweis von Homologien ist die Hauptmethode zur Ermittlung stammesgeschichtlicher Verwandtschaft und zur Aufstellung eines natürlichen Systems der Organismen. Die Übernahme der Homologiekriterien der Morphologie (REMANE 1952) in die Ethologie verdanken wir vor allem BAERENDS (1958) und WICKLER (1961, 1965). Homologien, die über das Gedächtnis weitergegeben werden, nennt WICKLER »Traditionshomologien« zum Unterschied zu den über das Genom als Informationsträger weitergereichten phyletischen Homologien. Im Gegensatz hierzu → *Analogie*, → *Konvergenz*.

homology □ The similarity of organs, structures and behaviour patterns due to their common evolutionary origin. The determination of homologies serves as the primary method for assessing evolutionary relationships and for illustrating the organization of natural systems. BAERENDS (1958) and WICKLER (1961, 1965) promote the use of the criteria applied to morphological homology (REMANE 1952) in all comparative studies of behaviour. WICKLER distinguishes homologies transmitted culturally (tradition homology) from those transmitted genetically (phylogenetic homology). In contrast to → *analogy*, → *convergence*.

homologie □ Similitude d'organes, de structures et de comportements fondée sur une communauté d'origine phylogénétique. La mise en évidence des homologies est la méthode principale pour l'élucidation des parentés phylogénétiques et l'établissement d'une classification naturelle des organismes. L'utilisation en éthologie comparée des critères d'homologie de la morphologie (REMANE 1952) est due particulièrement à BAERENDS (1958) et WICKLER (1961, 1965). Les homologies comportementales transmises par mémoire ont été appelées homologies de tradition par WICKLER pour les distinguer des homologies phylogénétiques dont la continuité est assurée par le génome comme vecteur d'information. En opposition à → *analogie*, → *convergence*.

Hoppeln → Gangart

E. hopping □ F. sautiller

Horde □ Im ursprünglichen Sinne eine kleine Gruppe von Menschen von Sammlern und Jägern mit eigenem Revier; heute nur noch bei Buschmännern, Hadzas, Pygmäen und Australiern. → *Klan.*

band □ Originally a small group of humans who are gatherers and hunters occupying a small territory of their own. Today only found in Bushmen, Hadzas, Pygmies, and Australian Aborigines. → *clan.*

horde □ Au sens originel du terme, un petit groupe humain vivant de la chasse et de la cueillette, avec un territoire propre, ce que nous ne trouvons plus aujourd'hui que chez les Boschimans, les Hadzas, les Pygmées et les Aborigènes de l'Australie. → *clan.*

Horten → Futterverstecken → Hamstern

Hospitalismus □ Krankhafter Zustand in der geistigen und körperlichen Entwicklung von Menschen, die als Kleinkinder unter Liebes- und Mutterentzug, meist in Heimen und Findelhäusern, aufwuchsen. Schon eine nur mehrmonatige Trennung von der Mutter kann zu einem Verlassenheitssyndrom, d. h. zu irreparablen Schädigungen führen. Ähnliches wurde bei Rhesusaffen festgestellt (SPITZ 1945, HARLOW und HARLOW 1962 a + b, TINBERGEN und TINBERGEN 1972).

hospitalism □ Disturbed emotional and bodily development of individuals that have grown up in orphanages without normal maternal affection. Separation of mother and child for a matter of months can lead to an isolation syndrome, i. e. to irreparable impairment. Also experimentally demonstrable in Rhesus monkeys (SPITZ 1945, HARLOW and HARLOW 1962 a + b, TINBERGEN and TINBERGEN 1972).

hospitalisme □ Etat pathologique du développement corporel et mental des individus qui, lorsqu'ils étaient petits enfants, ont été privés de contacts maternels et affectifs parce qu'ils étaient élevés dans des orphelinats ou qu'ils avaient effectué des séjours prolongés dans les hôpitaux. Seule une séparation de la mère de plusieurs mois peut conduire à un syndrome de solitude, c'est-à-dire à des dommages irréparables. Des phénomènes voisins ont été observés chez les Macaques (SPITZ 1945, HARLOW et HARLOW 1962 a + b, TINBERGEN et TINBERGEN 1972).

Hüftstöße □ Aus dem menschlichen Sexualverhalten abgeleitete Bewegungsweise, welche in ritualisierter Form im Kontext des Spottverhaltens, bei Tänzen und als Drohgebärde zum Aktionssystem des menschlichen Verhaltens gehört. Wir finden dieses Verhalten in verschiedenen Kulturen von den Buschmännern bis zu den Europäern. Bei Kindern tritt diese Bewegungsweise ähnlich wie das → *Gesäßweisen* oft spontan auf.

hip thrusting □ A behaviour pattern derived from human sexual behaviour, which occurs in ritualized form in the context of mocking, in dances and as a threat behaviour in the action system of human behaviour. The behaviour is present in various cultures from Bushmen to Europeans. In children this behaviour occurs spontaneously, similar to → *buttocks display.*

coups de reins □ Comportement dérivé du comportement sexuel chez l'Homme qui, sous une forme ritualisée, fait partie du contexte de taquinage. Il est également présent dans les différentes activités de danse et utilisé comme geste de menace. Nous rencontrons ce comportement dans les différentes cultures depuis les Boschimans jusqu'aux Européens. Chez les enfants, ce comportement apparaît spontanément, comme d'ailleurs la → *présentation du postérieur.*

Humanethologie □ Ein spezieller, rezenter Zweig der allgemeinen Verhaltensforschung, der versucht, das menschliche Verhalten mit Hilfe der an der Tierethologie gewonnenen Methoden zu untersuchen. Die Entdeckung der Ethologen, daß stammesgeschichtlich vorprogrammierte Gesetzmäßigkeiten das Verhalten von Tieren in entscheidender Weise bestimmen, eröffnet auch für die Erforschung menschlichen Verhaltens neue Perspektiven. Die Humanethologie hat sich zur Aufgabe gestellt, die stammesgeschichtlichen Grundlagen unseres Verhaltens zu erforschen und mit Hilfe des Kulturenvergleichs in den komplizierten Verhaltensweisen des Menschen den Anteil des Angeborenen zu eruieren. Die Dokumentation des Aktionssystems von taubblind geborenen Kindern spielt hierbei eine wichtige Rolle. Bei der Freilandarbeit benutzt der Humanethologe bevorzugt Spiegelobjektive, um Verhaltensweisen und soziale Interaktionen ungestellt und unbemerkt dokumentieren zu können (EIBL-EIBESFELDT und HASS 1966, SORENSON und GAJDUSEK 1966, KEITER 1969, CHAUVIN 1972, EIBL-EIBESFELDT 1973, EIBL-EIBESFELDT und LORENZ 1974, MORIN und PIATTELLI-PALMARINI 1974 – dort überall ausführliche Verzeichnisse von Einzelarbeiten).

human ethology □ A special and recent branch of ethology which endeavours to examine human behaviour with the aid of methods that were successful in studying animal behaviour from an ethological point of view. The discovery of ethologists that phylogenetically pre-programmed lawfulness in the behaviour of animals is an important determinant of animal behaviour, opens up new perspectives in the study of human behaviour. Human ethology tries to elucidate the phylogenetic bases of human behaviour, and to determine, with the help of a comparative study of cultures, the innate components of these complex behaviour patterns. The documentation of the action system of children that were born blind and deaf is important in this context. In field studies human ethologists preferably use a prism in their camera lens that allows filming at 90 degrees to the direction in which the camera is pointing, hence allowing the taking of natural, unposed behaviour (EIBL-EIBESFELDT and HASS 1966, SORENSON and GAJDUSEK 1966, KEITER 1969, CHAUVIN 1972, EIBL-EIBESFELDT 1973, EIBL-EIBESFELDT and LORENZ 1974, MORIN and PIATTELLI-PALMARINI 1974 – see these publications for more detailed references).

éthologie humaine □ Branche récente et spécialisée de l'éthologie qui a pour but d'étudier le comportement humain à l'aide des méthodes et des approches élaborées en éthologie animale. La découverte par les éthologistes que les bases fondamentales du comportement phylogénétiquement programmé déterminent d'une façon décisive le comportement des animaux, a ouvert de nouvelles perspectives pour les recherches sur le comportement humain. L'éthologie humaine s'est donné pour tâche d'élucider les bases phylogénétiques de notre comportement pour déterminer à l'aide de comparaisons interculturelles la part de l'inné dans les comportements hautement compliqués de l'Homme. La documentation sur le système d'action des enfants nés sourds et aveugles joue un rôle important dans ce contexte. L'éthologiste qui veut étudier le comportement humain sur le terrain utilise de préférence des caméras cinématographiques équipées d'objectifs à vision perpendiculaire ce qui lui permet d'obtenir des documents sur les comportements et interactions sociales d'une manière naturelle et inaperçue (EIBL-EIBELFELDT et HASS 1966, SORENSON et GAJDUSEK 1966, KEITER 1969, CHAUVIN 1972, EIBL-EIBESFELDT 1973, EIBL-EIBESFELDT et LORENZ 1974, MORIN et PIATTELLI-PALMARINI 1974 – voir ces publications pour bibliographie plus détaillée).

Hüpfscharren → Scharren
E. jump scratching □ F. grattage du sol en sautillant

Hüten → Herden

Hüteverhalten → Herden → Pflegeverhalten

I

Identifikation □ Gemeint ist hier in der Ethologie die Verbundenheit mit der Gruppe oder dem Familienverband. Man kann also bei der Identifikation von gruppenbindenden Mechanismen sprechen, wenn es sich um eine soziale Verantwortlichkeit handelt. Diese kann sich in totalitären Staatsformen zur Symbolidentifikation abwandeln.

identification □ Ethologically it implies the connection or relationship to the group or family. One may also speak of identification as a mechanism for promoting group cohesion if coupled with the additional concept of social consciousness. In totalitarian states a form of symbol identification has developed.

identification □ En éthologie, c'est une forme de liaison avec un groupe ou une organisation familiale. On peut assimiler cette identification à des mécanismes de cohésion d'un groupe, en particulier lorsqu'il s'agit d'une responsabilité sociale. Ce phénomène peut se modifier et dans les formes d'états totalitaires, il devient alors une identification avec des symboles.

Imponierbewegung → Imponierverhalten
E. display pattern movement □ F. mouvement de parade

Imponierbremsen → Scheinangriff
E. putting on the brakes display □ F. simulacre de menace

Imponieren → Imponierverhalten
E. display □ F. parade

Imponierfegen → Fegen
E. display polishing □ F. râclement d'intimidation

Imponierflug → Schauflug

Imponiergehaben → Imponierverhalten

Imponierlaufen → Imponierverhalten
E. display running □ F. course d'intimidation

Imponierschwimmen → Imponierverhalten
E. display swimming □ F. nage d'intimidation

Imponiertracht → Imponierverhalten
E. display costume □ F. costume (uniforme) de parade

Imponiertracht-Abbau → Imponierverhalten
E. simplification of dress □ F. uniformisation vestimentaire

Imponierveranstaltungen → Imponierverhalten
E. display manifestation □ F. manifestation de parade

Imponierverhalten □ Beeindruckende Verhaltensweisen gegenüber Rivalen oder gegenüber dem Fortpflanzungspartner, oft auch in Anwesenheit des Fortpflanzungspartners auf den oder die Rivalen bezogen. Im Imponierverhalten sind oft Angriffs- und Flucht- bzw. Verteidigungstendenzen überlagert. Dies ist auch beim Menschen der Fall, wie EIBL-EIBESFELDT (1971) zumindest bei den Waikas nachweisen konnte. Ein interessantes Imponierverhalten vieler Primaten, den Menschen eingeschlossen, gegenüber Artgenossen ist das → *Genitalpräsentieren.* Zur Unterstützung des Imponierverhaltens wurden spezielle Imponierorgane ausgebildet, die gleichzeitig auch als → *Auslöser* oder Signale Verwendung finden. So können z. B. die übergroße Schere bei den Winkerkrabben, die zu Klangkörpern umgebildeten Schwanzstacheln der Stachelschweine, die aufstellbare vergrößerte Rückenflosse bei *Emblemaria pandionis* sowie der Flügelspiegel bei Enten als Imponierorgane bezeichnet werden. Häufig sind auch Imponierveranstaltungen zur Einschüchterung des bzw. der Rivalen, die bei vielen Tieren in Form von Imponierlaufen und Imponierschwimmen besonders eindrucksvoll sind. Als Imponierveranstaltungen beim Menschen zur optischen Steigerung des Selbstwertes und zur Einschüchterung des Gegners wären insbesondere die Truppenparaden zu nennen, deren Wirkung durch Anlegung einer Imponiertracht, z. B. der Paradeuniform, sichtlich erhöht wird. Wie im Tierreich werden Waffen betont vorgezeigt und zur Schau getragen. Die Festbemalung bei Naturvölkern als Imponiertracht kann durchaus mit unseren Uniformen verglichen werden. Auf der anderen Seite entspricht die immer weiter fortschreitende Vereinheitlichung unserer Kleidung einem Imponiertracht-Abbau, um Reibungsflächen zu vermeiden. Das betrifft aber offenbar nicht die Frauen, welche ihre Reize betonen dürfen, da diese Herausforderung »bindende Mechanismen« aktiviert.

display behaviour □ Behaviour which may impress or intimidate the partner or rival and which is frequently directed at a rival or rivals in the presence of the mate. Both attack and flight tendencies may overlap in display behaviour, as shown in humans by EIBL-EIBESFELDT (1971) in his study of the Waika. An interesting display behaviour in many primates including man is the → *genital display* toward other conspecifics. Organs whose structure supports the effectiveness of behavioural display are called display organs. They may serve as → *releaser(s)* or signalizing mechanisms, as for example the enlarged chealopod of *Uca* crabs. Others include the enlarged sound producing spines in the porcupine's tail, the enlarged adjustable dorsal fin of *Emblemaria pandionis* or the speculum of the duck. Often parades and other display manifestations have been developed in order to intimidate rivals; display running and display swimming in many animals are particularly striking examples. In humans, display manifestations such as military parades also serve to increase self-esteem and to intimidate enemies; their effect is reinforced by the presentation of display costumes, e. g. parade uniforms. As in the animal kingdom, weapons are exhibited in a demonstrative manner. Festive body painting in certain primitive cultures has the same function as uniforms. On the other hand, the continuing trend to simplification of men's clothing helps to avoid sources of friction. In women this trend is not noticeable since their clothing often functions to emphasize stimuli which serve as bonding mechanisms.

comportement de parade □ Comportement démonstratif ou comportement d'intimidation envers les rivaux ou le partenaire sexual. Dans le comportement de parade, des tendances d'attaque et de fuite sont souvent superposées. Ceci est également le cas pour l'Homme, comme EIBL-EIBESFELDT (1971) a pu le démontrer chez les Waikas, Indiens de l'Amérique du Sud. Un comportement de parade intéressant chez de nombreux Primates, l'Homme y compris, est la → *présentation des organes génitaux* envers les congénères. Des organes d'intimidation se sont parfois spécialement développés pour renforcer le comportement de parade et peuvent, en même temps, fonctionner comme → *déclencheur(s)* et/ou signaux. A titre d'exemple, citons les pinces hypertrophiées chez les crabes du genre *Uca*, les longues épines transformées en organes de résonance des Porcs-épic, la nageoire dorsale élargie et érectile de nombreux Poissons, en particulier d'*Emblemaria pandionis,* ainsi que le miroir alaire des Canards. Souvent, les manifestations de parade sont destinées à l'intimidation du ou des rivaux ce qui est particulièrement évident, chez de nombreux animaux, sous forme d'une nage ou d'une course d'intimidation. Les manifestations de parade chez l'Homme sont surtout destinées à intimider les rivaux et servent essentiellement à la mise en valeur visuelle de sa propre force, comme les parades militaires dont l'efficacité est considérablement augmentée par les uniformes de parade. Comme dans le monde animal, les armes sont particulièrement mises en évidence. Les peintures corporelles chez un certain nombre de »peuples sauvages« en tant que costumes de parade sont tout à fait comparables à nos uniformes. D'autre part, une simplification de nos vêtements et leur uniformisation progressive est probablement destinée à diminuer la tension interindividuelle. Ce phénomène ne semble pourtant pas concerner les femmes qui, par contre, peuvent souligner leurs attraits naturels étant donné que de tels stimuli sont destinés à favoriser la formation du couple et constituent des mécanismes de liens.

Indikatorreaktion □ Bewegungsweisen, die nur bei wenigen Arten innerhalb einer Familie vorkommen, nannte SEITZ (1940–43) »Indikatorreaktionen«.

indicator response □ Movement patterns that only occur in a few species within a family; termed »indicator response« by SEITZ (1940–43).

indicateur comportemental □ Ce terme a été créé par SEITZ (1940–43) pour désigner un certain nombre de mouvements caractéristiques, présents seulement chez quelques espèces à l'intérieur d'une même famille.

Individualdistanz □ Der Mindestabstand zwischen Individuen einer Art; ein wichtiges Kriterium innerartlicher Bezugsformen. Viele gesellig lebende Vögel, wie z. B. Stare und Schwalben, die außerhalb der Fortpflanzungszeit große Schwärme bilden, aber den körperlichen Kontakt meiden, halten untereinander einen gewissen Abstand. Man spricht in diesem Fall auch von *Distanztieren* (HEDIGER 1942, 1950). Viele Koloniebrüter nisten in einer Distanz voneinander, in der sie sich gerade nicht mehr hakken können; man spricht in diesem Fall auch von *Hackdistanz.*

individual distance □ The minimum distance tolerated between individuals under normal social conditions; serves as an important criterion of intraspecific social relations. Many social birds, e. g. starlings and swallows, form large flocks outside the reproductive season, but nevertheless avoid bodily contact and maintain a characteristic distance (= *distance type animals*, HEDIGER 1942, 1950). Many colonial nesters nest at that distance from each other at which they can no longer peck each other; this distance is termed *peck distance.*

distance inter-individuelle □ Ecart minimum qui

doit être respecté entre individus appartenant à une même espèce et qui constitue une condition importante des relations intraspécifiques. De nombreux Oiseaux, par exemple les Etourneaux et les Hirondelles, forment d'importantes bandes en dehors de la période de reproduction, mais évitent le contact corporel et maintiennent entre eux une distance interindividuelle (*animaux du type distant*, HEDIGER 1942, 1950). Chez de nombreux Oiseaux nichant en colonie, les nids se trouvent entre eux à une distance telle que les Oiseaux couveurs ne peuvent plus s'atteindre avec leurs becs; on parle alors d'une *distance de becquetage*.

individualisierter Verband □ Tiergruppe, die durch das Band individueller Bekanntschaft zusammengehalten wird. Ihre soziale Organisation kann durch die Ausbildung einer → *Rangordnung* sehr kompliziert sein. Im Gegensatz hierzu → *anonymer Verband*.

individualized group □ Refers to a social group which is held together by bonds between individuals. The social organization of individualized groups may become complex through development of a → *rank order*. In contrast to the individualized group → *anonymous group*.

groupement individualisé □ Association d'individus dans laquelle le lien est maintenu par la reconnaissance individuelle. Son organisation sociale peut être très compliquée par la formation d'une → *hiérarchie sociale*. En opposition → *groupement anonyme*.

Infantilismus □ Rückfall in kindliche Verhaltensweisen, insbesondere im Zusammenhang mit der Werbung oder Demutsgesten. So rufen Hamster-♂♂ beim Werben wie Nestlinge, und der zärtlich werbende Menschenmann spricht in ausgesprochen kindlicher Weise unter betonter Verwendung von Diminutiven; oft treten früheste Stufen wieder auf: Die »Verhaltensuhr« läuft zurück bis zum Saugen an der Brust. Ein solches Regressionsverhalten steht beim Menschen offenbar im Dienste der Paarbindung. → *Mund-zu-Mund-Füttern*.

infantilism □ A return to child-like behaviour patterns especially in connection with courtship or submissive gestures. For example, male hamsters imitate the pup vocalizations during courtship and human males may speak in a distinctly child-like fashion using many diminutives; even earliest stages may reappear such as sucking at the breast. In humans, such behavioural regression apparently serves the pair bond. → *mouth-to-mouth feeding*.

infantilisme □ Retour à un comportement infantile, en particulier lors de la parade nuptiale ou des gestes de soumission. Ainsi, les Hamsters ♂♂ utilisent vis-à-vis de leurs ♀♀ les mêmes vocalisations que les nouveaux-nés; chez les humains, l'homme qui fait la cour à une femme parle souvent dans un langage enfantin, en utilisant de nombreux diminutifs. Souvent, des stades précoces réapparaissent ainsi, comme par exemple la tétée. Un tel comportement régressif chez l'Homme est probablement au service de l'attachement entre les partenaires d'un couple. → *nourrissage de bouche à bouche*.

Inferioritätsfärbung □ Insbesondere bei Fischen (z. B. Cichlidae oder Blennioidei), bei Chamaeleonidae und Cephalopoda spontan in Erscheinung tretende Verfärbung des schwächeren, unterlegenen Tieres (OHM 1958). Im Gegensatz hierzu → *Kampffärbung*.

submissive colouration □ A change of colouration in subordinate individuals especially among fish (e. g., Cichlidae or Blennioidei), in Chamaeleonidae and Cephalopoda (OHM 1958). Contrasts to → *fighting colour pattern*.

coloration d'infériorité, coloration de soumission □ Particulièrement chez les Poissons (p. ex. Cichlidae ou Blennioidei), chez les Chamaeleonidae et les Céphalopodes, une coloration indiquant l'infériorité ou la soumission d'un animal (OHM 1958). En opposition à → *coloration de combat* ou *de supériorité*.

Informationsaustausch → Verständigung
E. information exchange □ F. échange d'information

Informationsverhalten → Verständigung → Bienentänze
E. informational behaviour □ F. comportement d'information

Initiativgebärden □ Die Suche nach Kontakt durch verschiedene Gebärden und Bewegungen, z. B. Betteln durch → *Händepatschen*. Bei Naturvölkern konnte nachgewiesen werden, daß das Kleinkind nicht nur, wie auch bei uns, durch seine Erscheinungsform freundliche Kontakte auslöst (→ *Kindchenschema*), sondern aktiv deutliche Kontaktinitiative und Kontaktgebärden zeigt.

contact seeking behaviour □ Contact-oriented display and movements especially in small children, e. g. begging by → *hand patting*. Among primitive people it has been demonstrated that infants not only elicit contact behaviour from others (→ *baby schema*), but actively seek it themselves.

sollicitation gestuelle □ Recherche de contact à l'aide de mouvements et de gestes, en particulier chez les petits enfants, par exemple en quémandant par → *battement des mains*. Chez les différents peuples de culture primitive, comme d'ailleurs chez nous, il a pu être démontré que les bébés déclenchent non seulement des contacts amicaux par leur apparence (→ *schéma infantile*), mais qu'ils prennent des initiatives de contact en ce sens.

Innere Uhr → Zeitgeber
E. internal clock □ F. horloge interne

Innergruppenaggression → Aggression
E. intra-group aggression □ F. agressivité intra-groupe

Inquilinismus □ Zwischenartliche Beziehung, bei welcher eine sozialparasitische Insektenart ihren gesamten Lebenszyklus im Nest ihrer Wirtsart verbringt. Arbeiterinnen fehlen oder sind nur in verkümmertem Zustand vorhanden und zeigen außerdem ein degeneriertes Verhalten. Diese Bedingungen werden gelegentlich als »Dauerparasitierung« bezeichnet. Wir kennen ein solches Verhalten bei vielen Ameisen und wollen nur zwei Beispiele nennen: *Doronomyrmex pacis* als Eindringling und *Leptothorax acervorum* als Wirt sowie *Hagioxenius schmitzi* als Eindringling und *Tapinoma erraticum* als Wirt (Übersicht bei WILSON 1971, 1975).

inquilinism □ Interspecific relation in which a socially parasitic species of insect spends its entire life cycle in the nests of its host species. Workers either are lacking or, if present, are usually scarce and degenerate in behaviour. This condition is sometimes referred to loosely as »permanent parasitism«. Such a behaviour is known in many ants as shown by the following two examples: *Doronomyrmex pacis* as a parasite and *Leptothorax acervorum* as host, *Hagioxenius schmitzi* as a parasite and *Tapinoma erraticum* as host (overview by WILSON 1971, 1975).

inquilinisme □ Relation interspécifique dans laquelle un parasite social d'une espèce d'Insectes passe sa vie entière dans le nid d'une espèce-hôte. Les castes ouvrières manquent ou n'existent que sous une forme atrophiée et leur comportement est dégénéré. Ces conditions correspondent à ce qu'on appelle généralement un »parasitisme permanent«. Un tel comportement est connu chez de nombreuses Fourmis dont nous ne citerons que deux exemples: *Doronomyrmex pacis* comme parasite et *Leptothorax acervorum* comme hôte et *Hagioxenius schmitzi* comme parasite et *Tapinoma erraticum* comme hôte (vue d'ensemble dans WILSON 1971, 1975).

Instinkt □ TINBERGEN nennt Instinkt einen »hierarchisch organisierten, nervösen Mechanismus, der auf bestimmte vorwarnende, auslösende und richtende Impulse, sowohl innere wie äußere, anspricht und sie mit wohlkoordinierten, lebens- und arterhaltenden Bewegungen beantwortet«. Er unterscheidet Hauptinstinkte und untergeordnete Instinkte. Der Instinktbegriff ist heute überholt und umstritten (BARLOW 1974).

instinct □ TINBERGEN regarded instinct as »a hierarchically organized nervous mechanism which reacts to priming, releasing and directing stimuli of either endogenous or exogenous character. The reaction consists of a coordinated series of movements that contributes to the preservation of the individual and species«. TINBERGEN distinguishes primary and secondary instincts. Today the instinct concept has become controversial and obsolete (BARLOW 1974).

instinct □ TINBERGEN entend par instinct »un mécanisme nerveux hiérarchiquement organisé qui réagit à des stimuli avertisseurs, déclencheurs et directeurs, – qu'ils soient autochtones ou allochtones –, et qui y répond par des enchainements moteurs bien coordonnés au service du maintien de la vie et de l'espèce«. Il distingue des instincts majeurs et des instincts subordonnés. La notion d'instinct est aujourd'hui dépassée et contestée (BARLOW 1974).

Instinktbewegung → Erbkoordination → Triebhandlung
E. instinctive motor pattern □ F. mouvement instinctif

Instinkt-Dressur-Verschränkung → Trieb-Dressur-Verschränkung
E. instinct-learning intercalation □ F. interpénétration de l'inné et de l'acquis

Instinktverschränkung → Trieb-Dressur-Verschränkung
E. instinct intercalation □ F. intercalation d'instincts

instrumentale Aggression → Aggression
E. instrumental aggression □ F. agression instrumentale

Instrumentalverständigung □ Die intraspezifische Kommunikation, vor allem zwischen ♂ und ♀ bei verschiedenen Spechten, bezeichnet man so, wenn nicht Rufe, sondern nur die verschiedenen Klopffolgen dabei beteiligt sind. Die Abb. 64 – *Dendrocopus major* – zeigt einen Trommelwirbel, der auf weite Entfernung wirkt und das ♀ anlockt. Danach fliegt das ♂ zur Höhle und »zeigt« diese mit einem demonstrativen Klopfen in getakteter Schlagfolge (BLUME 1967).

Eine weitverbreitete Form der Instrumentalverständigung bei Insekten ist die mechanische Lauterzeugung, indem ein Körperteil gegen einen anderen gerieben wird; insbesondere bekannt von Orthopteren und Zikaden und gemeinhin als *Stridulation* bezeichnet.

mechanical communication □ Intraspecific communication among woodpeckers in which various pecking patterns are used as signals. The accompanying illustration (Fig. 64 – *Dendrocopus major*) depicts such a pattern which can attract a ♀ from a distance. When the ♀ is in the vicinity the ♂ flies to the nest hole and »shows« the exact location with another rhythmic pecking signal (BLUME 1967).

A common form of *instrumental communication* in insects is the production of sound by rubbing one part of the body surface against another; well known in Orthopterans and Cicadians, commonly called *stridulation*.

communication instrumentale □ Forme de communication intraspécifique, en particulier entre ♂ et ♀, chez un certain nombre d'espèces de Pics, s'établissant non par des vocalisations, mais par des séquences de tambourinage. La figure ci-contre (Fig. 64 – *Dendrocopus major*) montre une séquence rapide de tambourinage qui est audible à grande distance et attire la ♀. A l'approche de celle-ci, le ♂ s'envole vers le creux d'arbre choisi pour la nidification et l'indique à la ♀ à l'aide d'une séquence de tambourinage démonstrative et rythmée (BLUME 1967).

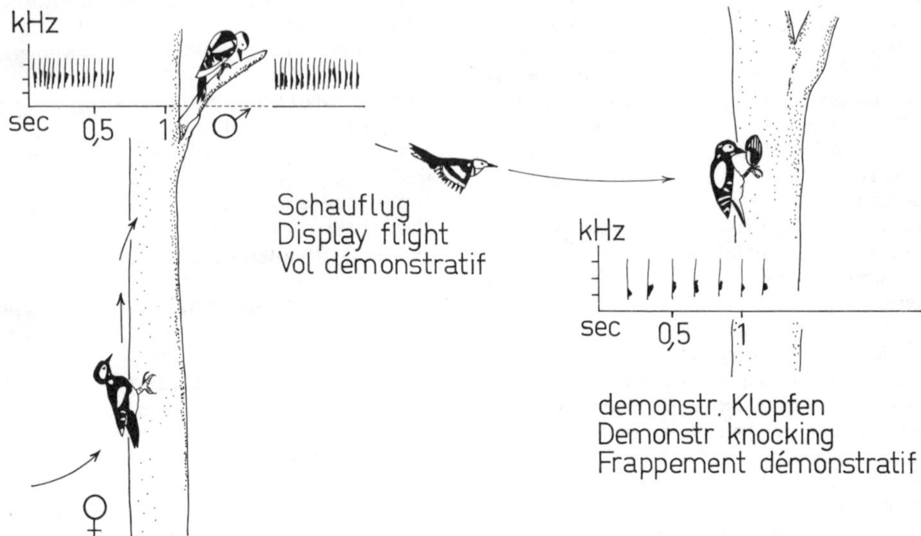

64

Une forme très répandue de communication instrumentale chez les Insectes est la production de sons qui consiste à frotter une partie de la surface corporelle contre une autre; connue surtout chez les Orthoptères et les Cicades et qu'on appelle *stridulation*.

Intentionsbewegung □ Angedeutete, nur begonnene Instinktbewegung mit Informationen für die Artgenossen, indem sie die Bereitschaft zu einer bestimmten Handlung anzeigt und über die »Stimmung« des Individuums Auskunft gibt (HEINROTH 1911).

intention movement □ A mere hint of or beginning part of a sequence of an innate behaviour pattern which contains information to a conspecific in that it indicates the readiness to perform a specific behaviour pattern, and which indicates the »mood« of the individual (HEINROTH 1911).

mouvement intentionnel □ Mouvement instinctif seulement esquissé ou amorcé et contenant des informations pour les congénères. Il indique la disponibilité à une action déterminée et donne des renseignements sur la »motivation momentanée« de l'individu (HEINROTH 1911).

irritative Aggression → Aggression
E. irritable aggression □ F. agression irritative

Isolationsexperiment → Kaspar-Hauser-Versuch
E. isolation experiment □ F. expérience d'isolement

isolierte Aufzucht → Kaspar-Hauser-Versuch
E. rearing in isolation □ F. élevage en isolement

J

Jagdspiele □ Spielverhalten, bei welchem → *Beutefanghandlung(en)* geübt werden. Oft bedienen sich die Tiere dazu verschiedener → *Ersatzobjekt(e)*. Die Katze spielt mit einem Wollknäuel, junge Löwen behandeln ihre Geschwister wie Beute.

mock hunting, hunting games □ A kind of play where prey-capture movements (→ *prey-catching behaviour*) are exercised; frequently the animal manipulates some → *substitute object,* e. g. a cat may play with a wool ball in this way and a young lion with its siblings.

jeux de prédation □ Comportement ludique pendant lequel les jeunes animaux s'exercent aux mouvements qui serviront à la capture des proies (→ *prédation*). Ils se servent souvent → *d'objets de remplacement*; les chatons utilisent une pelote de laine et les jeunes lions leurs frères et soeurs comme »proie«.

Jauchzen □ Ausdrucksverhalten mit Rufreihe bei Möwen. Der Jauchzer besteht bei der Silbermöwe aus drei Teilen: bei den ein oder zwei ersten Silben streckt die Möwe den Kopf etwas nach vorn, dann senkt sie ihn und quetscht einige hohe Töne heraus; schließlich wirft sie ihren Kopf mit einem Ruck nach oben, stößt eine Reihe von lauthallenden Schreien aus und reckt dabei mit weit geöffnetem Schnabel den Hals vorwärts (Abb. 65 – *Larus argentatus*). Die Möwen jauchzen beim Erscheinen von Artgenossen, Ehepartnern und der Jungen im Revier. Es gilt als Dominanzäußerung und Imponierruf und unterliegt

65

auch der → *Stimmungsübertragung* (GOETHE 1956, TINBERGEN 1958, 1959). Beim Menschen bedeutet Jauchzen Ausdruck höchster Freude.

oblique-cum-long-call □ In gulls, a demonstrative behaviour with accompanying vocalizations (greeting ritual). In the Herring Gull, the sequence occurs in three parts and consists of stretching the head first forward, then backward while uttering a few high notes. Then the head is jerked upward; a series of loud sounds are emitted followed by thrusting the neck and wide-open beak forward (Fig. 65 – *Larus argentatus*). Gulls perform this ceremony when greeting their acquaintances, mate or young on the territory. This ceremony seems to function in expressing dominance and is subject to → *mood induction* (GOETHE 1956, TINBERGEN 1958, 1959). In humans, the German term *Jauchzen* signifies »jubilation«.

»Jauchzen« □ Chez les Goélands et Mouettes, comportement expressif avec une série de cris qui constitue un rituel de présentation. Chez le Goéland argenté, ce comportement est composé de trois parties: pendant les deux premières syllabes, le Goéland avance sa tête, puis la baisse fortement de manière à ce que le bec se dirige presque vers l'arrière entre les pattes, en poussant quelques cris aigus; tout à coup, il rejette la tête vers le haut, le bec largement ouvert, le cou étiré en avant, et pousse une série de cris retentissants (Fig. 65 – *Larus argentatus*). Les Goélands et les Mouettes présentent ce comportement lors de rencontres avec des congénères et lors de l'arrivée du partenaire ou des jeunes dans le territoire. Ce comportement peut être considéré comme une expression de dominance et un cri d'intimidation. Il traduit également une → *induction allomimétique* (GOETHE 1956, TINBERGEN 1958, 1959). Chez l'Homme, le terme allemand *Jauchzen* est une expression de la plus grande joie (jubilation).

Jungefüttern □ Das Füttern der Jungen im allgemeinen. Es ist bei einigen in Dauerehe lebenden Vögeln aber insofern interessant, als sich aus diesem Jungefüttern eine Anzahl ritualisierter Verhaltensweisen im Dienste der Anpaarung und Partnerbindung entwickelt haben. → *Balzfüttern*, → *Begrüßungsfüttern*, → *Schnäbeln*.

feeding of young □ From this feeding of the young certain ritualized forms have developed which may serve as mating or bond maintenance mechanisms; such ritualized forms are common in birds with permanent pair bonds. → *courtship feeding*, → *partner feeding*, → *billing*.

nourrissage des jeunes □ L'alimentation des jeunes par leurs parents d'une manière générale; mais chez un certain nombre d'Oiseaux vivant en monogamie permanente, il est intéressant de constater qu'à partir du nourrissage des jeunes se sont développés nombre de comportements ritualisés au service de l'appariement et des liens entre les deux partenaires. → *nourrissage de parade nuptiale*, → *nourrissage de salut*, → *becquetage*.

K

Käfigverblödung □ LORENZ (1932) faßt unter dieser Bezeichnung alle feststellbaren geistigen Gefangenschaftserscheinungen und Bewegungshemmungen zusammen (FOX 1968).

captivity degeneration □ LORENZ (1932) includes here all disturbances in mental and physical performance resulting from conditions of captivity (FOX 1968).

syndrome de claustration □ LORENZ (1932) a résumé par le terme de *Käfigverblödung* ou syndrome du prisonnier tous les phénomènes d'amoindrissement mental et d'inhibition locomotrice dus à la captivité (FOX 1968).

Kainismus → Verwandtenfresserei
E. syngenophagy □ F. cainisme

Kainogenese → frühontogenetische Anpassung

Kameradenverteidigung → Beistandsverhalten

Kampfappetenz → Kampfinstinkt
E. appetence for fighting □ F. appétence au combat

Kampffärbung □ Besonders bei Fischen (z. B. Cichlidae), bei Chamaeleonidae und Cephalopoda spontan in Erscheinung tretende besondere Färbung eines im Kampf dominanten Tieres oder die Kampfbereitschaft anzeigend. Im Gegensatz hierzu → *Inferioritätsfärbung*.

fighting colour pattern □ Occurs in fish (e. g. Cichlidae), in Chamaeleonidae and in Cephalopods, prior to fighting and in the dominant individual during and after the fight; in contrast to → *submissive colouration*.

coloration de combat □ En particulier chez certains Poissons (par exemple Cichlidae), chez les Chamaeleonidae et chez les Céphalopodes, coloration particulière apparaissant d'une manière spontanée chez un individu dominant lors d'un combat ou indiquant la motivation au combat. Par opposition à la → *coloration d'infériorité*.

Kampfinstinkt □ Die angeborene Grundlage zum Kampf bzw. zur Aggression mit endogener Erregungsproduktion und dem entsprechenden → *Appetenzverhalten* (*Kampfappetenz* – LORENZ 1950,

1963, RASA 1971). Die Existenz einer solchen Kampfappetenz oder Kampfmotivation bzw. eines Aggressionstriebes ist heute aber umstritten (PLACK 1973). → *Trieb-Hypothese.*

fighting instinct □ The innate basis for fighting behaviour or aggression with its attendant fluctuation in motivational and appetitive components (*appetence for aggression* – LORENZ 1950, 1963, RASA 1971). However, the existence of such an appetence, motivation or drive for aggression is actually contested (PLACK 1973). → *drive-model.*

instinct de combat □ Fondement inné du combat ou de l'agression s'accompagnant d'une excitation endogène et du → *comportement d'appétence* adéquat (*appétence au combat* – LORENZ 1950, 1963, RASA 1971). L'existence d'une telle appétence ou motivation au combat ou encore d'une pulsion agressive est aujourd'hui fortement controversée (PLACK 1973). → *hypothèse de la pulsion agressive.*

Kampfintention □ Differenzierte Bewegungen, die die Bereitschaft zum Kampf dem Rivalen gegenüber zum Ausdruck bringen. Kampfintention und Fluchtintention können einander verschieden stark überlagern. → *Übersprungbewegung.*

fight intention movement □ Particular movements which communicate combat readiness to a rival. Both fight intention movements and flight intention movements may be present to varying degrees in any given context. → *displacement activity.*

mouvement intentionnel de combat □ Mouvements très différenciés indiquant la disposition au combat envers un rival. Les mouvements intentionnels de combat et de fuite peuvent coexister à des degrés divers dans une situation donnée. → *activité substitutive.*

Kampfkomment → Kommentkampf

Kampfspiele → Spielverhalten

E. play fighting □ F. combats ludiques

Kaspar-Hauser-Versuch □ In der Ethologie die Aufzucht einzeln isolierter Tiere in möglichst frühem Alter ohne jeden Kontakt. Was diese Tiere tun, können sie keinem Artgenossen nachgemacht haben, nicht von ihm gelernt haben. Die Benennung im deutschen Sprachgebrauch lehnt an eine menschliche Tragödie an. Kaspar Hauser erschien als etwa 17jähriger in Nürnberg, konnte kaum gehen und nicht sprechen, war lichtscheu und vertrug nur Wasser und Brot. Nachdem er zu sprechen gelernt hatte, gab er an, er habe, soweit seine Erinnerung reiche, allein in einem finsteren Gelaß gesessen, nie einen Menschen gesehen oder gehört, und wenn er aus dem Schlaf erwachte, habe Wasser und Brot neben ihm gestanden.

Kaspar-Hauser-experiment □ Refers to rearing animals in isolation from conspecifics from a very early age (isolation or deprivation experiment). Whatever the animal performs is therefore not attributable to imitation or learning from other conspecifics. The German term »Kaspar-Hauser-Versuch« is derived from the name of a youth in Nuremberg discovered at 17 years of age unable to speak, nearly unable to walk, shy of light and accustomed to only water and bread. After learning to speak, he supposedly said that he only remembered a dark room where he would awaken to bread and water, having never seen other humans.

expérience Kaspar-Hauser □ En éthologie, l'élevage d'animaux isolés dès la naissance ou à partir d'un très jeune âge et maintenus sans contact avec leurs congénères. Ce que font ces animaux n'a pas pu être imité ni appris des congénères. Le terme allemand est tiré de l'exemple d'un drame humain. Kaspar Hauser est apparu à Nuremberg à l'âge de 17 ans, ne sachant pas parler et à peine marcher; il craignait la lumière et ne pouvait s'alimenter que de pain et d'eau. Après avoir appris à parler, il raconta qu'il avait vécu, de si loin qu'il pouvait s'en souvenir, seul dans une pièce sombre, sans voir ni entendre un être humain. Chaque fois qu'il se réveillait, il trouvait auprès de lui du pain et de l'eau.

Katalepsie → Akinesis

katasematisches Verhalten → Demutsverhalten

Katzenbuckel → Buckelstellung

Katzenruf □ Bei der Silbermöwe ein langgezogener Ton, mit vorwärts und zuweilen abwärts gestrecktem Hals und weit geöffnetem Schnabel vorgetragen. Im Verhaltensinventar der Silbermöwe ein Lockruf, kann aber auch Feindschaft ausdrücken; bei Grenzkämpfen folgt ihm häufig der Angriff (TINBERGEN 1958).

mew call □ In the herring gull a long drawn-out tone given with the neck stretched forward and down and with the bill wide-open; serves as an enticement call, but may also express antagonism since it frequently follows a dispute on the territorial boundaries (TINBERGEN 1958).

miaulement □ Chez le Goéland argenté, vocalisation prolongée émise, le cou étiré en avant et vers le bas et le bec largement ouvert. Dans l'inventaire comportemental du Goéland argenté, c'est un cri d'appel, mais qui peut aussi exprimer une attitude agonistique; au cours d'affrontements à la limite de deux territoires, ce cri est souvent suivi d'une attaque (TINBERGEN 1958).

Kettenreflex □ SPENCER (1899) und LOEB (1905) nannten seinerzeit Instinkthandlungen »Kettenreflexe«. Diese Auffassung hat sich in manchen Lehrbüchern erhalten. Die Verhaltensforschung begann aber bereits 1937, sich davon zu entfernen. Auch die von BECHTEREW (1913) und PAWLOW (1927) begründete Reflextheorie des Verhaltens erklärt alles Verhalten aus bedingten und unbedingten Reflexen und behauptet, komplizierte Verhaltensabläufe seien »Kettenreflexe«.

chain reaction □ SPENCER (1899) and LOEB (1905) called instinctive movements »chain reactions« and this interpretation persisted in some manuals. Ethology, however, began diverging from this view in the 1930's, although the reflex theory of BECHTEREW (1913) and PAWLOW (1927) maintained that all behaviour could be explained as a sequence of conditioned and unconditioned reflexes and that complex behavioural responses were simply chain reactions.

réflexe en chaîne □ SPENCER (1899) et LOEB (1905) ont appelé en leur temps »réflexes en chaîne« les actions instinctives. Cette conception a persisté longtemps dans certains traités. L'éthologie a commencé, dès 1937, à s'éloigner de cette opinion. La théorie des réflexes de BECHTEREW (1913) et PAWLOW (1927) explique également tous les comportements par un enchainement de réflexes conditionnés et inconditionnés et prétend que les actions comportementales complexes sont des »réflexes en chaîne«.

Kindchenschema □ Die Kombination von Schlüsselreizen, die in summativer Wirkung beim Menschen und offensichtlich auch bei anderen Primaten bei Vorhandensein gewisser Proportionstypen, wie relativ dicker Kopf, Stirnwölbung, große Augen usw. → *Pflegeverhalten* auslösen (Abb. 66 – LORENZ 1943). Das Kindchenschema ist vielleicht ein angeborener Auslösemechanismus (→ *AAM*), der bei den meisten Wildtieren selektiv auf die eigene Art orientiert ist, bei Menschen und bei Wirten von Brutparasiten jedoch auch auf Jungtiere anderer Arten ansprechen kann.

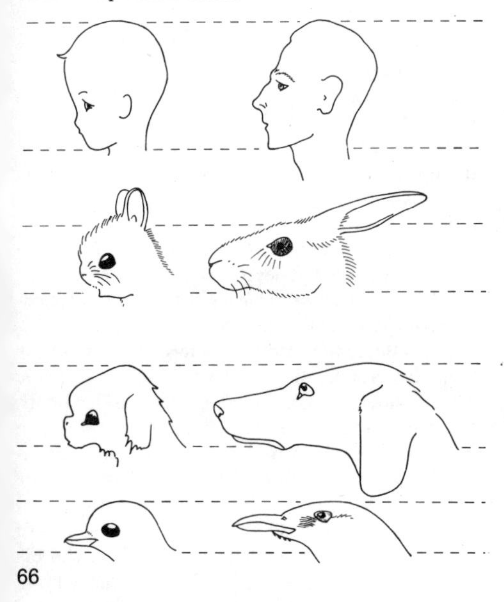

66

baby schema □ The combination of key stimuli that additively release → *epimeletic behaviour* in man and apparently other primates; these stimuli consist of a certain proportionality, e. g., relatively large head, forehead arch and eyes, etc. (Fig. 66 – LORENZ 1943). The child-like form is probably an innate releasing mechanism (→ *IRM*) whose properties differ from species to species, but which may release appropriate behaviour cross-specifically as it frequently does in man and in hosts of brood parasites.

schéma infantile □ La combinaison de certains stimuli-clés tels une tête relativement grande et ronde, un front bombé, de grands yeux, etc. (Fig. 66 – *schéma bébé*), déclenche, par sommation, un comportement parental de soins chez l'Homme et apparemment aussi chez d'autres Primates (LORENZ 1943). Ce schéma perceptif est probablement un mécanisme déclencheur inné (→ *MDI*) qui, chez la plupart des animaux sauvages, est orienté sélectivement vers les jeunes de leur propre espèce. Chez l'Homme et chez les hôtes des nidificateurs parasites, par contre, il peut viser aussi des jeunes hétérospécifiques.

Kinderfamilie □ Die Jungen einer Brut, die nach dem Schlüpfen zusammenbleiben, wie z. B. Raupen aus ein und demselben Gelege, die eine Assoziation bilden. DEEGENER (1918) spricht in diesem Fall von einem *Sympaedium*.

juvenile group □ Those young of a brood that stay together after hatching, e. g., certain caterpillars form associations of individuals only from the same egg mass. DEEGENER (1918) called this condition a *sympaedium*.

sympaedium □ Les jeunes d'une même couvée qui, après l'éclosion, restent ensemble dans une même »famille«. Par exemple, les chenilles d'une même ponte forment une association (DEEGENER 1918).

Kindkumpan → Kumpan
E. child companion □ F. compagnon juvénile

Kind-Mutter-Bindung → Mutter-Kind-Bindung
E. child-mother-bond □ F. lien enfant-mère

Kinese □ Man versteht darunter die Änderungen in der Art der Fortbewegung, die auf einer Änderung der Stärke des Reizes beruhen, aber nicht am Reiz ausgerichtet sind. Laufen die Tiere z. B. schneller, spricht man von *Orthokinese;* machen sie mehr oder größere Wendungen, spricht man von *Klinokinese* (FRAENKEL und GUNN 1961). Obwohl die Wendungen räumlich nicht auf den Reiz bezogen sind, führen diese Mechanismen die Tiere dennoch in eine bestimmte Reizgefälle-Richtung. Strudelwürmer z. B. machen um so weniger Wendungen und kriechen folglich um so längere Strecken geradeaus, je dunkler es wird. Auf diese Weise gelangen sie von hellen an dunkle Orte (LINDAUER 1973).

kinesis □ Refers to an orientation behaviour which depends on the stimulus intensity but not on the direction from which the stimulus is perceived. When the animals run faster one refers to *ortho-kinesis,* when the rate of turning increases it is called *klinokinesis* (FRAENKEL und GUNN 1961). Although the turns are not spatially dependent on the stimulus, these orienting mechanisms lead an animal in the direction of a certain stimulus gradient. Flatworms, for example, make fewer turns, hence crawl farther the darker the environment. In this way they move from lighter into darker places (LINDAUER 1973).

cinèse □ Modification des paramètres de la locomotion en reponse à une variation d'intensité d'un stimulus externe, sans composante d'orientation définie. Si les animaux accélèrent leur locomotion en ligne droite, on parle d'*orthocinèse;* s'ils augmentent la fréquence de leurs changements de direction, on parle de *clinocinèse* (FRAENKEL et GUNN 1961). Bien que ces changements de direction puissent être sans relation spatiale définie avec les »lignes de forces« du champ de stimulation, la combinaison des mécanismes d'orthocinèse et de clinocinèse amène néanmoins les animaux dans une zone définie du gradient de stimulation. Par exemple, les Planaires placées dans un gradient lumineux accomplissent d'autant moins de changements de direction et, par suite, font des portions de trajet rectilignes d'autant plus longues que l'éclairement est plus faible. De la sorte, elles se déplacent de régions éclairées vers des régions obscures de leur milieu (LINDAUER 1973).

Kinnheben □ Eine epigame Verhaltensweise bei Enten, die von beiden Partnern gemeinsam ausgeführt wird. Diese Bewegungsweise erinnert in ihrer Bedeutung an das → *Triumphgeschrei* der Gänse. Auf das → *Hetzen* der Ente hin hebt der Erpel den Schnabel, ohne ihn vorher gesenkt zu haben, bis weit über die Waagerechte, während sein Körper andeutungsweise tiefer ins Wasser einsinkt. Dieses Kinnheben wird öfter wiederholt, das ♀ führt synchron mit dieser Bewegung des ♂ sein abwechselndes Kinnheben aus (Abb. 67 – *Anas (Chanlelasmus) strepera*); der Pfeil deutet die Richtung der zwischen je zwei Kinnhebungen eingeschalteten Hetzbewegungen an (LORENZ 1941).

67

chin raising □ Epigamic behaviour in ducks in both partners together; functionally analogous to the → *triumph ceremony* in geese. In response to the → *inciting* of the ♀, the drake will raise its bill, without previously having lowered it, way above the horizontal, while its body overall becomes immersed deeper into the water. This chin-raising is repeated, and the ♀ performs this behaviour in synchrony with the ♂ in an alternating fashion (Fig. 67 – *Anas (Chanlelasmus) strepera*); the arrow indicates the direction of the inciting movement which is interspersed between any two chin-raising behaviour patterns (LORENZ 1941).

lever le menton □ Comportement épigame chez les Canards qui est exécuté simultanément par les deux partenaires. Ces mouvements rappellent par leur fonction le → *cérémonial de triomphe* chez les Oies. Après une → *provocation* de la part de la cane, le malard soulève son bec, sans l'avoir baissé auparavant, largement au-dessus de l'horizontale pendant que son corps est davantage immergé dans l'eau. Cette levée du menton est souvent répétée et la ♀ et le ♂ présentent alternativement ce comportement (Fig. 67 – *Anas (Chanlelasmus) strepera*); la flèche indique la direction d'une »provocation« qui est effectuée entre deux mouvements de levée du menton (LORENZ 1941).

Klammergriff □ Bei männlichen Fröschen und Kröten der Paarung dienende Instinkthandlung. Die ♂♂ umklammern die ♀♀ vom Rücken her, unterstützt von ihren → *Brunstschwielen,* deren Druck die Eiablage auslöst, und ergießen den Samen auf die ausgeschiedenen Eier. Der Klammerreflex kann so stark auftreten, daß z. B. Erdkröten zur Fortpflanzungszeit Steine, Holzstücke und sogar lebende Fische umklammern (SCHMIDT 1972).

clasping hold, amplexus □ An instinctive mating behaviour in frogs and toads in which the male, using his → *claspers,* grasps the female from the back. The pressure he exerts results in egg deposition and in the subsequent fertilization of released eggs. The clasping reflex can be so strong that toads, for example, may grasp stones, pieces of wood or even living fishes (SCHMIDT 1972).

étreinte □ Comportement instinctif chez les ♂♂ des Grenouilles et des Crapauds lors de la période de reproduction. Les ♂♂ enserrent les ♀♀ par l'arrière et exercent une pression à l'aide de leurs → *callosités tibiales,* déclenchant ainsi la ponte, puis expulsent le sperme permettant la fécondation. L'étreinte ou l'amplexus peut apparaître comme un mouvement réflexe très pulsionnel et il a été observé, chez *Bufo bufo* par exemple, que lors de la période de reproduction, des cailloux, des morceaux de bois et même des Poissons vivants ont été les victimes de ce déclenchement purement tactile (SCHMIDT 1972).

Klan □ Bezeichnung für eine meist exogame kleine Menschengruppe, die ihre Abstammung von einem gemeinsamen Vorfahren ableitet. In unserer heutigen westeuropäischen Gesellschaft versteht man darunter meist einen patrilinearen oder matrilinearen Familienklan, der zwar ideell noch besteht, seine Mitglieder aber leben geographisch verstreut. Besonders typisch für die Lebensweise der Pygmäen ist heute noch die Obereinheit → *Horde,* welche ein Revier innehat, und deren Aufgliederung in mehrere Klane als Untereinheiten mit jeweils einer älteren Respektsperson. Der Klan ist bei Pygmäen die Grundeinheit eines Lagers (HEYMER, in Vorbereitung).

Aus dem Tierreich kennen wir ein Beispiel vor allem von den Tüpfelhyänen, *Crocuta crocuta,* die in sozialen Verbänden leben, welche KRUUK (1968) Klans nannte und die bis zu hundert Tiere umfassen. Jeder

Klan, meist von einem erfahrenen ♀ angeführt, besitzt ein eigenes → *Revier,* patrouilliert regelmäßig seine Grenzen ab, welche außerdem an bestimmten Orten mit dem Duft der Analdrüsen markiert werden und gegen einen anderen fremden Klan verteidigt werden (VAN LAWICK-GOODALL 1970).

clan □ Term referring to a small usually exogamous group of humans that trace their origin from common ancestors. In our present day society this usually means a patrilinear or matrilinear family clan whose existence is only ideally and whose members may be scattered over a wide geographic area. In pygmy society a larger social unit is the → *band* which occupies a territory and is divided into several clans each with an elder head. The clan is the basic unit of the pygmy's camp (HEYMER, in preparation).

In the animal kingdom the Spotted Hyena, *Crocuta crocuta,* live in social groups termed clans by KRUUK (1968); these clans consist of up to 100 animals. Each clan, which is usually led by an experienced ♀, maintains its own → *territory,* patrols and defends the borders of this territory and marks the area with the scent secreted by the anal scent glands (VAN LAWICK-GOODALL 1970).

clan □ Chez l'Homme, appellation pour un petit groupe d'individus chez lesquels se pratique l'exogamie et descendant tous d'un ancêtre commun. Dans notre sociéte ouest-européenne actuelle, on entend par ce terme un clan familial patri- ou matrilinéaire dont l'existence est idéelle, étant donné que les membres dont il se compose sont géographiquement dispersés. Le mode de vie typique chez les Pygmées actuels de la forêt équatoriale est constitué par une unité supérieure, la → *horde,* occupant un territoire et à l'intérieur duquel elle est divisée en plusieurs unités inférieures, les clans, ayant chacune à sa tête une personne âgée et respectée. Le clan, chez les Pygmées, est en même temps l'unité de base d'un campement (HEYMER, en préparation).

Dans le monde animal, nous connaissons chez les Hyènes tachetées, *Crocuta crocuta,* des unités sociales organisées que KRUUK (1968) a appelées clans et qui peuvent comprendre une centaine d'animaux chacun. Chaque clan, généralement dirigé par une ♀ expérimentée, possède son propre → *territoire.* Les membres du clan patrouillent le long de ses limites et le marquent en commun dans des endroits déterminés à l'aide de leurs glandes anales. Ces territoires de clans sont vigoureusement défendus contre des clans étrangers (VAN LAWICK-GOODALL 1970).

Klangspektrogramm □ Die mit dem Tonband gewonnenen Aufnahmen von Lautäußerungen läßt man durch einen Klangspektrographen laufen, der auf Spezialpapier Amplitude, Frequenz und Schalldruck registriert. Die so entstehenden Klangbilder eignen sich zur Klanganalyse qualitativer und quantitativer Art. Zur Herstellung von Klangspektrogrammen sollten möglichst Originalaufnahmen Verwendung finden (TEMBROCK 1959, BUSNEL 1964, SHIOVITZ 1975).

sonogram □ Tape recordings of vocalizations can be run through a sound spectrograph which registers the amplitude, frequency and sound pressure, and prints this information as a »sound picture« on a special kind of paper. These sonograms, which are best made from original recordings, are useful for qualitative and quantitative analysis of the sound pattern (TEMBROCK 1959, BUSNEL 1964, SHIOVITZ 1975).

sonogramme □ Expressions vocales qui ont été enregistrées à l'aide d'une bande magnétique et qui, par la suite, sont introduites dans un spectrographe sonore permettant une impression, sur papier spécial, de l'amplitude, de la fréquence et de la pression acoustique. Ces images sonores sont destinées à une analyse qualitative et quantitative. Pour l'établissement d'un tel sonogramme, il est conseillé d'utiliser autant que possible des enregistrements originaux (TEMBROCK 1959, BUSNEL 1964, SHIOVITZ 1975).

Klapperzeremonie → Begrüßungsklappern

Kleptobiose → Beuteschmarotzer

Klinokinese → Kinese

Klinotaxis → Taxis

Knabberputzen = Putzknabbern → Körperpflegehandlungen

E. nibble-cleaning □ F. se nettoyer en mordillant

Kokettier-Ruf □ Bei manchen Enten der Gattungen *Aix* und *Lampronessa* vorkommender epigamer Ausdrucksruf, der in seiner Funktion dem → *Nickschwimmen* der Stockente, *Anas platyrhynchos,* gleichkommt (LORENZ 1941).

coquette call □ An epigamic vocalization in certain ducks, e. g., *Aix* and *Lampronessa,* that is functionally analogous to → *nod-swimming* in *Anas platyrhynchos* (LORENZ 1941).

cri de coquetterie □ Chez certains Canards des genres *Aix* et *Lampronessa,* cri épigame de valeur expressive qui, dans sa fonction, correspond à la → *nage de coquetterie* (Nickschwimmen) chez le Colvert, *Anas platyrhynchos* (LORENZ 1941).

Kokettierschwimmen → Nickschwimmen

Kollektivmimikry → Mimikry

E. collective mimicry □ F. mimétisme collectif

Komfortbewegung → Komfortverhalten

E. comfort movement □ F. mouvement de confort

Komfortverhalten □ Verhaltensweisen und Bewegungsformen im Dienste der »Behaglichkeit« und der Körperpflege. KORTLANDT (1940) sprach bereits von »Bequemlichkeitssuche«, BAERENDS (1950) führte dann den Begriff Komfortbewegungen ein. Sie sind im Tierreich weit verbreitet; generell lassen sich vier verschiedene Typen unterscheiden: 1. Putzbewegungen, 2. Bewegungen des Sichschüttelns, 3. Sichreiben, 4. Badebewegungen, während Sichstrecken und Gähnen vielleicht als → *Räkelsyndrom* zusammengefaßt werden können, da sie auf einer anderen physiologischen Faktorenkombination beruhen (TEMBROCK 1964).

comfort behaviour □ Behaviour patterns and movement coordinations that serve to enhance an animal's »comfort« and body care. By 1940 KORTLANDT had already introduced the term »comfort search« and BAERENDS introduced comfort movements in 1950. Widespread in the animal kingdom, this general category consists of 4 types: 1. cleaning movements, 2. selfshaking movements, 3. self-rubbing, 4. bathing movements. Stretching and yawning are more appropriately termed → *stretching syndrome* since they are based on a different combination of physiological factors (TEMBROCK 1964).

comportement de confort □ Comportements et mouvements au service du »bien-être« de l'individu et du soin corporel. KORTLANDT (1940) a déja parlé d'une »recherche du confort« et BAERENDS (1950) a créé l'expression de »mouvements de confort«. Ils sont très répandus dans le règne animal et d'une manière générale, on peut distinguer 4 types différents: 1. se nettoyer, 2. s'ébrouer, 3. se frotter, 4. se baigner. Il semble, par ailleurs, que l'étirement et le bâillement soient à considérer comme dépendants du → *syndrome d'étirement,* ces phénomènes étant basés sur une combinaison de facteurs physiologiques différents (TEMBROCK 1964).

Kommensalismus □ Form des Zusammenlebens verschiedener Arten. DEEGENER (1918) begrenzt diesen Begriff auf das Geschehnis, wenn z. B. ein Tier am oder im anderen lebt und sich an seiner Mahlzeit beteiligt, ohne als Parasit schädlich zu werden, ihm aber auch keinen Nutzen bringt. So ist z. B. der Springschwanz *Calobatinus grassei* in Reiterstellung auf dem Kopf eines Soldaten von *Bellicositermes natalensis* ein typischer Kommensale oder Mitesser. → *Freßgemeinschaft* oder Tischgenossenschaft ist etwas anderes. Grenzfälle von Kommensalismus sind gelegentlich schwer von Parasitismus und → *Symbiose* zu unterscheiden (FÜLLER 1958, WILSON 1975).

commensalism □ A form of community life of different species. DEEGENER (1918) used this term to depict organisms that live on or in other animals, partaking of the host's food without actually damaging the host and without being profitable to him. For example, the springtail, *Calobatinus grassei,* rides on the head of the termite, *Bellicositermes natalensis* as a commensal. Not to be confounded with → *feeding association.* In some border-line cases, commensalism is sometimes difficult to distinguish from parasitism and → *mutualism* (FÜLLER 1958, WILSON 1975).

commensalisme □ Forme de vie en commun d'organismes appartenant à des espèces différentes. DEEGENER (1918) désigne par ce terme le fait qu'un animal vit sur ou dans un autre animal et profite de la nourriture de celui-ci sans devenir un parasite nuisible et sans lui apporter un avantage, et il distingue nettement ce phénomène du → *symphagium.* Par exemple, le Collembole *Calobatinus grassei,* à cheval sur la tête d'un soldat du Termite *Bellicositermes natalensis,* est un commensal typique. Certains cas limite de commensalisme sont parfois difficiles à distinguer du parasitisme ou de la → *symbiose* (FÜLLER 1958, WILSON 1975).

Komment → Kommentkampf
E. ritual □ F. rituel

Kommentkampf □ Angeborene, nach festen Regeln ablaufende, gegen Artgenossen gerichtete Kampfweise. Bei Arten mit gefährlichen Waffen, die bei ihrem Einsatz den Gegner leicht töten können, haben sich meist ein spezieller Austragungsmodus und Hemmechanismen ausgebildet, die das Töten von Artgenossen verhindern, denn Beschädigungskämpfe würden hier zu starker Dezimierung der Art führen. Oft wird sogar der ganze Kampf zu einem Turnier (*Turnierkampf*) umgewandelt. Man spricht dann auch von einem Kampfkomment. Unter dem Ausdruck Komment (franz. *comment* = wie) versteht man die Art und Weise, etwas zu tun, das Einhalten einer bestimmten Regel.

ritualized fight □ Innate fighting patterns which are directed against conspecifics and which procede according to set rules. Species possessing dangerous weapons that easily could kill the opponent have usually developed a special manner of fighting as well as special inhibitory mechanisms which prevent an opponent's death. Otherwise, such fights could lead to a severe decimation of the population. Often the fight is entirely transformed into a duel or »fighting ritual«.

combat rituel □ Comportement inné et à stricte observance des règles dans le déroulement d'un combat entre congénères. Chez les espèces portant des armes dangereuses dont l'utilisation pourrait entraîner la mort de l'adversaire, il s'est développé un déroulement rituel dans le combat et des mécanismes inhibiteurs permettant d'éviter la mise à mort des congénères (*combat symbolique* – MONFORT-BRAHAM 1975). Il est évident que, dans ces cas, des combats sanglants pourraient aboutir à une forte décimation de l'espèce. Parfois même, tout le combat a pris la forme d'un rituel, – terme à ne pas confondre avec → *ritualisation* ce qui est souvent le cas. L'expression allemande *Kommentkampf* dérive du *Komment* estudiantin (combat au sabre réglementé des étudiants allemands au siècle dernier), terme emprunté du français *comment,* c'est-à-dire la manière selon laquelle on fait quelque chose.

Kommunikation → Verständigung

Kommunikationsbarriere → Verständigung
E. communication barrier □ F. barrière de communication

Kommunikationsmittel → Verständigung
E. means of communication □ F. support de communication

Kompaßorientierung → Orientierung
E. compass orientation □ F. orientation sur repères célestes
Komplexqualität □ Eine Zusammenfassung bestimmter Beschaffenheiten der inneren und äußeren Situation, z. B. Zustand des Hungers, Tageszeit, Wetter, sich darbietende Beute, Gelegenheit, die Beute auf gewohntem Wege zu ergreifen, in einem ganzheitlichen Erlebensbestand, und zwar in einem komplexen Ganzen, erlauben dem Tier, sich in verschiedenen Situationen verschieden zu verhalten (VOLKELT 1914, 1937).
quality of complexness □ A compilation of certain aspects of the internal and external milieu, e. g. hunger, daytime, weather, available prey, estimated prey capture probability, etc. into an unitary consideration or complex whole permitting the animal to behave differently under different situations. This was called quality of complexness by VOLKELT (1914, 1937).
qualité complexe □ Configuration de plusieurs particularités d'une situation interne et externe. Par exemple l'état d'inanition, l'heure de la journée, les conditions météorologiques, le mode de présentation de la proie et l'occasion de capturer la proie selon les processus usuels, perçus dans une totalité organisée, permettent à un animal de se comporter d'une manière différente, mais toujours adéquate, dans diverses situations. Telle est la définition de la qualité complexe selon VOLKELT (1914, 1937).
Konfliktverhalten □ Ist eine auslösende Reizsituation so beschaffen, daß verschiedene Dränge gleichzeitig aktiviert werden, wie etwa der Drang zum Angreifen und zum Flüchten, dann geraten diese geradezu entgegengesetzten Verhaltensweisen miteinander in Konflikt. Solche Spannungen führen oft zu → *Übersprungbewegung(en)*.
conflict behaviour □ If simultaneously releasing stimuli occur that activate contradictory tendencies, as for example the tendencies to attack and to flee, then the two opposing behaviours conflict and the resulting tension often leads to → *displacement activity*.
comportement conflictuel □ Si une situation déclenchante est telle que des pulsions contradictoires sont activées simultanément, comme par exemple la tendance à l'attaque et à la fuite, il en résulte alors un conflit. De telles tensions conduisent dans la plupart des cas à une → *activité substitutive*.
Konfusionseffekt □ Eine sehr auffällige → *Schutzanpassung* an Freßfeinde, wie z. B. der Fischschwarm. Der Zielmechanismus des Raubfisches wird durch die Vielzahl bewegter Beuteobjekte im Schwarm, der so einen Konfusionseffekt bildet, verwirrt.
confusion effect □ A striking → *protective adaptation* to predators, e. g. a fish school. The shear number of prey objects in a swarm distracts and confuses the aim mechanism of the predator.
effet de confusion □ Une → *adaptation protectrice* évidente à l'encontre des prédateurs potentiels, comme par exemple un banc de Poissons. La visée du prédateur est brouillée par la multiplicité des proies disponibles: c'est typiquement un effet de confusion.
Konstruktionsspiele → Spielverhalten
E. playing at building something (construction) □ F. jeux de construction
Kontaktabbruch → Hinterkopfzudrehen → Verneinung
E. cut-off behaviour □ F. rupture de contact
Kontaktaufforderung → Initiativgebärden
E. contact initiating gestures □ F. sollicitation de contact
Kontaktentzug □ Der Entzug des persönlichen Kontaktes in der frühkindlichen Entwicklung, z. B. längerer Krankenhausaufenthalt, der zu irreversiblen Schäden führen kann (SPITZ 1965, TINBERGEN u. TINBERGEN 1973 – → *Hospitalismus*). Ähnliches wurde bei isoliert aufgezogenen Rhesusaffen festgestellt (HARLOW u. HARLOW 1962 a + b).
contact withdrawal □ The withdrawal of personal contact in early childhood such as may occur during long hospitalization, may lead to irreversible damage (SPITZ 1965, TINBERGEN and TINBERGEN 1973 – → *hospitalism*). Contact withdrawal has been extensively studied in monkeys (HARLOW and HARLOW 1962 a + b).
privation de contact □ La privation de contact personnel durant le développement précoce de l'enfant, par exemple par un séjour prolongé à l'hôpital, peut entraîner des dommages irréversibles (SPITZ 1965, TINBERGEN et TINBERGEN 1973 – → *hospitalisme*). Des phénomènes comparables ont pu être mis en évidence chez les Macaques élevés dans l'isolement (HARLOW et HARLOW 1962 a + b).
Kontaktgebärden → Initiativgebärden
Kontaktgruß □ Man versteht darunter alle Formen des Sich-Begrüßens mit körperlichem Kontakt, vom Händedruck bis zum Nasereiben. Bei → *Kontakttieren*, die sich gut kennen, gibt es auch Körperberührung als Gruß. Bei Vögeln vor allem das → *Schnäbeln*. Bei vielen Säugern kennen wir verschiedene Arten des Sich-Begrüßens mit Körperkontakt, wie z. B. naso-nasale, naso-anale oder naso-genitale Begrüßung. Beim Menschen können die Nasen bei der Begrüßung aneinander gerieben werden. Oft ergreift der Grüßende auch nur eine Hand und reibt sich damit die Nase. Bei verschiedenen Völkern ersetzt das Nasenreiben unseren Kuß. Es kann auch als Riechkuß gedeutet werden (EIBL-EIBESFELDT 1970). Schon ANDREE (1889) erwähnt den Nasengruß als freundliches Beschnüffeln.
contact greeting □ All forms of greeting in which bodily contact occurs, e. g., hand shake, nose rubbing, etc. Certain → *contact-type animals* rub against each other as a sign of individual recognition. In birds the most frequent form of contact greeting is → *billing*. In many mammals there are various forms of greeting behaviour which involve body contact, e. g.

naso-nasal, naso-anal, or naso-genital greeting. Nose rubbing is a form of greeting in humans. Often the greeter may simply grasp a person's hand and rub it against the nose. In various cultures this behaviour replaces the kiss, and smelling the partner may be a part of this »kiss« (EIBL-EIBESFELDT 1970). As early as 1889 ANDREE described nose greeting as a friendly sniffing.

salut de contact □ Cette notion comprend toutes les formes de salut avec contact corporel, depuis la poignée de mains jusqu'au frottement du nez, etc. Chez certains → *animaux à contact corporel* quand il y a reconnaissance individuelle, il existe également des formes de salut avec contact corporel. Chez les Oiseaux, le salut de contact est surtout le → *becquetage.* Chez les Mammifères, nous connaissons nombre de saluts mutuels, comme par exemple le contact naso-nasal, naso-anal ou naso-génital. Chez les Hommes, les nez peuvent être frottés l'un contre l'autre en signe de salut. Parfois, l'individu qui salue prend simplement la main de l'autre pour y frotter son nez. Chez de nombreux peuples, ce comportement remplace notre baiser. Il peut être considéré comme un »baiser de flairage« (EIBL-EIBESFELDT 1970). ANDREE (1889) signalait déjà le salut avec le nez comme un flairage amical.

Kontaktinitiative → Initiativgebärden
E. contact initiation □ F. initiative de contact

Kontaktscheu □ Bei Tieren die Meidung engen körperlichen Kontaktes und Einhaltung einer → *Individualdistanz.* Bei Menschen die Meidung sozialen Kontaktes.

contact avoidance □ The avoidance or shyness of close bodily contact in animals and maintenance of an → *individual distance;* in humans the avoidance of social relations.

aversion du contact □ Chez les animaux, évitement du contact corporel et maintien d'une → *distance inter-individuelle.* Chez l'Homme, timidité ou tendance à se soustraire aux contacts sociaux.

Kontaktstreben → Initiativgebärden
E. contact-seeking behaviour □ F. recherche du contact

Kontakttiere □ Tiere, denen eine → *Individualdistanz* fehlt und die ein ausgeprägtes Bedürfnis nach körperlichem Kontakt haben.

contact-type animals □ Species in which an → *individual distance* is lacking and which have a distinct need for bodily contact.

animaux à contact corporel □ Animaux auxquels la → *distance inter-individuelle* fait défaut et qui ont un besoin prononcé d'un contact corporel mutuel.

Kontrastbetonung □ Erhalten bestimmte Arterkennungsmerkmale durch Selektionsdruck eine besondere Betonung zur besseren Differenzierung, so spricht man von einer Kontrastbetonung. So haben z. B. sympatrisch lebende, nah verwandte Heuschrecken deutlicher verschiedene Rufe als allopatrisch nah verwandte und sympatrische Leuchtkäfer verschiedene Blinksignale.

character enhancement □ The differentiation that occurs when certain distinctive characteristics of a species are under a selection pressure which further emphasizes this character. Sympatric grasshoppers, for example, have developed distinctly different vocalizations and sympatric fire-flies different blinking schemes, which is not always the case in allopatric species.

renforcement du contraste □ On parle de renforcement du contraste lorsque certains caractères de reconnaissance spécifique deviennent, sous la pression de la sélection naturelle, particulièrement distincts en vue d'un meilleur isolement reproductif. Ainsi, certaines espèces sympatriques de Criquets ont développé des chants nettement distincts; de même, les Lampyres sympatriques possèdent des signaux lumineux très différenciés, ce qui n'est pas toujours le cas chez les espèces allopatriques.

Konturverfremdung □ Von LILLI KOENIG (1973) benutzter Begriff zur Bezeichnung der Tarnstellung bei Eulen, die eine stark veränderte Sitzstellung einnehmen und die Federohren hochstellen, um sich ihrer Umgebung anzupassen und somit fast unsichtbar werden; eine ähnliche Art von → *Schutzanpassung* kennen wir auch bei Rohrdommeln.

contoure disguise □ Term coined by LILLI KOENIG (1973) to describe the camouflaged appearance of owls that by raising the ear tufts and altering their resting posture merge quite closely with surroundings; a similar form of → *protective adaptation* occurs also in the bittern.

effacement des contours □ Terme utilisé par LILLI KOENIG (1973) pour désigner une forme de camouflage chez les Chouettes qui adoptent en se perchant une position très modifiée et dressent leurs oreilles emplumées, s'adaptant à leur environnement et devenant ainsi pratiquement invisibles. Les Butors agissent de même. → *adaptation protectrice.*

Konvergenz □ Formenähnlichkeit ursprünglich ganz verschieden gestalteter Organe bzw. Organismen und auch Verhaltensweisen als Ergebnis stammesgeschichtlicher Anpassung an gleichartige Umweltbedingungen; z. B. Haie und Wale, Nektarvögel und Kolibris, altweltliche Wiederkäuer und neuweltliche Nager.

convergence □ Form similarity in organs, organisms, and behaviour patterns with different modes of origin; results from evolutionary development under similar environmental conditions; for example, sharks and whales, old world Nectariniidae and new world humming-birds, old world ruminants and new world rodents.

convergence □ Développement phylogénétique de ressemblances étroites entre les formes d'organes ou d'organismes originellement très différents, sous la pression de conditions d'environnement semblables. Par exemple, Requins et Baleines, Nectariniidae et Colibridae, Ruminants de l'ancien monde et Rongeurs du nouveau monde. Des faits de convergence ont aussi été décrits dans le domaine des traits de comportement.

Kopfhalstauchen □ Eine der häufigsten Badebewegungen bei Schwänen und Pelikanen, also ein → *Komfortverhalten,* wobei Kopf und Hals mit einer Schlangenbewegung öfter rhythmisch ins Wasser eintauchen (MEISCHNER 1959, PETZOLD 1964).

head-neck dipping □ One of the most frequent bathing movements in swans and pelicans in which the neck and head are dipped in a rhythmic snake-like fashion into the water. → *comfort behaviour* (MEISCHNER 1959, PETZOLD 1974).

plongée de la tête et du cou □ Un des mouvements de baignade les plus répandus chez les Cygnes et les Pélicans; c'est un → *comportement de confort* pendant lequel la tête et le cou plongent dans l'eau d'une manière rythmique et avec un mouvement serpentin (MEISCHNER 1959, PETZOLD 1964).

Kopfkratzen □ Als Komfortbewegung bei Vögeln als → *Hintenherumkratzen* und → *Vornherumkratzen.* Die meisten Säuger kratzen sich mit dem Hinterfuß am Kopf; andere Arten putzen sich mit den Vorderpfoten den Kopf (→ *Körperpflegehandlungen*). Beim Menschen kommt ein *Sich-am-Kopf-Kratzen* mit der Hand bei »Verlegenheit« vor (z. B. beim Überlegen oder während eines Vortrages) und ist wohl als → *Übersprungbewegung* zu deuten. Stellte man Schimpansen im Versuch ein Problem, so konnte ebenfalls ein solches Übersprung-Kopfkratzen beobachtet werden.

head scratching □ Comfort movement in birds separated into → *scratching over the wing* and → *direct scratching.* Many mammals scratch their head with the hind legs; other species groom their head with the fore-paws (→ *grooming movements*). In humans head scratching with the hand occurs as a → *displacement activity* during concentration or when perplexed, e. g., while delivering a lecture. If a chimpanzee is given a difficult problem to solve such head scratching may also be observed.

se gratter la tête □ Comportement de confort effectué chez de nombreux Oiseaux à l'aide d'une patte en passant soit par-dessus soit par-dessous l'aile (→ *grattage par-dessus l'aile* et → *grattage direct de la tête*). De nombreux Mammifères se grattent la tête à l'aide d'un membre postérieur; d'autres espèces se nettoient la tête avec les pattes antérieures (→ *soins corporels*). Chez l'Homme, on observe souvent un grattage de la tête avec la main pendant des moments d'embarras ou de perplexité (par exemple au cours d'un exposé); il peut être considéré comme une → *activité substitutive.* Il a été observé que les Chimpanzés en situation expérimentale de résolution de problème se grattaient souvent la tête de la même manière.

Kopfnicken □ Bei vielen Blenniiden-♂♂, wenn sie in Röhren sitzen, eine Imponierbewegung, gegenüber Rivalen mit besonders starker Intensität ausgeführt. Bei vielen Menschen bedeutet es Zustimmung.

head-nodding □ A behaviour found in ♂♂ of Blenniid fishes performed while resting in tube-like enclosures and when in the presence of rivals; in many humans signifies agreement.

hochement de tête □ Les ♂♂ des Poissons Blenniidae occupant un tube territorial dans la roche effectuent des mouvements verticaux de la tête, avec une intensité particulièrement forte, en présence de rivaux. Dans de nombreuses cultures humaines, ce mouvement de la tête signifie l'acquiescement.

Kopfpendeln → Suchautomatismus
E. head-wagging □ F. mouvement pendulaire de la tête

Kopfschütteln □ Bei Schwänen und Enten eine Komfortbewegung, die der zweiten Phase des Streckschüttelns (→ *Sich-Schütteln*) ähnelt, wobei aber nur der Kopf geschüttelt wird. Beim Menschen bedeutet Kopfschütteln in vielen Kulturen → *Verneinung.*

head-shaking □ A comfort movement in swans and ducks resembling the second phase of stretch-shaking (→ *shaking*) except that only the head is shaken; in humans implies → *negation.*

secouement latéral de la tête □ Chez les Cygnes et les Canards, mouvement de confort qui ressemble à la deuxième phase du *Streckschütteln* (s'ébrouer en s'étirant; → *s'ébrouer*), mais où la tête seule est agitée. Dans de nombreuses cultures humaines, le secouement latéral de la tête signifie refus ou → *négation.*

Koprophagie □ Orale Aufnahme von Kot, von vielen Säugern bekannt; nicht zu verwechseln mit → *Coecotrophie.*

coprophagy □ Oral intake of feces occurring in many mammals; not to be confused with → *caecotrophy.*

coprophagie, scatophagie □ Ingurgitation de matières fécales, connue chez de nombreux Mammifères; à ne pas confondre avec le phénomène de la → *caecotrophie.*

Kopulation □ Bei Einzellern und mehrzelligen Tieren der Vorgang des Verschmelzens (Hologamie) oder der Vereinigung der männlichen und weiblichen Geschlechtsorgane (Begattung) zur Herbeiführung einer Befruchtung oder Besamung. Bei vielen Vögeln, die keinen Penis haben, werden lediglich die Geschlechtsöffnungen bzw. Kloaken aufeinandergepreßt. Beim Menschen kennen wir neben der zur Befruchtung und Fortpflanzung führenden Kopulation noch die *Bindungskopulation,* welche ausschließlich im Dienste der Partnerbindung steht. Kopulation als aggressive Handlung → *Wutkopulation.*

copulation □ In uni- or multicellular animals the process of merging (hologamy) or uniting of male and female sex organs (mating) for fertilization or insemination. In many birds species without a penis, the sexual orifices, i. e., cloaca, are simply pressed together. In man, in addition to reproductive copulation, *bond-oriented copulation* may occur which exclusively serves pair-bond maintenance. Copulation as an aggressive action → *rage copulation.*

copulation □ Chez les Unicellulaires et les animaux pluricellulaires, le processus de fusion (hologamie) ou l'union des organes sexuels ♂ et ♀ (accouplement) permettant la fécondation ou l'insémination. Chez un certain nombre d'Oiseaux auxquels le pénis fait défaut, seuls les orifices génitaux ou cloaques sont mis en contact. Chez l'Homme, nous connaissons, outre l'accouplement destiné à la fécondation et la reproduction, un accouplement exclusivement au service du renforcement du lien entre partenaires (*accouplement de cohésion*). L'accouplement en tant qu'action agressive → *copulation de colère.*

Körperpflegehandlungen □ Verhaltensweisen der Reinigung der Haut, der Federn und der Haare, des gesamten Körpers schlechthin, die auch in den Kontext des Komfortverhaltens gehören. Bei Insekten sind Mundwerkzeuge und Extremitäten wichtige Putzorgane. Vögel beknabbern ihre Federn mit dem Schnabel, was MEISCHNER (1959) bei Pelikanen *Knabberputzen* und PETZOLD (1964) bei Schwänen *Putzknabbern* nannte. Auch Säuger beknabbern putzend ihr Fell mit den Zähnen und benutzen die Zunge zum Lecken; außerdem wird in beiden Tiergruppen die Hinterextremität häufig als Putzorgan eingesetzt, → *Vornherumkratzen* und → *Hintenherumkratzen.* Die Aufeinanderfolge der verschiedenen Einzelhandlungen während des Sich-Putzens sowie die Art und Weise der ausgeführten Putzhandlungen werden als Putzritus bezeichnet und können einen stammesgeschichtlichen Aussagewert haben, wie BÜRGER (1959) an Nagern und URSULA JANDER (1966) an Tracheaten herausfanden.

Die intraspezifische soziale Körperpflege oder auch Fremdputzen hatte wohl ursprünglich nur zur Aufgabe, jene Körperstellen zu reinigen, die der Partner selbst nicht erreichen kann. In ritualisierter Form hat das Sich-Einander-Putzen eine wichtige bindende Funktion zwischen Paarpartnern und/oder Mitgliedern einer Gruppe. Vögel benutzen für die soziale Körperpflege, das Gefiederkraulen, den Schnabel, viele Säuger die Zunge oder die Zähne; dies tun auch die Prosimier, die aber dabei noch zusätzlich die Hände auf den Partner legen oder in dessen Fell verkrallen.

Bei allen übrigen Primaten, den Menschen eingeschlossen, vorkommende Körperpflegehandlung ist das Lausen, das mit den Händen ausgeführt wird. Man kann sich selbst lausen (Sichlausen); Einander-Lausen ist bei Primaten und Menschen Ausdruck einer gewissen Vertrautheit und festigt das Band; es hat soziale Bedeutung. Schreiende Kinder bei primitiven Völkern beruhigt man oft, indem man sie laust. Lausen besagt auch nicht unbedingt, daß man nach Läusen sucht, sondern hat rein symbolische Bedeutung. Das englische *grooming* sagt mehr aus als Lausen, da es alle Putzhandlungen mit einschließt. Eine besondere Form zwischenartlicher Körperpflegehandlungen finden wir in der → *Putzsymbiose.*

preening movements (birds), **grooming movements** (mammals) □ Behaviour patterns pertaining to the cleaning the skin, feathers or hairs of the entire body. They also belong into the context of comfort behaviour. In insects mouth parts and extremities are important cleaning organs. Birds nibble their feathers with their bills (*nibble-preening* or *nibble-cleaning* – MEISCHNER 1959, PETZOLD 1964). Mammals also nibble-groom their fur with their teeth and lick it with their tongue. Furthermore, the hind legs are often used for cleaning (scratching), → *direct scratching* and → *scratching over the wing.* The sequence of various individual movement patterns of self-grooming as well as the manner in which the specific grooming patterns are executed are called a grooming (preening) ritual and can tell us about the phylogenetic relationships between species, as was found by BÜRGER (1959) in rodents and URSULA JANDER (1966) in tracheates.

The intra-specific, social grooming or the grooming of others seems to have originally served to clean parts of the body that the animal could not reach by itself. In its ritualized form mutual grooming seems to have an important function in maintaining bonds between partners and/or members of a group. Birds use their beaks for preening their feathers, many mammals use their tongues, as do the prosimians, which, however, place their hands on the partner in addition or actually hold on to its fur.

In all other primates, including humans, grooming patterns include the German *Lausen* (searching for lice), which is done with the hands. This can be done to one's self as well as others and is in primates and humans an indication and expression of a certain degree of familiarity, strengthens bonds, hence has a social function. Crying children are often calmed in primitive peoples by delousing them. Delousing in this context does not necessarily imply that one is looking for lice, it may have a purely symbolic meaning. The English term grooming implies more than delousing, since it includes all cleaning patterns. A special form of interspecific grooming is found in → *cleaning symbiosis.*

soins corporels □ Activités destinées au nettoyage de la peau, des plumes ou des poils ou du corps dans son ensemble et faisant partie du comportement de confort *sensu lato.* Chez les Insectes, les pièces buccales et les membres sont des organes de nettoyage importants. Les Oiseaux se nettoient le plumage en le mordillant avec leur bec et les Mammifères mordillent leur pelage avec les dents ou le lèchent avec leur langue. D'autre part, chez les Oiseaux, le membre postérieur est souvent utilisé comme organe de nettoyage (→ *grattage par-dessus l'aile,* → *grat-*

tage direct de la tête). Parmi les Mammifères, selon les groupes, les membres antérieurs et postérieurs sont utilisés. La succession des différents mouvements au cours de l'auto-nettoyage et la manière selon laquelle ces mouvements sont exécutés, constituent le rituel de nettoyage et peuvent donner des renseignements phylogénétiques, comme ont pu le démontrer BÜRGER (1959) chez les Rongeurs et URSULA JANDER (1966) chez les Trachéates Hexapodes.
Le soin corporel social et intraspécifique (allogrooming) était probablement destiné, à l'origine, à nettoyer les parties corporelles que l'individu ne pouvait pas atteindre lui-même. Sous une forme ritualisée, l'*allo-grooming* ou nettoyage mutuel possède une fonction importante pour la cohésion entre partenaires d'un couple et/ou membres d'un groupe. Les Oiseaux utilisent leur bec lors des soins corporels sociaux, de nombreux Mammifères leur langue ou leurs dents. Parmi les Primates, les Prosimiens utilisent encore leurs dents comme peigne, mais, de plus, posent leurs mains sur le partenaire ou s'agrippent dans son pelage.
Chez tous les autres Primates, l'Homme y compris, l'une des activités de soins corporels est l'épouillage qui est essentiellement effectué à l'aide des mains. On peut s'épouiller soi-même ou mutuellement ce qui est, notamment chez l'Homme, l'expression d'une certaine intimité et renforce le lien interindividuel. Chez les peuples primitifs (comme par exemple les Boschimans ou les Pygmées), les petits enfants qui pleurent sont épouillés pour les calmer. Chez tous les Singes, l'épouillage mutuel correspond à des soins corporels sociaux. L'expression anglaise *grooming* dit beaucoup plus que le terme épouillage, car elle inclut le nettoyage. S'épouiller ne veut pas dire que des poux sont effectivement recherchés; ce comportement a plutôt une signification symbolique. Une forme particulière de relations de soins corporels interspécifiques existe sous forme de → *symbiose de nettoyage*.

Koten □ Ausscheidung des Kotes. Dem Koten liegen reflektorische Vorgänge zugrunde (ALTMANN 1969).

defecation □ Excretion of feces; controlled by reflex processes (ALTMANN 1969).

défécation □ Elimination des matières fécales. La défécation est sous le contrôle de processus réflexes (ALTMANN 1969).

Kotfressen → Koprophagie

Kotzeremoniell □ Unter dem von PILTERS (1956) geprägten Begriff versteht man ein bei Säugetieren aus mehreren Komponenten bestehendes Appetenzverhalten zwischen zwei Partnern (ALTMANN 1969). Ein besonders eindrucksvolles Zeremoniell beschrieb HENDRICHS (1971) vom Dikdik, *Rhynchotragus kirki* (Abb. 68). A) ♀ uriniert; B) ♀ kotet; C) ♂ schnuppert am Kot des ♀; D) ♂ uriniert scharrend; E) ♂ hockt sich nach Drehung um 180° hin und kotet; F) ♂ kotet nochmals nach erneuter

68 A 68 B 68 C 68 D 68 E 68 F

Drehung um 180°.

defecation ceremony □ Appetitive behaviour with several components in mammals between two partners described by PILTERS (1956) and ALTMANN (1969). HENDRICHS (1971) describes a particularly impressive example in the dikdik, *Rhynchotragus kirki* (Fig. 68). A) ♀ urinates; B) ♀ defecates; C) ♂ sniffs feces of ♀; D) ♂ urinates scratching; E) ♂ turns 180° crouching and defecates; F) ♂ defecates again after another turn of 180°.

cérémonial de défécation □ Ce terme a été créé par PILTERS (1956) et définit chez les Mammifères un comportement d'appétence composé de plusieurs éléments entre deux partenaires (ALTMANN 1969). Un exemple particulièrement impressionnant a été décrit par HENDRICHS (1971) chez le Dikdik, *Rhynchotragus kirki* (Fig. 68). A) ♀ urine; B) ♀ défèque; C) ♂ renifle les matières fécales de la ♀; D) ♂ urine et gratte le sol; E) après avoir effectué une rotation de 180°, le ♂ s'incline et défèque; F) ♂ effectue une nouvelle rotation de 180° et défèque à nouveau.

Kreuzgang □ Wechselschritt, bei dem Vorder- und Hinterbein der gleichen Körperseite sich entgegengesetzt bewegen. Der Vorderlauf wird erst dann wieder nach vorn geführt, wenn der dazugehörige Hinterlauf auf dem Boden aufgesetzt hat (Abb. 69 A – *Capra falconeri*). Die natürliche Fortbewegung des Menschen entspricht ebenfalls den Gesetzmäßigkeiten des Kreuzgangs (Ab. 69 B). → *Gangart*, → *Paßgang*.

cross gait □ Alternating step in which the front and rear leg of each side move in opposition to each other. Progression in anterior limbs occurs when the corresponding rear leg is on the ground (Fig. 69 A –

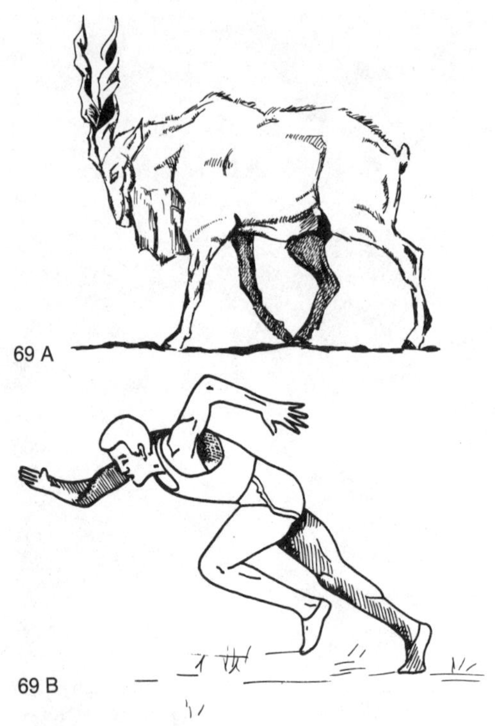

69 A

69 B

Capra falconeri). The natural locomotion in man is also a sort of cross gait (Fig. 69 B). → *locomotory pattern*, → *pacing*.

allure croisée □ Chez les Mammifères, allure locomotrice au cours de laquelle le membre antérieur gauche avance en même temps que le membre postérieur droit tandis que le membre antérieur droit et le membre postérieur gauche servent de points d'appui, et vice versa (Fig. 69 A – *Capra falconeri*). La locomotion naturelle de l'Homme est également du type allure croisée (Abb. 69 B). → *schème locomoteur*, → *amble*.

Kriegergruß → Drohgruß
E. warrior greeting □ F. salut du guerrier

Kronismus → Verwandtenfresserei

Kulturenvergleich → Humanethologie
E. cross-cultural studies or comparison □ F. comparaison inter-culturelle

Kulturethologie □ Eine spezielle Arbeitsrichtung der allgemeinen vergleichenden Verhaltensforschung (Ethologie) und insbesondere der Humanethologie, die sich mit den ideellen und materiellen Produkten des Menschen, seiner Kultur, deren Entwicklung, ökologischer Bedingtheit und ihrer Abhängigkeit von angeborenen Verhaltensweisen sowie mit entsprechenden Erscheinungen bei Tieren vergleichend befaßt. KOENIG (1970, 1975) versucht nachzuweisen, daß gewisse fundamentale Gesetzlichkeiten der organischen Natur auch für die Kulturentwicklung des Menschen gelten.

cultural ethology □ A special branch of ethology concerned with the intellectual and material products of mankind and his culture in relation to their development, their ecological relevance, their dependence on innate behaviour patterns and their relation to corresponding situations in animals. KOENIG (1970, 1975) attempted to show that certain fundamental principles of nature apply also to culture development.

éthologie culturelle □ L'un des axes de recherches de l'éthologie, notamment humaine, orienté vers l'étude des productions matérielles et intellectuelles de l'Homme, de sa culture et du développement de celle-ci, des influences écologiques, des relations de la culture avec les *conduites innées*, de l'existence de phénomènes comparables chez les animaux (KOENIG 1970, 1975).

Kulturfolger □ Die Schwalben und vor allem die Mauersegler als ursprüngliche Felsbrüter, aber auch andere Vogelarten, die durch Häuser- und Städtebau des Menschen neue geographische Räume besiedeln konnten und sich dadurch eher ausbreiten als zurückgehen, werden Kulturfolger genannt.

commensals of civilization □ Animals, esp. birds, which have been able to extend their habitat in the presence of man; for example, swallows and swifts which originally nested in cliffs now do so in houses and buildings.

commensaux de civilisation □ Les Hirondelles et les Martinets qui nichaient initialement dans des biotopes rocheux – mais aussi d'autres Oiseaux – ont pu peupler de nouvelles aires géographiques grâce à la construction de maisons et de villes par l'Homme; ils ont ainsi étendu leur répartition géographique et sont plutôt en progression qu'en régression.

Kumpan □ Objekt mit bestimmten Eigenschaften in der Umgebung eines Tieres, an das dieses in einer bestimmten Phase und innerhalb kurzer Frist durch → *Prägung* irreversibel gebunden wird. Für ein neugeborenes Jungtier, besonders für Nestflüchter, kann sich damit die Bindung an die Eltern vollziehen; bietet sich stattdessen der Mensch in dieser sensiblen Phase an, so kann er der *Elternkumpan* werden. Umgekehrt handelt es sich beim *Kindkumpan* im tierischen und menschlichen Verhalten um die Jungen und Kinder schlechthin. Es sind nur dann nicht die eigenen Jungen, wenn man diese experimentell vertauscht oder Brutschmarotzer im Spiele sind. Die zusammen aufgewachsenen Jungen, z. B. einer Brut oder auch verschiedener Bruten, bilden oft auch nach der Trennung von ihren Eltern einen geschwisterlichen Zusammenhalt oder *Sympaedium* und erkennen einander als *Geschwisterkumpan*. Der *Geschlechtskumpan* ist in der Regel ein Artgenosse des anderen Geschlechts oder ein durch Prägung gebundenes Ersatzobjekt. Der Kumpan ist aber nicht immer irreversibel das gleiche »Objekt«, sondern auch ein Partner in einem bestimmten Funktionskreis. So war z. B. für eine von LORENZ handaufgezogene Dohle der Mensch der Eltern- und Geschlechtskumpan; sie flog jedoch mit Nebelkrähen als sozialen und Flugkumpanen; schließlich nahm sie junge Dohlen als Kindkumpane an. Die Problemstellung des Kumpans wurde von LORENZ (1935) ursprünglich an Vögeln untersucht.

companion □ An organism or object in the environment of an animal which has specific significance in some or all functional systems to that individual. The preference for a specific companion may be, but does not have to be, acquired during a short, sensitive period in early life. If, during this time, an animal is exposed to a human being, the latter will become the *parent-companion*. On the other hand, we speak of *child-companion* with respect to animal and human behaviour, when not only the own but all other young are regarded as such. In other words, parental behaviour may be directed at young other than one's own, as is the case with brood parasites or young that were exchanged experimentally. The young of mixed species raised together by a parent(s) form a sibling bond, hence are called *sibling-companions*. The *sexual-companion* as a rule is a conspecific or a substitute object acquired as a result of imprinting. The companion is not, however, »irreversibly« the same object, but is also a partner in a particular functional system. Thus, a jackdaw handraised by LORENZ considered humans as parent- or sexual-companions, but a hooded crow was the social and flight-companion. Finally, it accepted young jackdaws as child-companions. The problems dealing with the companion concepts were originally investigated by LORENZ (1935) in birds.

compagnon □ Objet ou organisme ayant des propriétés déterminées dans l'environnement d'un animal auquel il sera, au cours d'une phase sensible, imprégné et donc lié d'une manière irréversible. Pour un animal nouveau-né, surtout chez les nidifuges, c'est ainsi que s'établit le lien avec les parents. Par contre, si lors de cette phase sensible, l'Homme se présente, il peut alors devenir le *compagnon parental*. Inversement, le *compagnon juvénile* dans le comportement animal et humain est représenté par les propres jeunes ou les enfants d'une manière générale en rapport et en liaison avec leurs parents. Le compagnon juvénile, cependant, ne sera pas la propre progéniture dans le cas où celle-ci a été expérimentalement échangée ou dans le cas des nidificateurs parasites. Les jeunes qui ont grandi ensemble, par exemple ceux d'une même couvée ou même de couvées différentes, forment souvent après la séparation de leurs parents une *congrégation fraternelle* ou *sympaedium* et se reconnaissent entre eux comme frères et sœurs *(compagnon fraternel)*. Le *compagnon sexuel* est en règle générale un congénère du sexe opposé, mais il peut être aussi un objet de remplacement jouant ce rôle et auquel l'individu est lié par imprégnation. Le compagnon n'est cependant pas toujours et d'une manière irréversible un même »objet«, mais aussi un partenaire dans un cycle fonctionnel déterminé. Par exemple, un Choucas élevé à la main par LORENZ considérait l'Homme comme le compagnon parental et le compagnon sexuel, mais il préférait s'envoler avec des Corneilles mantelées qui étaient donc les compagnons sociaux de vol, et il finit par accepter de jeunes Choucas comme compagnons filiaux ou juvéniles. Tout le problème du cycle fonctionnel du compagnon a été étudié initialement sur des Oiseaux (LORENZ 1935).

Kundschafterinnen □ Bienen, die ausfliegen und nach neuen Futter- und Sammelquellen Ausschau halten. Nach der Rückkehr in den Stock übermitteln sie die Nachricht auf der Wabe tanzend den Stockgenossen. → *Rundtanz*, → *Schwänzeltanz*, → *Sicheltanz*.

scouting bees □ Bees that search for new supplies of food; when returning home they transmit their news to their hivemates with a dance. → *round dance*, → *waggle dance*, → *sickle dance*.

abeilles exploratrices □ On appelle ainsi les Abeilles qui s'envolent de leur ruche à la recherche de nouveaux lieux de récolte. De retour à la ruche, elles transmettent leurs informations à leurs congénères en effectuant des danses sur les rayons. → *danse circulaire*, → *danse frétillante*, → *danse en croissant*.

Kurzhochwerden □ Eine epigame Verhaltensweise bei Schwimmenten, die nach dem einleitenden → *Sichschütteln* ausgeführt wird. Diese Verhaltensweise ist der → *Abaufbewegung* als Werbehandlung gleichwertig. Der Erpel reißt zunächst unter einem lauten Pfiff den Kopf mit eingezogenem Kinn nach hinten und oben und krümmt gleichzeitig den Steiß mit stark gesträubtem Bürzelgefieder aufwärts, so daß der ganze Vogel eigenartig kurz und hoch wird (Abb. 70 A – *Anas platyrhynchos*; B – *Poecilonetta bahamensis*; diese Extremstellung wird mehrere Sekunden aufrechterhalten) LORENZ (1941).

70 A 70 B

head up-tail up □ An epigamic behaviour pattern in dabbling ducks occurring after the introductory *self-shaking (→ shaking)*. This behaviour pattern is comparable to the → *down-up-movement* as a courtship behaviour. The drake pulls up his head sharply while producing a whistling sound, while bending his chin down towards his breast. At the same time he bends his tail region up steeply, a movement that is enhanced by the conspicuously spread-out tail feathers. These combined movements result in the bird appearing at once much shorter as well as higher (Fig. 70 A – *Anas platyrhynchos*; B – *Poecilonetta bahamensis*; this extreme position is maintained for several seconds) LORENZ (1941).

tête haute-queue haute □ Un comportement épigame chez les Anatidae qui succède au mouvement d'introduction *(→ s'ébrouer)*. Ce comportement possède la même signification épigame que le → *mouvement de bas en haut.* Le malard se rengorge, ceci accompagné d'un sifflement très fort, et rejette la tête vers le haut en arrière, le menton rétracté; en même temps, il recourbe fortement le croupion dont les plumes sont très ébouriffées. L'Oiseau, dans son ensemble, a ainsi curieusement un aspect court et haut (Fig. 70 A – *Anas platyrhynchos*; B – *Poecilonetta bahamensis*; cette position extrême est maintenue pendant quelques secondes) LORENZ (1941).

Kurzschlußhandlung → Affekthandlung

Kurzzeitgedächtnis → Gedächtnis

E. short term memory □ F. mémoire immédiate

Kußfüttern □ Die ritualisierte Form des Fütterns beim Liebeswerben der Menschen. → *Mund-zu-Mund-Füttern.*

kiss-feeding □ A ritualized form of feeding occurs in the courtship behaviour in man. → *mouth-to-mouth feeding.*

baiser de nourrissage □ Une forme ritualisée de nourrissage que l'on observe lors du comportement de parade nuptiale chez les humains. → *nourrissage de bouche à bouche.*

L

Labyrinthversuch □ Im einfachsten Fall arbeitet man mit einem T- oder Y-Labyrinth. Es handelt sich dabei um das Erlernen der richtigen Entscheidung an den Orten der Gabelung oder der Verzweigung in einem System von vielen Gängen, in dem schließlich nur einer zum Ziel führt. Die richtige Entscheidung wird am Ziel meist mit Futter belohnt. Höhere Anforderungen stellen Labyrinthe, die mehrfache Entscheidungen hintereinander erforderlich machen (Abb. 71 – Schimpansin Julia wählt zwischen zwei deutlich getrennten Systemen). Ihre Konstruktion kann zwei- oder dreidimensionale Systeme beinhalten. Labyrinthversuche spielen in der Lernpsychologie eine wichtige Rolle. Für die Lernleistung ist die Zahl der Durchgänge wichtig, die ein Tier benötigt, um das Ziel in einem fehlerfreien Durchgang zu erreichen (BUCHHOLTZ 1973, RENSCH 1973).

maze experiment □ The T or Y maze are the simplest examples in which learning of the correct path at each choice point eventually brings the animal to its goal. This goal is usually rewarded with food. The difficulty of the maze is increased when the animal is faced with a series of decisions in two or three dimensional systems (Fig. 71 – ♀ chimpanzee Julia choosing between two different systems). Maze experiments are important for learning psychologists who access learning capacity according to the complexity of the maze which can be executed without error (BUCHHOLTZ 1973, RENSCH 1973).

71

expérience du labyrinthe □ Dans le cas le plus simple, le labyrinthe est en forme de T ou de Y. L'animal doit apprendre à choisir la bonne direction au niveau d'une bifurcation ou du croisement de plusieurs couloirs dont un seul mène au but. Le choix correct est en général récompensé par de la nourriture placée à l'arrivée. Des labyrinthes plus compliqués exigent de l'animal plusieurs décisions consécutives (Fig. 71 – la ♀ Chimpanzé Julia choisit entre deux systèmes bien différenciés). Leur topographie peut être bi- ou tridimensionnelle. Les expériences en labyrinthe sont un moyen d'investigation important en psychologie de l'apprentissage. La performance se mesure au nombre de passages nécessaires pour trouver le but sans erreurs (BUCHHOLTZ 1973, RENSCH 1973).

Lächeln □ Eine von allen Menschen bekannte Ausdrucksform, die somit zu den → *Universalien* gehört (KOEHLER 1954 a + b). EIBL-EIBESFELDT (1973 a) konnte inzwischen nachweisen, daß auch taubblind geborene Kinder ebenso lächeln wie normale und es sich somit um ein angeborenes Verhaltensmuster handelt. Lächeln ist u. a. auch eine *Beschwichtigungsgebärde* zur Abwendung aggressiven Verhaltens, und man lächelt auch beim zwischenmenschlichen Sich-einander-Begrüßen (*Begrüßungslächeln*). Auf der anderen Seite ist ironisches Lächeln offenbar aggressionsauslösend. Weiter ist Lächeln eine Vorstufe zum → *Lachen* (EIBL-EIBESFELDT 1970, 1972, 1973 b, VAN HOOFF 1972, SCHEYGROND *et al.* 1973).

smiling □ An expression common to all humans which consequently belongs to → *universals* (KOEHLER 1954 a + b). EIBL-EIBESFELDT (1973 a) has demonstrated that children born blind and mute also smile and that it is an innate behaviour pattern. Smiling serves among other things as an appeasement display to divert aggressive behaviour. Smiling also occurs in human salutation (*greeting smile*). On the other hand, ironic smiling may elicit aggressive gestures. Furthermore smiling may be considered a preliminary of → *laughing* (EIBL-EIBESFELDT 1970, 1972, 1973 b, VAN HOOFF 1972, SCHEYGROND *et al.* 1973).

sourire □ Forme d'expression faciale universelle (→ *universaux*) existant dans toutes les cultures humaines (KOEHLER 1954 a + b). EIBL-EIBESFELDT (1973 a) a pu démontrer que les enfants nés sourds et aveugles sourient de la même manière que les enfants normaux: il s'agit donc d'un comportement inné. Le sourire exerce une fonction d'apaisement et il empêche l'agression. On sourit également en se saluant (*sourire de salut*). D'autre part, le sourire ironique peut déclencher des réactions agressives. Enfin, le sourire est également une amorce du → *rire* (EIBL-EIBESFELDT 1970, 1972, 1973 b, VAN HOOFF 1972, SCHEYGROND *et al.* 1973).

Lachen □ Eine von allen Menschen bekannte Ausdrucksform, die somit zu den → *Universalien* gehört. Lachen beim Spiel ist ein Zeichen freundlicher Absicht; außerdem ist Lachen eine milde Form von Aggression (Auslachen). EIBL-EIBESFELDT (1972, 1973) spricht vom Lachen als ritualisierter Aggressionsform. – »Lachen« nennt man auch den Alarmruf der Silbermöwe (TINBERGEN 1958).

laughing □ An expression common to all humans – → *universals*. Laughing during playing is an indication of friendly intention, but may also represent mild aggression. EIBL-EIBESFELDT (1972, 1973) considers laughing as ritualized aggression. – The alarm call of the Herring Gull is also called laughing (TINBERGEN 1958).

rire □ Forme d'expression universelle existant dans toutes les cultures humaines (→ *universaux*). Au cours du jeu, le rire est l'indice d'une intention amicale, mais par ailleurs, il a valeur d'une agression atténuée. EIBL-EIBESFELDT (1972, 1973) considère le rire comme une forme d'agression ritualisée. – D'autre part, on appelle »rire«, le cri d'alarme du Goéland argenté (TINBERGEN 1958).

Lahmstellen → Verleiten

Lalldialog → Lallen

E. babbling dialogue □ F. dialogue de lallation

Lallen □ Lautäußerungen ohne Sinnesbedeutung für die Wortsprache in der Vorsprachenphase beim Kind. Die Lallphase beginnt etwa im Alter von $2^1/_2$ Monaten (DECROLY 1934, PIERON 1968). Das Kind lallt vor allem beim Erkundungsverhalten und Spielen mit Objekten. Der Säugling führt mit Erwachsenen einen freundlichen *Lalldialog*, der meist vom Kind ausgelöst wird. Das vor der Wortsprache auftretende Lallen scheint eine angeborene Fähigkeit zu sein und scheint in den verschiedenen Kulturen gleich.

babbling □ Sounds made by the child of pre-speech age. Babbling begins at about $2^1/_2$ months and occurs mainly during explorations and while playing with objects (DECROLY 1934, PIERON 1968). The infant often carries on a *babbling dialogue* with an adult, and this is usually elicited by the infant. Babbling appears to be an innate trait which apparently is identical from culture to culture.

lallation □ Vocalisations caractéristiques de la période préverbale de l'enfant, n'ayant pas de signification dans le langage parlé (DECROLY 1934, PIERON 1968). Cette phase commence à l'âge de 2 mois $^1/_2$. La lallation accompagne surtout le comportement d'exploration et le jeu avec des objets. Le nourrisson établit avec les adultes un *dialogue de lallation* amical qui est le plus souvent déclenché par l'enfant. Il semble que la lallation précédant l'apparition du langage parlé soit une capacité innée, identique dans toutes les cultures humaines.

Landmarken □ Markante Gegenstände in der Natur, wie einzeln stehende Bäume, Baumgruppen usw., die der optischen Orientierung dienen. Auch viele Insekten benutzen Landmarken zu ihrer Orientierung, und VON FRISCH und LINDAUER (1954) konnten nachweisen, wie bedeutsam Landmarken für das Auffinden des Weges sein können, indem z. B. bei Bienen ein Waldrand als Leitlinie wichtiger ist als der Himmelskompaß.

landmarks □ Natural objets such as trees standing alone, tree groups which are useful for optical orientation. Many insects use landmarks in their orientation, and VON FRISCH and LINDAUER (1954) studied the significance of this means of orientation in bees, e. g., the forest edge rather than the sky compass may be used as a reference.

repères topographiques □ Repères naturels, comme par exemple un arbre isolé, un groupe d'arbres etc., servant à l'orientation visuelle. De nombreux Insectes utilisent de tels repères pour leur orientation et VON FRISCH et LINDAUER (1954) ont démontré leur importance chez les Abeilles pour lesquelles la lisière d'une forêt peut être un indice plus important pour le retour au gîte que la position du soleil.

Langzeitgedächtnis → Gedächtnis

E. long-term memory □ F. mémoire à long terme

Laß-los-Ruf oder **-Reaktion** □ Bei Anuren, wenn ein ♂ ein anderes ♂ oder ein nicht paarungsbereites ♀ mit dem → *Klammergriff* ergreift; manche Arten zittern, was ebenfalls ein Loslassen auslöst.

let-go call or **response** □ In frogs and toads when a ♂ clasps another ♂ or a non-receptive ♀, this call or vibrations of the body are emitted.

cri ou **réaction spasmodique** □ Chez les Anoures, cri ou réaction spécifique émis lorsqu'un ♂ a serré un autre ♂ ou une ♀ non réceptive; quelques espèces présentent des mouvements de vibration.

Latenzzeit → Reiz

E. latency □ F. temps de latence

Lauerjäger □ Ein Tier, das seine Beute nicht aktiv sucht, sondern von einer Warte aus erwartet bzw. auflauert.

ambush predator □ An animal that awaits its prey and ambushes rather than stalks it.

prédateur à l'affût □ Un animal qui ne recherche

pas activement sa proie, mais attend celle-ci sans bouger, embusqué sur un perchoir ou dans une cachette.

Laufscharren → Scharren
E. alternating scratching □ F. grattage alternatif du sol

Laufschlag □ Im Paarungszeremoniell bei mehreren Unterfamilien der Bovidae schlägt das ♂ beim Verfolgen mit dem gestreckten Vorderlauf nach dem ♀ (Abb. 72 – *Litocranius walleri*). Unmittelbar nach diesem Laufschlag folgt meist der Ansprung zur Kopulation. Den Laufschlag zwischen die Hinterbeine nannte WALTHER (1958) zunächst »Laufeinschlag«. Da aber beide Formen innerhalb einer Art vorkommen können, wird nur noch Laufschlag verwendet (WALTHER 1966, 1968).

72

foreleg-kick □ During courtship in several subfamilies of Bovidae the ♂, when pursuing the ♀, strikes her with his outstretched forelimbs (Fig. 72 – *Litocranius walleri*). This is usually followed by mounting and copulation. WALTHER (1958) originally distinguished as a separate form the *Laufeinschlag* (striking between the back legs). Since both forms may occur in the same species only *Laufschlag* now is used for both (WALTHER 1966, 1968).

Laufschlag □ Pendant le comportement de parade nuptiale chez plusieurs sous-familles de Bovidae, le ♂ donne à la ♀, lors de la poursuite, des coups avec la patte antérieure tendue (Fig. 72 – *Litocranius walleri*). Immédiatement après ce coup de patte se produit la monte pour l'accouplement. Un coup de patte administré entre les pattes postérieures de la ♀ a été initialement appelé *Laufeinschlag* par WALTHER (1958), mais étant donné que les deux formes peuvent se produire à l'intérieur de la même espèce, WALTHER (1966, 1968) n'a conservé que le terme *Laufschlag*. DUBOST (1975), en français, utilise également le terme de *Laufschlag*.

Lausen → Körperpflegehandlungen
E. grooming □ F. épouillage

Lautinventar → Ethogramm
E. vocal inventory □ F. vocabulaire sonore

Lebensraum → Aktionsraum

Leerlaufhandlung □ Wenn eine Instinkthandlung längere Zeit nicht zur Auslösung gelangt, erniedrigt sich der Schwellenwert der zu ihrer Auslösung notwendigen Reize. Die Schwellenerniedrigung kann insofern einen Grenzwert erreichen, daß die Reaktion schließlich ohne nachweisbaren Reiz zum Durchbruch kommt, d. h. im Leerlauf abläuft. Seit LORENZ (1937) das Beispiel seines Stars beschrieb, sind inzwischen viele solcher Leerlaufhandlungen bekannt geworden. BASTOCK, MORRIS und MOYNIHAN (1953) haben vorgeschlagen, den Begriff durch *Überflußhandlung* zu ersetzen, da man ja nie genau das absolute Fehlen eines auslösenden Reizes nachweisen könne. Da aber der Begriff Leerlauf nichts impliziert, können wir ihn beibehalten.

vacuum activity □ When an instinctive behaviour has not occurred for an extended period of time, the threshold value for releasing stimuli is lowered. This reduction in threshold may reach a point at which the response occurs without a demonstrable stimulus, and hence the term vacuum-activity. Since LORENZ (1937) first described the phenomenon in starlings many other cases have been described. BASTOCK, MORRIS and MOYNIHAN (1953) have suggested that the term be replaced with *overflow activity* since one cannot be certain that no stimulus is present. Since the term vacuum activity does not refer to stimuli there is no reason to discard it.

activité à vide □ Lorsqu'une activité instinctive n'a pas été déclenchée pendant un certain temps, le seuil des stimuli nécessaires à son déclenchement s'abaisse. Le processus peut aboutir à une situation limite dans laquelle la »réaction« se déroule sans que la présence d'un stimulus puisse être démontrée, c'est une activité à vide. Depuis les observations de LORENZ (1937) sur son Etourneau, de nombreuses observations d'activités à vide ont été faites. BASTOCK, MORRIS et MOYNIHAN (1953) ont proposé le terme de »débordement« puisqu'il n'est jamais possible de prouver l'absence complète d'un stimulus déclencheur. Mais comme la notion d'activité à vide n'a pas d'implication théorique, ce terme peut être conservé.

Lerndisposition □ Erblich vorbestimmtes Lernvermögen, welches die funktionelle Potenz der Informationsaufnahme und ihrer Verwertung bestimmt, d. h. die genetisch bedingte Lerndisposition ist für jedes Individuum vorbestimmt. Es werden dadurch den verschiedenen Lernmöglichkeiten eindeutige und nicht überschreitbare Grenzen für das Einzeltier gesetzt (BUCHHOLTZ 1973, RENSCH 1973).

learning disposition □ That genetic aspect of learning capacity which governs the functional potential for information uptake and analysis. This learning disposition is an individual characteristic which sets definite limits to the various types of learning in an individual (BUCHHOLTZ 1973, RENSCH 1973).

prédisposition à l'apprentissage □ Capacité héréditaire d'apprentissage qui détermine le potentiel individuel de réception et d'analyse des informations. La prédisposition génétique à l'apprentissage est propre à chaque individu. Elle impose des limites définies et infranchissables aux différentes modalités d'apprentissage (BUCHHOLTZ 1973, RENSCH 1973).

Lernen □ In der Ethologie und in der Humanpsychologie sehr weit gefaßter Begriff. Der Lernvorgang zergliedert sich in Teilmechanismen, wie Aufnahme und Speicherung von Informationen, welche für eine adäquate Handlung abrufbar sind (→ *Gedächtnis*) und auch über Generationen weitergegeben werden können (FOPPA 1968, BUCHHOLTZ 1973, RENSCH 1973, ANGERMEIER 1976). Das *Lernrezept* beim Gimpel, *Pyrrhula pyrrhula,* z. B. besteht darin, daß die Jungen nur den Gesang des Vaters nachahmen. NICOLAI (1959) ließ ein Gimpel-♂ von Kanarienvögeln aufziehen, und es sang danach wie ein Kanarienvogel und gab dies auch an seine Jungen und über diese an die Enkel weiter. → *Tradition.*

learning □ A broad concept of ethology and psychology which may be divided into various phases such as information uptake, storage and retrieval. The latter phase is essential if the information is to be effectively used, e. g., transmitted to later generations (FOPPA 1968, BUCHHOLTZ 1973, RENSCH 1973, ANGERMEIER 1976). In the bullfinch, *Pyrrhula pyrrhula,* the *learning program schedule* is the young birds' habit of imitating only the song of their father. NICOLAI (1959) once had a bullfinch ♂ raised by canaries. This bird later sang like a canary and passed on this song to it's young and these in turn passed it on to their young. → *tradition.*

apprentissage □ En éthologie et en psychologie humaine, terme d'une très grande extension. Le processus d'apprentissage peut être décomposé en plusieurs mécanismes partiels tels que la réception et le stockage des informations auxquelles l'individu pourra faire appel pour l'exécution d'une action adéquate (→ *mémoire*) et qui pourront même être transmises d'une génération à l'autre (FOPPA 1968, BUCHHOLTZ 1973, RENSCH 1973, ANGERMEIER 1976). Chez le Bouvreuil, *Pyrrhula pyrrhula,* par exemple, la *formule d'apprentissage* consiste pour les jeunes à n'imiter que le chant de leur père. NICOLAI (1959) a fait élever un Bouvreuil ♂ par des Canaris; ce Bouvreuil chanta ensuite comme un Canari et transmit par imitation son chant à ses jeunes de la première et de la deuxième génération. → *tradition.*

lernpsychologisches Modell → Aggression
E. learning-theory model □ F. modèle de l'agression acquise

Lernrezept → Lernen
E. learning program schedule □ F. formule d'apprentissage

Lestobiose □ Das Leben von kleinen Arten sozialer Insektenkolonien in den Nestwänden weitaus größerer Arten. Die kleine Art dringt in die Kammern der größeren Art ein, um dort die Brut zu erbeuten oder in deren Nahrungsspeichern zu räubern. Am besten bekannt ist das Beispiel der Diebsameisen der Gattung *Solenopsis* (Übersicht bei WILSON 1971, 1975).

lestobiosis □ The relation in which colonies of a small species of social insects nest in the walls of the nests of a larger species and enter the chambers of the larger species to prey on brood or to rob the food stores. The best known example are thief ants of the genus *Solenopsis* (overview by WILSON 1971, 1975).

lestobiose □ La vie des colonies de petites espèces d'Insectes sociaux dans les parois du nid d'espèces plus grandes. Les petites espèces pénètrent alors dans les chambres de la grande espèce pour y effectuer la prédation parmi la progéniture et pour faire des dégâts dans les stocks de nourriture. L'exemple le plus connu sont les Fourmis voleuses du genre *Solenopsis* (vue d'ensemble dans WILSON 1971, 1975).

Lex-Heinze-Stimmung → Mit-Tretinteresse
E. Lex-Heinze-mood – reaction of a ♀ against mating □ F. motivation Lex-Heinze – jalousie d'une ♀ vis-à-vis d'un accouplement

Lichtkompaßorientierung → Orientierung
E. light compass orientation □ F. orientation sur la boussole lumineuse

Lichtrückeneinstellung → Orientierung
E. dorsal light response □ F. réaction photo-dorsale

Lichtrückenorientierung → Orientierung
E. dorsal light orientation □ F. orientation photo-dorsale

Lichtstreß □ Mit Hilfe einer starken Lichtquelle als Reiz im Zentrum des Behälters untersuchte RYALL (1958) das »emotionale« Verhalten von Ratten, welches er an der Frequenz von Defäkationen, Putzbewegungen als Übersprungbewegungen, Umherlaufen usw. überprüfte. Mit Hilfe von verschiedenen Drogen konnte er die Lichtstreßwirkung beliebig hemmen bzw. fördern.

light stress □ RYALL (1958) studied emotional behaviour in rats by using a bright light in the center of the cage. Reaction to this light stress, measured as defecation frequency, displacement grooming, ambulation, etc., could be inhibited or facilitated with various drugs.

stress lumineux □ A l'aide d'une source lumineuse puissante installée comme stimulus au centre d'une cage, RYALL (1958) a effectué des recherches sur le comportement »émotionnel« des Rats, mesuré par la fréquence des défécations, des mouvements de toilette substitutifs et de l'activité locomotrice. A l'aide de diverses drogues, il a pu à volonté inhiber ou stimuler l'effet stressant de la lumière dans cette situation.

lineares Rangordnungssystem → Rangordnung
E. linear hierarchy, linear rank order □ F. hiérarchie linéaire

Locken → Bodenpicken → Futterlocken
E. enticement □ F. comportement d'appel
Lorenz-Freud'sches Triebmodell → Aggression
E. Lorenz'-Freudian drive model □ F. modèle de l'agression pulsionnelle
Lotsenfische □ Fische der Gattung *Naucrates,* die man fast nie allein sieht. Begleiten sie räuberische Haie, so schwimmen sie meist in Höhe der Rückenflossen. Lotsenfische putzen auch ihre Begleiter, wie z. B. das Maul der Manta.
pilot fish □ Fish of the genus *Naucrates* which accompany sharks, mantas, etc., and serve as »cleaners« for them; they are often seen swimming near the dorsal fin of sharks.
poissons pilotes □ Poissons du genre *Naucrates* qui ne se rencontrent presque jamais seuls. Si ces Poissons accompagnent des Requins prédateurs, ils nagent au niveau de la nageoire dorsale. Les Poissons pilotes peuvent aussi accomplir la fonction de nettoyage de leurs compagnons, par exemple au niveau de la bouche de la Raie manta.
Luftkopulation □ Die Begattung im Fluge, eine bei vielen Insekten häufige Ercheinung (Abb. 73 A – Libellulidae der Gattung *Orthetrum*). Bei Vögeln konnte die Begattung im Fluge erstmals vom Mauersegler (Abb. 73 B – *Apus apus*) nachgewiesen werden (ROTHGÄNGER 1973). Inzwischen ist dieses Verhalten auch von anderen Seglern, *Apus melba* und *Apus pallidus,* bekannt (eigene Beob.).

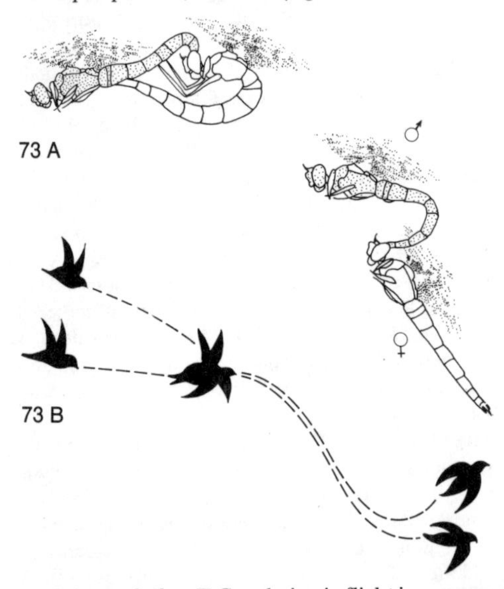

aerial copulation □ Copulation in flight is common in insects (Fig. 73 A – Libellulidae of the genus *Orthetrum*). Evidence for aerial copulation in birds has first been presented by ROTHGÄNGER (1973) in the swift (Fig. 73 B – *Apus apus*). In the meantime, this behaviour has also been observed in *Apus melba* and *Apus pallidus* (pers. obs.).
accouplement aérien □ Phénomène répandu chez de nombreux Insectes (Fig. 73 A – Libellulidae du genre *Orthetrum*). Parmi les Oiseaux, l'accouplement pendant le vol a été signalé pour la première fois chez le Martinet noir *Apus apus* (Fig. 73 B) par ROTHGÄNGER (1973). Depuis, ce comportement a également été reconnu chez d'autres Martinets, par exemple *Apus melba* et *Apus pallidus* (obs. pers.).
Lunarorientierung = Mondorientierung → Orientierung
E. lunar orientation □ F. orientation lunaire
Lunarperiodik □ Bei vielen, vor allem in der Gezeitenzone lebenden marinen Organismen, deren Fortpflanzungszyklus auf die Zeiten der Spring- und Nipptiden (Springfluten) beschränkt ist und mit den Mondphasen teilweise sehr genau übereinstimmt. Die dabei auftretenden physiologischen Rhythmen entsprechen also den lunaren Zyklen von ungefähr 29,5 Tagen, in anderen Fällen den semilunaren Zyklen von 14–15 Tagen (KORRINGA 1957, BÜNNING 1963).
lunar periodicity □ In many marine animals living in the tidal zone the reproduction cycles are limited to the periods of spring tide and may coincide precisely with the moon phases. The physiological rhythms involved then correspond to the lunar cycles of about 29.5 days, in other cases to the semilunar cycles of 14 to 15 days (KORRINGA 1957, BÜNNING 1963).
périodicité lunaire □ Chez de nombreux animaux marins vivant dans la zone de balancement des marées, le cycle reproducteur est limité à la période de grande marée; ce cycle correspond souvent d'une manière précise aux phases de la lune. Les rythmes physiologiques en jeu coincident soit avec le cycle lunaire de 29,5 jours environ, soit avec le cycle semilunaire de 14 à 15 jours (KORRINGA 1957, BÜNNING 1963).

M

Mach-mit-Verhalten → Stimmungsübertragung
E. allomimetic behaviour □ F. comportement allomimétique
Magneteffekt □ Bei Untersuchungen von Flossenrhythmen bei Fischen stellte VON HOLST (1937, 1939) fest, daß sich Flossenrhythmen einander etwa gleich stark beeinflussen können. Sehr oft behält jedoch eine Flosse (oder ein Flossenpaar) ihren konstanten Rhythmus bei und zwingt der anderen Flosse ihren Rhythmus auf, was VON HOLST Magneteffekt nannte.
magnet effect □ The research of VON HOLST (1937, 1939) on fin rhythms in fishes suggests that each fin is capable of influencing adjacent fins to adapt its rhythm. In many cases, however, a certain fin (or pair of fins) maintains a constant rhythm and induces the other fins to follow this rhythm; he termed this ma-

gnet effect.

effet inducteur □ VON HOLST (1937, 1939), au cours de ses recherches sur les rythmes natatoires des Poissons, a trouvé que les rythmes des différentes nageoires peuvent s'influencer mutuellement avec une même intensité; pourtant, une nageoire (ou une paire de nageoires) déterminée peut maintenir son rythme propre, entraînant ainsi le rythme d'une autre nageoire.

Magnetfeldorientierung □ Die Orientierung nach dem Erdmagnetfeld bei Ausfallen des Himmelskompasses. Der Mechanismus der Magnetfeldorientierung ist nur mangelhaft bekannt, und spezifische Magnetorezeptoren sind noch nicht entdeckt worden. SCHNEIDER (1961) gelang es, die Orientierung von Maikäfern, *Melolontha vulgaris,* mit Hilfe von Hufeisenmagneten zu irritieren; im normalen Magnetfeld behielten sie ihre einmal eingeschlagene Flugrichtung bei. Termiten der Gattungen *Reticulitermes, Heterotermes* und *Coptotermes* richten ihren Galerienbau nach den Feldlinien des Erdmagnetfeldes aus (BECKER 1971), und Bienen orientieren danach ihren Wabenbau; außerdem kann der → *Schwänzeltanz* durch das Erdmagnetfeld gestört werden. Weitere zahlreiche Beispiele kennen wir von Vögeln, und überzeugende Belege wurden vor allem von MERKEL und WILTSCHKO (1965) sowie WILTSCHKO und WILTSCHKO (1975 a + b) erbracht (MARTIN und LINDAUER 1973 – ausführliche Abhandlung mit wichtigen Literaturangaben).

magnetic field orientation □ Animal orientation using the earth's magnetic field when sky-compass orientation fails. The mechanism for magnetic field orientation is poorly understood, and specialized magnetic receptors have not been discovered. SCHNEIDER (1961) succeeded in displacing the orientation of lady bugs, *Melolontha vulgaris,* with horse shoe magnets; in a normal magnetic field they retained their initial flight direction preference. Termites in the genus *Reticulitermes, Heterotermes* and *Coptotermes* construct their chambers along lines of the earth's magnetic field (BECKER 1971); bees also construct their combs in this fashion. Moreover, the → *waggle dance* can be influenced by the magnetic field. Numerous additional examples have been reported in birds, with especially convincing results from MERKEL and WILTSCHKO (1965), WILTSCHKO and WILTSCHKO (1975 a + b) – (see MARTIN and LINDAUER 1973 for detailed discussion and literature).

orientation par rapport au champ magnétique terrestre □ Ce mode d'orientation est possible en l'absence d'indication de la »boussole céleste«. Mais les mécanismes en sont encore mal connus: en particulier, on n'a pas encore découvert de récepteurs sensoriels magnétospécifiques. SCHNEIDER (1961) a réussi à perturber l'orientation de Hannetons communs, *Melolontha vulgaris,* à l'aide d'aimants en fer à cheval; mais dans le champ magnétique naturel, ces Insectes ne modifient pas une direction de vol antérieurement choisie. Chez les Termites (genres *Reticulitermes, Heterotermes, Coptotermes*), la construction des galeries suit les lignes de force du champ magnétique terrestre (BECKER 1971). Celui-ci influence également la construction des rayons de la ruche d'Abeilles; en outre, la → *danse frétillante* peut être perturbée, dans sa composante directionnelle, par ce même agent physique. De nombreux autres exemples concernent les migrations des Oiseaux. Des preuves convaincantes ont été surtout établies par les travaux de MERKEL et WILTSCHKO (1965) et par ceux de WILTSCHKO et WILTSCHKO (1975 a + b). Voir la revue de MARTIN et LINDAUER (1973) pour une exposition détaillée des phénomènes, ainsi que pour des données bibliographiques importantes.

Männerbalzplatz □ Bekannte Plätze, auf denen die ♂♂ zur gemeinsamen Balz zusammenkommen und untereinander wilde Kämpfe ausführen. Die ♀♀ kommen meist zur Begattung heran, und bei solchen Arten fehlt meist auch die Paarbildung *sensu stricto.* Aus der Vogelwelt kennen wir Männerbalzplätze von vielen Vogelarten, insbesondere von Kampfläufern, *Philomachus pugnax,* von Birkhähnen, *Lyrurus tetrix,* und Paradiesvögeln. Sie gibt es aber auch bei Säugern, manchen Bovidae und bei einigen Fledermausarten.

lek, male courtship arena □ Sites where ♂♂ congregate for social courtship displays and fighting. The ♀♀ come to these sites for mating; in most species a true pair bond does not develop. Many bird species including ruffs, *Philomachus pugnax,* grouse, *Lyrurus tetrix,* and paradise birds, as well as certain mammals including various bovids and bats utilize leks or courtship arenas. The area defended by individual ♂♂ within a lek or communal display area is called *court. Lek* was first described by SELOUS (1907).

arène, terrain de parade nuptiale collective □ Endroits précis où les ♂♂ se réunissent pour une parade nuptiale en commun et sur lesquels ils effectuent entre eux d'importants combats. Les ♀♀, deleur côté, s'approchent pour l'accouplement et, chez ces espèces, le processus de la formation des couples *sensu stricto* fait généralement défaut. Parmi les Oiseaux, nous connaissons de telles arènes surtout chez les Chevaliers combattants, *Philomachus pugnax,* les Coqs de Bruyère, *Lyrurus tetrix,* ainsi que chez les Oiseaux Paradisiers. Parmi les Mammifères, de tels terrains de parade nuptiale collective sont connus chez un certain nombre de Bovidae et quelques espèces de Chauves-Souris. Le terme anglais *lek* (SELOUS 1907) est fréquemment utilisé en français (BROSSET 1974).

Manteln □ Ausdruck aus der Jägersprache für die Haltung der Greifvögel, wenn sie beim Rupfen die weit gespreizten Flügel über die Beute halten (Abb. 74 – *Accipiter gentilis* mit *Sciurus vulgaris* als Beute).

»manteln« □ A term used by falconers for the pos-

74

ture of a trained bird as it spreads its wings wide when about to pluck the prey (Fig. 74 – *Accipiter gentilis* with *Sciurus vulgaris* as a prey).

se couvrir □ Expression du langage des fauconniers désignant l'attitude des Rapaces tenant leur proie avec les griffes, les ailes étendues en forme de toit (Fig. 74 – *Accipiter gentilis* avec *Sciurus vulgaris* comme proie).

Markierung □ Die Kennzeichnung von Wegen und Abgrenzung eines Reviers; das Markieren von Geschlechtspartnern und Gruppenmitgliedern. → *Duftmarkierung,* → *Gruppenduft,* → *Harnmarkierung,* → *Harnspritzen,* → *Reviermarkierung.*

marking □ The designation of trails and delineation of a territory; marking of the mate or members of a group. → *scent marking,* → *group smell,* → *urine marking,* → *urine spraying,* → *territory marking.*

marquage □ Signalisation des chemins et délimitation d'un territoire; marquage du partenaire sexuel ou des membres appartenant au même groupe. → *marquage odorant,* → *odeur sociale,* → *marquage à l'urine,* → *aspersion d'urine,* → *marquage du territoire.*

Markierungstrommeln → Duftmarkierung
E. marking drumming □ F. tambourinage de marquage

Massenbewegung □ Bei Fisch-, Vogel- und Säugerembryonen beobachten wir ein vom Kopf bis zum Körperende voranschreitendes Heranreifen der Bewegungen. Es handelt sich dabei um regelmäßig wiederkehrende, von kurzen Ruhepausen unterbrochene Bewegungsausbrüche (COGHILL 1929).

»embryonic mass action« □ In embryos of fish, birds and mammals one observes a progressive maturing of motor function. Movement first occurs at the rostral end of the embryo and later progresses to the caudal end; mass action is characterized by regularly alternating periods of movement and rest (COGHILL 1929).

»mouvement d'ensemble progressif« □ Chez les embryons des Poissons, des Oiseaux et des Mammifères, il existe une maturation des mouvements progressant de la tête vers la partie caudale. Ces phases motrices sont régulièrement récurrentes et sont interrompues par de brèves phases de repos (COGHILL 1929).

Massensiedlungseffekt □ Erscheinungen wie die herabgesetzte Fruchtbarkeit, die bei Übervölkerung eines Raumes zu beobachten ist. Mehlkäfer, *Tenebrio molitor,* fressen bei Übervölkerung ihre Eier auf, bei Säugern führt sie zu Streß und Populationszusammenbruch, noch lange bevor sich Nahrungsmangel bemerkbar macht. Hervorgerufen wird der Streß durch zu hohe Populationsdichte, verbunden mit einer erhöhten adreno-corticalen Aktivität und zunehmenden antagonistischen Auseinandersetzungen (LOUCH 1956, LEVI 1971, VON HOLST 1973). In diesem Sinne hat dieser Streß einen dichteregulierenden Wert.

overpopulation effect □ Various phenomena, observed in overcrowded populations, lead to lower reproductive rates. In the flour beetle, *Tenebrio molitor,* overcrowding results in eating of their own eggs while mammals may undergo stress and population collapse long before the food supply is exhausted. Overpopulation is thought to evoke stress coupled with elevated adreno-cortical activity and increasing agonistic confrontations (LOUCH 1956, LEVI 1971, VON HOLST 1973). Such stress may have a density regulating function.

effet de surpopulation □ Ce terme désigne certains phénomènes, comme par exemple une diminution de la fertilité lors d'une surpopulation dans un espace donné. Dans une telle situation, chez *Tenebrio molitor,* les imagos mangent leurs propres œufs. Chez les Mammifères, ceci conduit vers une situation de stress et un effondrement de la population, souvent bien avant qu'un véritable manque de nourriture se fasse sentir. Ce stress est déclenché par une trop grande densité de population liée à une activité adréno-corticale très élevée et une augmentation des comportements agonistiques (LOUCH 1956, LEVI 1971, VON HOLST 1973). Dans ce sens, un tel stress possède donc une signification de régulation de la densité de population.

Maulbrüter □ Fische, vor allem aus der Familie der Buntbarsche (Cichlidae), bei denen die ♀♀ die eben gelegten Eier ins Maul nehmen und dort ausbrüten. Die Besamung der Eier erfolgt im Maul des ♀ durch nachträgliche Aufnahme des Spermas; um das zu erreichen, wurden innerhalb der Familie verschiedene Methoden entwickelt (WICKLER 1962, zusammenfassend dargestellt). Auch die Jungen finden später im Maul der Mutter Platz und Schutz.

mouth breeding fish □ Fish, especially of the Cichlid family, in which the ♀♀ take the eggs in their mouth following spawning and incubate them there. The fertilization of the eggs occurs in the mouth of the ♀ by subsequent uptake of sperm; for this different behaviour patterns have evolved within the Cichlid family (summarized by WICKLER 1962). The fry may return to the mouth of the ♀ for protection as well.

incubateurs buccaux □ Poissons appartenant surtout à la famille des Cichlidae chez lesquels les ♀♀, juste après la ponte, prennent les œufs dans leur bouche pour les incuber. La fécondation des œufs s'effectue dans la bouche de la ♀ par absorption ultérieure du sperme; pour cela, des méthodes différentes ont été développées à l'intérieur de cette famille (vue d'ensemble chez WICKLER 1962). Les alevins trouvent également abri et protection dans la bouche de leur mère.

Maulkampf □ Antagonistische Auseinandersetzungen bei Fischen mit Hilfe des Maules. Die Substratbrüter unter den Cichliden packen sich dabei fest an den Lippen und ermitteln in einem Schiebekampf oder durch *Maulzerren* den Sieger (Abb. 75 A – *Tilapia zillii*). Einen ähnlichen Kampf kennen wir

combat buccal □ Comportement antagoniste et combat chez un certain nombre de Poissons à l'aide de leur bouche. Chez les Cichlidae, les espèces qui n'incubent pas dans la bouche se tiennent face à face et se saisissent mutuellement par la bouche en poussant et en tirant pour désigner le vainqueur (Fig. 75 A – *tiraillement mutuel de la bouche* chez *Tilapia zillii*). Nous connaissons un combat semblable chez *Emblemaria pandionis* (Blennioidei, Salariinae). Chez les Cichlidae incubateurs buccaux, les rivaux ne se saisissent pas par la bouche, mais se tiennent face à face et s'administrent des claques avec la bouche largement ouverte (Fig. 75 B – *choc à bouche ouverte* chez *Tilapia nilotica*) APFELBACH (1967).

Maulklatschen → Maulkampf

E. mouth clashing □ F. choc à bouche ouverte

75 A

75 B

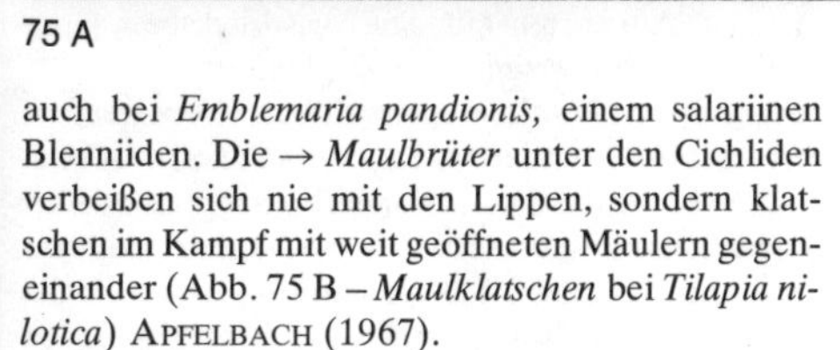

auch bei *Emblemaria pandionis,* einem salariinen Blenniiden. Die → *Maulbrüter* unter den Cichliden verbeißen sich nie mit den Lippen, sondern klatschen im Kampf mit weit geöffneten Mäulern gegeneinander (Abb. 75 B – *Maulklatschen* bei *Tilapia nilotica*) APFELBACH (1967).

mouth fight □ Antagonistic behaviour pattern in fish using their mouths for fighting. Substrate nesting Cichlids fight by grasping each other lip to lip and wrestling and shoving back and forth (Fig. 75 A – *mouth pulling* in *Tilapia zillii*). A similar fighting behaviour is found in *Emblemaria pandionis,* a salariine Blenniid. The → *mouth breeding fish* among Cichlids do not lock lips as the substrate nesters do, but rather butt and strike at each other with open mouth during fighting (Fig. 75 B – *mouth clashing* in *Tilapia nilotica*) APFELBACH (1967).

Maulzerren → Maulkampf

E. mouth pulling □ F. tiraillement mutuel de la bouche

Mäuseln □ Aus der Jägersprache übernommene Bezeichnung für den Beutesprung *(Mausfangsprung)* der Caniden und Feliden (Abb. 76 – Chowchow Susi von Konrad Lorenz).

mouse pounce (stiff-legged jumps) □ A term for the pouncing attack of Canids and Felids on mice and other small prey animals (Fig. 76 – Konrad Lorenz's chow-chow Susi).

bond sur la proie □ L'expression allemande *Mäuseln* est tirée du langage des chasseurs qui désigne le comportement des Canidae et Felidae au moment de la capture d'une proie du type Souris ou Mulot (Fig. 76 – Chow-chow Susi de Konrad Lorenz).

Menotaxis → Taxis

76

Merkmalsträger → Funktionskreis
E. perceptual cue bearer □ F. porteur d'information perceptible

Merkwelt → Funktionskreis
E. perceptual field □ F. champ perceptif

Milchtritt □ Bei vielen Säugerordnungen während des Saugens eine Tretbewegung nach der Mutter, meist mit dem Vorderlauf, aber auch der Hinterlauf kann benutzt werden.

treading, kneading □ Milk elicitation movement in many mammals. They use either their front or hind feet to paw at the teat during nursing.

coup de patte de tétée □ Chez de nombreux Mammifères, on observe pendant la tétée une série de coups de patte dirigés vers la mère, le plus souvent avec les pattes antérieures. DUBOST (1974), en français, utilise le terme de *Milchtritt.*

Milieutheorie □ Die Milieutheorie geht davon aus, daß der Mensch sozusagen als *tabula rasa* geboren wird und erst über Lernprozesse seine Verhaltensweisen erwerben muß. Eine in der Pädagogik und den politischen Wissenschaften weit verbreitete Ansicht: nur die Umwelt forme den Menschen (MONTAGU 1962, SKINNER 1971). Engegen diesen Ansichten hat die vergleichende Verhaltensforschung an Tieren und Menschen entdeckt, daß es in genau feststellbaren Bereichen des Verhaltens eine → *Vorprogrammierung* gibt. Die Verhaltensweisen reifen im Laufe der ontogenetischen Entwicklung heran und treten erst dann in Erscheinung, wenn sie benötigt werden. Probleme gibt es, wenn man sich nicht einig ist, was man unter Verhalten versteht.

learning theory □ This theory maintains that the human brain at the time of birth is a *tabula rasa* which then acquires all its information, i. e., behaviour, through learning processes. This opinion is widespread in the fields of education and political science and is reflected in the idea that only the environment forms the man (MONTAGU 1962, SKINNER 1971). Ethology has opposed this view by showing the existence of certain → *pre-programmed behaviour* elements in man and animals which mature during ontogenetic development and make their appearance at the appropriate time. Problems arise when there is no agreement as to what constitutes the behaviour in question.

théorie du milieu □ Empirisme psychologique; la théorie du milieu part du principe que l'Homme naît comme une *tabula rasa* et que l'enfant acquiert ses différents comportements grace à des processus d'apprentissage. Cette opinion est très répandue dans les sciences pédagogiques et politiques. Seul l'environnement formerait l'Homme (MONTAGU 1962, SKINNER 1971). Contrairement à cette opinion, les recherches en éthologie comparée sur les animaux et sur l'Homme ont pu démontrer qu'il existe une → *programmation* dans des domaines bien déterminés du comportement. De tels comportements programmés se développent au cours de la maturation ontogénétique et n'apparaissent qu'au moment adéquat de la vie. Cette divergence d'opinion n'existe que parce qu'il n'y a pas identité de définition du comportement.

Mimese → Schutzanpassung, → Untergrundangleichung
E. camouflage □ F. camouflage

Mimik □ Gesichtsausdrücke bei höheren Säugern und Menschen, die als Ausdrucksbewegung über die Stimmung des entsprechenden Individuums Auskunft geben und als Signal bei der innerartlichen Verständigung verwendet werden. → *Drohmimik.*

facial expression □ Expressive movements in higher mammals and man which convey mood or intentions of an individual and are used as signals in intraspecific communication. → *threat face.*

mimique faciale □ Chez les Mammifères supérieurs et l'Homme, mouvement expressif ayant une valeur d'information sur la motivation de l'individu concerné, et qui est utilisé en tant que signal dans la communication inter-individuelle. → *mimique de menace.*

Mimikerkennen □ Das richtige Deuten und Verstehen einer Mimik, eines Ausdrucksverhaltens. Bei Rhesusaffen konnte ein angeborenes Mimikerkennen nachgewiesen werden, was auch beim Menschen der Fall zu sein scheint, da wir automatisch auf die Mimik eines Mitmenschen ansprechen und es sich um durch angeborene Auslösemechanismen determinierte Antworten auf eine Ausdrucksbewegung handelt. → *Universalien,* → *Drohmimik.*

recognition of facial expression □ The correct interpretation or understanding of facial expressions. In the rhesus monkey this mood perception has an innate basis; in humans certain mechanisms may also exist that determine the response to various facial expressions in fellow humans. → *universals,* → *threat face.*

reconnaissance d'une mimique □ Interprétation correcte et compréhension d'une expression faciale. Chez le Macaque Rhésus, on a pu démontrer qu'il existe une faculté innée de reconnaissance de la mimique; ceci est probablement aussi le cas chez l'Homme, étant donné que nous réagissons d'une manière automatique aux expressions faciales de nos congénères et qu'il s'agit là d'une réponse à un mouvement expressif déterminé par un mécanisme déclencheur inné. → *universaux,* → *mimique de menace.*

Mimikry □ Nachahmung, Täuschung, Signalfälschung und → *Schutzanpassung* in der Natur, und zwar im Tier- wie im Pflanzenreich und sogar in »sinnvollen« Wechselbeziehungen zwischen beiden. Die Nachahmung erfolgt meist in der Form und Färbung.

Viele Insekten tarnen sich als Blüten, wie z. B. verschiedene Fangschrecken. Kleine Zikaden, wie z. B. *Ityrea gregorie* (Fulgoridae), sitzen zu mehreren so an einem Pflanzenstengel zusammen, daß sie einen

77 A

ganzen Blütenstand nachahmen (Abb. 77 A), was man auch *Kollektivmimikry* nennt.
Umgekehrt ahmen Pflanzen Tiere nach und profitieren davon. So ähneln beispielsweise die Blüten vieler Ragwurzarten der Gattung *Ophrys* schwirrenden Bienen- oder Hummel-♀♀ (Abb. 77 B – *Ophrys insectifera*-Blüte); die davon angezogenen ♂♂ bestäuben dann beim Kopulationsversuch diese Blüten (Abb. 77 C – Langhornbienen-♂ der Gattung *Eucera*).
Andere Mimikry-Formen zielen auf Freßschutz: So sind harmlose Zweiflügler (Diptera) und auch manche Lepidoptera, wie z. B. der Hornissenschwärmer, häufig Nachahmer von stachel- und gifttragenden Wespen, Bienen oder Hummeln (Hymenoptera) (Abb. 77 D – *Vespa crabro,* E – *Aegeria apiformis*). Viele Vögel lernen Wespen nach wenigen Freßversuchen an ihrer Färbung zu erkennen, meiden sie fortan und lassen auch die gleich aussehenden Dipteren oder Schwärmer in Ruhe. Die Hymenopterenmimikry bringt ihnen also einen Freßschutz. Man

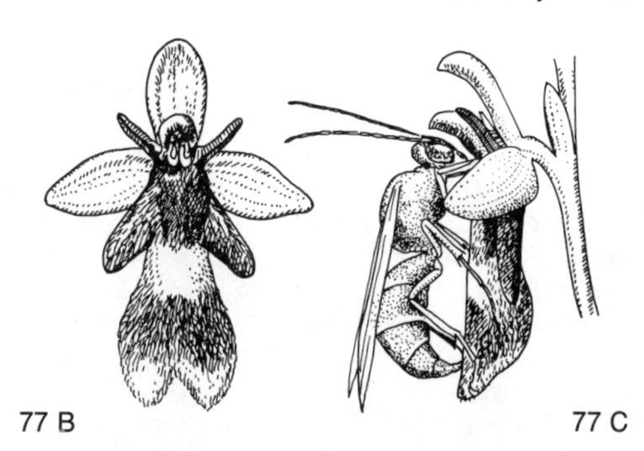
77 B 77 C

spricht in diesem Falle auch von *Bates'scher Mimikry*. Solche Mimikry-Fälle sind im Insektenreich weit verbreitet, einerseits innerhalb der Lepidopteren zwischen verschiedenen Familien, und weiter vor allem zwischen schlechtschmeckenden Coleopteren und genießbaren Schaben, Heteropteren und auch

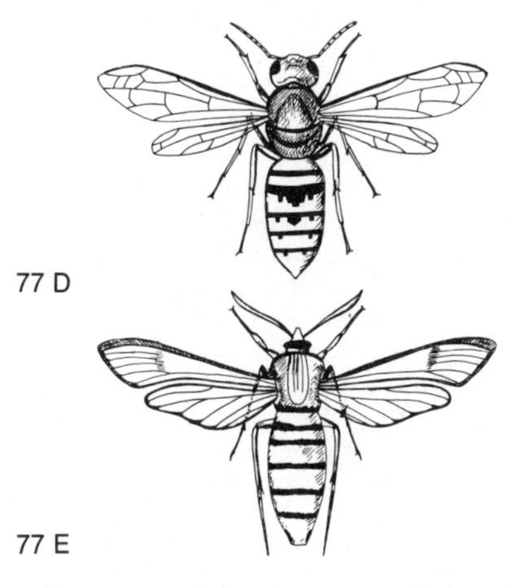
77 D

77 E

Orthopteren (Abb. 77 F – Laufkäfer der Gattung *Tricondyla,* G – Heuschrecke der Gattung *Condylodera*).
Neben Körpermerkmalen wie Form und Färbung

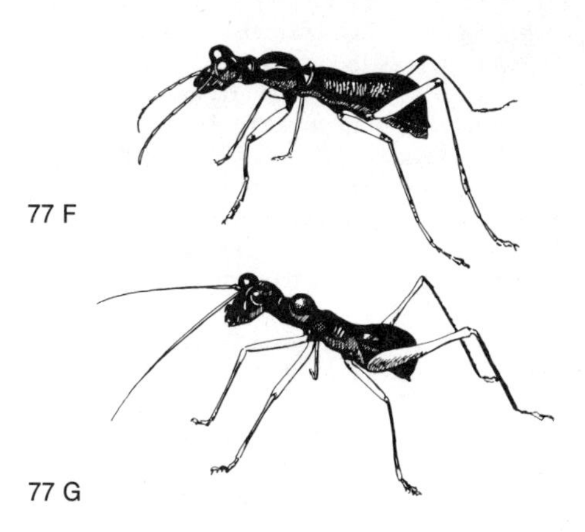
77 F

77 G

gibt es auch Nachahmung von Verhaltensweisen; bei verwandtschaftlich fernstehenden Arten sind normalerweise auch die Verhaltensmechanismen sehr unterschiedlich. Treten aber wirkungsvolle Wechselbeziehungen auf, so kommt es oft auch zur Anpassung oder besser Nachahmung von Verhaltensweisen, was man als *Ethomimikry* bezeichnet *(→ Putzernachahmer)*. Die Nachahmung der Geschlechter untereinander innerhalb ein und derselben Art bezeichnet man allgemein als → *Automimikry*. Ausführlich abgehandelt finden wir die Mimikry-Probleme bei WICKLER (1968) und dort auch ausführliche Literaturhinweise.

mimicry □ Imitation, deception, disguise and → *protective adaptation* in nature, both in the animal and plant kingdom and frequently beneficial to both. Imitation mainly concerns form or colour.
Many insects are disguised as blossoms, as for example mantises (Mantodea). Small cicadas, e. g., *Ityrea gregorie* (Fulgoridae) arrange themselves in a group on a plant stem so that they resemble a flower cluster (Fig. 77 A). This is termed *collective mimicry*.
The reverse situation in which the plant resembles the animal is possible. For example, many species of fly orchids in the genus *Ophrys* resemble wing-beating bee and bumblebee ♀♀ (Fig. 77 B – blossom of *Ophrys insectifera*); this attracts ♂♂ and they pollinate the flowers while attempting to copulate with them (Fig. 77 C – ♂ bee of genus *Eucera*).
Other forms of mimicry concern protection from predators. Many harmless flies (Diptera) and several Lepidopterans mimic stinging or poisonous wasps, bees and bumblebees (Hymenoptera) (Fig. 77 D – *Vespa crabro*, E – *Aegeria apiformis*). Insectivorous birds recognize these unpalatable forms from their colour and avoid them as well as their mimics. Hence mimicry of Hymenopterans may bring protection from predation; this is often termed *Batesian Mimicry*. Such cases of mimicry are frequent in the insect kingdom, on the one hand among Lepidopterans between different families, and on the other hand especially between unpalatable Coleopterans and palatable cockroaches, Heteropterans and also Orthopterans (Fig. 77 F – Carabid of the genus *Tricondyla*, G – grasshopper of the genus *Condylodera*).
In addition to morphological characters such as form and colour, behavioural mimicry also occurs; in distantly related species behaviour patterns are generally very different. Sometimes, however, effective interactions may lead to an adaptation or imitation of behaviour patterns called *ethomimicry (→ cleaner fish imitator)*. The imitation of one sex by the other within the same species is commonly called → *automimicry*. WICKLER (1968) gives a detailed discussion of different mimicry problems as well as exhaustive references.

mimétisme □ Imitation, feinte, reproduction de faux signaux et → *adaptation protectrice* dans la nature chez les animaux et les plantes et parfois même sous forme d'interactions significatives entre les deux. Le mimétisme se produit le plus souvent dans la forme ou dans la coloration.
De nombreux Insectes miment des fleurs, comme par exemple un certain nombre de Mantes (Mantodea). Chez les petites Cigales de l'espèce *Ityrea gregorie* (Fulgoridae), tout un groupe d'individus se dispose autour d'une tige végétale de manière à mimer une inflorescence complète (Fig. 77 A); dans de tels cas, on peut parler d'un *mimétisme collectif*.
En sens inverse, plusieurs plantes imitent des animaux et en profitent: ainsi, les fleurs d'un certain nombre d'Orchidées du genre *Ophrys* ressemblent étrangement à des ♀♀ d'Abeilles et de Bourdons (Fig. 77 B – fleur d'*Ophrys insectifera*); les ♂♂ de ces espèces, attirés par ces leurres, essaient de s'accoupler et effectuent en même temps et involontairement la pollinisation de ceux-ci (Fig. 77 C – Abeille ♂ du genre *Eucera*).
D'autres formes de mimétisme ont pour but un effet aposématique. Ainsi, certains Diptères inoffensifs et quelques Lépidoptères imitent souvent des Hyménoptères venimeux tels qu'Abeilles, Guêpes, Bourdons (Fig. 77 D – *Vespa crabro*, E – *Aegeria apiformis*). De nombreux Oiseaux apprennent par les expériences d'essai et d'erreur à reconnaître ces Hyménoptères dangereux par leur coloration et à les éviter, mais ils évitent en même temps les Diptères et Lépidoptères qui leur ressemblent. Ce mimétisme d'Hyménoptères protège donc les Diptères et Lépidoptères mimes contre leurs prédateurs et constitue un *mimétisme aposématique*. En pareil cas, on parle aussi de *mimétisme batésien*. Un tel mimétisme est très répandu parmi les Insectes, d'une part entre différentes familles parmi les Lépidoptères, et d'autre part entre Coléoptères de mauvais goût et des Blattes, Hétéroptères ou Orthoptères comestibles (Fig. 77 F – Carabide du genre *Tricondyla*, G – Sauterelle du genre *Condylodera*).
A côté de l'imitation de caractères morphologiques tels que forme ou coloration, il existe aussi un mimétisme du comportement. Chez des espèces phylogénétiquement éloignées, les mécanismes comportementaux sont également très différents, mais lorsqu'il y a apparition d'interactions efficaces, il peut se produire une adaptation ou plus exactement une imitation de comportements ce qui peut être considéré comme un *éthomimétisme (→ mime du poisson nettoyeur)*. L'imitation d'un sexe par l'autre à l'intérieur d'une même espèce est généralement appelée → *automimétisme*. Les problémes du mimétisme en général ont été traités par WICKLER (1968); voir aussi pour bibliographie exhaustive à ce sujet.

Mitesserei → Kommensalismus

Mit-Tretinteresse □ PETZOLD (1964) beschrieb bei Schwänen, wie ein zweites ♀ einem Paar beim Paarungsvorspiel interessiert zusah, dann ganz nah herankam und die Bewegungen des angepaarten ♀ synchron mitmachte. Dieses Verhalten ist das glatte

Gegenteil der bei Vögeln oft zu beobachtenden *Lex-Heinze-Stimmung,* d. h. des Tretneids eines zweiten ♀ gegenüber einer Begattung.

interest in participation with a copulating pair □ PETZOLD (1964) reported an instance of a ♀ swan which approached a swan pair during their precopulatory displays and performed the ♀ courtship movements sychronously with the paired ♀. This behaviour can be contrasted with the *Lex-Heinze-mood* (reaction of a single ♀ against mating) which is often seen in birds.

comportement de »participation copulatoire« □ PETZOLD (1964) a observé chez les Cygnes une deuxième ♀ observant avec intérêt un couple lors de la parade nuptiale. Elle s'approcha alors très près et commença à exécuter avec un synchronisme parfait tous les mouvements de la parade nuptiale de l'autre ♀. Ce comportement traduit une motivation radicalement opposée à celle qui sous-tend le comportement de jalousie d'une ♀ vis-à-vis d'un accouplement (dit *motivation Lex-Heinze*) et que l'on rencontre chez de nombreux Oiseaux.

Mnemotaxis → Taxis

Mondorientierung → Orientierung

E. lunar orientation □ F. orientation lunaire

Mosaikbewegung → ambivalentes Verhalten

E. mosaic movement □ F. mouvement mosaique

Motivation → Handlungsbereitschaft

Motivationsanalyse □ Die Analyse von Gesetzmäßigkeiten im Auftreten von Instinkthandlungen einer Tierart. Die Motivationsanalyse befaßt sich mit der Frage nach der Art und Zahl der Antriebsquellen für ein bestimmtes Verhalten. Man beginnt meist damit, eine bestimmte Verhaltensweise mit einer beliebig gewählten anderen zu vergleichen, – später können andere hinzukommen. Daraus ergibt sich dann die Frage: Haben die Bereitschaften zu diesen beiden Verhaltensweisen gemeinsame Ursachen oder nicht? Nach der von TINBERGEN (1939) vorgeschlagenen Methode sollen folgende Kriterien geprüft werden: Ist die Bewegungsform beider Verhaltensweisen ähnlich oder nicht? Werden die beiden Verhaltensweisen in gleichen oder ähnlichen Außensituationen ausgelöst oder nicht? Treten die beiden Verhaltensweisen sehr häufig oder sehr selten zusammen auf? Gute Beispiele für diese Arbeitsweise geben vor allem die Arbeiten von MORRIS (1958), WIEPKEMA (1961) und HEILIGENBERG (1964).

motivational analysis □ The study of regularities in the occurrence of instinctive behaviours in an animal species. Motivational analysis deals with the number and type of drive factors mediating a particular behaviour pattern. First a particular behaviour pattern is compared with another chosen behaviour to determine if the two behaviours have a common causation; then additional behaviours may be compared with the original one. TINBERGEN (1939) suggested the following criteria be used in the comparison: Is the movement pattern in both behaviours similar? Are the two behaviours released by the same or similar external stimuli? Do the behaviours occur very frequently or very seldom together? Good examples of this type of work are found in MORRIS (1958), WIEPKEMA (1961) and HEILIGENBERG (1964).

analyse motivationnelle □ C'est l'analyse des lois qui régissent l'apparition des actes instinctifs dans une espèce animale. Ce type d'analyse tend à dégager le nombre et la nature des sources d'incitation à l'accomplissement d'un comportement déterminé. En général, on commence par comparer une action donnée avec une autre, choisie plus ou moins arbitrairement; d'autres comparaisons de ce type peuvent intervenir par la suite. Il s'agit alors de savoir si les dispositions à l'origine des deux activités que l'on compare possèdent, ou non, des causes communes. D'après la méthode proposée par TINBERGEN (1939), les critères suivants doivent entrer en ligne de compte: les formes des mouvements déployés dans l'un et l'autre cas sont-elles semblables ou non? Les deux activités sont-elles évoquées dans une même situation ou dans des situations voisines, ou non? Ces deux comportements surviennent-ils fréquemment, ou rarement, en même temps? De bons exemples de la mise en œuvre de cette démarche sont fournis, notamment, par les travaux de MORRIS (1958), de WIEPKEMA (1961) et de HEILIGENBERG (1964).

Motivationswechsel □ Wenn eine Verhaltensweise im Laufe der Evolution eine neue oder zusätzliche biologische Funktion bekommt, muß sie den inneren Antrieb wechseln oder einen eigenen Antrieb bekommen, d. h. die Verhaltensweise löst sich von ihrer ursprünglichen Motivation und entwickelt eigene motivierende Mechanismen, was man auch als *Emanzipation* bezeichnen kann. Einen solchen Motivationswechsel beschreibt WICKLER (1966) am Beispiel der zur Grußgebärde gewordenen weiblichen Präsentierbewegung der Paviane.

change in motivation □ When a behaviour pattern takes on a different or additional function in the course of evolution, the underlying drive must be changed or replaced, i. e., the new behaviour is released from its original motivation and develops its own motivating mechanisms; this process can be termed *emancipation.* WICKLER (1966) describes an example of this process in the genital presentation display of the baboon.

changement de motivation □ Lorsqu'un aspect du comportement, au cours de l'évolution, assume une fonction nouvelle ou supplémentaire, l'incitation interne correspondante doit changer ou se créer *de novo.* Cet aspect du comportement se désolidarise donc de sa motivation originelle et développe un mécanisme motivationnel qui lui est propre. Cette *émancipation* est bien illustrée par un exemple étudié par WICKLER (1966): la présentation du postérieur par la ♀, qui devient, chez le Babouin, un mode de salutation.

Muldescharren → Ausmulden

Mund-zu-Mund-Füttern □ Bei vielen Völkern und Kulturen übliche Fütterung des Kleinkindes mit vorgekauter Nahrung, wobei das Kind den Mund leicht öffnet und die Mutter oder auch eine andere Person, z. B. die größere Schwester, mit der Zunge etwas Nahrung in den Mund des Kindes schiebt (Abb. 78 A – Papua-Frau aus Neuguinea). Dieses Verhalten gibt es auch unter erwachsenen Ituri-Pygmäen nach der Jagd (Abb. 78 B). Weiter werden bei manchen Völkern junge Haustiere so gefüttert. Außerdem tritt es bei sich Liebenden in ritualisierter Form als Kußfüttern und als Zärtlichkeitsfüttern wieder auf (Abb. 78 C – europäisches Liebespaar). Ein ähnliches Verhalten ist von Menschenaffen bekannt (EIBL-EIBESFELDT 1970, 1972, 1973). Weiterhin gibt es Mund-zu-Mund-Füttern bei den meisten Vögeln, und es entspricht hier dem normalen → *Jungefüttern* sowie dem → *Balzfüttern* und dem → *Begrüßungsfüttern* zwischen Paar-Partnern (WICKLER 1969). Bei sozialen Insekten → *Trophallaxie.*

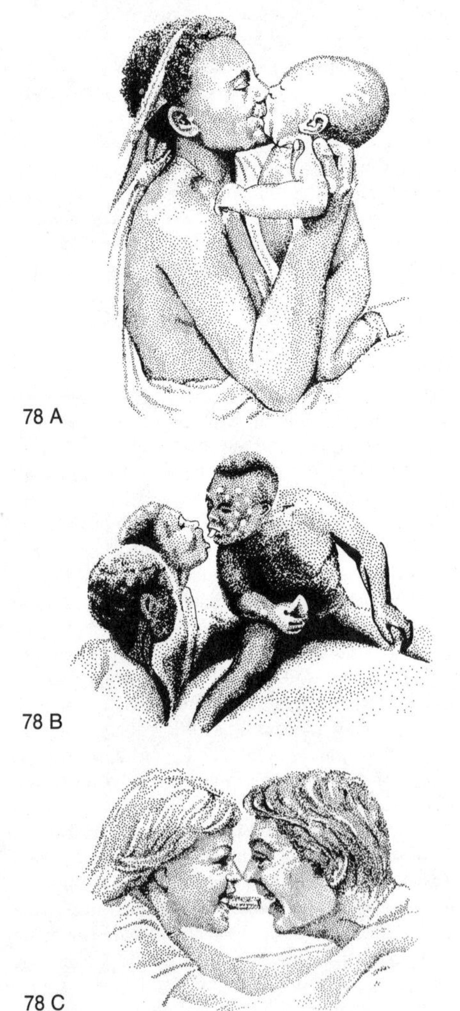

78 A

78 B

78 C

mouth-to-mouth feeding □ In many cultures mothers feed the infant pre-chewed food by inserting it in the infant's mouth with their tongue (Fig. 78 A – Papua woman from New Guinea). This behaviour is also practiced by adult Ituri Pygmies after the hunt (Fig. 78 B). In addition certain people in these cultures may use this means of nourishing young domestic animals. It also occurs in a ritualized form as kiss-feeding in human couples of lovers (Fig. 78 C). A similar behaviour is seen in the anthropoid apes (EIBL-EIBESFELDT 1970, 1972, 1973). Mouth-to-mouth feeding occurs in many birds and corresponds in this case to normal → *feeding of young,* and also to → *courtship feeding* and → *greeting feeding* between partners of a pair (WICKLER 1969). In social insects → *trophallaxis.*

nourrissage de bouche à bouche □ Manière de nourrir les jeunes enfants connue chez de nombreux peuples et cultures. La nourriture prémâché par la mère ou par une grande sœur est poussée avec la langue dans la bouche entrouverte de l'enfant (Fig. 78 A – femme Papou de la Nouvelle Guinée). Ce comportement a été également observé chez les Pygmées adultes de l'Ituri après la chasse, lors du partage de la viande (Fig. 78 B). Chez d'autres peuples, les jeunes animaux domestiques sont nourris de la sorte. D'autre part, ce comportement existe sous une forme ritualisée en tant que baiser de nourrissage chez les amoureux humains (Fig. 78C). Le nourrissage de bouche à bouche a été fréquemment observé chez les Singes Anthropoides. EIBL-EIBESFELDT (1970, 1972, 1973). Il existe également chez un grand nombre d'Oiseaux et correspond dans la plupart des cas au → *nourrissage des jeunes,* mais aussi au → *nourrissage de parade nuptiale* et au → *nourrissage de salut* entre les partenaires d'un couple (WICKLER 1969). Chez les Insectes sociaux → *trophallaxie.*

Mutterbindung → Mutter-Kind-Beziehung
E. mother bonding, bond to the mother □ F. lien maternel

Mutterfamilie □ Die Mutter allein bildet mit den Kindern eine Familie. Diese Art des sozialen Zusammenhaltes nannte DEEGENER (1918) *Gynopaedium.*

maternal family □ A group composed of the mother and her children. DEEGENER (1918) called this form of social structure *gynopaedium.*

famille maternelle □ Forme de famille comprenant seulement la mère et ses enfants. Une telle association sociale a été désignée par DEEGENER (1918) sous le terme de *gynopaedium.*

Mutter-Kind-Beziehung □ Die Beziehung und Bindung zwischen Mutter und Kind. Sie wird primär durch eine Reihe angeborener Reaktionen hergestellt, über welche BOWLBY (1958) im einzelnen berichtet. Die Bindung der Mutter an ihr Kind ist in erster Linie vom Brutpflegetrieb motiviert. Das Kind stellt für die Mutter nicht den Artgenossen mit Heimvalenz dar, sondern ist primär ein Betreuungsobjekt. Bei allen Säugern senden die Jungtiere bestimmte Signale aus, die Betreuung auslösen. Sie können geruchlich, akustisch oder optisch sein, wobei innerhalb der Primaten optische Signale zunehmend an Bedeutung gewinnen (EIBL-EIBESFELDT 1970, 1973, HASSENSTEIN 1973).

mother-child-relationship □ The relationship and bond between mother and child. This bond develops primarily through a series of innate responses which are described in detail by BOWLBY (1958). The mother's bond is motivated mainly by an epimeletic drive. The child does not represent a companion to share the home, but rather is an object that elicits care. In all mammals the young give certain signals that release care responses. The signals may be olfactory, acoustic or visual; in the primates visual stimuli are most important (EIBL-EIBESFELDT 1970, 1973, HASSENSTEIN 1973).

relation mère-jeune □ Relation et lien entre mère et enfant qui s'établissent primairement grâce à une série de réactions innées, décrites en détail par BOWLBY (1958). L'attachement de la mère à son jeune est fondé avant tout sur une pulsion épimélétique. L'enfant ne représente pas seulement pour la mère un congénère quelconque de l'environnement parental; c'est avant tout un objet de protection et de soins. Chez tous les Mammifères, les jeunes émettent des signaux déterminés, qui déclenchent les conduites maternelles. Ces signaux peuvent être d'ordre olfactif, auditif ou visuel; chez les Primates cependant, les composantes visuelles prennent une importance croissante (EIBL-EIBESFELDT 1970, 1973, HASSENSTEIN 1973).

Mutter-Kind-Bindung → Mutter-Kind-Beziehung
E. mother-child bond □ F. cohésion mère-enfant

N

Nachahmer → Mimikry
E. imitator □ F. mime

Nachahmung → Mimikry
E. mimicry □ F. mimétisme

Nachbeißen □ Bei Raubtieren auftretende Verhaltensweise, die regelmäßig beim Beutetöten nach dem ersten Zubeißen eine Schüttelbewegung ausführen und dann die Zähne leicht lösen und einmal nachbeißen (EIBL-EIBESFELDT 1956, LEYHAUSEN 1965).

repeated bites □ Occur in carnivores during killing of the prey whereby the animal first bites into the prey, shakes it, relaxes the jaws and then bites, once again (EIBL-EIBESFELDT 1956, LEYHAUSEN 1965).

morsure finale □ Comportement observé chez les Mammifères prédateurs qui, après la première morsure, secouent la proie, puis desserrent légèrement les dents pour mordre à nouveau (EIBL-EIBESFELDT 1956, LEYHAUSEN 1965).

Nachfolgeobjekt → Nachfolgereaktion
E. imprinting model □ F. objet de poursuite

Nachfolgereaktion □ Die angeborene Disposition junger Nestflüchter, gleich nach dem Schlüpfen bewegten Objekten nachzulaufen. Ist dieses bewegte Objekt *(Nachfolgeobjekt)* nicht die Mutter, sondern eine Attrappe, irgendein Objekt, auf welches die Jungen dann folgen, als ob es das Elterntier wäre, so führt dieses Verhalten zur → *Objektprägung.*

following response □ The innate disposition of young nidifugous birds to approach and follow a moving object shortly after hatching. If this moving object *(imprinting model)* is not the mother, but a model used to elicit the following response in the young as if it was the parent, this behaviour leads to → *object imprinting.*

réaction de suivre □ Chez les Oiseaux nidifuges, la tendance innée à suivre, dès la naissance, tous les objets en mouvement. Lorsque l'objet en mouvement *(objet de poursuite)* n'est pas la mère, mais un leurre ou un objet quelconque lequel est suivi comme si c'était un parent, ce comportement conduit alors à une → *empreinte sur objet aberrant.*

Nachläuferinnen □ Bienen, die der heimgekehrten Kundschafterin beim → *Rundtanz* bzw. → *Schwänzeltanz* nachlaufen und die Meldung und Information über die Entfernung und Himmelsrichtung der Futterquelle entgegennehmen. ESCH (1964) erreichte auch, daß Nachläuferinnen interessiert einer »tanzenden« Holzattrappe folgten, den damit angedeuteten Futterplatz aber suchten sie nicht auf.

follower bees □ Bees which meet the forager bees at the hive and follow them in the → *round dance* or → *waggle dance* hence obtaining the information concerning location of the food source. ESCH (1964) pointed out that follower bees will follow a »dancing« wooden model but do not subsequently search for a food source.

abeilles suiveuses □ Abeilles qui suivent l'exploratrice rentrée à la ruche pendant qu'elle exécute la → *danse circulaire* ou la → *danse frétillante* et qui recueillent ainsi l'information sur la source de nourriture et sa position par rapport à la ruche. ESCH (1964) a obtenu que des ouvrières suivent les évolutions d'un leurre en bois »dansant«, mais celles-ci ne se rendaient pas à la source de nourriture prétendument indiquée.

Nach-»Nichts«-Schnappen → Reflexverspätung
E. objectless snapping □ F. happer à vide

Nachtänzerinnen → Nachläuferinnen

Nachttänzerinnen □ Bienen, die auch in der

Nacht um 3h22 einen klaren → *Schwänzeltanz* ausführten und – mit einem Fehler von nur 12,5° – den Azimut der Sonne anzeigten, den sie um diese Zeit nie gesehen haben konnten (LINDAUER 1954).

night dancers □ Bees that performed the → *waggle dance* at 3:22 AM with a miscalculation of the sun's azimuth of only 12.5° even though they could not have seen the sun at that time (LINDAUER 1954).

danseuses nocturnes □ Abeilles qui, à 3h22 du matin, exécutaient une → *danse frétillante* caractéristique en indiquant avec une erreur de 12,5° seulement l'azimut du soleil qu'elles ne pouvaient voir à cette heure (LINDAUER 1954).

Nackenbiß □ Bei vielen Säugern und Vögeln (Schwäne) beißt das ♂ während der Begattung in den Hals bzw. in den Nacken des ♀. Bei vielen Feliden ist der Nackenbiß eine perfektionierte Beutefanghandlung. Dieser *Tötungsbiß* gleicht jenem Griff, mit dem Raubtiere und Nager ihre Jungen tragen und löst deshalb bei leicht verletzten Tieren oft die → *Tragstarre* aus.

neck-bite □ In a variety of mammals and birds (e. g. swans) the ♂ bites the female's neck during copulation. The same movements occur in many predators when attacking their prey *(killing bite)*. This type of predatory neck bite resembles the grip used by certain mammals and rodents to carry their young. → *limp posture*.

morsure à la nuque □ Chez de nombreux Mammifères et Oiseaux (Cygnes), le ♂ mord le cou ou la nuque de sa partenaire pendant l'accouplement. Par ailleurs, chez un grand nombre de Felidae, la morsure à la nuque constitue un acte de capture de proie perfectionné *(morsure fatale ou léthale)*. Cette morsure qui ressemble à la prise par laquelle certains Carnivores et Rongeurs portent leurs jeunes, déclenche souvent chez les proies légèrement blessées → *l'immobilisation de transport*.

Nahorientierung → Orientierung
E. orientation within the home range □ F. orientation proche

Nahrungsaustausch → Trophallaxie
E. food exchange □ F. échange de nourriture

Nahrungsspeichern → Futterverstecken
E. food hoarding □ F. amassement de nourriture

Nasengruß → Kontaktgruß
E. nose greeting □ F. salut nasal

Nasenreiben → Kontaktgruß
E. nose rubbing □ F. friction nasale

Necken → Scherzpartnerbeziehung
E. teasing □ F. taquiner

Nekrophoresie □ Der Heraustransport von toten Koloniemitgliedern aus dem Nest, ein hochentwickeltes stereotypes Verhalten bei Ameisen (WILSON 1971, 1975).

necrophoresis □ Transport of dead members of the colony away from the nest. A highly developed, stereotype behaviour in ants (WILSON 1971, 1975).

nécrophorésie □ L'évacuation du nid de membres morts dans une colonie; un comportement hautement développé et stéréotypé chez les Fourmis (WILSON 1971, 1975).

Nestablösung → Ablösungszeremoniell
E. relieving ceremony □ F. relève au nid

Nestausmulden → Ausmulden

Nestgeruch → Gruppenduft
E. nest odour □ F. odeur du nid ou de la ruche

Nestplatzzeigen □ Ein bei vielen Vögeln und Fischen zum Werbeverhalten gehörendes Element, wobei das ♂ dem frischgeworbenen ♀ den Nestplatz bzw. die Höhle zeigt, um es in »Neststimmung« zu bringen, d. h. in einen Motivationszustand, der es dem ♀ auch erlaubt, das vom ♂ gezeigte Nest in Besitz zu nehmen.

»showing the nest« behaviour □ A component of courtship behaviour in many birds and fish in which the ♂ draws the ♀ toward a potential nest or nesting hole for the purpose of bringing the ♀ in a nesting mood, i. e. a motivational state (broodiness) inducing the ♀ to occupy the nest site shown by the ♂.

indication du nid □ Chez de nombreux Oiseaux et Poissons, élément de la parade nuptiale pendant laquelle le ♂ montre à la ♀ qu'il vient de courtiser l'emplacement du futur nid ou la cavité de nidification ce qui stimule la motivation reproductrice de la ♀; cette dernière est ainsi placée dans un état motivationnel qui lui permet de prendre possession du nid ou du lieu de nidification préalablement choisi par le ♂.

Neststimmung → Nestplatzzeigen
E. nesting mood □ F. motivation reproductrice

Neugierverhalten □ Eine einem inneren Antrieb folgende Verhaltensweise, neue Gegenstände oder neue Situationen zu untersuchen, zu prüfen und zu erkunden *(→ Erkundungsverhalten)*. Das Neugierverhalten ist bei Tieren nur an eine kurze Entwicklungsphase in der Jugend gebunden, die mit der fortschreitenden phylogenetischen Entwicklungsstufe, vor allem bei vielen Säugern, zwar plastischer wird, im Prinzip aber bleibt die Kluft zwischen jungem und altem Tier erhalten. Beim Menschen bleibt das Neugierverhalten bis an die Grenze des Greisenalters erhalten (EIBL-EIBESFELDT und LORENZ 1974).

curiosity behaviour □ Behaviours associated with the examination and exploration of new objects, areas or situations and arising from an inner drive *(→ exploratory behaviour)*. Curiosity behaviour in many animals is associated with the juvenile phase of development; in more phylogenetically advanced groups, e. g., mammals, the dichotomy in curiosity behaviour between young and adult animals is not so marked. In humans curiosity tendencies remain intact up into old age (EIBL-EIBESFELDT and LORENZ 1974).

comportement de curiosité □ Un comportement obéissant à une pulsion endogène et consistant dans l'exploration et l'examen des objets nouveaux ou des situations nouvelles (→ *comportement exploratoire*).

Chez les animaux, ce comportement de curiosité n'est lié qu'à une courte phase du développement juvénile. Au fur et à mesure de l'élévation du niveau phylogénétique, cette phase de développement devient plus étendue, particulièrement chez les Mammifères, mais en principe, l'écart entre l'animal jeune et l'animal adulte reste évident. Chez l'Homme, par contre, le comportement de curiosité reste présent d'une manière permanente jusqu'à la vieillesse (EIBL-EIBESFELDT et LORENZ 1974).

Nickschwimmen □ Bei manchen Schwimmenten eine epigame Ausdrucksbewegung (Abb. 79 A – Entengattung *Anas*), die HEINROTH (1910) ursprünglich *Kokettierschwimmen* nannte (LORENZ 1941). Bei vielen Blenniiden ist das Nickschwimmen ein Ausdruck eines Konfliktes zwischen Weiterschwimmen und Umkehren. Der Putzernachahmer *Aspidontus taeniatus* (Abb. 79 B) zeigt ausschließlich eine ritualisierte Form des Nickschwimmens sowohl im innerartlichen als auch im zwischenartlichen Verkehr (WICKLER 1963).

nod swimming □ An epigamic behaviour found in several dabbling ducks (Fig. 79 A – genus *Anas*) originally referred to as *coquette swimming* by HEINROTH (1910), LORENZ (1941). In certain Blenniid fish the nod swimming shows a conflict between the tendency of progressing forward or retreating. A ritualized form of nod swimming occurs in the intra- and interspecific encounters of the saber-toothed blenny *Aspidontus taeniatus* (Fig. 79 B), WICKLER (1963).

nage de coquetterie □ Chez un certain nombre de Canards (Fig. 79 A – genre *Anas*), mouvement expressif du comportement épigame décrit par HEINROTH (1910) sous le terme de *Kokettierschwimmen* (LORENZ 1941). Chez de nombreux Poissons Blenniidae, cette forme de nage exprime un conflit entre la poursuite du même trajet et la tendance à faire demi-tour. La Blennie *Aspidontus taeniatus* (Fig. 79 B) qui mime le Poisson nettoyeur, présente exclusivement une forme ritualisée de la nage de coquetterie, aussi bien dans ses relations intra- qu'interspécifiques (WICKLER 1963).

Nistmaterialüberreichen □ Das Überreichen von Nistmaterial bei vielen Vogelarten als eine beschwichtigende Grußgebärde bei der Brutablösung. → *Ablösungszeremoniell.*

presentation of nest material □ This behaviour is found in many birds as an appeasement gesture when the mate arrives to take over incubation. → *relieving ceremony.*

offrande de matériaux de nidification □ Chez de nombreux Oiseaux, comportement par lequel les partenaires d'un couple s'offrent mutuellement, lors du relais au nid, des matériaux de nidification. Ce geste constitue un salut à valeur apaisante. → *cérémonie de relève.*

O

Oben-unten-Orientierung = Lichtrückenorientierung → Orientierung
E. above-below orientation □ F. orientation vers le haut et vers le bas

Oberlichtorientierung = Lichtrückenorientierung → Orientierung

Objektorientierung → Orientierung
E. orientation towards objects, object fixation □ F. orientation vers un objet

Objektprägung □ Bei Menschen und Tieren die Prägung auf ein nicht-adäquates, also falsches Objekt, auf eine falsche Person oder auf eine fremde Art *(Fehlprägung)*. Rhythmische Rufe und die verschiedensten bewegten Objekte lösen beim frischgeschlüpften Graugansgössel und anderen Nestflüchtern die Folgereaktion *(→ Nachfolgereaktion)* aus. Es läuft ebenso gut einem Menschen wie einer Gans oder einem vor ihm hergezogenen Kistchen nach. Folgt es einem solchen Objekt auch nur kurze Zeit, so bleibt es dabei. Lief das Gössel z. B. einem Menschen nach, konnte man es später nicht mehr dazu bringen, der wirklichen Mutter nachzulaufen (LORENZ 1935). Es ist bezüglich seiner Folgereaktion auf den Menschen geprägt. → *Prägung.*

object imprinting □ In man and animals the imprinting to a biologically inadequate, i. e. wrong object, to an inappropriate person or false species *(inappropriate imprinting)*. Rhythmic calls and diverse moving objects release the → *following response* in a young gosling and other precocial animals shortly after hatching. It follows a human as readily as it would follow a goose or moving box. If it follows such an object even for a short time, it will stay with it. Once the gosling has followed a person, it will not later follow its own mother (LORENZ 1935). With respect to the following reaction, it has become imprinted to man. → *imprinting.*

empreinte sur objet aberrant □ Chez l'Homme et chez les animaux, l'imprégnation sur un objet inadéquat, sur une personne inappropriée ou sur une espèce étrangère *(fausse empreinte)*. Des cris rythmiques et des objets très divers en mouvement déclenchent chez l'oison de l'Oie Cendrée, juste après l'éclosion, la → *réaction de suivre.* Le jeune animal suit aussi bien un être humain, une Oie ou une quelconque boîte tirée devant lui. S'il suit un tel objet, même pendant un temps très court, il lui restera attaché. Par exemple si l'oison suit un être humain, il ne sera plus possible de lui faire suivre sa vraie mère (LORENZ 1935); du point de vue de la réaction de suivre, il s'est ainsi attaché à l'Homme. → *empreinte.*

objektübertragene Handlung → Ersatzobjekt

E. redirected activity □ F. activité de redirection

Obstruktionsmethode □ Zur Messung der Triebstärke, um festzustellen, ob die Schwelle eines zweiten Reizes schwankt, der gerade eben die Beantwortung eines ersten Reizes unterdrückt.

obstruction method □ A means of measuring drive strength by determining if the threshold of response to a second stimulus has changed following failure to respond to an initial stimulus.

méthode d'obstruction □ Méthode servant à mesurer la force des pulsions en examinant les variations éventuelles du seuil d'inhibition de la réponse à un stimulus appétitif par un autre stimulus nociceptif ou dissuasif.

ödipale Phase → Ödipuskomplex

E. oedipal stage □ F. phase oedipienne

Ödipuskomplex □ Erotische Bindung des Kindes an den andersgeschlechtlichen Elternteil, die infolge des ambivalenten Konfliktes mit dem zugleich geliebten, gehaßten und gefürchteten gleichgeschlechtlichen Elternteil verdrängt wird (positiver Ödipuskomplex). Beim »negativen« oder »umgekehrten« Ödipuskomplex wird die Rivalität mit dem gleichgeschlechtlichen Elternteil durch eine erotische Bindung an eben diesen Elternteil ersetzt, was z. B. bei Jungen zu einer unbewußten homosexuellen Passivität gegenüber dem Vater führt. Um das fünfte Jahr erlebt das Menschenkind eine kritische Phase (*ödipale Phase;* FREUD), die für sein ganzes späteres Geschlechtsleben von Bedeutung ist. Es reifen die sexuellen Antriebe, und das geschlechtsspezifische Verhalten wird festgelegt.

Oedipus complex □ Erotic involvement of a male child with his mother accompanied by hostile feelings toward the father. In females, called the Electra complex, it is involvement with the father and resentment toward the mother. Negative Oedipus complex is one erotic involvement with the parent of the same sex and resentment toward the opposite sex. This can lead to homosexual preferences in people who resolved the Oedipus complex negatively. Approximately during the fifth year the child goes through a critical period (*oedipal stage;* FREUD) which is important for his later sexuality. The sexual drive and sex-specific role, male or female, are determined during this period.

complexe d'Oedipe □ Attachement érotique de l'enfant au parent du sexe opposé, refoulé à cause du conflit ambivalent avec le parent du même sexe qui est à la fois aimé, haï et craint (complexe d'Oedipe positif). Le complexe d'Oedipe est dit »négatif« ou »inverse« lorsque la rivalité avec le parent du même sexe est remplacée par l'attachement érotique à celui-ci; par exemple chez les garçons, il s'établit une passivité homosexuelle inconsciente à l'égard du père. Vers la cinquième année, l'enfant humain passe par une phase critique (*phase oedipienne;* FREUD) qui est importante pour l'orientation ultérieure de sa vie sexuelle. Les pulsions sexuelles commencent à mûrir et le rôle sexuel – mâle ou femelle – s'établit alors.

Ohrenstellung □ Bei vielen Säugern eine sehr wichtige Ausdrucksmimik, die Auskunft gibt über die Stimmung des Tieres.

ear position □ In many mammals the position of the ears is an important means of expressing and signaling mood.

position des oreilles □ Chez de nombreux Mammifères, mimique expressive très importante indiquant l'état motivationnel de l'individu.

Ökotrophobiose → Trophallaxie

Omegastellung → Rangordnung

E. omega position □ F. position omega

Ontogenese des Verhaltens → Ethogenese

E. ontogeny of behaviour □ F. ontogenèse du comportement

Orientierung □ Das Sichzurechtfinden im Raum (Raumorientierung). Die Primärorientierung kontrolliert die Körperstellung, die Sekundärorientierung kontrolliert die Einstellung auf einen aus der Umgebung stammenden Reiz. Wenn der Reiz nur die Aktivität der Bewegung steigert, spricht man von einer → *Kinese,* werden aber reizbezogene Wendungen ausgeführt, so handelt es sich um eine → *Taxis.* Bei einer Orientierung ohne richtende Außenreize spricht MITTELSTAEDT (1973) von einer idiothetischen Orientierung, im Gegensatz zur Orientierung nach Fremdinformation (Sonnenazimut), die er allothetische Orientierung nennt. Unter dem Begriff Objektorientierung versteht man Orientierungshandlungen schlechthin, die sich auf ein Objekt richten, in der Regel ein Beuteobjekt. Die Mechanismen wurden von MITTELSTAEDT (1953, 1954) besonders eingehend am Beispiel der Fangschrecken untersucht.

Die Orientierung und das Zurechtfinden in der unmittelbaren Umgebung des Lebensraumes (Nahorientierung) wird durch Duftstraßen (Ameisen), akustische Raumorientierung (Fledermäuse) oder elektrische Ortung (Mormyridae) erreicht. Bienen orientieren sich nach dem Sonnenstand (Kompaßorientierung) oder mit Hilfe des polarisierten Lichtes (Lichtkompaßorientierung); es genügt ein kleiner Ausschnitt aus dem blauen Himmel, um den von Wolken verdeckten Sonnenstand zu errechnen (zusammenfassend VON FRISCH 1965). Der Strandfloh *Talitrus saltator,* der sich auch nach der Sonne orientiert, kann sich nachts nach dem Mond orientieren (Mondorientierung), um den richtigen Fluchtweg zu finden (PAPI und PARDI 1959). Besonders ausgeprägt ist bei Zugvögeln die Fähigkeit, sich nach

Himmelskörpern zu orientieren (Sonnen-, Mond- und Sternorientierung); hinzu kommt aufgrund rezenter Arbeiten von WILTSCHKO und WILTSCHKO (1975 a + b) die →*Magnetfeldorientierung*.
Der prinzipielle Unterschied zwischen Nahorientierung und Fernorientierung ist nicht die Entfernung Ausgangsort – Ziel, sondern die Tatsache, daß bei der Nahorientierung das Ziel permanent wahrnehmbar ist; ist dagegen das Ziel nicht ständig wahrnehmbar, so spricht man von einer Fernorientierung, selbst dann, wenn die Entfernung Ausgangsort – Ziel nur gering ist.
Viele Fische und Wasserkäferlarven (Dytiscidae) orientieren sich mit Hilfe ihrer Augen nach dem einfallenden Licht. Die Wasserkäferlarven schwimmen z. B. zum Luftholen nach oben; beleuchtet man das Becken von unten, dann schwimmen die Tiere nach unten, als sei es die Wasseroberfläche, und ersticken dort (Lichtrückenorientierung). Außerdem haben viele Fische die Tendenz, den Rücken zum Licht hin zu richten (Lichtrückeneinstellung), und so schwimmen sie oft in Höhlen mit von unten kommendem Reflexlicht »Bauch nach oben«.
Umfassende Literatur zum Problem der Orientierung im allgemeinen finden wir vor allem bei KRAMER (1952, 1957, 1959), VON FRISCH (1965), SCHÖNE (1972, 1973, 1974) und LINDAUER (1974 – Internat. Symposium Orientierung der Tiere im Raum, Band I und II).

orientation □ The perception of one's position in space (space orientation). Primary orientation controls the body posture whereas secondary orientation controls the response to a stimulus from the environment. If this stimulus only increases activity, it is called a → *kinesis,* however, if the organism moves in some manner in relation to the stimulus, it is called a → *taxis.*
MITTELSTAEDT (1973) considered orientation in the absence of external stimuli as idiothetic orientation in contrast to orientation utilizing environmental information (e. g. sun azimuth) which he refers to as allothetic orientation. Object orientation concerns exclusively orientation in relation to an object, usually a prey object. MITTELSTAEDT (1953, 1954) has intensively studied the mechanisms of object orientation using the praying mantis as a model.
Orientation within the home area can occur by means of scent trails (ants), echolocation (bats), or electrical discharges (Mormyridae). Bees orient by using the sun's position (sky-compass orientation), or by using polarized light (light-compass orientation); a small opening in the clouds suffices to permit computation of the sun's position (summarized by VON FRISCH 1965). The tidal sand flea, *Talitrus saltator,* which may orient to the sun during the day, is also able to use the moon at night for assessing the correct escape route (lunar orientation; PAPI and PARDI 1959). Of particular note is the ability of migratory birds to orient according to celestial bodies (sun, lunar and star orientation); in addition, recent studies of WILTSCHKO and WILTSCHKO (1975 a + b) have also demonstrated → *magnetic field orientation.*
The principal difference between local orientation and remote orientation is not the distance between origin and goal itself, but rather that the goal is continuously within sight or perception in the former type. Remote orientation refers to orientation where the goal is not continuously in sight, even if there is only a short distance between origin and goal.
Many fishes and larvae of water beetles (Dytiscidae) orient according to the impinging light source. *Dytiscus* larvae, which must surface to breath, will drown if the aquarium is lighted from beneath since they move toward the source of light (dorsal light orientation). Moreover, many fish swim with their back toward the light (dorsal light response); inside caves where the light is reflected from the floor, these fish swim upside-down.
Comprehensive bibliographies on orientation are found in KRAMER (1952, 1957, 1959), VON FRISCH (1965), SCHÖNE (1972, 1973, 1974) and LINDAUER (1974 – Internat. Symposium Orientation of Animals in Space, Vol. I and II).

orientation □ Le fait d'assumer une position définie dans l'espace (orientation spatiale). L'orientation primaire contrôle la position du corps tandis que l'orientation secondaire contrôle la direction du corps vis-à-vis d'une stimulation exogène. Si le stimulus augmente seulement l'activité motrice, on parle d'une → *cinèse;* si, par contre, l'animal effectue des mouvements d'orientation par rapport à la source de stimulation, il s'agit d'une → *taxie.*
Lors d'une orientation en absence de stimuli externes particuliers, MITTELSTAEDT (1973) parle d'une orientation idiothétique en opposition à une orientation selon des informations exogènes (azimuth du soleil) qu'il appelle orientation allothétique. Par la notion de l'orientation vers un objet déterminé, on entend des mouvements orientés vers un objet, généralement une proie. Les mécanismes de ce comportement ont également été étudiés minutieusement par MITTELSTAEDT (1953, 1954) chez les Mantidae.
L'orientation dans l'environnement immédiat du domaine vital est assurée par différentes modalités, comme par exemple des traces odorantes (Fourmis), l'orientation acoustique (Chauves-Souris) ou encore l'orientation électrique (Mormyridae). Les Abeilles s'orientent selon la position du soleil (orientation sur la boussole lumineuse) ou à l'aide de la lumière polarisée par réflexion sur le ciel bleu; il leur suffit d'une petite partie du ciel visible pour leur permettre de calculer la position du soleil lorsqu'il est caché par les nuages (résumé VON FRISCH 1965). La Puce de mer *Talitrus saltator* qui s'oriente le jour par rapport au soleil, possède la capacité de s'orienter la nuit par rapport à la position lunaire pour retrouver une direction de fuite adéquate (orientation lunaire; PAPI

et PARDI 1959). Chez les Oiseaux migrateurs, la capacité de l'orientation astrotaxique (solaire, lunaire, stellaire) est particulièrement bien connue, à laquelle s'ajoute, selon de récentes recherches effectuées par WILTSCHKO et WILTSCHKO (1975 a + b) → *l'orientation par rapport au champ magnétique terrestre.*
Ce qui distingue particulièrement l'orientation proche de l'orientation lointaine, ce n'est pas la distance sujet-but, c'est plutôt le fait que ce dernier est perceptible de façon continuelle durant tout le trajet; en revanche, si le but doit être perdu de vue durant certains segments du parcours, il s'agit d'une orientation lointaine, même si la distance à parcourir est faible.
De nombreux Poissons et larves de Coléoptères aquatiques (Dytiscidae) s'orientent à l'aide de leurs yeux vers la lumière incidente. Ces larves de Coléoptères aquatiques, par exemple, viennent à la surface de l'eau pour respirer; si dans des conditions expérimentales l'aquarium est éclairé par le bas, ces larves nagent aussi vers le bas, comme s'il s'agissait de la surface de l'eau, et s'asphyxient au fond de l'aquarium (orientation photo-dorsale). D'autre part, de nombreux Poissons ont tendance à s'orienter le dos vers la lumière (réaction photo-dorsale) et ils nagent souvent le ventre en l'air dans les grottes, orientant leur dos vers la lumière réfléchie qui alors vient d'en bas.
On trouvera une bibliographie exhaustive sur les problèmes de l'orientation d'une manière générale surtout dans KRAMER (1952, 1957, 1959), VON FRISCH (1965), SCHÖNE (1972, 1973, 1974) et LINDAUER (1974 – Symposium International sur l'orientation des animaux dans l'espace, Vol. I et II).

Orientierungsbewegung □ Freie Ortsbewegung bei freilebenden Tieren und Phytoflagellaten; → *Orientierung,* → *Taxis.* Orientierende Bewegung bei festgewachsenen Tieren und bei Pflanzen; → *Tropismus.*

orientation response □ Movement from one site to another in freely moving animals and phytoflagellates; → *orientation,* → *taxis.* Orientating movement in sessile animals and plants; → *tropism.*

mouvement d'orientation □ Chez les animaux libres et les Phytoflagellés, il s'agit d'une → *taxie* (voir aussi → *orientation*); chez les animaux fixes et les plantes, on parle de → *tropisme.*

Orientierungsflug □ Der Orientierungsflug dient zum Kennenlernen der Umgebung. Der Bienenwolf, *Philanthus triangulum,* absolviert bei jedem Verlassen seines Nestes einen neuen Orientierungsflug und prägt sich den Standort genau ein. Legt man einen Kiefernzapfenkranz um das Nest, während der Bienenwolf sich darinnen befindet, so prägt sich das Tier diese neue Situation beim Verlassen des Schlupfloches ein, bevor es den Ort verläßt. Verändert man die geographische Disposition dieses Kiefernzapfenkranzes während seiner Abwesenheit, so findet das Tier nach der Rückkehr seinen Nestplatz nicht mehr wieder oder nur nach sehr langem Suchen. Es fliegt in die Mitte des versetzten Kiefernzapfenkranzes, als wäre dort der Eingang (Abb. 80). Das gleiche Verhalten findet man bei den Sandwespen der Gattung *Ammophila* (TINBERGEN 1932, BAERENDS 1941).
Setzt man Bienen, die noch nie ihren Stock verlassen haben, weiter als in 50 m Abstand davon aus, so finden sie nicht mehr nach Hause. Hat aber eine Biene von sich aus den Stock verlassen und einen Flug zur Orientierung in der Umgebung absolviert, so findet sie dann leicht aus einigen hundert Metern zurück, nach mehreren Orientierungsflügen sogar aus mehreren tausend Metern (VON FRISCH 1965, zusammenfassend dargestellt).

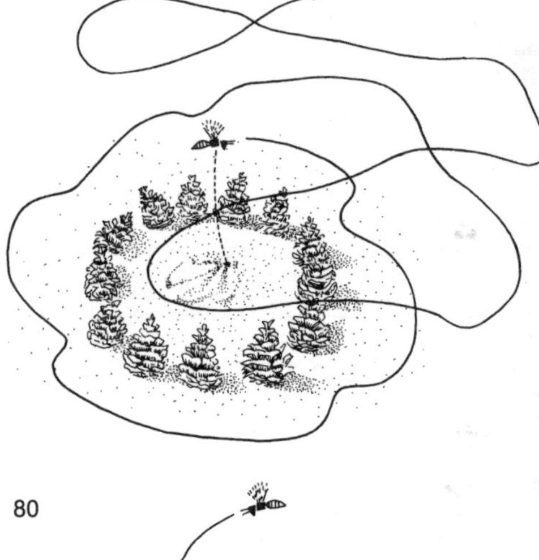

80

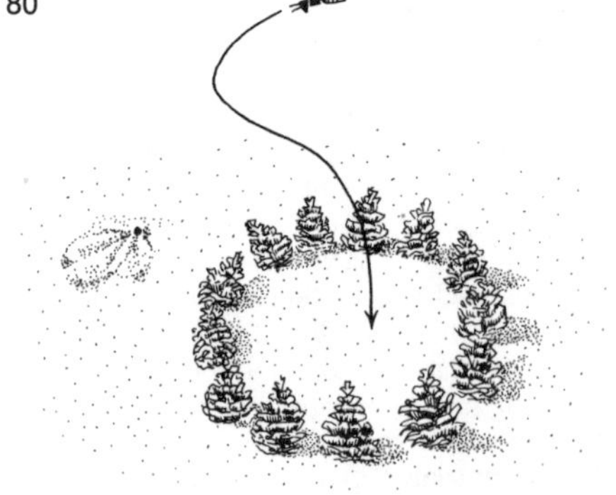

orientation flight □ During the orientation flight, the animal learns the landmarks of its environment. The bee-hunting digger wasp *Philanthus triangulum,* for example, makes a new orientation flight each time it leaves the nest in order to memorize the site. If one places pine cones in a circle around the nest while the digger wasp is inside, the animal will learn these new cues upon leaving the nest entrance, before it leaves the area. If one then displaces the pine cones during the animal's absence, it is unable to find the nest entrance or will find it only after a long time. The wasp will fly into the center of the displaced circle of cones, where the entrance was located before relative to the surrounding cones (Fig. 80). The same behaviour is found in the wasps of the genus *Ammophila* (TINBERGEN 1932, BAERENDS 1941). If one removes a bee, which has never been out of the hive, more than 50 meters from the hive he will not be able to return home. However, if the bee has had only one previous opportunity to fly in the vicinity of the hive he can find his home from several hundred meters distance. After several orientation flights he may successfully home from several thousand meters (summarized by VON FRISCH 1965).

vol d'orientation □ Le vol d'orientation sert à la reconnaissance de l'environnement. Le »Loup des Abeilles«, *Philanthus triangulum,* effectue, à chaque sortie de son nid, un nouveau vol d'orientation pour mémoriser le site de ce dernier. Lorsqu'on dispose des pommes de pin en cercle autour du nid pendant que le Philanthe se trouve à l'intérieur, celui-ci mémorise cette nouvelle disposition à la sortie de son trou avant de quitter les lieux. Si l'on modifie l'emplacement du cercle de pommes de pin pendant son absence, l'animal, à son retour, ne retrouve plus son nid ou seulement après des recherches prolongées. L'insecte se dirige au centre du cercle de pommes de pin déplacé comme si l'entrée se trouvait là (Fig. 80). Le même comportement s'observe chez les Guêpes du genre *Ammophila* (TINBERGEN 1932, BAERENDS 1941).
Si l'on transporte des Abeilles qui n'ont jamais quitté leur ruche à une distance supérieure à 50 m, elles seront incapables de retrouver leur gîte. Si, par contre, une Abeille a pu sortir d'elle-même de la ruche pour effectuer un vol d'orientation dans les environs, elle est capable de retrouver facilement son chemin jusqu'à une distance de quelques centaines de mètres. Après plusieurs vols d'orientation, elle revient même à la ruche quand elle en est éloignée de plusieurs kilomètres (vue d'ensemble chez VON FRISCH 1965).

Orthokinese → Kinese

Ortsgedächtnis □ Die Erinnerung an einen bestimmten Ort, eine bestimmte Stelle, als Beitrag zum Heimfindevermögen. → *Orientierung.*

topographic memory □ Memory of a given location or site as a component of migratory or homing behaviour. → *orientation.*

mémoire topographique □ La mémorisation d'un lieu précis, condition de la capacité de retour au gîte. → *orientation.*

Östrus → Brunst
E. estrus □ F. oestrus

P

Paarbildung □ Verhaltensweisen zwischen zwei künftigen Partnern, die zur Bildung eines Paares führen. Man spricht in diesem Fall oft vom Paarungsvorspiel. Während der Paarbildung treten Elemente des Drohens, Beschwichtigungsgebärden und Infantilismen auf. → *Werbeverhalten.*

pair formation □ The behavioural interaction of two future mates which leads to pair formation. This may also be referred to as mating foreplay. During pair formation elements of threat, appeasement and infantile behaviours may occur. → *courtship behaviour.*

appariement □ Comportement de deux futurs partenaires, aboutissant à la formation d'un couple. On peut parler aussi de jeu prénuptial. Au cours de l'appariement, apparaissent des manifestations de menace, d'apaisement et d'infantilisme. → *comportement de parade nuptiale.*

Paarbindung → Partnerbindung

Paarfüttern → Balzfüttern → Begrüßungsfüttern → Futterübergabe
E. feeding between pair-partners □ F. nourrissage entre partenaires d'un couple

Paarsitzen → Partnerbindung
E. spatial bond □ F. connexion spatiale

Paarungsaufforderung □ Meist vom ♀ ausgehende Verhaltensweisen und die Einnahme besonderer Körperhaltungen, die zur Paarung auffordern. → *Kopulation.*

invitation to mate □ Body postures and behaviour of the ♀ which lead to mating. → *copulation.*

invite à l'accouplement □ Comportements et postures assumés par la ♀, en vue d'inciter le ♂ à l'accouplement. → *copulation.*

Paarungskreisen □ Ein Element aus dem Paarungszeremoniell bei einigen Bovidae und Cervidae (BACKHAUS 1958, WALTHER 1958). Die Partner stehen umgekehrt parallel nebeneinander und kreisen so eng umeinander. Savannenarten führen das Kreisen eng auf der Stelle laufend aus, Waldarten verfolgen einander im kleinen Kreis um eine Vegetationsinsel herum. Die so entstehenden Spuren werden in der Jägersprache auch *Hexenkreise* genannt (KURT 1968).

display circling □ An element of the courtship ceremony in several Bovidae and Cervidae in which the partners stand head to back beside each other (BACKHAUS 1958, WALTHER 1958). Savanna dwelling animals circle around each other within a small area, whereas forest animals may circle an island of vegetation. These circular tracks are called *witches circles* in the hunter's language (KURT 1968).

parade nuptiale circulaire □ Elément du comportement de formation du couple chez quelques Bovidae et Cervidae (BACKHAUS 1958, WALTHER 1958). Les partenaires sont placés tête-bêche, côte à côte, et se poursuivent mutuellement en rond. Les espèces de savane le font en restant pratiquement sur place et les espèces de forêt se poursuivent généralement en petits cercles autour d'un îlot de végétation. Les traces qui en résultent sont appelées, en langage de chasseur, *ronds de sorcière* (KURT 1968).

Paarungsvorspiel → Paarbildung
E. mating foreplay, precopulatory behaviour □
F. jeu prénuptial

Partnerbindung □ Das Leben in fester Dauer-Einehe. Die Paar-Partner halten ständigen Seh- oder Hörkontakt und versuchen, sich gegenseitig nicht zu verlieren. Die Partnerbindung setzt die Fähigkeit eines individuellen Erkennens voraus. Beispiele hierzu sind uns von vielen Wirbeltieren bekannt, und zwar gleichermaßen von Fischen, Vögeln und Säugern. Bei Tieren mit Partnerbindung sitzt das ♂ immer nur mit einem ganz bestimmten ♀, seinem ♀, zusammen, was SEIBT und WICKLER (1972) bei Untersuchungen am Verhalten der Garnele *Hymenocera picta* als *Paarsitzen* bezeichneten (WICKLER und SEIBT 1972). Ein ähnlich ausgeprägtes Paarsitzen gibt es auch beim Großen Mara (Abb. 81 – *Dolichotis patagonum;* DUBOST und GENEST 1974, GENEST und DUBOST 1974).

partner bonding □ Refers to the long term attachment between two individuals which maintain continuous visual or auditory contact and try not to lose each other. One prerequisite for a partner-bond is individual recognition; examples are common in vertebrates including fish, birds and mammals. Partner-bonded animals always maintain a close spatial relationship. The ♂ will sit only with a given ♀, i. e. his ♀, and in close proximity to her. For this behaviour the term *Paarsitzen* or *spatial bond* was proposed by SEIBT and WICKLER (1972) in their observations on the shrimp *Hymenocera picta* (see also WICKLER and SEIBT 1972). A similar spatial bond is found in maras (Fig. 81 – *Dolichotis patagonum;* DUBOST and GENEST 1974, GENEST and DUBOST 1974).

81

lien entre partenaires □ Attachement entre partenaires d'un couple qui sont en permanence en contact visuel ou auditif et essaient de ne pas se perdre. Un tel lien entre deux partenaires implique une capacité de reconnaissance individuelle. Nous en connaissons de nombreux exemples chez les Vertébrés, aussi bien chez les Poissons que chez les Oiseaux et les Mammifères. Chez les animaux où existe un lien entre les membres d'un couple, les deux partenaires se trouvent presque toujours l'un près de l'autre. Le ♂ se tient seulement près d'une ♀ déterminée, sa ♀, ce que SEIBT et WICKLER (1972) ont appelé *Paarsitzen* ou *connexion spatiale,* lors de leurs recherches sur le comportement de la Crevette *Hymenocera picta* (WICKLER et SEIBT 1972). Un comportement semblable a été observé chez le Grand Mara (Fig. 81 – *Dolichotis patagonum;* DUBOST et GENEST 1974, GENEST et DUBOST 1974).

Partnereffekt → Stimmungsübertragung
E. partner effect □ F. effet du partenaire

Paßgang □ Im Gegensatz zum → *Kreuzgang* werden beim Paßgang beide Beine einer Körperseite gleichzeitig vorgesetzt (Abb. 82 – *Ammodorcas clarkei*), so daß eine schaukelnde Fortbewegung entsteht. Wir finden den Paßgang als normal übliche → *Gangart* vor allem bei Tylopoden, beim Elefant, bei der Giraffe und bei einigen Antilopen.

82

pacing □ In contrast to the → *cross gait* both legs on one side of the body are moved forward at the same time (Fig. 82 – *Ammodorcas clarkei*). This results in a swaying gait. Pacing is the usual form of locomotion in Tylopodes, in elephants, giraffes and in some antelopes.

amble □ Par opposition à → *l'allure croisée;* un animal qui va l'amble avance en même temps les deux pattes épilatérales (Fig. 82 – *Ammodorcas clarkei*): il en résulte un tangage dans la locomotion. Ce type de locomotion se rencontre normalement surtout chez les Tylopodes, chez l'Eléphant, la Girafe et un certain nombre d'Antilopes.

passive Verbreitung → Phoresieverhalten
E. passive distribution □ F. répartition passive

Pfeilgruß □ Eine aggressive Begrüßungsform bei Indianer-Stämmen des Amazonasgebietes. Die Ankommenden werden von den Indianern mit Pfeilen beschossen, aber man zielt dabei so gut, daß die Pfeile genau vor ihnen in den Boden treffen. Manche Besucher haben das als feindselig mißverstanden. Nach EIBL-EIBESFELDT (1970) ist diese Begrüßung durchaus mit unserem Salutschießen vergleichbar.

arrow greeting □ An aggressive form of greeting among certain Indian tribes of the Amazon region. The visitor is greeted with a barrage of arrows which land just in front of him. Many visitors have misunderstood this as a hostile gesture. EIBL-EIBESFELDT (1970) suggests that arrow greeting is comparable to our gun salute.

salut par flèches □ Une forme de salut agressif chez les Indiens d'Amazonie. Les arrivants sont salués par des flèches tirées en leur direction, mais les Indiens visent si bien que les flèches s'enfoncent dans le sol juste devant l'arrivant. Ce comportement de salut a souvent été interprété à tort comme hostile. Selon EIBL-EIBESFELDT (1970), cette forme de salut est comparable à nos salves d'honneur.

Pflegeverhalten □ Verhaltensweisen der Betreuung und Überwachung des Nachwuchses bei Tieren und Menschen; auch *Hüteverhalten* genannt. Das gegenseitige Sich-Beknabbern und Sich-Belekken, auch Fremdputzen genannt. Meist werden solche Verhaltensweisen durch Unterlegenheitsgesten und Infantilismen ausgelöst und dienen der Ableitung von Aggression. Beim Menschen dienen solche Pflegehandlungen, wie z. B. das Einander-Lausen, der Partnerbindung, vielleicht auch der Gruppenbindung. Auch Kindchenmerkmale (→ *Kindchenschema*) lösen Pflegehandlungen aus. → *Infantilismus.*

care-giving behaviour □ Behaviour of guarding and protecting of young in animals and man. Mutual nibbling and licking, also called allogrooming. Submissive gestures and infantile behaviour often elicit *epimeletic behaviour* which serves to divert aggression. In humans care-giving behaviour and grooming strengthen the pair-bond, perhaps also the group bond. Child-like characteristics (→ *baby schema*) also elicit care-giving tendencies. → *infantilism.*

comportement épimélétique □ Comportement de soins et de surveillance de la progéniture chez les animaux et chez l'Homme. Le mordillement et le léchage entre individus est aussi appelé épouillage mutuel. Ces différents comportements de soins sont généralement déclenchés par des gestes de soumission ou des gestes infantiles et servent à dévier l'agression. Chez l'Homme, les comportements de soins, comme par exemple le toilettage réciproque, ont pour fonction de resserrer le lien entre deux partenaires et peut-être aussi à renforcer les liens au sein d'un groupe. Les caractères infantiles (→ *schéma bébé*) déclenchent également ce comportement de soins. → *infantilisme.*

Pfotentrommeln □ Bei vielen Nagern ist das Auf-den-Boden-Trommeln mit den Vorderpfoten ein Imponierverhalten. Bei Hamstern und Mäusen mit den Hinterpfoten, die vorher über die Lateraldrüsen geführt werden, ein Markierungsverhalten (*Markierungstrommeln*). → *Duftmarkierung.*

drumming with paws □ In many rodents the forepaws are beaten on the ground during threat display. The hamsters and mice may mark the ground by drumming with the hind paws after rubbing them across their scent glands (*marking drumming*). → *scent marking.*

tambourinage avec les pattes □ Chez de nombreux Rongeurs, le fait de tambouriner vivement le sol avec les pattes antérieures, constitue un comportement d'intimidation. Chez les Hamsters et les Souris qui l'exécutent avec les pattes postérieures après les avoir passées sur les glandes latérales, le tambourinage constitue un comportement de marquage (*tambourinage de marquage*). → *marquage odorant.*

Phallusdrohen → Genitalpräsentieren
E. phallic threat □ F. menace phallique

Pheromone □ Hormone, die eine humorale Korrelation zwischen Individuen einer Art vermitteln. Hierzu gehören vor allem die Sexuallockstoffe, die besonders bei staatenbildenden Insekten von Bedeutung sind. Man spricht auch von *Ektohormonen, Soziohormonen* oder *Sozialhormonen*. Der hormonartige Stoff, der in bestimmten Drüsen gebildet wird, wird im Gegensatz zu den eigentlichen Hormonen nach außen abgegeben. Er entfaltet daher seine Wirkung nicht im Erzeugertier, sondern beeinflußt andere Individuen der gleichen Art. Das bekannteste Beispiel ist die sog. Königinnensubstanz der Honigbiene, die von den mächtig entwickelten Oberkiefer-Drüsen der Königin abgesondert wird und die Entwicklung weiterer Königinnen im Stock hemmt. Auch bei der Regelung der Kastenzugehörigkeit staatenbildender Insekten spielen Pheromone eine wichtige Rolle. Im weiteren Sinne dienen Pheromone auch der sozialen → *Verständigung*. Zu ihnen gehören Sexualstoffe, Duftstoffe zum Markieren, Alarmstoffe, wie sie von verschiedenen Schwarmfischen bekannt sind, und auch Orientierungspheromone und Rekrutierungspheromone (WILSON 1971, 1975).

pheromones □ Hormones which transmit information between individuals of a species. Of particular significance here are the sex attractants prevalent among social insects. These hormone-like substances, sometimes called ectohormones, sociohormones or social hormones are produced in distinct glands and released outside the organism. Hence the pheromone produces its effect on other individuals of the same species. The best known example is the so-called "royal substance" produced in the highly developed supramaxillary glands of the queen in honey bees. Secretion of this substance prevents the development of additional queens. Pheromones also play an important role in caste determination in social insects. Moreover, they are important in other aspects of social → *communication*. Among these are sexual attractants, marking substances, alarm substances as known from various fishes, as well as orientation and recruitment pheromones (WILSON 1971, 1975).

phéromones □ Hormones transmettant des informations entre individus à l'intérieur d'une même espèce. Font surtout partie de cette catégorie d'hormones les substances d'attraction sexuelle, jouant un rôle particulier parmi les Insectes sociaux. Ces phéromones sont également appelées éctohormones ou hormones sociales. Ces substances sont produites dans des glandes déterminées et en opposition aux hormones proprement dites, elles sont sécrétées vers l'extérieur. Elles ne produisent donc pas d'effet chez l'animal producteur, mais seulement chez un autre individu de la même espèce. Un bon exemple est constitué par la substance royale de l'Abeille mellifère produite dans des glandes mandibulaires géantes de la reine et qui est sécrétée dans le but d'empêcher le développement d'autres reines dans la même ruche. Les phéromones jouent également un rôle important dans la régulation de l'appartenance aux différentes castes chez de nombreux Insectes sociaux. Au sens plus large, les phéromones servent également à l'intercommunication sociale (→ *communication*). Elles comprennent des substances sexuelles, des substances odorantes au service du marquage, des substances d'alarme connues chez de nombreux Poissons vivant en bancs, mais nous connaissons également des phéromones d'orientation et des phéromones de recrutement (WILSON 1971, 1975).

Phoresieverhalten □ Verhalten von Tieren, die andere Arten, meist größere Organismen, als vorübergehende Transportmittel benutzen. Es handelt sich um einen Sondertypus zwischenartlicher Beziehungen, der weit verbreitet ist. Besonders häufig ist Phoresie bei aas- und dungbewohnenden Organismen, wie gewissen Nematoden, Milben, Pseudoskorpionen und manchen Käferlarven, die auf diese Weise von einem Futterplatz an einen anderen gelangen. Ein besonders eindrucksvolles Beispiel von Phoresie liefert uns der Schiffshalter *(Echeneis),* der sich meist an Haie heftet, aber auch an Schiffsrümpfe (daher der deutsche Name) und sogar an Taucher. Das transportierende Tier wird auch als *Tragwirt* bezeichnet. Eine Zusammenfassung des Phoresie-Problems aus ethologischer Sicht finden wir bei SCHALLER (1960).

phoretic behaviour □ Behaviour of animals using other animals, generally of larger size, as a temporary means of transport. It constitutes a special form of interspecific interaction which is widespread among animals. Phoresy occurs frequently in animal organisms living in carrion and dung, e. g. certain nematods, mites, pseudoscorpions, and some Coleopteran larvae which in this manner move from one feeding site to another. A particularly striking example is given by the fish genus *Echeneis* which generally use scales as carrier hosts, but also cling to the hull of ships and even to human skin divers. The transporting animal may be designated the *carrier host.* The phoresy problem has been summarized from an ethological point of view by SCHALLER (1960).

comportement phorétique □ Comportement connu chez un certain nombre d'animaux utilisant d'autres espèces, généralement des êtres organisés plus grands, pour se laisser transporter passsivement d'une manière temporaire. Il s'agit d'un type très particulier de relations interspécifiques, répandu dans le monde animal. La phorésie est très fréquente chez les organismes vivant dans la charogne et dans le fumier, tels que certains Nématodes, Acariens, Pseudoscorpions et larves de Coléoptères qui se laissent ainsi transporter d'un emplacement alimentaire à un autre. Un exemple particulièrement impressionnant de phorésie est connu chez les Rémoras, Poissons du genre *Echeneis,* qui utilisent normalement des Requins comme moyen de transport, mais s'accrochent aussi aux coques des bateaux et même aux plongeurs scaphandriers. L'animal transporteur peut être appelé *hôte phorétique.* Un résumé des problèmes de la phorésie, vus sous l'angle éthologique, est donné par SCHALLER (1960).

Polyethismus → Arbeitsteilung

Präferenzversuch → Wahlversuch

E. preference test □ F. expérience de choix simultané

Prägung □ LORENZ (1935) beschrieb das Phänomen der Prägung und hob einige Kennzeichen hervor, die diesen Lernvorgang vom normalen Assoziationslernen unterscheiden. HESS hat diese Unterschiede weiter ausgearbeitet und (1973) zusammengefaßt. – Eine Prägung findet immer nur in einer bestimmten sensiblen Periode statt. Verstreicht diese Zeit, kann das Tier nicht mehr geprägt werden. Diese sensible Periode muß aber nicht unbedingt auf die ersten Lebenstage oder -wochen beschränkt sein. Für die Folgereaktion bei Enten liegt sie zwischen der 13. und 16. Stunde nach dem Schlüpfen.

Die erworbene Kenntnis des auslösenden Objektes wird zeitlebens behalten, während sonst gerade das Vergessen ein wesentliches Merkmal ist. Das Gelernte wird aber nicht nur behalten, sondern das Objekt der Prägung auch zeitlebens bevorzugt; die Prägung ist also irreversibel. Als Merkmale des Prägungsobjektes greift das geprägte Tier nur überindividuelle, artkennzeichnende Merkmale heraus.

Geprägt wird immer nur eine bestimmte Reaktion auf ein bestimmtes Objekt (Abb. 83 – sexuell auf den Menschen geprägtes ♂ von *Tragelaphus imberbis*). Für eine von LORENZ (1935) aufgezogene Dohle war der Mensch *Elternkumpan* und *Geschlechtskumpan* (→ *Kumpan* = als Partner in einem bestimmten Funktionskreis). Sie flog jedoch mit Nebelkrähen (*Flugkumpan*) und nahm schließlich junge Dohlen als *Kindkumpan* an. Die Festlegung des Objektes einer Triebhandlung kann im Falle sexueller Prägung zu einem Zeitpunkt erfolgen, zu dem die betreffende Triebhandlung noch gar nicht ausgereift ist.

imprinting □ LORENZ (1935) described the phenomenon of imprinting and emphasized several criteria which distinguish this learning process from normal association learning. HESS (1973) has further clarified these differences. – Imprinting takes place only during a specific sensitive period. If this time passes

83

the animal can no longer be imprinted. This sensitive period does not have to be restricted absolutely to the first few days or weeks of life. For the following reaction the sensitive period in ducklings is between 13 and 16 hours after hatching.
The acquired knowledge of releasing object is retained for life, while normally, forgetting is common after learning. What is learned is not only retained, but the imprinting object is also preferred during the rest of the animal's life; imprinting is then irreversible. During imprinting the animal learns only supra-individual species-specific characteristics.
Only specific reactions become imprinted to a particular object (Fig. 83 – ♂ of *Tragelaphus imberbis* sexually imprinted to man). A jackdaw raised by LORENZ (1935) regarded humans as *parent* and *sexual companions* (→ *companion* is a partner in a specific functional cycle). The bird flew around with hooded crows as *flight companions* and accepted young jackdaws as *child companions*. The determination of the object for an instinctive activity can take place, as in the case of sexual imprinting cited above, at a time when the appropriate behaviour pattern is not yet matured and thus has not yet been performed by the animal.

empreinte, imprégnation □ LORENZ (1935) a décrit le phénomène de l'empreinte et en définit divers critères qui différencient ce processus d'apprentissage de l'apprentissage associatif conventionnel. HESS (1973) a continué à préciser ces différences. L'empreinte a toujours lieu au cours d'une période »sensible« déterminée. Cette période passée, l'animal ne peut plus être imprégné. Cette période sensible n'est pas obligatoirement limitée aux premiers jours qui suivent la naissance. Dans la réaction de suivre des Canards, elle est située entre la treizième et la seizième heure après l'éclosion.
La connaissance acquise de l'objet déclencheur subsiste toute la vie, alors que l'oubli est fréquent après un apprentissage conventionnel. Ce qui est appris par empreinte est non seulement retenu, mais l'objet de l'empreinte est toujours préféré durant tout le reste de la vie de l'animal. L'empreinte est donc »irréversible«. Durant la période d'empreinte, l'animal n'apprend que les caractéristiques supra-individuelles distinctives de l'espèce.
Seules s'imprégnent des réactions spécifiques sur un objet déterminé (Fig. 83 – ♂ de *Tragelaphus imberbis* sexuellement imprégné sur l'Homme). Dans le cas d'un des Choucas élevés par LORENZ (1935), l'Homme était à la fois *parent* et *compagnon sexuel* (→ *compagnon*). Le Choucas vola toutefois avec des Corneilles mantelées (*compagnons de vol*) et accepta finalement de jeunes Choucas comme *compagnons-enfants*. La détermination de l'objet d'une activité instinctive peut se manifester, comme dans le cas de l'imprégnation sexuelle citée précédemment, même si le mode de comportement approprié n'est pas encore installé et n'a pas encore été exécuté par l'animal.

Prägungsobjekt → Objektprägung
E. imprinting object □ F. objet d'empreinte

Präsentierbewegung →Imponierverhalten
E. presentation movement □ F. mouvement de présentation

Prellsprung □ Unter den Paarhufern, vornehmlich bei Vertretern der Gazellinae und Antilopinae, aber auch bei einigen Hirscharten, kommt bei Feindverdacht, überraschenden Störungen, auf der Flucht und im Spiel eine eigenartige Bewegungsweise vor, die man als Prellsprung bezeichnet (WALTHER 1948, 1966, JUNGIUS 1969, HEIDEMANN 1973). Mit steil gehobenem Hals, waagerecht bis senkrecht gestelltem Schwanz stößt das Tier sich steifbeinig vom Boden ab und federt so in 2 bis 12 Sprüngen auf und ab. Der Streckengewinn ist gering und beträgt meist nur einige Zentimeter. SCHALLER (1967) mißt dem Prellsprung vier mögliche Sozialfunktionen bei: 1. Das Geräusch der aufstoßenden Hufe kann in der Nähe befindlichen Artgenossen als Alarmsignal dienen. – 2. Die hohen Sprünge mit dem Präsentieren der hellen Analfläche und des Rumpfes sind ein weithin sichtbares visuelles Signal. – 3. Die Sprünge ermöglichen dem Tier die Überwachung der Störquelle im unübersichtlichen Gelände bzw. geben ihm Gelegenheit, den sich verbergenden Feind ausfindig zu machen. – 4. Durch das Aufstoßen der Läufe und die Sekretion der Klauendrüsen wird eine Duftmarkierung hinterlassen, die den nachfolgenden Artgenossen als Information dient. – Es sei noch erwähnt, daß der Prellsprung im Zusammenhang mit nicht sicherer Feindidentifikation und Schrecken als Ausdruck hoher Erregung vorkommt. Man findet ihn hauptsächlich bei Bewohnern offenen Geländes mit langen Beinen. In diesem Zusammenhang ist der Nachweis des Prellsprungs bei einem Nager (Abb. 84 – *Dolichotis patagonum*) interessant (DUBOST und GENEST 1974).

84

stotting □ An unusual behaviour which occurs among Artiodactylids, especially gazelles and antelopes and also deer, during fright, sudden disturbance or play (WALTHER 1948, 1966, JUNGIUS 1969, HEIDEMANN 1973). It consists of jumping from 2 to 12 times straight into the air with all four legs held stiff, the neck erect and the tail held in a horizontal to vertical position. The animal, however, gains perhaps only a few centimeters of viewing advantage. SCHALLER (1967) proposed four possible social functions for this behaviour. 1. The sound of the hoofs striking the ground could serve as an alarm signal for conspecifics in the immediate vicinity. – 2. The higher jumps expose the bright anal area and rump which may constitute a visual signal over a greater distance. – 3. The jumping allows the animal to monitor the source of disturbance in an area where the visibility is poor, e. g., offers the opportunity to detect a predator. – 4. The striking of hoofs and the secretion of the hoof scent glands may provide information for conspecifics who pass by. It should be mentioned that stotting occurs in cases where the animal is frightened and uncertain of the source or nature of disturbance, i. e., predator. This behaviour is primarily seen in open grassland species, and its occurrence in a different order, i. e., rodents (Fig. 84 – *Dolichotis patagonum*) is noteworthy (DUBOST and GENEST 1974).

stotting □ Parmi les Artiodactyla, en particulier chez les Gazellinae et Antilopinae, mais aussi chez un certain nombre de Cervidae, il existe un mode de locomotion particulier appelé »stotting« qui se manifeste lors de la détection d'un prédateur, lors de dérangements inattendus, pendant la fuite ainsi qu'au cours du jeu (WALTHER 1948, 1966, JUNGIUS 1969, HEIDEMANN 1973). Le cou dressé verticalement, la queue tendue à l'horizontale ou à la verticale, l'animal effectue des bonds, les quatre pattes tenues roidies quittant le sol en même temps. En sautant ainsi, l'animal gagne peu de terrain, parfois pas plus de quelques centimètres. SCHALLER (1967) suppose que le »stotting« peut avoir quatre fonctions sociales différentes. 1. Le bruit que produisent les sabots en battant le sol peut être un signal d'alarme pour des congénères se trouvant à proximité. – 2. Ces sauts particulièrement hauts, en association avec la présentation du miroir anal clair et parfois du ventre clair, peuvent être un signal visuel perceptible à grande distance. – 3. Les sauts peuvent faciliter la surveillance ou la détection de la source de dérangement dans un terrain accidenté et donnent à l'animal la possibilité de déceler un ennemi. – 4. Les coups de sabots sur le sol et la sécrétion des glandes unguéales peuvent produire un marquage olfactif qui donne des informations aux congénères qui suivent. D'autre part, le »stotting« qui apparaît comme une réaction phobique en cas d'identification incertaine d'un prédateur, est alors l'expression d'une grande excitation. Le »stotting« existe surtout chez les espèces possédant des pattes relativement longues et habitant des biotopes ouverts. Il est intéressant de signaler dans ce contexte l'existence du »stotting« chez un Rongeur (Fig. 84 – *Dolichotis patagonum*). Le terme anglais *stotting* est utilisé en français (DUBOST et GENEST 1974).

Prestigesitten □ Die Demonstration des eigenen Wertes im menschlichen Verhalten: größeres und schöneres Auto, Haus usw. als der Nachbar oder Kollege. Gastgeber und Gäste versuchen sich zuweilen gegenseitig an Gastlichkeit zu überbieten (EIBL-EIBESFELDT 1970). Eine abartige Form von Prestigesitten hat sich bei den Kwaikutl-Indianern entwikkelt, die bei den »Potlatsch« genannten Festen ihre eigene Überlegenheit gegenüber den Gästen derart demonstrieren, daß schließlich der Gast als armer Teufel verspottet wird (BENEDIKT 1934). Man darf aber nicht vergessen, daß Prunkentfaltung dem Gast eine hohe Rangstellung zuweist und somit gleichzeitig eine → *Achtungserweisung* darstellt.

prestige customs □ The demonstration of one's own worth in human social behaviour; for example a bigger and more attractive automobile, house, etc., than the neighbours have. Often the host and his guest attempt to outdo each other in their hospitality (EIBL-EIBESFELDT 1970). The Kwaikutl Indians have a ceremonialized form of prestige called the »potlatsch« in which the superiority of the host is carried to such an extent that the guest may be ridiculed in the end as a poor man (BENEDIKT 1934). At the same time this display of lavishness to the guest affords him a high rank and hence represents a form of recognition (→ *showing »respect«*).

manifestations de prestige □ Démonstration de sa propre valeur dans le comportement humain, par exemple par la possession d'une voiture ou d'une maison plus grande et plus belle que celle du voisin ou du collègue. Hôtes et invités essaient parfois de se surpasser mutuellement en hospitalité (EIBL-EIBESFELDT 1970). Une autre forme extrême de manifestation de prestige se rencontre chez les Indiens Kwaikutl qui, pendant leurs fêtes appelées »potlatsch«, démontrent à l'extrême leur propre supériorité envers les invités, allant jusqu'à se moquer d'eux et à les considérer comme de pauvres diables (BENEDIKT 1934). Il est à remarquer cependant que de grandes démonstrations envers un invité lui donnent en même temps une position hiérarchique supérieure et que les manifestations de prestige sont, dans ce sens, à considérer comme un → *témoignage de respect.*

Propriozeptoren □ Einrichtungen, die auf körpereigene Reize ansprechen, wie z. B. freie Nervenendigungen an den Gelenkhäuten und Muskeln, die zusammen mit den Chordotonalorganen dazu dienen, Lageveränderungen der gegeneinander beweglichen Teile des Chitinpanzers wahrzunehmen. Im Gegensatz hierzu → *Exterozeptoren.*

proprioceptors □ Receptors responding to stimuli from the body. They consist of free nerve endings terminating in articular cuticle and muscle which with the chordotonal organs transmit information about the relative position of different parts of the body. In contrast to → *exteroceptors.*

propriocepteurs □ Dispositifs répondant à des stimulations autochtones du corps. Il s'agit d'extrémités libres du système nerveux situées au niveau des cuticules articulaires et des muscles et qui, en connexion avec les organes chordotonaux, servent à la perception des changements de position des différentes parties cuticulaires mobiles. En opposition → *extérocepteurs.*

Propulsus □ Verhaltenselement während der Begattung bei Wiederkäuern. Nach der *Immissio* der Penisspitze *intra vaginam* wird das Glied augenblicklich maximal gestreckt und dabei ganz in die Vagina eingeführt. Im Moment der stärksten Streckung erfolgt die Ejakulation. Gleichzeitig werden Hüft- und Kniegelenk gestreckt, das Lumbosakralgelenk gebeugt und die Bauchmuskeln zusammengezogen, so daß der Stier, der Bock oder der Hirsch einen Sprung (Abb. 85 – Propulsus von *Cervus nippon*) zur Kuh hin macht und sich dabei mit den Hinterläufen völlig vom Boden abhebt (SAMBRAUS 1971, 1973).

propulsus □ Behaviour pattern during copulation observed in ruminants. After the immission of the tip of the penis *intra vaginam* the penis is immediately stretched to the maximum and at the same time completely introduced into the vagina. At the moment of the strongest extension ejaculation occurs. At the same time the hip joint and the knee joint are stretched, the lumbo-sacral articulation bended and the ventral muscles contracted. In this situation the bull, buck or deer performs a jump towards the ♀ (Fig. 85 – *Cervus nippon*) and takes off completely from the ground with its hind legs (SAMBRAUS 1971, 1973).

propulsus □ Elément comportemental lors de l'accouplement observé chez les Ruminants. Après l'immission de la pointe du pénis *intra vaginam,* le pénis est étendu au maximum et alors complètement introduit dans le vagin. L'éjaculation se produit

85

lorsque l'extension est maximale. Les articulations iliaque et du genou sont alors tendues et l'articulation lombo-sacrale fléchie. Les muscles ventraux sont contractés ce qui permet au taureau, au bouc ou au cerf d'effectuer un saut (= *propulsus*) contre la ♀ (Fig. 85 – *Cervus nippon*) et de décoller ainsi complètement du sol (SAMBRAUS 1971, 1973).

proteanisches Verhalten □ Ein täuschendes Verhalten oder Ablenkungsmanöver, so genannt nach dem griechischen Meergott PROTEUS, welcher durch dauernden Gestaltswechsel seinen Verfolgern entging (CHANCE und RUSSELL 1959). → *Verleiten.*

protean behaviour □ A diversionary tactic or behaviour named after PROTEUS, the Greek sea god who escaped his pursuers by continually changing his form (CHANCE and RUSSELL 1959). → *distraction display.*

manoeuvre protéenne □ Comportement de feinte appelé protéen d'après le nom du dieu marin grec PROTEE qui, grâce à de multiples changements de forme et d'aspect, sut échapper aux importuns (CHANCE et RUSSELL 1959). → *manoeuvre de diversion.*

Prüfwittern → Harnmarkierung → Kotzeremoniell
E. sniffing test □ F. flairage de contrôle

Pseudopenis □ Bei der Fleckenhyäne *Crocuta crocuta* haben die ♀♀ einen Pseudopenis, der dem echten Penis des ♂ stark ähnelt und ebenso wie dieser ausgeschachtet und erigiert werden kann (Abb. 86). Bei der Begrüßung belecken sich zwei einander bekannte Tiere ausdauernd die Genitalien, und ein Teil des männlichen Paarungsvorspiels ist zur sozialen Grußgeste beider Geschlechter geworden (WICKLER 1969).

pseudo-penis □ In the spotted hyena *Crocuta crocuta* the ♀♀ have a »pseudo-penis« which clearly re-

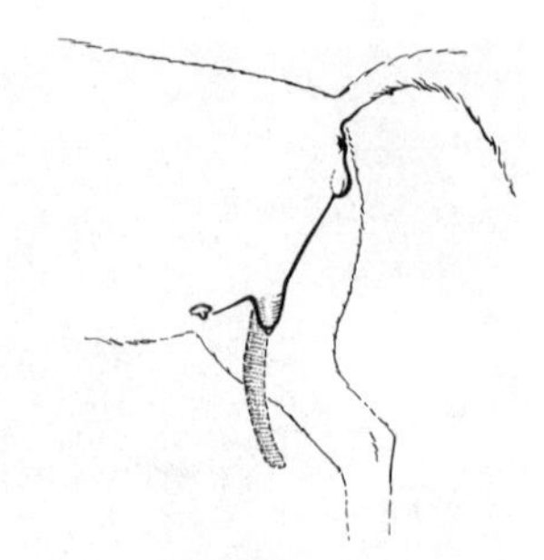

86

sembles the male's penis and which can be similarly extruded and erected (Fig. 86). During the greeting ceremony two animals familiar with each other may lick the other's genitalia; here, a part of the male's precopulatory behaviour has developed into a social greeting display (WICKLER 1969).

pseudo-pénis □ Chez l'Hyène tachetée *Crocuta crocuta,* les ♀♀ possèdent un pseudo-pénis qui ressemble étrangement à celui du ♂ et qui est également érectile (Fig. 86). Lors de la rencontre de deux animaux qui se connaissent individuellement, ceux-ci se lèchent longuement et mutuellement les organes génitaux. Une partie de la parade nuptiale du ♂ s'est donc transformée chez les deux sexes en un comportement de salutation (WICKLER 1969).

Pseudoart □ Die rasche, kulturell gesteuerte Evolution des Menschen birgt eine Gefahr. Kulturen bilden sich schnell und kapseln sich aufgrund ihrer besonderen Kulturmuster von anderen ab; und die verschiedenen Kulturen oder Nationen verhalten sich zueinander wie »biologische Arten«. ERIKSON (1966) bezeichnete sie deshalb auch als *Pseudospecies.*

pseudo-species □ The rapid cultural evolution of man represents a potential danger since cultures tend to differentiate rapidly and isolate themselves from each other. They begin to behave toward each other as »biological species«; hence ERIKSON (1966) refers to cultures and nations as pseudo-species.

pseudo-espèce □ L'évolution rapide et culturellement dirigée de l'Homme contient un danger latent. En effet, les différentes cultures se forment vite et s'isolent entre elles sur la base de modèles culturels différents; elles se comportent alors entre elles comme des »espèces biologiques« séparées. ERIKSON (1966) a utilisé dans ce contexte, pour les cultures ou nations, l'expression pseudo-espèce.

Pumpen □ Eine paarungseinleitende Verhaltensweise bei den meisten Anatinen. Erpel und Ente vollführen dabei rhythmische alternative Pumpbewegungen mit dem Hals (Abb. 87 – *Anas platyrhynchos*). Halspumpen in Form von rhythmischen Halsbewegungen vor und zurück ist u. a. auch eine typische Bewegungsweise während des Schwimmens beim *Coscoroba*-Schwan und beim Teichhuhn, *Gallinula chloropus.* Halspumpen kommt auch beim → *Nickschwimmen* vor.

pumping □ A pre-copulatory display in most of the Anatinae in which the drake and duck perform rhythmic and alternate pumping or thrusting with the neck (Fig. 87 – *Anas platyrhynchos*). A rhythmic forward-backward movement of the neck during swimming (*nodding*) also occurs in the Coscorobid swans or the coot, *Gallinula chloropus.* Neck-pumping or nodding also occurs during → *nod-swimming.*

mouvement de pompage □ Comportement épigame et prénuptial chez la plupart des Anatinae. Le malard et la cane exécutent des mouvements alternatifs verticaux et rythmiques avec leur cou (Fig. 87 – *Anas platyrhynchos*). Un autre mouvement de pompage, mouvement rythmique du cou en avant et en arrière pendant la nage, existe entre autres chez le Cygne *Coscoroba* et la Poule d'eau, *Gallinula chloropus.* Le mouvement de pompage existe aussi pendant la → *nage de coquetterie.*

87

Pumpsaugen → Saugtrinken

Punktattrappe → Attrappe

E. eye spot model □ F. leurre à taches oculaires

Pupillenreaktion □ Die Erweiterung oder Verengung der Pupillen dient der Regulierung des ins Auge eindringenden Lichtes und scheint ein bei Säugern weit verbreitetes Phänomen zu sein. Menschen reagieren bei gleichbleibender Beleuchtung auf ihnen vorgeführte Bilder mit Erweiterung der Pupillen, wenn ihnen etwas gefällt oder ihr besonderes Interesse erweckt; es ist somit ein Indiz für eine orthosympathische Aktivierung (z. B. Freude beim Anblick eines potentiellen Geschlechtspartners). Die Pupille zieht sich dagegen zusammen, wenn die Person etwas wahrnimmt, was sie ablehnt; dies kann als parasympathische Aktivierung angesehen werden (bei heftiger und unangenehmer Gemütserregung und als Reaktion auf einen Unlustreiz). Der *Pupillentest* wurde von HESS (1965) auch bei Homosexuellen durchgeführt, die auf Bilder von Männern mit Pupillenerweiterung reagierten, während die Pupille heterosexueller Männer sich vor allem bei Mädchenbildern erweitert (HESS 1975, HESS *et al.* 1965).

pupillary response □ The expansion and contraction of the pupils serve to regulate the amount of light entering the eye. This mechanism is widespread among mammals. Humans react to pictures that please or interest them with pupillary dilation even though no differences in light intensity occur. This represents an index of orthosympathetic activation as for example during viewing of a potential sex-partner. Conversely, the pupils contract if the person perceives something unpleasant; this can be considered parasympathetic activation, as occurs, e. g., in intense anger or repugnance. With this *pupillary test* HESS (1965) demonstrated that homosexuals react to pictures of men with dilation whereas heterosexual males show dilation primarily to pictures of girls (HESS 1975, HESS *et al.* 1965).

réaction pupillaire □ Elargissement ou rétrécissement des pupilles servant à la régulation de l'éclairement admis dans l'œil; cette réaction semble être très répandue parmi les Mammifères. Sous un éclairement invariable, les Hommes réagissent par une dilatation de la pupille lorsqu'une image perçue leur plait ou les intéresse particulièrement, ce qui constitue un indice d'une activation orthosympathique (par exemple le plaisir pris à contempler un partenaire sexuel pontentiel). Les pupilles, par contre, se contractent lorsque la personne perçoit quelque chose de désagréable ou sans intérêt, signe d'une activation parasympathique (rencontré dans les émotions violentes ou désagréables ou en réponse à un stimulus nociceptif). HESS (1965) a fait de nombreuses recherches à ce sujet et a effectué notamment ce test chez des homosexuels qui présentent une dilatation de la pupille lorsqu'on leur montre des photos de sujets du sexe masculin; les hommes hétérosexuels, par contre, réagissent avec une dilatation des pupilles lorsqu'on leur montre des photos de jeunes filles (HESS 1975, HESS *et al.* 1965).

Pupillentest → Pupillenreaktion
E. pupillary test □ F. test pupillaire

Putzappetenzschwimmen → Putzsymbiose
E. cleaning appetence swimming □ F. nage d'appétence de nettoyage

Putzaufforderungsstellung → Putzsymbiose
E. cleaning invitation posture □ F. posture d'invite de nettoyage

Putzdrang □ Bei Vögeln findet man nach dem Baden einen ausgesprochen hohen Putzdrang, mit einem zunächst steil ansteigenden *spezifischen Aktionspotential (→ SAP),* das dann langsam abklingt. Bei Fischen, die sich zu Putzstationen begeben und geduldig warten, bis der »Kunde« vor ihnen fertig geputzt ist, kann man wohl auch vom Vorhandensein eines Putzdranges sprechen.

»cleaning drive« □ After bathing, birds have a strong tendency to preen; this reflects an initial steeply rising *specific action potential (→ SAP)* which wanes somewhat later. In fish which must wait at a »cleaning station« to be cleaned one can also speak of a »cleaning drive«.

pulsion de nettoyage □ Chez les Oiseaux, après le bain, il existe une tendance à se nettoyer très marquée avec, au début, un *potentiel d'action spécifique (→ PAS)* en augmentation rapide qui, par la suite, diminue lentement en intensité. Chez les Poissons qui se rendent dans une »station de nettoyage« et attendent patiemment que »le client« devant eux leur cède la place, on peut sans doute aussi penser à l'existence d'une pulsion correspondante.

Putzergarnele → Putzsymbiose
E. cleaner shrimp □ F. crevette nettoyeuse

Putzernachahmer □ Der Säbelzahnschleimfisch *Aspidontus taeniatus* (Abb. 88 A), ein kleiner Räuber, hat Form, Färbung und Fortbewegungsverhalten des echten Putzers *Labroides dimidiatus* (Abb. 88 B) kopiert und profitiert auch davon, indem er den Putzkunden kleine Stücke aus Haut, Flossen und Kiemen beißt. Er wird allgemein als Putzernachahmer bezeichnet. → *Putzsymbiose.*

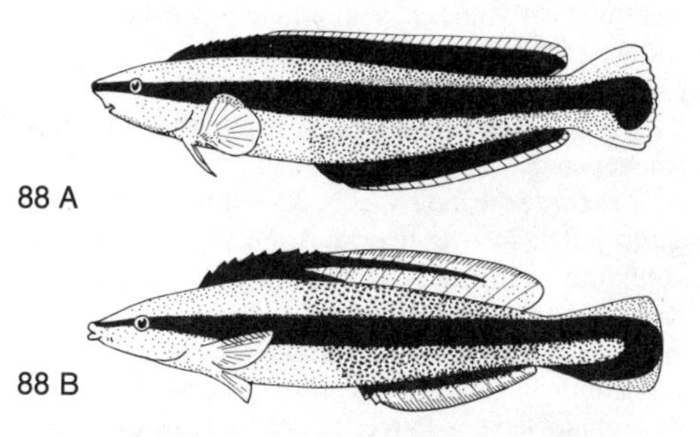

cleaner fish imitator □ The sabertoothed blenny *Aspidontus taeniatus* (Fig. 88 A) has copied the form, coloration and locomotory behaviour of the true cleaner fish, *Labroides dimidiatus* (Fig. 88 B), and profits from the mimicry by biting small pieces out of the skin, the fins and the gills of the fish expecting to be cleaned. → *cleaning symbiosis.*

mime de poisson nettoyeur □ La Blennie *Aspidontus taeniatus* (Fig. 88 A), un petit prédateur, imite la forme, la coloration et même le comportement locomoteur du vrai Poisson nettoyeur *Labroides dimidiatus* (Fig. 88 B) et profite de ce mimétisme. Il approche, comme le nettoyeur, les Poissons qui veulent se faire nettoyer et leur arrache de petits morceaux de la peau, des nageoires et des branchies. → *symbiose de nettoyage.*

Putzertanz → Putzsymbiose
E. cleaner fish dance □ F. danse du poisson nettoyeur

Putzgespräch → Wechselgespräch
E. grooming talk □ F. dialogue épimélétique

Putzknabbern = Knabberputzen → Körperpflegehandlungen
E. nibble-cleaning □ F. se nettoyer en mordillant

Putzritus → Körperpflegehandlungen
E. cleaning ritual □ F. rituel de nettoyage

Putzstation → Putzsymbiose
E. cleaning station □ F. station de nettoyage

Putzstimmung → Putzdrang
E. cleaning mood □ F. motivation de nettoyage

Putzsymbiose □ Eine zwischenartliche Beziehung, die allen Beteiligten nützt. Dem Putzkunden werden die Ektoparasiten entfernt, welche ihrerseits dem Putzer als Nahrung dienen. Solche Putzsymbiosen gibt es nicht nur unter Fischen, von denen sie am bekanntesten sind, sondern auch zwischen Garnelen der Gattungen *Stenopus* und *Periclimontes* als »Putzer«-Garnelen und Fischen als Kunden, weiter an Land zwischen Vögeln als Putzern und Säugern, Krododilen und Schildkröten als Kunden (HEDIGER 1953, EIBL-EIBESFELDT 1955, 1959, WICKLER 1956, VON WAHLERT 1961, ABEL 1971). Zwischen Putzer und Putzkunde haben sich Signale im Dienste der zwischenartlichen Verständigung entwickelt. Bei Fischen hat der Putzer, wie z. B. *Labroides dimidiatus,* einen *Putzertanz* entwickelt, den man auch als Wippschwimmen bezeichnet (WICKLER 1963) und den er nicht nur vor Kunden, sondern auch vor beliebigen anderen Objekten, ja sogar vor dem Taucher ausführt. Dieser Putzer hat auch einen Nachahmer (→ *Putzernachahmer*). Fische, die geputzt werden wollen, geben dies ihrerseits durch eine bestimmte *Putzaufforderungsstellung* (Abb. 89 – *Labroides dimidiatus* putzt *Odonus niger*) bekannt; sie stehen dabei vollkommen still, halten das Maul offen und spreizen Kiemendeckel und Flossen. Durch ein weiteres Signal, meist eine kurze ruckartige Bewegung, gibt der Putzkunde bekannt, wenn er weiterschwimmen will. An Standorten von Putzerfischen finden sich oft eine Vielzahl von Kunden ein, und so haben sich regelrechte Putzstationen entwickelt. LIMBAUGH *et al.* (1961) beobachteten in 6 Stunden über 300 Kunden bei einem einzigen Putzer. Der Mittelmeerputzer *Symphodus melanocercus* kann aber auch sein Revier bzw. seine *Putzstation* verlassen und Brutkolonien anderer Fische aufsuchen, z. B. von *Chromis chromis,* um dort in der Kolonie mit einem *Putzappetenzschwimmen* auf sich aufmerksam zu machen (HEYMER 1972).

89

cleaning symbiosis □ An interspecific relationship which is profitable to all participants. The cleaners deliver their »customers« from the ectoparasites on which they feed. Cleaning symbiosis is found not only between fish where it is most common, but also between shrimps of the genera *Stenopus* and *Periclimontes* as cleaners and fish as customers as well as between birds as cleaners and mammals, crocodiles and turtles as customers (HEDIGER 1953, EIBL-EIBESFELDT 1955, 1959, WICKLER 1956, VON WAHLERT 1961, ABEL 1971). A set of signals has arisen between cleaners and their customers which permits interspecific communication. The cleaner fish *Labroides dimidiatus,* e. g., performs a *cleaner fish dance* (WICKLER 1963) which is demonstrated not only to customers, but also to objects or even divers. This cleaner fish is imitated by a predator (→ *cleaner fish imitator*). Fish which wish to be cleaned assume a *cleaning invitation posture* (Fig. 89 – *Labroides dimidiatus* cleaning *Odonus niger*). They remain motionless, open their mouth and spread the fins and gill covers. By another signal, generally a short jerky movement, the customer indicates to the cleaner that he wants to swim away. The place where a cleaner fish stays is frequently visited by a number of other fish which want to be cleaned; in this manner regular cleaning stations may develop. LIMBAUGH *et al.* (1961) observed over 300 customers visiting a single cleaner fish during 6 hours. The Mediterranean cleaner fish *Symphodus melanocercus* may leave his home base or *cleaning station* and look for the breeding grounds of other fish, e. g., *Chromis chromis,* where he attracts attention with a particular cleaner fish display (*cleaning appetence swimming;* HEYMER 1972).

symbiose de nettoyage □ Relation interspécifique utile à tous les participants. Le »client« se fait débarrasser de ses ectoparasites qui, à leur tour, constituent une grande partie de la nourriture du nettoyeur. Des relations de nettoyage existent non seulement entre différentes espèces de Poissons où elles sont le plus connues, mais aussi chez des Crevettes des genres *Stenopus* et *Periclimontes* qui nettoient des Poissons et chez différentes espèces d'Oiseaux qui nettoient des Mammifères, des Crocodiles et des Tortues (HEDIGER 1953, EIBL-EIBESFELDT 1955, 1959, WICKLER 1956, VON WAHLERT 1961, ABEL 1971). Entre le nettoyeur et le »client« se sont développés des signaux de communication inter-spécifique. Le Poisson nettoyeur *Labroides dimidiatus,* par exemple, effectue une véritable *danse d'invite au nettoyage* qu'il pratique d'ailleurs non seulement devant ses »clients«, mais aussi devant différents objets artificiels et même devant un plongeur. Ce Poisson nettoyeur possède, d'autre part, un mime (→ *mime de poisson nettoyeur*). Les Poissons qui veulent se faire nettoyer indiquent cette disposition par une posture particulière (*Putzaufforderungsstellung*): ils se tiennent parfaitement immobiles, ouvrent largement la bouche et leurs opercules branchiaux et étalent leurs nageoires (Fig. 89 – *Labroides dimidiatus* nettoyant *Odonus niger*). Un autre signal, en général un mouvement bref et saccadé, informe le nettoyeur que le »client« veut continuer son chemin. Les Poissons qui veulent se faire nettoyer se rendent dans les territoires des nettoyeurs et dans ce contexte, se sont développées de véritables *stations de nettoyage.* LIMBAUGH *et al.* (1961) a observé que, pendant 6 heures, plus de 300 poissons se sont présentés devant un même nettoyeur. Le nettoyeur méditerranéen *Symphodus melanocercus* possède également un territoire, mais il peut le quitter pour se rendre dans les colonies reproductrices d'autres Poissons tels que *Chromis chromis;* il a développé en conséquence une nage particulière qui signale sa présence (HEYMER 1972).

Putztanz □ Bei diesem Verhalten schüttelt die Biene ihren Körper in schneller Bewegung nach rechts und links hin und her. Andere Bienen werden nach Meinung mancher Beobachter dadurch veranlaßt, die Haare der Tänzerin an Körperstellen, die ihr selbst schwer zugänglich sind, mit ihren Mandibeln zu putzen. HAYDAK (1929, 1945) spricht nach der Art der Bewegung von *shaking dance,* MILUM (1947, 1955) nach dem angestrebten Ziel von *grooming dance* (vergleiche auch LINDAUER 1948 und VON FRISCH 1965).

grooming dance □ Bees which shake their body rapidly back and forth are thought to attract other bees for cleaning the bristles on areas of their body which they, themselves, cannot reach. HAYDAK (1929, 1945) designated this behaviour as a »shaking dance«, whereas MILUM (1947, 1955) in reference to the presumed function called it »grooming dance« (see also LINDAUER 1948 and VON FRISCH 1965).

danse de nettoyage □ Chez l'Abeille, mouvement latéral et rapide du corps, alternativement vers la droite et la gauche. Selon un certain nombre d'observateurs, les autres Abeilles qui se trouvent à proximité, se sentent sollicitées et se précipitent alors sur la danseuse pour nettoyer avec leurs mandibules les poils aux endroits du corps que cette dernière ne peut atteindre elle-même. HAYDAK (1929, 1945) qui a décrit ce phénomène a appelé ce comportement *shaking dance,* en raison des modalités de cette manœuvre. MILUM (1947, 1955) parle de *grooming dance* en se référant à sa fonction (voir aussi LINDAUER 1948 et VON FRISCH 1965).

Putzverhalten → Körperpflegehandlungen E. grooming behaviour (mammals), preening behaviour (birds) □ F. comportement de nettoyage

---

R

Radfahrer-Reaktion □ Von GRZIMEK (1944) gebrauchter Ausdruck für eine Handlung am Ersatzobjekt. Man versteht darunter auch die Weitergabe nach »unten« einer von »oben« subierten Bedrohung (der volkstümliche Ausdruck »nach oben buckeln und nach unten treten« ist in unserer Gesellschaft weit verbreitet). Ein Fuchs, der von einem ranghöheren bedroht wird, in dessen Nähe aber zufällig ein rangtieferer steht, reagiert das aktivierte Defensivverhalten an diesem ab und schnappt nach ihm. In der Rangordnung der menschlichen Gesellschaft weit verbreitetes Phänomen. Der Ausdruck hat sich in der ethologischen Literatur nicht eingebürgert.

redirected aggression to an inferior □ The German term *Radfahrer-Reaktion* (bicyclist response) was used by GRZIMEK (1944) for a response directed toward a substitute, in particular the transfer of a threat received from a superior to a subordinate. A fox threatened by a higher-ranking member of the group may release his activated defensive behaviour on a lower ranking individual by snapping at him. Similar phenomena are widespread in human society. The term »bicyclist response« has not become accepted in the ethological literature.

répression vers l'inférieur □ L'expression allemande de *Radfahrer-Reaktion* (réaction de cycliste) a été utilisée par GRZIMEK (1944) pour désigner une activité sur un objet de remplacement. On entend par ce terme la transmission vers le »bas« dans un contexte hiérarchique, de menaces subies par le »haut«. Ainsi, un Renard qui a été menacé par un congénère hiérarchiquement supérieur, essaie d'agresser un animal de rang inférieur si ce dernier se trouve à proximité de lui. Dans les hiérarchies sociales humaines, ce comportement est largement répandu. L'expression de GRZIMEK n'a pourtant pas trouvé de résonance dans la bibliographie éthologique.

Radstellung □ Begattungsstellung bei Libellen, die durch die im Tierreich einmalige Anordnung des Kopulationsorgans des ♂ am 2. Abdominalsternit zustande kommt (Abb. 90 A – *Orthetrum coerulescens,* Anisoptera; B – *Agrion lindeni,* Zygoptera). Das männliche Kopulationsorgan ist mit den inneren Geschlechtsorganen nicht verbunden und muß demzufolge vor der Begattung vom Genitalporus des 9. Sternits mit Sperma aufgefüllt werden (HEYMER 1966, 1973).

»wheel posture« □ Copulation posture in dragonflies which is determined by the unique position of the male copulatory organ on the second abdominal sternite (Fig. 90 A – *Orthetrum coerulescens,* Anisoptera; B – *Agrion lindeni,* Zygoptera). The male's organ is not connected to the internal sex organs. Hence, sperm must be transferred from the genital pore in the 9th sternite to the copulatory organ prior to copulation (HEYMER 1966, 1973).

accouplement tête-bêche □ Posture d'accouplement propre aux Libellules, unique dans le monde animal, qui résulte d'une disposition particulière de l'organe copulateur ♂ (Fig. 90 A – *Orthetrum coerulescens,* Anisoptera; B – *Agrion lindeni,* Zygoptera). En effet, celui-ci est situé sur le 2ème sternite abdominal et il n'y a pas de voies de communication avec les organes sexuels internes. Par conséquent, il doit,

90 A

90 B

avant l'accouplement, être chargé de spermatozoïdes provenant du pore génital du 9ème sternite (HEYMER 1966, 1973).

Rahmenhandlungen □ Unter diesem Begriff (HASSENBERG 1965) sind einige heterogene Handlungen (vor allem Putz-, Kratz- und Freßbewegungen) zusammengefaßt, denen gemeinsam ist, daß sie als nicht situationsbedingtes (also syndromfremdes, allochthones) Verhalten die Bewegungsabläufe der Ruheappetenz begleiten und »umrahmen«. → *Übersprungbewegung.*

structural activities □ A number of heterogenous behaviours (especially grooming, scratching, and feeding movements) are grouped under this term (HASSENBERG 1965) which refer to activities accompanying and »surrounding« the appetence for rest, and requiring no other specific context for their occurrence. → *displacement activity.*

activités d'encadrement □ HASSENBERG (1965) a décrit sous le terme allemand de *Rahmenhandlungen* un certain nombre d'activités hétérogènes, en particulier de nettoyage, de grattage et d'alimentation, qui accompagnent et »encadrent« les activités d'appétence au repos. → *activité substitutive.*

Rahmenkoordinationen □ Unter Rahmenkoordinationen werden Haltung und Bewegung einzelner Körperstrukturen verstanden, die mehr oder weniger häufig zugleich mit den → *Grundkoordination(en)* auftreten. Mimik, Kopfhaltung und → *Ohrenstellung* zeigen Zusammenhänge mit dem Miktions- und Defäkationsverhalten. Jedoch sind diese Verhaltenselemente nicht so fest verankert, daß sie immer und obligatorisch auftreten wie die Grundkoordinationen.

associated movements □ Refers to the postures and movements which occur in conjunction with certain → *basic movement pattern(s)*. In mammals, certain body and head postures as well as → *ear position,* for example, are associated with urinating and defecating; however, these behaviour elements are not that firmly fixed that they always occur in these situations. In this sense they differ from basic movement coordinations.

mouvements d'accompagnement □ Postures et mouvements de parties du corps qui apparaissent plus ou moins fréquemment en même temps que les → *coordinations motrices de base.* Une certaine mimique, une certaine position de la tête ou des oreilles, indiquent des relations avec le comportement de miction et de défécation. Ces éléments comportementaux ne sont cependant pas aussi invariables et obligatoires dans leur apparition que les coordinations motrices de base.

Räkelsyndrom □ Es handelt sich um einen stoffwechselphysiologisch bedingten Verhaltensablauf in einem Zusammenhang mit Funktionen des Kreislaufs. Die bekanntesten Bewegungsformen sind das *Gähnen* und das *Sich-Strecken.* Diese Verhaltensformen unterscheiden sich prinzipiell von vielen anderen Bewegungsweisen dadurch, daß sie nicht durch Außenbedingungen aktiviert oder ausgelöst werden (PEIPER 1956, TEMBROCK 1954, 1961, LUTTENBERGER 1975). Ihnen kann demnach kein → *AAM* zugeordnet werden. Gähnen und Streckbewegungen sind oft miteinander gekoppelt oder folgen aufeinander. Daher darf auch ein gemeinsamer Ursachenkomplex angenommen werden. Gähnen kann aber auch ansteckend wirken, insbesondere beim Menschen (Abb. 91 A – Bayaka-Pygmäe aus Zentralafrika).

Beim *Sichstrecken* (Streckbewegung) folgt die Impulswelle, die physiologisch auf dem Wege vom Atemzentrum her aktiviert wird, im allgemeinen der *Vorn-Hinten-Regel.* Es beginnt mit dem Gähnen, dann folgt das Vornstrecken und danach das Hintenstrecken, wobei die motorische Bewegungswelle auch auf den Schwanz übergreifen kann. Neben diesem *Sukzessivstrecken* kann man auch ein *Simultanstrecken* beobachten, bei dem alle Komponenten annähernd gleichzeitig aktiviert werden (Abb. 91 B – Mara, *Dolichotis patagonum*).

Vögel strecken sich im Gegensatz zu den Säugern meist einseitig (*Seitabstrecken*). Eine Körperseite bleibt in Ruhelage, auf der anderen wird der Flügel weit ausgebreitet und gleichzeitig mit dem entsprechenden Bein nach hinten weggestreckt (Abb. 91 C – Graugänse, *Anser anser*). PETZOLD (1964) zählt dieses Verhalten bei Schwänen zum Komfortverhalten. So wie es HEINROTH (1930) beschrieb, müßte es aber in den Kontext des Räkelsyndroms gehören.

stretching syndrome □ Certain behaviour sequences associated with metabolic regulation and circulatory function. The best known such patterns are *yawning* and *stretching*. These forms of behaviour differ from others principally in that they do not require external stimuli for elicitation (PEIPER 1956, TEMBROCK 1954, 1961, LUTTENBERGER 1975). Hence they cannot be classified as dependent on innate releasing mechanisms (→ *IRM*). Yawning and stretching frequently follow each other; therefore, a common causal complex may be postulated. Yawning, especially in humans, may also have a contagious effect (Fig. 91 A – Bayaka Pygmy from Central Africa).

During stretching movements the impulse set into action by the breathing center in general follows the *rostral-caudal rule*. Yawning occurs first followed by stretching of the fore-limbs, then the hindlimbs and possibly the tail. In addition to *successive stretching, simultaneous stretching* has also been observed in which all motor components may be activated at the same time (Fig. 91 B – Mara, *Dolichotis patagonum*).

In contrast to mammals, birds stretch on one side at a time (*lateral stretching*). While one side of the body remains at rest the wing and leg on the other side are stretched caudally (Fig. 91 C – greylag geese, *Anser anser*). PETZOLD (1964) considered this behaviour in swans as a comfort movement; however, it may also be considered to be part of the stretching syndrome (HEINROTH 1930).

syndrome d'étirement □ Processus de comportement sous-tendu par des facteurs métaboliques et en rapport avec la fonction circulatoire. Les formes les plus connues sont le *bâillement, l'étirement latéral* chez les Oiseaux, et *l'étirement complet* du corps chez les Mammifères, y compris l'Homme. Ces phénomènes comportementaux se distinguent de beaucoup d'autres mouvements principalement par le fait qu'ils ne sont pas activés ou déclenchés par des conditions extérieures (PEIPER 1956, TEMBROCK 1954, 1961, LUTTENBERGER 1975). Il n'est donc pas possible de les associer à un mécanisme déclencheur inné (→ *MDI*). Le bâillement et les mouvements d'étirement sont souvent simultanés ou se déroulent en succession immédiate. Il est donc possible de supposer qu'ils dépendent d'un même ensemble de causes. Le bâillement, lui, peut avoir des effets contagieux, en particulier chez l'Homme (Fig. 91 – Pygmée Bayaka de l'Afrique Centrale).

Pendant le *mouvement d'étirement,* la vague d'impulsion activée physiologiquement par le centre respiratoire, obéit en général à la *règle de progression antéro-postérieure*. Elle commence par un bâillement, suivi d'un étirement antérieur, puis d'un étirement postérieur d'où le mouvement moteur peut s'étendre à la queue. A côté de ce phénomène qu'on appelle

91 A

91 B

91 C

l'étirement successif, on peut observer un *étirement simultané* au cours duquel toutes les composantes sont activées à peu près en même temps (Fig. 91 B – Mara, *Dolichotis patagonum*).

Contrairement aux Mammifères, les Oiseaux ne s'étirent que d'un seul côté à la fois (*étirement latéral*). Un côté du corps reste en position de repos pendant que de l'autre côté, l'aile est dépliée et simultanément la patte étendue en arrière (Fig. 91 C – Oies cendrées, *Anser anser*). PETZOLD (1964) considère ce comportement chez les Cygnes comme un mouvement de confort, mais selon la description de

HEINROTH (1930), il s'inscrirait mieux dans le contexte du syndrome d'étirement.

Rammelzeit → Brunst

E. »shag time« – slang term for the breeding season in hares and rabbits □ F. période de parade nuptiale et d'accouplement chez les Lièvres et les Lapins

Rangdemonstration □ Verhaltensweisen zur Aufrechterhaltung und Festigung des Ranges innerhalb einer Gruppe, die nicht immer vom α-Tier ausgehen, sondern auch von anderen Gruppenmitgliedern, aber wohl immer gegen rangniedere Individuen eingeleitet werden. → *Aufreitdrohung,* → *Wutkopulation.*

rank demonstration □ Behaviour patterns for the maintenance and strengthening of the rank positions within a group. The actions are not always initiated by the α-animal, but are invariably directed against lower ranking individuals. → *mounting threat,* → *rage copulation.*

démonstration de dominance □ Modes de comportement permettant le maintien et la consolidation d'une position hiérarchique à l'intérieur d'un groupe social. Ces manifestations ne sont pas toujours induites par l'animal α, mais souvent aussi par d'autres membres du groupe, mais elles sont presque toujours dirigées vers des individus inférieurs. → *chevauchement de menace,* → *copulation de colère.*

Rangkämpfe → Rangordnung

E. dominance fighting □ F. combats de dominance

Rangordnung □ Das Phänomen der Rangordnung hat als erster SCHJELDERUP-EBBE (1922) an Hühnern untersucht; heute spricht man bei Hühnern vorwiegend von → *Hackordnung.* Am Futterplatz genießen einige Hennen gewisse Vorrechte. Sie dürfen zuerst zur Futterstelle und hacken andere, rangniedere Hennen. Es ist genau festgelegt, wer wen hacken darf. Es gibt neben einfachen linearen Rangordnungen auch komplizierte *Dreiecksverhältnisse,* in denen zwar α über β und β über γ, γ aber über α dominiert. Rangordnung gibt es bei Tieren vorwiegend in individualisierten Verbänden, wo sie die soziale Organisation aufrecht erhält, während sie in anonymen Verbänden fehlt. Die Rangordnung entwickelt sich innerhalb einer Gruppe aufgrund gelegentlicher Kämpfe. Jedes Gruppenmitglied merkt sich im Laufe der Auseinandersetzungen, wer ihm über- und wer ihm unterlegen ist, und richtet sein Verhalten danach. In einer einmal festgelegten Rangordnung wird im allgemeinen selten gekämpft. Die Rangordnung wird durch kurzes Drohen, sog. → *Rangdemonstration(en),* aufrecht erhalten. Genügt ein α-Tier jedoch nicht mehr seinen Anforderungen, werden Rangkämpfe vor allem von den nächsttieferen Individuen inszeniert, die versuchen, sich den hohen Rang zu erkämpfen.
Der zukünftige Rang eines Tieres wird zuweilen auch von der Rangstellung der Mutter bestimmt, also über den Weg der Tradition weitergegeben, wie man bei Rhesusaffen und japanischen Makaken nachgewiesen hat. Man kann in einem solchen Falle von einer *Rangordnungstradition* sprechen, die es auch in der menschlichen Gesellschaft gibt (KAWAI 1958, KOFORD 1963).
Das ranghöchste Tier eines Systems oder einer Gesellschaft nennt man α-Individuum, die darauffolgenden β, γ, δ usw. Die niedrigste Position eines Tieres ist die Omegastellung, benannt nach dem letzten Buchstaben im griechischen Alphabet Ω. Ein solches Omegaindividuum wird in der Regel von allen übrigen Gruppenmitgliedern dominiert, gehackt oder gebissen und verjagt.

rank order, hierarchy system □ The occurrence of rank order in animal social behaviour was first studied by SCHJELDERUP-EBBE (1922) in chickens. Today the term → *peck order* replaces rank order for chickens and this phenomenon is often seen at the feeding station where certain hens feed first, pecking at any lower ranking hens that approach. Besides simple linear rank orders there are also more complicated *triangular relationships* in which α dominates over β and β dominates over γ, but γ dominates over α. Rank order occurs predominantly in individualized groups, where it promotes social organization, but is not found in anonymous groups. The rank order develops within a group through occasional quarrels. Each member of the group determines in the course of these altercations who is superior and who is subordinate to him and his subsequent behaviour reflects this. In a stable rank order very little fighting occurs; rather, a → *rank demonstration,* e. g. brief threats, reinforces the rank order structure. Should the α animal fail to sustain his position, dominance fighting occurs among the next lower ranking animals and results in a new α animal.
Sometimes the rank order among animals is closely related to the rank order of their mother. Such a *tradition of ranking order* has been described in the rhesus monkey and the Japanese macaques. Similar tradition of ranking order occurs in human society (KAWAI 1958, KOFORD 1963).
The highest ranking animal of a system or society is called α individual, the next following ones β, γ, δ, etc. The lowest rank in a social hierarchy is the omega position according to the last letter in the Greek alphabet Ω. An omega animal is normally dominated by all other members of the group which peck or bite at it and chase it away.

hiérarchie sociale □ Ce phénomène a été décrit pour la première fois par SCHJELDERUP-EBBE (1922) à la suite de ses recherches effectuées sur des Poules; chez ces animaux, on parle aujourd'hui plutôt de → *hiérarchie de becquetage.* A l'emplacement de la nourriture, un certain nombre de Poules jouissent de certains privilèges et peuvent ainsi se diriger les premières vers la mangeoire et donner des coups de bec à d'autres Poules de rang inférieur. Il est exactement

établi qui a le droit d'attaquer qui. Il existe, à côté des hiérarchies strictement linéaires, des relations plus compliquées, comme par exemple la *hiérarchie triangulaire* dans laquelle α domine β et β domine γ alors que γ domine α. Une hiérarchie sociale existe surtout dans des groupements individualisés où elle maintient une certaine organisation sociale. Par contre, dans les groupements anonymes, il n'y a pas de hiérarchie. La hiérarchie à l'intérieur d'un groupe s'établit et se maintient à l'aide de luttes occasionnelles. Chaque membre du groupe retient au cours des différents affrontements quel individu lui est supérieur ou inférieur, et règle son comportement en conséquence. Dans une hiérarchie une fois établie, les combats sont rares. Elle est maintenue par de simples attitudes de menace ou → *démonstration(s) de dominance.* Dans le cas où un animal α ne satisfait plus aux exigences de son statut, des combats de dominance s'engagent, déclenchés surtout par des individus subdominants qui saisissent ainsi l'occasion de s'emparer du rang suprême. La future position hiérarchique d'un animal est parfois déterminée par celle que détient sa mère. Elle est donc transmise par voie de tradition (*hiérarchie sociale par tradition*) comme cela a été prouvé chez les Macaques Rhésus et les Macaques japonais. Ce phénomène se rencontre aussi dans les sociétés humaines (KAWAI 1958, KOFORD 1963).
L'individu le plus élevé dans une société est appelé individu α, suivi de β, γ, δ etc. La dernière position hiérarchique d'un animal à l'intérieur d'un groupe social est la position oméga, ainsi appelée selon la dernière lettre de l'alphabet grec Ω. En règle générale, un individu oméga est dominé par tous les autres membres du groupe et est, de ce fait, menacé, mordu et pourchassé.

Rangordnungsdemonstration → Rangdemonstration

Rangordnungskämpfe = Rangkämpfe → Rangdemonstration

Rangordnungstradition → Rangordnung
E. tradition of ranking order □ F. hiérarchie sociale par tradition

Ranz → Brunst
E. courtship period in foxes and badgers □ F. période de parade nuptiale chez le Renard et le Blaireau

Raubfeind □ Nicht das Raubtier schlechthin, sondern stets in Zusammenhang mit der Beute. Für Mäuse z. B. sind Katze und Bussard typische Raubfeinde.

natural predator □ Refers not to predators in general, but to specific animals which attack a given species, i. e. the cat and the buzzard are natural predators of mice.

prédateur potentiel □ Il ne s'agit pas de prédateurs dans l'absolu, mais par rapport à une proie déterminée. Pour les Souris, par exemple, le Chat et la Buse sont des prédateurs potentiels typiques.

Raumorientierung → Orientierung
E. space orientation □ F. orientation dans l'espace

Raum-Zeit-Gefüge □ In der Verhaltensforschung die Analyse und Prüfung eines Tieres, das zu einer bestimmten Zeit und an einem bestimmten Ort ein bestimmtes Verhalten zeigt. → *Motivationsanalyse.*

spatio-temporal structure □ The analysis and testing of an animal's behaviour in relation to its time and place of occurrence. → *motivational analysis.*

système spatio-temporel □ En éthologie, examen et analyse des activités d'un animal qui, à un moment et à un endroit déterminés, montre un comportement défini. → *analyse motivationnelle.*

Rauschzeit → Brunst
E. reproduction period in the wild boar □ F. période de reproduction chez le Sanglier

Reafferenz → Reafferenzprinzip

Reafferenzprinzip □ Erklärungsprinzip für Regelungsvorgänge zwischen der vom Körper in Gang gesetzten Motorik und den zugeordneten Meldungen der Sinnesorgane (Rezeptoren), das von VON HOLST und MITTELSTAEDT (1950) eingeführt wurde. Generell werden Meldungen, die von der Peripherie (den Rezeptoren) zum Zentrum (ZNS) verlaufen, als *Afferenzen* und alle Impulse, die zentrifugal verlaufen und die Motorik in Gang setzen, als *Efferenzen* bezeichnet. Das Reafferenzprinzip fordert für jede vom Zentrum her spontan aktivierte Motorik eine Efferenzkopie, die den vorgesehenen Sollwert der Bewegung fixiert und erst gelöscht wird, wenn von den ausführenden Systemen die entsprechende Rückmeldung kommt, die man als Reafferenz bezeichnet. Deckt sie sich mit der Efferenzkopie, so wird der Bewegungsimpuls eingestellt. Wird jedoch die Ausführung durch eine »unerwartete« Störung beeinträchtigt, dann bleibt die Kopie erhalten und der »Befehl« zur Bewegung bestehen, bis sie ausgeführt ist. Solche zusätzlichen Meldungen aus der Peripherie werden als Exafferenzen bezeichnet.

reafference principle □ A concept developed by VON HOLST and MITTELSTAEDT (1950) to explain the regulation and interaction of internal signals and sensory signals in directing and in coordinating bodily movements. In general, signals from the periphery (sense organs) running to the processing unit (CNS) are termed *afferences,* and all impulses that run centrifugally and set the motor units in action are called *efferences.* The reafference principle requires storage of an efference copy of each spontaneous activation of a motor unit by the processing unit; this efference copy fixes the *Sollwert* or reference value of parameters required to execute the movement and this information guides the response until the reafference from the motor unit to the processing unit indicates accordance with the *Sollwert* (reference value). If during execution of the movement an »unexpected« disturbance is encountered, the efference copy is reexamined in relation to the new information and the executing maneuvers are corrected until the goal is reached. All such additional inputs from the periphery are termed exafferents.

principe de réafférence □ Systématisation théorique (élaborée par VON HOLST et MITTELSTAEDT 1950) de la régulation des comportements moteurs, en fonction des informations, ou *afférences,* venant de la périphérie (organes récepteurs) et acheminées vers les centres nerveux; les impulsions centrifuges activant une conduite sont des *efférences.* Pour chaque manifestation motrice, le principe de réafférence postule la formation d'une copie d'efférence, qui spécifie les paramètres du mouvement »prévu« et de sa trajectoire théorique. Cette trace centrale du message moteur ne s'efface pas avant que ne parvienne au système nerveux central une information en retour (proprioceptive, visuelle . . .) relative à l'exécution du mouvement, à sa trajectoire réelle: cette information en retour est la réafférence. Si elle est conforme à la copie d'efférence, elle l'annule et l'incitation motrice centrale prend fin. Par contre, si l'exécution du mouvement est entravée par une perturbation »inattendue«, la copie d'efférence se maintient, l'ordre moteur persiste et le mouvement subit des corrections guidées par les exafférences, ou informations extérieures additionnelles qui tiennent compte de la perturbation.

Reaktionsbereitschaft → Handlungsbereitschaft
E. responsiveness □ F. état réactionnel

Reaktionskette □ Programmierte Handlungsfolge, auch Handlungskette genannt. Verhalten, welches vor allem im sozialen Bereich auftritt, wo viele Verhaltensabläufe aus einer festgelegten Kette aufeinanderfolgender Einzelhandlungen bestehen, die stets weitgehend in gleicher Weise ablaufen; → *Balzkette,* → *Kommentkampf.* Die langwährende und höchst komplizierte Paarungszeremonie der Weinbergschnecke *Helix pomatia,* die ihren Höhepunkt im Ausstoßen der beiden Liebespfeile erreicht, hat schon SZYMANSKI (1913) als echte Reaktionskette erkannt (siehe auch MEISENHEIMER 1921). Reaktionsketten treten auch in anderen Bereichen des Verhaltens eines Tieres auf, wie z. B. bei der Nahrungssuche der Honigbiene oder beim Beutefangverhalten vieler Arten; besonders schön belegt beim Bienenwolf *Philanthus triangulum* (TINBERGEN 1935, 1964, Instinktlehre).

reaction chain □ A pre-programmed sequence of responses or reactions especially among behaviours occurring in a social context; many single responses follow each other in a certain sequence or chain which seldom varies in order; → *courtship chain,* → *ritualized fight.* The extended and highly complicated mating ceremony of the snail, *Helix pomatia* which reaches its zenith with the expulsion of the "love darts", was recognized as a response chain by SZYMANSKI (1913); see also MEISENHEIMER (1921). Response chains also occur in other behaviour contexts, e. g., in the honey bees' food searching or in prey catching of various species. TINBERGEN (1935, 1964, »The Study of Instinct«) describes a particularly beautiful example in the bee-hunting digger wasp, *Philanthus triangulum.*

réaction en chaîne □ Comportement et activité programmés apparaissant surtout dans le domaine social et dont les différents éléments comportementaux forment une sorte de chaîne qui se déroule généralement de la même manière; → *Balzkette,* → *combat rituel.* Le cérémonial d'appariement extrêmement complexe chez *Helix pomatia,* par exemple, dont le point culminant est atteint lors de la projection synchronisée de deux dards »d'amour«, a déjà été reconnu comme une véritable réaction en chaîne par SZYMANSKI (1913); voir aussi MEISENHEIMER (1921). Des réactions en chaîne apparaissent également dans d'autres domaines du comportement animal, comme par exemple lors de la recherche alimentaire chez l'Abeille mellifère ou lors du comportement de capture de proie chez un grand nombre d'espèces; particulièrement bien décrit chez le Loup d'Abeilles, *Philanthus triangulum* (TINBERGEN 1935, 1964, »Etude de l'Instinct«).

Reaktionsnorm □ In der Verhaltensforschung die Norm des Reagierens eines Tieres auf richtunggebende Reize, wobei die Form der Bewegung, die Bewegungsnorm, infolge der Erbkoordination unabhängig vom jeweiligen Reizangebot bleibt.

reaction norm □ The normal pattern of reaction of an animal to directional stimuli in which the form of the movement remains independent of the respective stimulus.

norme de réaction □ On désigne ainsi les propriétés invariables d'un mouvement en réponse à des stimulations directionnelles (forme, déroulement spatio-temporel de ce mouvement) qui seraient indépendantes du stimulus déclencheur et spécifiées par une »coordination motrice héréditaire«.

Reaktionsträgheit → Anfangsreibung

Reaktionszeit = Latenzzeit → Reiz
E. reaction time □ F. temps de latence

Reflex □ Eine nicht willkürlich auslösbare Reaktion des tierischen und menschlichen Organismus auf einen Reiz, die in einem weitgehend festgelegten *Reflexbogen* abläuft. Der Reflexbogen besteht aus folgenden Elementen: dem Rezeptor als Aufnahmeorgan für den Reiz, der afferenten Leitungsbahn, die Rezeptorenerregungen zu Synapsen und Ganglienzellen im ZNS leitet, der efferenten Bahn, über die Erregungsimpulse zum Effektor laufen, in dem dann die Antwortreaktion erfolgt. Stets sind an einem Reflex zahlreiche gleichwertige Reflexbögen beteiligt. Die Gesamtheit ihrer Synapsen und Ganglienzellen im ZNS nennt man Reflexzentrum.

reflex □ An involuntary response of an organism to a stimulus in which the response proceeds in a clearly prescribed fashion or *reflex arc.* Reflex arcs consist of the following elements: the receptor organ which responds to the stimulus; the afferent pathway which conducts the action potential from the receptor to the synapses and ganglion cells of the CNS; the efferent pathway carrying the impulse to the effector organ which produces the appropriate response. Always the reflex itself consists of numerous equivalent reflex arcs. The totality of their synapses and ganglion cells in the CNS is termed the reflex center.

réflexe □ Réaction d'un organisme animal ou humain en réponse à une stimulation et qui n'est pas déclenchable à volonté. Elle se déroule selon un *arc réflexe* déterminé, constitué des éléments suivants: du récepteur qui capte le stimulus, de la voie afférente qui achemine les messages sensoriels vers les synapses et les neurones centraux, de la voie efférente par laquelle les impulsions motrices sont conduites vers l'effecteur qui exécute la réponse. Dans un réflexe, il y a toujours plusieurs arcs réflexes équivalents engagés. L'ensemble de leurs synapses et de leurs neurones centraux constitue un centre réflexe.

Reflexbogen → Reflex
E. reflex arc □ F. arc réflexe

Reflexverspätung □ Beim Beutefangverhalten bei Fröschen und Kröten tritt das letzte Glied, der Zungenschlag, einmal durch den adäquaten Reiz der Beutewahrnehmung ausgelöst, mit einer gewissen Verzögerung *(Tardus)* auf und wird nicht ortskorrigiert. HINSCHE (1935) sprach von einem »T-Phänomen« des *Schnappreflexes,* nach ihm auch LESCURE (1965). Bewegt sich die Beute noch in dieser Phase, so schnappt der Frosch oder die Kröte daneben und verfehlt die Beute (*Nach-Nichts-Schnappen*). Der Zungenschlag ist eine echte, verhältnismäßig einfache → *Erbkoordination* (EIBL-EIBESFELDT 1951).

delayed reflex □ The tongue strike is the final component of the prey catching behaviour of frogs and toads. It follows the stimulus which releases it with a certain delay, and once released, it is not subject to correction should the prey have moved in the meantime. HINSCHE (1935) termed this the »T-phenomenon« of the *snapping reflex;* this term is also used by LESCURE (1965) (*objectless snapping*). The tongue strike is a true example of a relatively simple instinctive response or → *fixed action pattern* (EIBL-EIBESFELDT 1951).

retardement du réflexe □ Dans le comportement prédateur chez les Grenouilles et les Crapauds, la dernière composante – le coup de langue –, une fois déclenchée par la stimulation adéquate de perception de la proie, se déroule avec un certain retard et ne subit pas de correction directionnelle. On parle aussi, en pareil cas, de »phénomène T« (HINSCHE 1935, LESCURE 1965). Si la proie se déplace pendant ce temps, la Grenouille ou le Crapaud manquent alors leur but (*happer à vide*). Le coup de langue est une → *coordination héréditaire* véritable, mais relativement simple (EIBL-EIBESFELDT 1951).

Refraktärstadium □ Das einer abgeleiteten Erregung folgende Stadium, in dem das erregbare System auf einen zweiten, gleichgroßen Reiz nicht oder mit verminderter Amplitude antwortet.

refractory period □ The interval following excitation of a system during which a subsequent stimulus of equal magnitude elicits no response or a response reduced in amplitude.

période réfractaire □ Situation qui suit une excitation et dans laquelle le système excitable ne répond plus ou avec une amplitude affaiblie à une seconde stimulation équivalente.

Regelkreis □ Zusammengesetztes System, bei dem ein bestehender Wert oder »Istwert« (Ausgangswert) fortlaufend mit einer vorgegebenen Größe oder »Sollwert« (Führungsgröße) verglichen und letzterer permanent angepaßt wird. → *Rückkoppelung.*

feedback circuit □ A self regulating system in which the actual value (»Istwert«) of a parameter is continuously compared with and corrected towards a predetermined reference value (»Sollwert«). → *feedback.*

boucle de régulation □ Système cybernétique dans lequel la valeur de la sortie est continuellement comparée à une grandeur prédéterminée. → *rétroaction.*

Regressionsverhalten → Infantilismus
E. behavioural regression □ F. comportement régressif

Reifung □ In der Ethologie versteht man darunter die Heranreifung von angeborenen Verhaltensweisen, die nicht gelernt zu werden brauchen und zur adäquaten Zeit auftreten. Didaktische Beispiele für die Reifung eines Zielmechanismus bei Hühnerküken finden wir bei HESS (1956) und über Reifung der Selektivität des *angeborenen Auslösemechanismus* (→ *AAM*) der Nachfolgereaktion bei den Cichlidae bei KUENZER (1962, 1968).

maturation □ In Ethology maturation refers to innate behaviour patterns which appear at a given point in the animal's life cycle without prior learning. HESS (1956) has described the maturation of a goal-directed response in chicks, and KUENZER (1962, 1968) reported the maturation of selection of the *innate releasing mechanism* (→ *IRM*) in the following response of young Cichlids.

maturation □ En éthologie, la maturation est l'apparition, au moment adéquat, de comportements innés dont l'apprentissage n'est pas nécessaire. Des exemples instructifs sont donnés par HESS (1956) pour les mécanismes de picorage chez les Poussins et par KUENZER (1962, 1968) pour la sélectivité du *mécanisme déclencheur inné* (→ *MDI*) de la »réaction de suivre« chez les Cichlidae. A ne pas confondre avec le sens donné en biologie.

Reiz □ Äußere oder innere reversible Zustandsänderung, welche ab einer gewissen → *Reizschwelle* eine oder mehrere Sinneserregung(en) auslöst. Zur Aufnahme der verschiedenen Reize mechanischer, physischer oder chemischer Art hat der tierische Organismus in der Regel verschiedene, den Bedingungen angepaßte Sinnesorgane ausgebildet. Jedes dieser Rezeptororgane besitzt seine eigenen adäquaten Reize, auf welche es bei relativ niedriger Reizschwelle anspricht. Gegenüber einem adäquaten Reiz spielt das Sinnesorgan die Rolle eines Transduktors, d. h. die speziellen Reizarten werden in nervöse Erregung transformiert und in dieser Form an das zentrale Nervensystem weitergeleitet. Die für ein Sinnesorgan bzw. für einen Rezeptor typische Reizart wird als adäquater Reiz den inadäquaten Reizen gegenübergestellt. Die Zeit zwischen Reizbeginn und Reizantwort bezeichnet man als Latenzzeit. → *Schlüsselreiz.*

stimulus □ A reversible change in the external or internal environment which, beyond a certain → *stimulus threshold,* elicits one or more sensory excitations. Animals have developed certain sense organs which are adapted for receiving the various mechanical, physical or chemical stimuli. Each such receptor organ responds to its own »adequate stimulus« at a relatively low threshold. In reference to the adequate stimulus, the receptor is a transducer of the stimulus energy into nerve impulses which are then transmitted to the CNS. In contrast to the special »adequate stimulus« mode for this receptor, all other types of stimuli are termed inadequate. The time between stimulus onset and response is termed latency. → *key stimulus.*

stimulus (pl. **stimulus** ou **stimuli**) □ Modification réversible du milieu extérieur ou intérieur, engendrant, au-dessus d'un certain → *seuil absolu de réponse*, une (ou des) excitation(s) sensorielles. En général, les organismes animaux ont développé des organes de réception sensorielle adaptés à différents types de stimulus (mécaniques, physiques, chimiques), chacun ayant sa gamme propre de stimulus adéquats, auxquels il répond à des seuils particulièrement bas. Un organe sensoriel joue, à l'égard d'un stimulus adéquat, le rôle d'un transducteur, c'est-à-dire qu'il le met dans une forme propre à son acheminement vers le système nerveux central (notion de codage sensoriel). Le stimulus adéquat typique pour un organe sensoriel déterminé est opposé au stimulus inadéquat. Le temps séparant le début d'une stimulation et celui de la réponse est appelé temps de latence. → *stimulus-clé.*

Reizadaptation → Reizgewöhnung

Reizbarkeit □ Die Fähigkeit aller Lebewesen, auf Einwirkungen der Außenwelt mit bestimmten Reaktionen (Reizantwort) zu antworten.

irritability □ The ability of all living organisms to react to influences from the external environment with a particular response.

irritabilité □ Faculté, pour tout organisme, de répondre à l'aide de réactions déterminées aux influences de l'environnement.

Reizgefälle □ Eine Folge von Reizen mit abnehmender Wirksamkeit zur Auslösung einer Reaktion. Ist ein Tier z. B. darauf dressiert, auf einen Ton von 1000 Hz zu reagieren, so wird es ohne vorhergehende Dressur auch auf niedrigere und höhere Frequenzen reagieren. Die Reaktion wird jedoch schwächer, je weiter der jeweilige Reiz vom Dressurreiz entfernt ist.

stimulus generalization □ A graded series of stimuli of decreasing effectiveness in eliciting a response. E. g., an animal is trained to respond to a tone of 1000 cps, but will also respond to lower and higher frequencies without prior training. The response becomes weaker the farther the stimulus is removed from the training stimulus.

gradient de stimulation □ Succession de stimuli d'une intensité décroissante pour déclencher une réaction donnée. Si, par exemple, un animal a été conditionné pour répondre à un son de 1000 Hz, il répond également à des fréquences plus basses et plus hautes sans être dressé. Cependant, la réaction de l'animal s'affaiblit en fonction de la différence entre le signal donné et le stimulus de conditionnement.

Reizgewöhnung □ Gegenüber dauernden oder regelmäßig-rhythmischen Reizen tritt bald ein Nicht-mehr-Reagieren (Reizgewöhnung) ein. Die reizspezifische Empfindlichkeitsminderung betrifft die Sinnesorgane selbst und die primären Sinneszentren; sobald nicht mehr gereizt wird, stellt sich die Empfindlichkeit bald wieder her.

stimulus habituation □ Eventually following repeated presentations of a stimulus the given response wanes. This stimulus-specific waning occurs in the sense organ and the primary sensory centers; when the stimulus is withdrawn the sensitivity rapidly returns.

accoutumance aux stimulations □ Des stimuli durables ou réguliers ne provoquent généralement plus de réaction au bout d'un certain temps; il s'est donc établi une accoutumance. La diminution de la sensibilité spécifique à la stimulation concerne les organes sensoriels eux-mêmes et les centres sensoriels primaires. Après l'arrêt de la stimulation continue, la sensibilité initiale s'établit de nouveau.

Reizschwelle □ Die Reizintensität, bei der als Folge des Reizes eine bestimmte, charakteristische Antwortreaktion feststellbar ist. Da die Reizschwelle im umgekehrten Verhältnis zum funktionellen Zustand des jeweiligen Objektes steht, kann sie zu dessen Charakterisierung verwendet werden. Beispielsweise ist die Erregbarkeit eines Nervs umso höher, je niedriger seine Reizschwelle ist.

stimulus threshold □ The strength of a stimulus which just results in a measurable characteristic response. Since the stimulus threshold is inversely related to the functional status of the receptor, the stimulus threshold can be used to describe the receptor. For example, the excitability of a nerve is increased as the stimulus threshold declines.

seuil absolu de réponse □ Intensité minimale de

la stimulation conduisant à une réaction caractéristique déterminée. Comme le seuil absolu de réponse varie en sens inverse de l'excitabilité momentanée du système stimulé, on peut l'utiliser pour évaluer cette excitabilité: celle d'un nerf, par exemple, est d'autant plus grande que le seuil absolu de réponse est plus bas.

Reizsummenregel □ Regel von der summativen Wirkung der Schlüsselreize bei der Auslösung einer artspezifischen Instinktbewegung (SEITZ 1940). So sind z. B. die summativ auf das Bettelverhalten junger Silbermöwen wirkenden Reize die rote Farbe und der Kontrast des Fleckes am Schnabel, schmale Schnabelform und Steilstellung des Schnabels. Diese Bedingungen erfüllt der fütternde Altvogel mit seinem gelben Schnabel und dem roten Fleck (TINBERGEN 1949).

law of heterogeneous summation □ Rule describing the additive effect of key stimuli in releasing a species specific instinctive movement (SEITZ 1940). For example, in the herring gull the red colour and contrast of a beak spot, the narrow bill form and the steep oblique position of the beak all summate to release begging responses from the young. These conditions are fulfilled during feeding by the adult bird with the red spot on its yellow beak (TINBERGEN 1949).

règle de la sommation hétérogène des stimuli □ Règle exprimant le caractère plus ou moins strictement obligatoire de l'addition des stimuli-clés pour le déclenchement d'un mouvement instinctif et propre à l'espèce (SEITZ 1940). Par exemple, les stimuli sommatifs pour le comportement de sollicitation des poussins du Goéland argenté sont la couleur rouge et le contraste de la tache sur le bec, la forme mince du bec et la position à peu près verticale de celui-ci. Ces conditions sont remplies par les parents dont le bec jaune porte une tache rouge (TINBERGEN 1949).

Rekrutierungsverhalten □ Verhaltensweisen von Kundschafterinnen bei Ameisen zur Gewinnung von Neulingen zur Futtereinbringung oder zum Umzug zu neuen Nestplätzen. Bei vielen Arten erfolgt die Rekrutierung allgemein durch Spurpheromone (zusammenfassend bei WILSON 1971). Daneben gibt es bei einigen Arten eine individuelle Rekrutierung durch *Tandemlauf* (HINGSTON 1929). *Bothroponera tesserinoda* z. B. wirbt um Neulinge, indem sie eine Nestgenossin kurz ruckartig mit den Mandibeln zieht und sich dann herumdreht und ihr Hinterende bietet. Die Folgerin läuft dann in engem Fühlerkontakt hinter der Führerin her. Die führende Ameise wird durch Betrillern von Abdomen und Hinterbeinen gereizt, und die Nachfolgerin ist durch ein Oberflächenpheromon und durch mechanische Reize an die vorausgehende Ameise gebunden. Bei einigen *Leptothorax*-Arten wird beim Tandemlauf sogar ein spezifisches Rekrutierungspheromon aus der Giftdrüse eingesetzt (MASCHWITZ *et al.* 1974, HÖLLDOBLER 1976 – dort weitere Literatur).

recruitment behaviour □ That behaviour of worker ants that wins recruits for carrying food or for moving to a new nest. In many species the recruitment is generally accomplished through trail marking pheromones (see WILSON 1971). In addition, some forms recruit by means of tandem running (HINGSTON 1929). *Bothroponera tesserinoda,* for example, recruits by grasping a nest-mate with her mandibles and pulling backwards. She then turns and presents her rear end to the nestmate who promptly follows her. The leader ant is stimulated on the abdomen and rear legs by the antenna of the follower ant. The follower ant is attracted to the leader by a »surface« pheromone and by the mechanical stimulation of touching the leader. In certain *Leptothorax* species the tandem running is associated with the production of a specific recruiting pheromone from the poison gland (see MASCHWITZ *et al.* 1974 and HÖLLDOBLER 1976 for further references).

comportement de recrutement □ Comportement des exploratrices chez les Fourmis pour recruter de nouveaux individus en vue d'un ramassage collectif de nourriture ou pour un déménagement vers un autre nid. Chez de nombreuses espèces, le recrutement s'effectue à l'aide d'un phéromone directeur (résumé chez WILSON 1971), mais chez quelques espèces, il existe un recrutement individuel à l'aide d'une marche en tandem (HINGSTON 1929). *Bothroponera tesserinoda,* par exemple, recrute les nouveaux individus en tirant un congénère d'abord rapidement avec ses mandibules, puis se retourne et présente son abdomen. La fourmi qui suit reste en étroit contact antennaire avec l'individu qui la précède. Ainsi l'individu qui précède est constamment stimulé par un tapotement de son abdomen et des pattes postérieures, et l'individu qui suit est lié par des stimulations mécaniques et un phéromone spécifique. Chez certaines espèces du genre *Leptothorax,* lors de la marche en tandem, un phéromone spécifique au recrutement est sécrété par la glande venimeuse (voir MASCHWITZ *et al.* 1974 et HÖLLDOBLER 1976 pour bibliographie exhaustive).

Revier □ Ein Abschnitt innerhalb des für die betreffende Tierart typischen *Lebensraums,* der als Revier oder Territorium gegen Artgenossen verteidigt und in bestimmter Weise abgegrenzt wird. Außerdem sollte der Revierbesitzer kontinuierlich über diesen Raumabschnitt orientiert sein, damit die Möglichkeit besteht, jeden Eindringling innerhalb kurzer Zeit zu stellen und zu vertreiben. Revierkämpfe treten vorwiegend während der Reviergründung auf oder beim Eindringen fremder Artgenossen. Selten sind Revierkämpfe mit den Nachbarn nach der Stabilisierung und Kennzeichnung der Grenzen. Das Revier kann Besitz eines einzelnen Tieres sein, das keinen anderen Artgenossen oder nur keine gleichgeschlechtlichen Artgenossen duldet; es kann aber auch Besitz einer Familie oder einer Gruppe sein, wo nur familienfremde bzw. gruppenfremde Artgenossen abgewiesen werden. Bei vielen Primaten ist *Gruppenterritorialität* ein verbreitetes Merkmal, und die Analogien zum menschlichen Verhalten sind sehr auffällig; z. B. Hordenterritorien bei Buschmännern, Hadzas und Pygmäen, Stammesterritorien bei Waldpflanzern, und nicht zuletzt unsere Provinzen und Nationalstaaten (→ *Pseudospeziation*). ALTUM (1868) war der erste, der den Begriff des Reviers definierte und die Bedeutung des Vogelgesanges für die Revierverteidigung richtig interpretierte. Das besetzte Gebiet wird in der Regel in besonderer Weise gekennzeichnet (→ *Reviermarkierung*). Die Revierbindung ist je nach Art in der Zeit variabel, meist abhängig von der Fortpflanzung; häufig hat sich ein hochentwickeltes Heimfindevermögen ausgebildet (JEWELL und LOIZOS 1966, KLOPFER 1969, HOLLOWEY 1974, STOKES 1974, WILSON 1975).

territory □ A subdivision of the → *home range* of an animal species. The territory is set apart from the surround in some manner and defended against other conspecifis. The territory owner keeps continuously informed about this area in order to be able to detect and to chase immediately any intruder. Territorial fights occur primarily during formation of the territory and later during trespassing by strange individuals. Territorial disputes seldom occur with neighbours after the boundaries have been marked. The territory may belong to a single individual who will not tolerate other individuals or same sex individuals; it may also belong to a family or a group where non-members are rejected. In many primate species *group territoriality* is a widely distributed characteristic. Analogies to human behaviour are quite obvious; e. g. band territories in bushmen, hadzas and pygmies; tribal territories in forest dwellers and finally provinces and nation states (→ *pseudospeciation*). ALTUM (1868) first defined the concept of territory and correctly interpreted the significance of bird song in territorial defense. Usually the territory is marked in a special fashion (→ *territorial marking*). The territorial bond may vary according to the species or the season; it is particularly strong during the reproductive period. Frequently a highly developed homing ability is present (JEWELL and LOIZOS 1966, KLOPFER 1969, HOLLOWEY 1974, STOKES 1974, WILSON 1975).

territoire □ Surface comprise à l'intérieur du → *domaine vital* typique d'un animal qui est délimitée d'une certaine manière et défendue envers les congénères. L'occupant du territoire se tient continuellement informé sur cette zone pour être en mesure de repérer et de chasser tout intrus en peu de temps. Des combats territoriaux apparaissent souvent lors de l'établissement du territoire ou lors de l'intrusion des congénères inconnus. De tels combats, par contre, sont rares avec les voisins, une fois les territoires stabilisés et les limites marquées. Le territoire peut être la propriété d'un individu seul qui ne laisse pénétrer aucun autre congénère ou aucun congénère du même sexe. Il peut aussi être la propriété d'une famille ou d'un groupe d'animaux d'où seuls les individus étrangers à la famille ou les congénères étrangers au groupe sont refoulés. Chez de nombreux Primates, le *territoire de groupe* est très répandu. Les analogies dans le comportement humain sont très apparentes, comme par exemple les territoires d'une horde chez les Boshimans, les Hadzas et les Pygmées, les territoires des tribus de planteurs et enfin nos provinces et nos états nationaux (→ *pseudo-spéciation*). ALTUM (1868) a été le premier à définir la notion de territoire et à interpréter correctement la signification du chant des oiseaux dans le contexte de la défense du territoire. Le terrain occupé est généralement marqué d'une manière spécifique (→ *marquage du territoire*). L'attachement au territoire est variable selon les espèces et dans le temps et dépend souvent de la période de reproduction. Dans la plupart des cas, il s'est développé une capacité de retour au gîte (JEWELL et LOIZOS 1966, KLOPFER 1969, HOLLOWEY 1974, STOKES 1974, WILSON 1975).

Revierbindung → Revier
E. territorial bond □ F. attachement au territoire

Reviergesang → Reviermarkierung
E. territorial song □ F. chant territorial

Revierkampf → Revier
E. territorial fight □ F. combat territorial

Reviermarkierung □ Die Kennzeichnung und Abgrenzung des Reviers. Säugetiere markieren meist durch Duftstoffe, indem sie Drüsensekrete und auch Harn und Kot an bestimmten Punkten des Reviers absetzen. Vögel und auch Brüllaffen kennzeichnen ihre Reviere durch Gesang und andere Lautäußerungen. Viele Fische, besonders bunte Meeresfische, kennzeichnen ihren Besitz optisch durch einen → *Signalsprung*. Optische Reviermarkierung gibt es auch bei Libellen. Bei den schwach elektrischen Fischen dient wahrscheinlich die ständige Aussendung der elektrischen Signale der Revierbegrenzung. In der Literatur wird der Ausdruck Markierung für alle diese Möglichkeiten benützt, obwohl akustische, optische und elektrische Signale keine »Marken« hinterlassen, wenn der Revierbesitzer abwesend ist. Man sollte vielleicht in den letzteren Fällen von Reviersignalisierung sprechen, was in dem Wort Signalsprung schon ausgedrückt wird, obwohl dieses Verhalten auch dem »Anlocken« der paarungswilligen ♀♀ dient. Damit aber würde Markierung schlechthin hinfällig, denn man könnte dann ebenso von Duftsignalen sprechen, denn das Wort Signal antizipiert nicht, daß es nicht permanent ist.

marking of territory □ The posting and designation of a territory. Mammals usually scent mark their territories with gland secretions, urine or feces at various spots. Birds and howler monkeys designate their territorial boundaries with song or other vocalizations. On the other hand many fish mark their territorial space optically with special conspicuous movements (→ *signal jump*). Visual marking also occurs in dragonflies. The weak electric fields continuously emitted by electric fish probably serve for marking of territory. In the literature acoustic, optic and electrical signals are all termed marking although no actual »marks« are left behind when the territory owner is absent. One perhaps might better refer to these cases as territorial signalling. This sense is inherent in the term »signal jump« although this particular term also refers to the solicitation of the ♀ during courtship in certain fish. Carrying this idea even further, scent signalling rather than marking could be used since the word signalling itself does not imply that there is no permanency.

marquage du territoire □ Mode de délimitation d'un territoire. Les Mammifères marquent leur territoire à l'aide de substances olfactives en déposant des sécrétions glandulaires, de l'urine ou des matières fécales en des endroits déterminés. Les Oiseaux, mais aussi les Singes hurleurs, signalent leur territoire à l'aide de chants et de cris particuliers. De nombreux Poissons, en particulier ceux de coloration vive, manifestent visuellement la possession d'un territoire à l'aide d'un → *saut-signal.* Une signalisation optique du territoire existe aussi chez les Libellules. Chez certains Poissons électriques, l'émission permanente de signaux électriques de faible tension sert probablement à la délimitation de leur territoire. Dans la bibliographie, l'expression marquage est utilisée pour toutes ces possibilités bien que les signaux acoustiques, visuels et électriques ne laissent à proprement parler aucune trace ou »marque« en l'absence de l'occupant. On pourrait, dans les cas cités ci-dessus, parler d'une signalisation du territoire ce qui est déjà indiqué par le saut-signal chez les Poissons bien que ce dernier comportement serve également à l'attraction des ♀♀. Finalement, l'expression de »marquage« serait pratiquement superflue: on pourrait évidemment parler aussi de signaux odorants, étant donné que le mot signal n'implique pas qu'il n'est pas permanent.

Revierverhalten → Revier → Reviermarkierung

E. territorial behaviour □ F. comportement territorial

rhythmisches Brustsuchen → Suchautomatismus

Richtungsweisung □ Verhaltensweisen von Sammelbienen, die ihre Stockgenossen über die Entdeckung einer guten Futterquelle und über die Richtung zu diesem Ziel informieren. In der Regel tanzen die Bienen im finsteren Stock auf vertikaler Wabenfläche. Dabei transponiert die Tänzerin den Winkel zur Sonne auf den Winkel zum Lot. Manchmal tanzen sie auch auf horizontaler Fläche, z. B. auf dem Anflugbrettchen vor dem Stock oder (im Versuch) im horizontal gelegten Stock. Dann weist die Richtung des → *Schwänzeltanz(es)* direkt nach dem

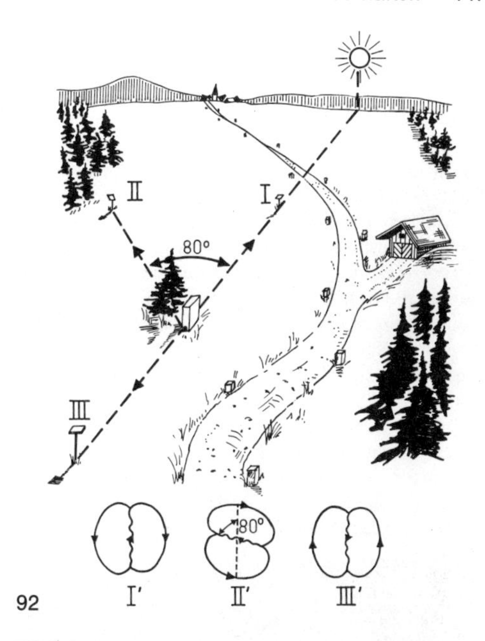

92

Ziel, indem die Tänzerin denselben Winkel zur Sonne einhält wie zuvor beim Flug vom Stock zur Futterquelle (Abb. 92 – Richtungsweisung mit drei verschiedenen Beispielen und den entsprechenden Tänzen I, II, III).

indication of direction □ Behaviour patterns of forager bees who inform their conspecifics of the discovery of a food source and the direction leading to it. Usually the bee dances inside the hive on the dark vertical surface of the comb and portrays the angle of the goal to the sun as the angle of her waggle dance in relation to the plumb line. Sometimes she dances on a horizontal surface, e. g., on the hive's landing ramp or in an experimentally rotated hive, and in this case, she indicates the direction of the food source in a → *waggle dance* by maintaining the same angle to the sun which she had earlier used in flying from the hive to the food source (Fig. 92 – indication of direction with three different examples and the corresponding dances I, II, III).

indication de direction □ Comportements des Abeilles butineuses informant les compagnes de leur ruche sur la direction d'une source de nourriture. En règle générale, les Abeilles dansent à l'intérieur de la ruche dans l'obscurité sur des rayons verticaux. Pendant cette danse, l'Abeille transpose l'angle de la direction du but par rapport au soleil en un angle par rapport à la verticale. Quelquefois, les butineuses dansent aussi sur un plan horizontal, comme par exemple à l'entrée de la ruche ou dans une ruche expérimentalement couchée à l'horizontale. Dans ces cas, la direction du parcours de la → *danse frétillante* indique directement le but et la danseuse maintient le même angle par rapport au soleil qu'auparavant pendant le vol de la ruche vers le lieu de la provende

(Fig. 92 – indication de direction dans trois situations différentes avec les danses correspondantes I, II, III).

Riechgähnen □ Ein dem → *Flehmen* nah verwandtes Verhalten, welches bei den meisten Bovidae anzutreffen ist. Nach Lecken von Fruchtwasser, Embryonalhüllen nach der Geburt und Lecken der Jungen hebt das Muttertier leicht den Kopf, beleckt intensiv Muffel und Nüstern, und gähnt dann. Bei diesem »Gähnen« sind die Nasenrücken leicht gerümpft, die Augen fast oder ganz geschlossen und die Zunge gaumenwärts eingerollt (Abb. 93 – *Syncerus caffer* ♀). Nach 1–2 sec wird das Maul geschlossen. Auch Aasgeruch kann Riechgähnen auslösen. Vermutlich wird hierbei das Jacobson'sche Organ betätigt. HALDER und SCHENKEL (1972) betrachten das Riechgähnen als vorwiegend weiblich-mütterliches primitives Verhalten, während sie das Flehmen als männliche, höher spezialisierte Form der Rezeption mittels des Jacobson'schen Organs ansehen.

93

sniff-yawning □ Behaviour pattern in many Bovidae which is closely related to → *Flehmen* (lip-curl). After licking of amnion fluid, licking of the placenta after parturition and licking of the offspring, the mother animal raises the head, licks the snout and the nostrils followed by yawning. In this kind of yawning the top of the nose is somewhat wrinkled, the eyes are almost or completely closed and the tongue is rolled upwards (Fig. 93 – *Syncerus caffer* ♀). After 1–2 sec the mouth is closed. Cadaverous odour may also elicit sniff-yawning. It is probable that in sniff-yawning Jacobson's Organ is set in action. HALDER and SCHENKEL (1972) consider sniff-yawning as a predominantly female-maternal variation of receptory activity involving Jacobson's Organ, while Flehmen would have to be considered as the more specialized male variation.

bâillement de flairage □ Comportement apparenté au → *Flehmen* et connu chez un certain nombre de Bovidae. Après léchage du liquide amniotique, des enveloppes embryonnaires après la naissance et après léchage des jeunes, la mère soulève légèrement la tête, se lèche d'une manière intense le museau et les narines, puis bâille. Lors d'un tel bâillement, le dos du nez paraît retroussé, les yeux sont presque ou complètement fermés et la langue est enroulée vers le palais (Fig. 93 – *Syncerus caffer* ♀). Après 1 à 2 sec, la bouche est refermée. L'odeur de charogne peut également déclencher le bâillement de flairage. L'organe de Jacobson est probablement mis en activité lors de ce comportement. HALDER et SCHENKEL (1972) considèrent le bâillement de flairage comme un comportement primitif, principalement féminin-maternel. Par contre, le Flehmen serait principalement masculin et représenterait une forme de réception plus spécialisée.

Riechkuß → Kontaktgruß

E. smell-kiss, sniffing kiss □ F. baiser de flairage

Riechstoffe = Duftstoffe → Duftmarkierung

ritualisiertes Futterbetteln → Balzfüttern → Begrüßungsfüttern

E. ritualized begging □ F. sollicitation ritualisée de nourriture

Ritualisierung □ HUXLEY (1914) nannte die Verwendung nichtsexueller Verhaltensweisen in der Balz von Tieren – in Anlehnung an kulturelle Vorgänge beim Menschen – ein »Ritual«. Seither und mit HUXLEY (1923) spricht man generell bei der Veränderung von Verhaltensweisen im Dienste der Signalbildung (Signalfunktion) von Ritualisierung. HUXLEY (1966) definierte weiter: »Ritualisierung ist eine durch natürliche Auslese hervorgebrachte adaptive Ausrichtung von Ausdrucksverhalten, zu der Lernprozesse hinzukommen können«. Ritualisierung impliziert also die Änderung einer Verhaltensweise mit Signalwirkung unter dem Selektionsdruck besserer Verständigung in Richtung auf größere Deutlichkeit und Unzweideutigkeit des Signals für den Empfänger (WICKLER 1967, 1970). → *Abdomenschwingen,* → *Futterlocken.*

ritualization □ In 1914 HUXLEY referred to animal courtship displays in terms of the characterization of human social conventions as a »ritual«. Since then the term has been used generally to refer to the modification of a behaviour pattern to serve a communicative function (HUXLEY 1923). The same author (1966) refined the definition: »Ritualization is an adaptive device arising through natural selection in which learning processes have been combined with behavioural displays«. Ritualization, therefore, implies the natural selection for improved communicability in the signal properties of a behaviour pattern through increases in the clarity and distinctiveness of the signal for the receiver (WICKLER 1967, 1970). → *abdominal up-and-down movement,* → *food-calling.*

ritualisation □ HUXLEY (1914) a considéré l'utilisation de comportements non sexuels dans la parade nuptiale des animaux comme un »rituel«, par comparaison avec des processus culturels humains. Depuis lors et avec HUXLEY (1923), on parle généralement de ritualisation lorsqu'il y a changement de signification d'un comportement au service d'une fonction de signalisation. HUXLEY (1966) a défini par ailleurs: »La ritualisation est une ré-orientation adaptative du comportement vers des fonctions expressives. Elle s'opère par sélection naturelle qui peut être accompagnée de processus d'apprentissage«. La ritualisation implique donc la modification d'un comportement à fonction de signalisation; elle s'effectue sous la pression sélective d'une meilleure communication et tend vers une plus grande clarté et vers une moindre ambiguïté du signal pour celui qui le reçoit (WICKLER 1967, 1970). → *mouvement pendulaire de l'abdomen,* → *attraction alimentaire.*

Rivalenkampf = Revierkampf → Revier

Rolligkeit → Brunst

E. state of oestrus in Felid ♀♀ □ F. état d'oestrus chez les ♀♀ des Felidae

Rückkoppelung □ Informationsverarbeitung, bei der das Ergebnis der Wirkung eines Stellgliedes erneut zur Aktion eben dieses Stellgliedes benützt wird. Beim Vorliegen einer Rückkoppelung findet ein Kreisen von Informationen statt (*feedback*).

feedback □ Information processing in which the result of the action of a certain system component serves as an initiator of new action in the system. Feedback involves circular transfer or cycling of information.

rétroaction □ Transmission d'information en boucle fermée, permettant la réutilisation des effets d'une instance pour solliciter celle-ci à nouveau (*feedback*).

Rückmeldung → Reafferenzprinzip

Rucktanz □ Er kommt vor, wenn die Sammelbiene, die auf der Wabe da und dort ihren Honigblaseninhalt verfüttert, zwischendurch kurze gerichtete Schwänzelläufe macht (Abb. 94 – Rucktanzschema einer Honigbiene; die Punkte zeigen Futterübergabe an). Der Rucktanz ist eher ein Ausdruck der Tanzstimmung als ein wirksames Signal (VON FRISCH 1965).

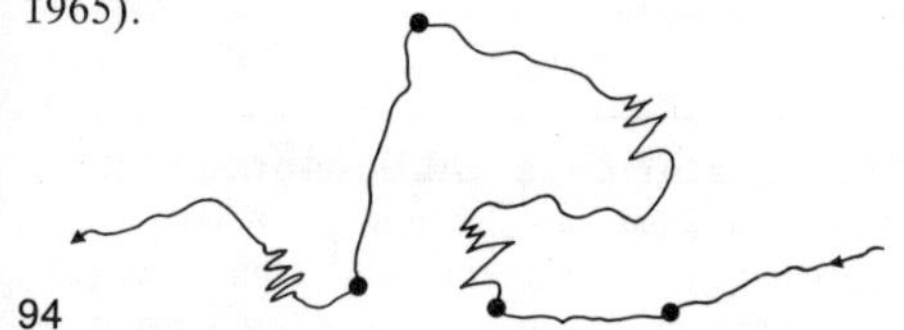

94

jerk dance □ This dance occurs when the forager bee, while depositing the contents of her pollen sacs on the comb, makes occasional short runs wagging her abdomen (Fig. 94 – jerk dance pattern of a honeybee; the points indicate food transfers). The jerk dance is more an expression of a mood or tendency to communicate (dance) than an actual signal (VON FRISCH 1965).

danse saccadée □ Cette danse a lieu lorsqu'une Abeille butineuse distribue ça et là le contenu de son jabot. Elle effectue alors, entre les différentes rencontres trophallactiques, de courtes danses frétillantes directionnelles (Fig. 94 – schéma de la danse saccadée d'une Abeille; les points indiquent les emplacements où elle effectue des transferts de nourriture). La danse saccadée est plutôt l'expression d'une disposition à danser qu'un signal d'information valable (VON FRISCH 1965).

Rückwärtsspritzen → Harnmarkierung → Harnspritzen

E. backwards urine spraying □ F. giclement d'urine vers l'arrière

Rudelgeruch → Gruppenduft

Ruf des Verlassenseins □ Lautäußerungen im Dienste des Wiederfindens nach Verlust des Elternkumpans und/oder Geschwisterkumpans. Bei Tieren weit verbreitet und auch vom Menschensäugling bekannt. Von LORENZ erstmals in seiner Bedeutung erkannt und beschrieben.

distress call, »lost call« □ Vocalization which serves to bring the separated young animal once again into contact with its parent figure or siblings. This type of vocalization is wide spread in the animal kingdom and is present in the human infant as well. LORENZ was the first to recognize and describe the significance of the distress call.

cri de détresse □ Vocalisation au service de la reprise de contact du jeune animal avec ses parents et/ou ses compagnons syngéniques. Chez les animaux, ce phénomène est très répandu et également connu chez le nourrisson humain. Il a été reconnu pour la première fois dans sa signification et décrit par LORENZ (1935).

Ruhestellung → Schlafverhalten

E. resting posture □ F. posture de repos

Ruheverhalten → Schlafverhalten

Rumpellauf □ Eine auffällige Bewegungsweise von erfolgreich heimkehrenden Sammelbienen, die hastig trippelnd auf der Wabe vorwärts laufen und die im Wege stehenden Stockgenossen anrempeln. Diese erfahren dadurch nur, daß etwas los ist, und können auf diese Weise zum Ausflug veranlaßt werden (SCHMID 1964, VON FRISCH 1965).

bumping run □ A conspicuous movement pattern in foraging bees returning to the hive with pollen; they excitedly bound forward onto the comb trampeling other workers in their path. This simply alerts the others and may motivate them to make excursion flights (SCHMID 1964, VON FRISCH 1965).

marche tapageuse □ Mouvement apparent chez les Abeilles butineuses, de retour à la ruche. Ces Abeilles, très excitées, avancent rapidement sur les rayons en bousculant les compagnes qui se trouvent sur leur chemin. Celles-ci n'obtiennent pas de renseignements précis par ces manœuvres, mais sont tout simplement informées que quelque chose se passe et peuvent ainsi être incitées à quitter la ruche (SCHMID 1964, VON FRISCH 1965).

Rundtanz □ Mit raschen, wippelnden Schritten läuft die Sammelbiene in so engem Kreise, daß oft nur eine Zelle im Inneren liegt. Auf den 6 angrenzenden Zellen rennt sie herum, wobei sie bald mit einer plötzlichen Wendung kehrt macht und in entgegengesetzter Richtung weiterläuft. Zwischen zwei Wendungen liegen oft ein bis zwei volle Kreise (Abb. 95 – Rundtanzschema der Honigbiene). Der Rundtanz bedeutet, daß es in der Umgebung des Stockes reichlich Futter gibt. Bis 25 m Entfernung wird nur rundgetanzt, von 25–100 m gibt es Übergänge (→ *Sicheltanz*), und liegt das Ziel 100 m und weiter, so gibt es nur noch den → *Schwänzeltanz*. Diese Entfernungsangaben variieren bei *Apis mellifica* von Unterart zu Unterart.

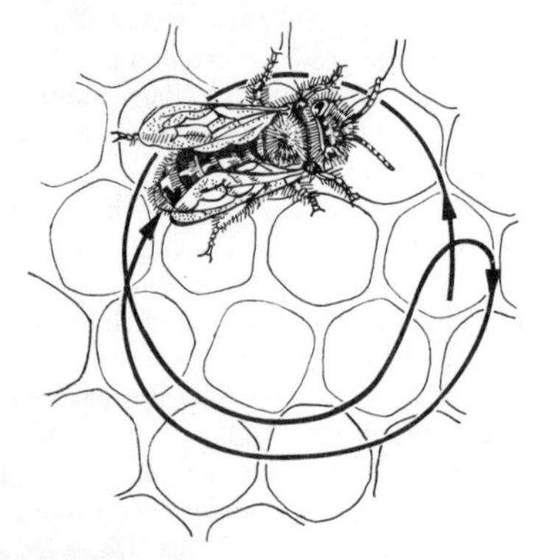

95

round dance □ With rapid rocking steps the forager bee moves in circles which are often so tight that only one comb cell lies inside. She runs around the six adjacent cells, suddenly turning and running in the opposite direction. One or two full circles may separate each reversal of direction (Fig. 95 – round dance pattern of the honeybee). The round dance signifies that a rich source of food lies in the vicinity of the hive. Up to a distance of 25 meters only the round dance is performed; from 25 to 100 m a transitional dance (→ *sickle dance*) is performed, and food sources over 100 m elicit the → *waggle dance*. These distances vary in *Apis mellifica* from one subspecies to another.

danse circulaire □ L'Abeille butineuse marche rapidement en rond en décrivant un cercle de très faible rayon de manière à ce qu'il n'y ait qu'une seule cellule à l'intérieur. Elle marche alors sur les 6 cellules environnantes pour effectuer subitement un demi-tour sur elle-même et avancer en direction opposée. Entre deux demi-tours, il y a parfois un à deux tours complets (Fig. 95 – schéma de la danse circulaire de l'Abeille mellifère). La danse en rond signale la présence, autour de la ruche, d'une abondante source de provende, sans indication de direction. La danse en rond est effectuée lorsque la nourriture se trouve à moins de 25 m de la ruche. Entre 25 et 100 m de distance, l'information est donnée par la → *danse en croissant* et si la nourriture se trouve au-delà de 100 m, l'information n'est transmise que par la → *danse frétillante*. Ces indications de distance varient chez *Apis mellifica* d'une sous-espèce à l'autre.

Rütteltanz □ Tanzverhalten, bei welchem die Biene mit einer anderen Kontakt aufnimmt und mit ihrem Hinterleib in dorsoventraler Richtung eine rasche Vibrationsbewegung ausführt (Abb. 96). Nach ALLEN (1959 a + b) sind Rütteltänze gegenüber der Königin sehr häufig vor dem Schwärmen, nach HAMMANN (1957) sehr oft auch vor dem → *Hochzeitsflug*. Die Bedeutung dieses Verhaltens ist allerdings nicht geklärt (siehe auch VON FRISCH 1965).

shaking dance □ Here the worker bee initiates contact with another bee by rapidly shaking or undulating its abdomen in a dorso-ventral direction (Fig. 96). According to ALLEN (1959 a + b) the shaking dance occurs very frequently in response to the queen; HAMMANN (1957) reports that it occurs often before the → *nuptial flight*. The significance of this behaviour, however, has not been determined (see also VON FRISCH 1965).

danse vibrante □ Forme de danse par laquelle une Abeille recherche le contact d'une congénère en effectuant avec son abdomen des mouvements de vibration rapide en direction dorsoventrale (Fig. 96). Selon ALLEN (1959 a + b), ces danses vibrantes sont fréquentes envers la reine juste avant l'essaimage et selon HAMMANN (1957), elles sont aussi effectuées souvent avant le → *vol nuptial*. La signification exacte de ce comportement n'est cependant pas élucidée (voir aussi VON FRISCH 1965).

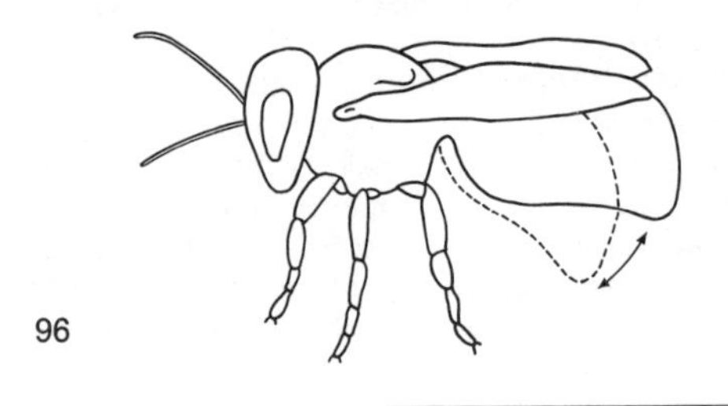

96

S

SAP, spezifisches Aktionspotential □ HINDE (1954) hat im Anschluß an THORPE (1951) vorgeschlagen, den Begriff der »reaktionsspezifischen Energie«, die den spezifischen Verhaltensablauf nach LORENZ von innen her aktiviert, durch den am Verhalten selbst meßbaren Begriff des »SAP« zu ersetzen. THORPE hat dieses SAP wie folgt umschrieben: Der Status eines Tieres ist abhängig von seiner Bereitschaft, die Verhaltensweisen eines Instinktes vorzugsweise gegenüber allen anderen Verhaltensformen abzuhandeln. Diese spezifische Bereitschaft verringert sich oder endet, wenn die → *Endhandlung* des betreffenden Instinktes vollzogen wurde.

SAP, specific action potential □ HINDE (1954) following the lead of THORPE (1951), suggested that the LORENZian concept of »reaction-specific energy« as an internal activator of behaviour be replaced by SAP which can actually be measured from the behaviour. THORPE described the SAP as follows: An animal's SAP status is dependent on its readiness or tendency to perform a particular instinctive behaviour pattern in preference to all other behaviours. This specific tendency declines or ends after the → *consummatory act* of the instinctive behaviour.

PAS, potentiel d'action spécifique □ HINDE (1954), à la suite de THORPE (1951), a proposé de remplacer par cette notion celle de »l'énergie spécifique d'une réaction« qui, d'après LORENZ, active de l'intérieur de l'organisme le déroulement d'un acte spécifique. Le PAS aurait l'avantage d'être directement mesurable par observation du comportement. THORPE décrit comme suit ce PAS: l'état d'un animal dépend de sa disposition à actualiser les éléments d'une conduite instinctive donnée, de préférence à toutes les autres possibles. Cette disposition spécifique s'atténue ou prend fin quand → *l'acte consommatoire* correspondant a été accompli.

Saugtrinken □ Vögel, die so trinken, tauchen den Schnabel ein, pumpen mit deutlich sichtbaren Bewegungen des vorderen Ösophagus, bis sie sich satt getrunken haben, ohne den Schnabel aus dem Wasser herauszunehmen. Ein solches Saugtrinken (auch Pumpsaugen genannt) findet man bei den Pteroclidae, bei Columbidae (mit Ausnahme von *Didunculus*) und den Astrildidae. Dieses Verhalten ist entgegen früheren Annahmen kein Verwandtschaftskriterium, sondern eine Verhaltensanalogie (WICKLER 1961).

»Saugtrinken« □ *A pumping mode of drinking.* Birds which drink in this manner dip their beaks into the water and pump the water upward with clearly visible movements of their anterior oesophagus without removing their bill from the water. This type of drinking is found in the Pteroclidae, Columbidae (except in *Didunculus*) and in the Astrildidae. Contrary to earlier assumptions, this behaviour pattern is not a criterion of phylogenetic relationship, but a behavioural analogy (WICKLER 1961).

boire en aspirant □ Les Oiseaux qui boivent de cette manière trempent leur bec dans le liquide et pompent très visiblement à l'aide de la partie antérieure de l'oesophage. Ils boivent ainsi jusqu'à satiété sans retirer leur bec de l'eau. Ce mode de prise de boisson se rencontre chez les Pteroclidae, les Columbidae (à l'exception du genre *Didunculus*) et les Astrildidae. Ce comportement constitue une analogie comportementale et ne peut servir en aucune manière en tant que critère d'affinité phylétique, comme on l'avait supposé naguère (WICKLER 1961).

Schalenappetenz □ Das Suchverhalten der Paguridae nach einem neuen Schneckenwohnhaus, wenn das eigene zu klein geworden ist. Kann durch Wegnehmen der Wohnschale auch experimentell ausgelöst werden.

shell-searching behaviour □ The searching behaviour of Pagurids for a new snail-shell when they have outgrown the present one. This behaviour can be elicited experimentally by removing the shell.

recherche des coquilles □ Comportement de recherche d'une coquille d'escargot chez les Paguridae dès que la leur est devenue trop petite. Ce comportement peut être déclenché expérimentalement en sortant le Pagure de son habitation.

Schallokalisation → Echo-Orientierung
E. sound localization □ F. localisation sonore

Schampräsentieren □ Bei Buschmädchen beobachtete EIBL-EIBESFELDT (1972) eine tiefe Verbeugung wie beim Zukehren des Gesäßes *(→ Gesäßweisen),* wobei die Scham in Richtung des zu Verspottenden deutlich gezeigt wird (Abb. 97 – !ko-Buschmädchen aus der Kalahari). Anschließend wird meist frech gelacht. Es handelt sich möglicherweise um eine alte primatenabgeleitete Form weiblichen → *Genitalpräsentieren(s)*, die EIBL-EIBESFELDT (1973) als Schampräsentieren vom frontalen → *Schamweisen* unterscheidet. Beide Verhaltensweisen werden oft in wechselnder Folge ausgeführt.

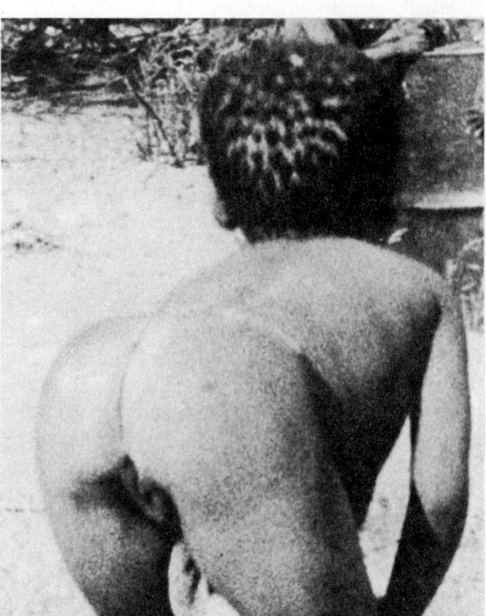

97

vulva presentation □ In Bushman girls EIBL-EIBESFELDT (1972) observed a type of deep bow, similar to the → *buttocks display,* in which the genitals are exhibited toward a person who is being ridiculed (Fig. 97 – Bushman girl from the Kalahari Desert). This is followed by raucous laughter. This display is similar to a form of → *genital presentation* found in female primates and is distinguished by EIBL-EIBESFELDT (1973) from the → *pubic presentation* which often follows or preceeds it.

présentation vulvaire □ EIBL-EIBESFELDT (1972)

a observé un comportement de moquerie chez les filles Boshiman qui présentent leur vulve à la personne dont elles veulent se moquer. Ceci se fait en posture très inclinée comme lors de la → *présentation du postérieur* (Fig. 97 – jeune fille Boshiman du Kalahari). Ensuite, on observe souvent un rire insolent. Il s'agit probablement d'une forme ancienne de → *présentation des organes génitaux* féminins dérivée d'un comportement existant chez les Primates infrahumains et que EIBL-EIBESFELDT (1973) tient à distinguer de la → *présentation pubienne* frontale. Les deux comportements sont souvent exécutés en alternance.

Schamschürzenwippen → Schamweisen
E. loin cloth flashing □ F. balancement du tablier pubien

Schamweisen □ Die Scham wird bei diesem Spottverhalten frontal gewiesen, und zwar meist mit leicht vorgestreckten Hüften und einem Arm angewinkelt auf die Hüfte gestützt; mit der anderen Hand wird die Schamschürze leicht angehoben, ein Fuß vorgesetzt und der Schenkel auffällig vorgeschoben. Manchmal wird auch die Schamspalte mit beiden Händen auseinandergezogen. Dieses Verhalten ist offenbar im Gegensatz zum → *Schampräsentieren* kulturritualisiert (EIBL-EIBESFELDT 1972). Eine abgewandelte Form des Schamweisens finden wir im Schamschürzenwippen. Dieses Verhalten ist eine Art Spott-Tanz, bei dem die Buschmädchen ihr Schamschürzchen nicht mit der Hand anheben; sie wippen vielmehr mit den Hüften im Tanzrhythmus, so daß die Schürzchen auf und nieder fliegen (Abb. 98 – !ko-Buschmädchen aus der Kalahari). Man kann auch kurz von Schürzenwippen sprechen.

pubic presentation □ Frontal genital display in human females. In this mocking display one hand rests on the hips, which are usually tilted forward toward the subject. The other hand is used to lift the loin cloth apron while one leg is shoved a step forward. Occasionally the genital orifice is spread apart with the hands. This behaviour, in apparent contrast to → *vulva presentation,* is culturally ritualized (EIBL-EIBESFELDT 1972). A derived form of this display is found in loin cloth flashing. This behaviour is a type of ridiculing dance in which Bushman girls expose their genitals by flapping their loin cloth up and down with a rhythmic action of the hips rather than with the hand (Fig. 98 – Bushman girl from the Kalahari Desert).

présentation pubienne □ Comportement de moquerie, au cours duquel le pubis est présenté frontalement, souvent en avançant légèrement la hanche, une main posée dessus, l'autre main soulevant légèrement le tablier pubien pendant qu'un pied est avancé et une cuisse mise en évidence. Parfois, la fente pubienne est écartée à l'aide des deux mains. Ce comportement, contrairement à la → *présentation vulvaire,* est probablement une ritualisation culturelle (EIBL-EIBESFELDT 1972). Une forme modifiée de la présentation pubienne est le balancement du tablier pubien: ce comportement est une sorte de danse de moquerie au cours de laquelle les jeunes filles Boshiman effectuent des mouvements rythmiques avec leurs hanches et balancent ainsi leur petit tablier pubien. Le tablier n'est pas soulevé avec la main (Fig. 98 – jeune fille Boshiman du Kalahari).

98

Scharren □ Vögel scharren entweder abwechselnd mit einem und dem anderen Bein, wie z. B. die Hühner, oder springen beidbeinig vorwärts und sofort wieder zurück, wobei die Zehen über den Boden fegen, wie z. B. Drosseln und Finken. Das erste nennt man *Laufscharren* und das zweite *Hüpfscharren* (WICKLER 1970). Ob ein Vogel überhaupt scharrt, hängt wohl davon ab, wo er sein Futter sucht. Wohl alle Galli scharren, aber nicht die Hokkos (*Crax*), die ihre Nahrung auf Bäumen suchen (NICOLAI mündl.). Es scharren auch viele Vögel, die zwar nicht am Boden leben, aber dort ihr Futter suchen. Nach KÖNIG (1951) scharren von hüpfenden Ahnen abstammende Singvögel hüpfend, auch wenn sie sekundär laufen, wie die Bartmeise, *Panurus biarmicus.*

Säuger scharren mit den Vorder- oder Hinterextremitäten stets alternativ, und bei manchen Gruppen gibt es ein einseitiges Vorderlaufscharren, oft bei aggressiver Stimmung. Während wir im Deutschen zwischen *Scharren* (meist nach Futter) und *Kratzen* (Sich-Kratzen, jemanden kratzen) unterscheiden, ist das im Französischen nicht der Fall.

scratching behaviour □ Birds scratch on the substrate either by using first one and then the other leg, e. g. the chicken, or they spring forward with both legs at the same time pulling the claws backwards over the surface as in, e. g., thrushes and finches. The former type is referred to as *run-scratching* or *alternating scratching* and the latter as *hop-scratching* behaviour (WICKLER 1970). Whether a bird utilizes substrate scratching behaviour depends on where he typically feeds. Most Gallinacious birds, with the exception of the *Crax* which feeds in trees (NICOLAI verbal com.), show substrate scratching behaviour. According to KÖNIG (1951) birds evolving from *jump scratching* forms will also exhibit this type of substrate scratching behaviour even though they may secondarily have acquired running capability, e. g. the Bearded Titmice *Panurus biarmicus*.
Mammals scrape or scratch with fore and/or hind feet in alternation. Some groups only show foreleg scraping with one leg, sometimes with aggressive intention.

grattage du sol □ Les Oiseaux grattent le sol de deux manières, soit alternativement avec l'une ou l'autre patte, comme par exemple les Poules, soit simultanément avec les deux pattes; dans ce dernier cas, les individus effectuent un petit saut en avant et immédiatement après en arrière. Lors du mouvement en arrière, les griffes frôlent le sol, comme le font par exemple les Turdidae et Fringillidae. La première manière de gratter le sol est appelée *grattage alternatif du sol* et la deuxième *grattage en sautillant* (WICKLER 1970). Le grattage du sol chez une espèce d'Oiseau dépend de l'endroit où elle cherche sa nourriture. Nous observons ce comportement chez tous les Galliformes, mais il est absent chez les Hoccos (*Crax*) qui cherchent leur nourriture sur les arbres (NICOLAI verbalement). Le grattage s'observe chez de nombreux Oiseaux qui ne vivent pas habituellement au sol, mais y cherchent leur nourriture. KÖNIG (1951) a remarqué que les Passereaux descendant d'ancêtres sauteurs effectuent le grattage en sautillant, même si leur mode de locomotion s'est secondairement transformé en marche, comme par exemple chez *Panurus biarmicus*.
Chez les Mammifères, le grattage du sol est en général effectué alternativement avec les membres antérieurs ou postérieurs. Il existe cependant chez certains groupes un grattage unilatéral avec une patte antérieure, parfois avec une intention agressive.

Schauflug □ Man kennt von vielen Greifvögeln besonders auffällige Flüge, die wie beim Mäusebussard (Abb. 99 A – *Buteo buteo*) und beim Habicht (Abb. 99 B – *Accipiter gentilis*) als *Imponierflug* zu deuten sind und der Demonstration und dem Zur-

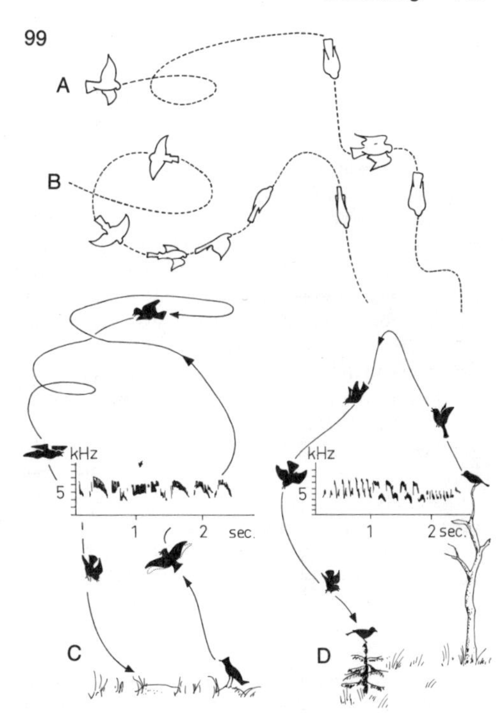

Schau-Stellen dienen, da sie vorwiegend aufgeführt werden, wenn Artgenossen das Brutrevier durchfliegen.
Bei Singvögeln, vor allem Bewohnern offener Landschaften, wie z. B. bei einigen Alaudidae, Motacillidae *(Anthus)* und den Viduinae, kennen wir Gesangsflüge (Abb. 99 C – *Alauda arvensis;* D – *Anthus trivialis*). Diese Demonstrationen dienen der Anlockung der Weibchen und vor allem der Signalisierung des Reviers gegenüber Artgenossen. Es sind im strengen Sinne keine Balzflüge, da sie nicht nur während der Paarbildung ausgeführt werden (BLUME 1967, NICOLAI 1969).

display flight □ When conspecifics intrude in their mating territory, birds of prey, e. g. the buzzard (Fig. 99 A – *Buteo buteo*) and the hawk (Fig. 99 B – *Accipiter gentilis*), make particularly striking *demonstrative flights*.
In song birds, especially in birds of open areas, e. g. some Alaudidae, Motacillidae *(Anthus)* and the Viduinae, we know of song display flights (Fig. 99 C – *Alauda arvensis;* D – *Anthus trivialis*). These displays serve to attract females and to indicate territorial possession to conspecifics. Strictly speaking they are not courtship flights, since they are not only performed during courtship (BLUME 1967, NICOLAI 1969).

vol de parade □ Comportement d'un grand nombre de Rapaces, comme la Buse (Fig. 99 A – *Buteo buteo*) et l'Autour (Fig. 99 B – *Accipiter gentilis*), qui effectuent des vols particulièrement démonstratifs. Le but de ces *vols démonstratifs* est probablement l'intimidation, car ils sont souvent exécutés lors

du passage de congénères dans le territoire de reproduction.
Chez les Passereaux, principalement les habitants de paysages ouverts, par exemple chez quelques Alaudidae, Motacillidae *(Anthus)* et chez les Viduinae, nous connaissons des vols démonstratifs avec chant (Fig. 99 C – *Alauda arvensis;* D – *Anthus trivialis*). Ces démonstrations servent à la fois à l'attraction de la ♀ et surtout à la signalisation du territoire vis-à-vis des congénères ♂♂. Ce ne sont pas des vols de parade nuptiale *sensu stricto* étant donné qu'ils ne sont pas seulement exécutés pendant la formation du couple (BLUME 1967, NICOLAI 1969).

Scheinangriff □ Ritualisiertes Drohverhalten, bei dem die Angriffe nur angedeutet sind und kurz vor dem Feind gestoppt werden. Ein gutes Beispiel von Scheinangriff ist z. B. das bei Dachs und Eichhörnchen bekannte *Imponierbremsen.* Das Tier springt dabei auf den Gegner zu, aber zu kurz, und bremst den Angriffssprung betont ab, wobei es mit den Beinen hart auf die Unterlage schlägt, oft noch gleichzeitig ruft und die Haare aufrichtet (EIBL-EIBESFELDT 1957). Wahrscheinlich kann man auch das → *Hassen* der Singvögel als Scheinangriff deuten.

sham attack □ A ritualized threat behaviour in which the feigned attack is stopped just short of the opponent. The animals springs at its opponent, but stops short of the goal coming down hard on the ground and often screaming and erecting the hair. This *putting on the brakes display* is described by EIBL-EIBESFELDT (1957) for the squirrel and the badger. The → *mobbing* in birds may also be considered as a sham attack.

pseudo-attaque □ Comportement de menace ritualisée où les attaques ne sont qu'ébauchées et sont stoppées juste devant l'adversaire. Un bon exemple de pseudo-attaque est le simulacre de menace chez le Blaireau et l'Ecureuil. L'animal fait un *simulacre d'attaque* vers le rival, mais nettement trop court, et freine son élan d'une manière très accusée. Souvent, l'animal achève son saut en se recevant avec force sur ses pattes. En même temps, il peut pousser des cris et hérisser son pelage (EIBL-EIBESFELDT 1957). Probablement, le → *houspillage* des Passereaux peut également être considéré comme une pseudo-attaque.

Scheinbeißen □ Bei vielen Säugern beim Spielverhalten eine am Partner ausgeführte Beißintention ohne jeden Ernstbezug. Im menschlichen Flirtverhalten zum → *Beißkuß* ritualisiert. Bei Kleinkindern im Alter der → *Fremdenfurcht* beobachtet man beim Schäkern mit fremden Personen ein Alternieren zwischen Zuwenden als Spielaufforderung und Wegwenden als ritualisierte Flucht, wobei eine Beißintention in die Mutterbrust ausgeführt wird (Abb. 100 – Bayaka-Pygmäen-Säugling aus Äquatorialafrika).

pseudo-biting, play-biting □ In many mammals during play without any serious intent, directed to-

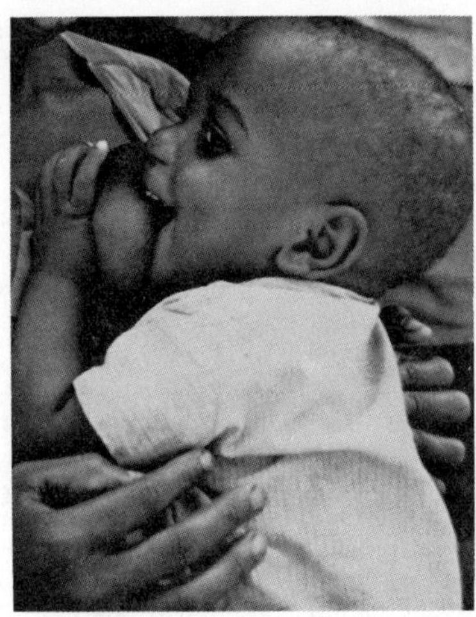

100

wards the partner. Ritualized into the → *love-bite* in human courtship behaviour. In small children during the phase of → *fear of strangers* it can be seen when they tease or flirt with strangers. They alternatively turn their head to the stranger as an invitation to play and away as a ritualized escape while performing a biting intention to the mother's breast (Fig. 100 – Bayaka Pygmies infant from Equatorial Africa).

pseudo-morsure □ Mouvement intentionnel observé chez de nombreux Mammifères au cours du jeu. Dans le comportement humain, lors du flirt, il est ritualisé en un → *baiser mordant.* Chez les enfants, à l'âge où ils ont → *peur des étrangers,* on observe au cours du badinage avec des personnes peu connues une alternance entre une invite au jeu et une fuite ritualisée en tournant la tête pour éviter le contact visuel: c'est lors de la fuite ritualisée que l'enfant exécute une morsure intentionnelle orientée vers la poitrine maternelle (Fig. 100 – nourrisson de Pygmées Bayaka de l'Afrique Equatoriale).

Scheinhinken → Verleiten
E. feigned limping □ F. boiterie feinte

Scheinpicken □ Ein bei vielen Vögeln während des Drohens oder in einer Kampfpause plötzlich auftretendes Auf-den-Boden-Picken, ohne daß dabei Nahrung aufgenommen wird. Schon von HEINROTH (1911) beobachtet, der auch den Ausdruck prägte. TINBERGEN (1940) und KORTLANDT (1940 a + b) analysierten diese Vorgänge genauer. TINBERGEN interpretierte dieses Verhalten als eine → *Übersprungbewegung.* Bei Möwen beobachtet man anstelle des Scheinpickens ein → *Grasrupfen.*

sham pecking □ A type of pecking at the ground in certain birds in which no food is consumed. This occurs during threat displays or during a pause in fighting and was first noted by HEINROTH (1911). Both TINBERGEN (1940) and KORTLANDT (1940 a + b) analysed the phenomenon more carefully and TINBERGEN considered it a → *displacement activity.* In gulls one finds → *grass pulling* instead of sham pecking.

pseudo-picorage □ Chez un grand nombre d'Oiseaux, comportement qui survient lors des actions de menace ou lors d'un arrêt du combat; il consiste en un picorage subit du sol pendant lequel aucune nourriture n'est prise. Ce phénomène a été observé par HEINROTH (1911) qui créa le terme. TINBERGEN (1940) et KORTLANDT (1940 a + b) l'ont analysé d'une manière plus précise. TINBERGEN le considère comme une → *activité substitutive*. Chez les Goélands, on observe à la place du pseudo-picorage un comportement de → *»désherbage«*.

Scheinputzen □ Demonstratives Verhalten im Kontext der Werbung bei vielen Entenvögeln. Viele Erpel putzen bei der Balz scheinbar ihre Flügel, was als → *Konfliktverhalten* oder → *Übersprungbewegung* gedeutet wird. Die Schwungfedern werden dabei nur flüchtig berührt, als würde das Tier mit dem Schnabel auf den prächtigen Spiegel weisen (Abb. 101 A – *Anas querquedula* ♂). Dieses Verhalten wurde zu einer → *Demonstrationsbewegung* ritualisiert, welche beim Mandarinenerpel besonders auffällig wirkt (Abb. 101 B – *Aix galericulata*). Diese Bewegung ist mit dem → *Antrinken* fest gekoppelt. Scheinputzen kommt als Übersprungbewegung auch bei anderen Vogelarten vor, ohne demonstrativen Charakter zu besitzen (LORENZ 1941).

101 A

101 B

sham preening □ Display behaviour in many species of Anatidae. Many drakes apparently preen their wings during courtship and this is interpreted as a form of → *conflict behaviour* or → *displacement activity*. The flight feathers are only lightly touched as if to draw attention to the colourful speculum (Fig. 101 A – *Anas querquedula* ♂). This behaviour pattern has been ritualized into a → *demonstrative movement* which is particularly conspicuous in the mandarin drake (Fig. 101 B – *Aix galericulata*). It is closely coupled with → *display drinking*. Sham preening occurs in other bird species as a displacement movement without any demonstrative character (LORENZ 1941).

pseudo-nettoyage □ Comportement démonstratif lors de la parade nuptiale chez de nombreuses espèces de la famille des Anatidae. Les ♂♂ font semblant de nettoyer leurs ailes lors de la parade nuptiale, activité qui est considérée comme un → *comportement conflictuel* ou une → *activité substitutive*. Les rémiges ne sont que légèrement touchées comme si les malards voulaient indiquer avec leur bec leur magnifique miroir alaire (Fig. 101 A – *Anas querquedula* ♂). Ce comportement s'est ritualisé en un → *mouvement démonstratif*, particulièrement apparent chez *Aix galericulata* (Fig. 101 B). Ce mouvement est étroitement lié avec le → *»Antrinken«*. Le pseudo-nettoyage existe aussi chez d'autres Oiseaux comme mouvement substitutif sans pour autant avoir la signification d'un caractère démonstratif (LORENZ 1941).

Scheinschlafen → Übersprungbewegung
E. pseudo-sleeping □ F. pseudo-sommeil

Scherzpartnerbeziehung □ Das Necken, Spotten und miteinander Scherzen hat über die erzieherische Wirkung hinaus noch eine aggressionsableitende Funktion. Scherzpartner, die gut befreundet sind, dürfen sich eine Reihe von Freiheiten herausnehmen. Sie dürfen einander necken und herausfordern, wobei die Anlässe meist fingiert sind, also keinerlei Verstöße eines Partners gegen die Gruppennorm vorliegen. Solches Verhalten dient einer harmlosen Aggressionsableitung innerhalb der Gruppe und ist in diesem Sinne als Ventilsitte aufzufassen (HEINZ 1967, EIBL-EIBESFELDT 1972, 1973).

joking (kidding) relationship □ Teasing, mocking and joking behaviour may serve to divert aggression in addition to its role as an educational mechanism. Persons who are close friends may tease each other over a variety of matters. Often the reason for such teasing is invented and there is no actual violation of a group norm. This type of behaviour serves to divert intra-group aggression and may be regarded in this sense as a release valve (HEINZ 1967, EIBL-EIBESFELDT 1972, 1973).

relations de taquinerie □ Les taquineries, les moqueries et les plaisanteries mutuelles ont, en plus de leur effet éducatif, une fonction de dérivation de l'agressivité. Les partenaires, liés par une amitié particulière, peuvent se permettre une série de libertés. Dans ce contexte, ils peuvent se taquiner et se défier, les raisons alléguées étant souvent fictives ou simulées, mais cette taquinerie ne constitue pas une atteinte à la norme du groupe. Un tel comportement a une fonction de dérivation anodine de l'agressivité à l'intérieur d'un groupe et, à cet égard, il peut être considéré comme une soupape (HEINZ 1967, EIBL-EIBESFELDT 1972, 1973).

Schilddachformation □ Das Gruppieren, Sich-aneinander-Schmiegen und die Köpfe-ineinander-Stecken bei Kaiserpinguinen während starker und kalter Stürme. Emanzipierte Junge tun dies untereinander. Dieses Verhalten dient der Thermoregulation, und die Schilddachformation ist ständig in Bewegung, indem die Randtiere nach innen drängen und dabei die Innentiere nach außen schieben (PREVOST 1961).

penguin huddling behaviour □ During severe storms Emperor penguins huddle together and form a sort of roof by holding their heads together. Juveniles do this among themselves. The group is in a continuous state of flux in which the outer animals push toward the center and the inner ones are pushed outwards. This behaviour serves thermo-regulation (PREVOST 1961).

formation en carapace de »tortue« □ Chez le Manchot Empereur, mode de regroupement dense d'un grand nombre d'individus lors de tempêtes fortes et froides: les Oiseaux s'agglomèrent les uns aux autres et accolent leurs têtes qui forment une sorte de toit protecteur. Les jeunes émancipés »font la tortue« entre eux, indépendamment de leurs parents. Ce comportement sert à la thermorégulation. La formation en »tortue« est constamment en mouvement, car les animaux de la périphérie se pressent vers le centre, poussant vers l'extérieur les animaux se trouvant au milieu (PREVOST 1961).

Schildwache → Wachesitzen
E. sentinel □ F. sentinelle

Schlafverhalten □ Die ein- oder mehrfach sich wiederholenden Restitutionsphasen im Aktivitätsrhythmus eines Tieres führen zu einem Ruhe- bzw. Schlafverhalten. Nach TEMBROCK (1961) ist dieses Verhalten nicht einfach eine »Erschöpfung«, sondern stellt vielmehr einen integrierenden Bestandteil des Verhaltensinventars des Tieres dar. Es wird in artspezifischem Zeitrhythmus unter arttypischen Vorbereitungen und mit charakteristischen Stellungen ausgeführt (HASSENBERG 1965, JAVANOVIC 1973).

sleeping behaviour □ The recurring phases of restitution in the circadian activity cycle of an animal which leads to rest or sleep. TEMBROCK (1961) maintains that sleeping behaviour is not simply a consequence of exhaustion, but rather represents an integration of the animal's behaviour repertoire. Sleeping behaviour may occur once or several times in the activity cycle according to the species. The preparatory movements and the posture assumed are also species specific (HASSENBERG 1965, JAVANOVIC 1973).

comportement de sommeil □ Les phases de restauration en recurrence simple ou multiple dans le rythme journalier d'activité d'un animal conduisent au comportement de repos ou de sommeil. Selon TEMBROCK (1961), ce comportement ne se réduit pas à un »épuisement«, mais constitue un élément parfaitement intégré dans le répertoire comportemental d'un animal. Le comportement de repos et de sommeil est précédé de préparatifs caractéristiques; il suit un rythme circadien spécifique et met en jeu des postures propres à chaque espèce (HASSENBERG 1965, JAVANOVIC 1973).

Schlagen → Fegen

Schlittenfahren □ Das Auf-dem-Bauch-Rutschen der Pinguine zur schnelleren Fortbewegung; dabei wird mit den Füßen abgestoßen (PREVOST 1961).

sliding, tobogganning □ Penguins often slide on their stomach, propelling themselves forward with their feet. They are able to move faster in this manner. This behaviour is also called tobogganning (PREVOST 1961).

tobogganning □ Mode de locomotion rapide chez les Manchots qui consiste à se laisser glisser sur la glace, couché sur le ventre, en se propulsant vigoureusement avec les pattes. PREVOST (1961), en français, utilise le terme anglais *tobogganning*.

Schloß-Schlüssel-Verhältnis → Schlüsselreiz
E. lock-key relationship □ F. relation clé-serrure

Schlüsselreiz □ Der zum Ansprechen eines angeborenen Auslösemechanismus (→ *AAM*) notwendige Reiz, der über eben diesen AAM eine artspezifische Instinktbewegung in Gang setzt. RUSSELL (1943) nannte solche entscheidenden Reize *Signalreize* und später auch *Wahrnehmungssignale*. KOEHLER (1964) wies jedoch darauf hin, daß im Deutschen Schlüsselreiz der gebräuchlichere Ausdruck ist. Dem Ausdruck liegt die Vorstellung zugrunde, daß die auslösenden Merkmale analog einem Schlüssel, der ein Schloß öffnet (*Schloß-Schlüssel-Verhältnis*) auf eine dem Zentralnervensystem vorgeschaltete Apparatur (AAM) wirken, die eine Entladung der zentralen Impulse zur unpassenden Zeit verhindert und sie erst beim Eintreffen der spezifischen Schlüsselreize auslöst. Für jeden Funktionskreis gelten andere Schlüsselreize, werden andere Merkmale vom Tier beachtet. → *übernormaler Auslöser*, → *Reiz*.

key stimulus □ The stimulus which is adequate to activate an innate releasing mechanism (→ *IRM*) which then sets in motion a species specific action pattern is termed a key stimulus. RUSSELL (1943) referred to such definitive stimuli as *signal stimuli* or *perceptual stimuli*. KOEHLER (1964) pointed out that »key stimulus« is in wider usage. This term is based on the idea that the releasing cues of a behaviour are analogous to a key which opens a lock (*lock-key relationship*). Thus a perceptual releaser (IRM) activates motor responses specific to that IRM. It is the latter which prevents the release of CNS impulses in inappropriate situations and which activates the motor patterns only in the right situation. Each functional cycle has its own key-stimuli; the animal thus attends only to those cues appropriate at a given time. → *su-*

pra-normal releaser, → *stimulus.*

stimulus-clé □ Stimulus nécessaire à l'activation d'un mécanisme déclencheur inné (→ *MDI*); par ce MDI, le stimulus-clé déclenche un mouvement instinctif spécifique. RUSSELL (1943) a appelé de tels stimuli décisifs *stimuli-signaux* et plus tard *signaux perceptifs.* KOEHLER (1964), cependant, a signalé qu'en allemand stimulus-clé est l'expression la plus utilisée ce qui aujourd'hui est également le cas en français. Cette expression est basée sur un modèle d'après lequel les caractères déclencheurs fonctionnent d'une manière analogue à une clé ouvrant une serrure (*relation clé-serrure*). Le stimulus-clé agit sur un mécanisme central (MDI) précédant le système nerveux et empêchant une décharge des impulsions centrales à un moment inadéquat. Ce mécanisme ne transmet des ordres moteurs qu'en présence des stimuli-clés spécifiques. Pour chaque type d'activité, il existe des stimuli-clés différents, traduisant la prise en considération de diverses catégories de signaux. → *déclencheur supra-normal,* → *stimulus.*

Schmollen □ Verhalten bei Menschen: Unterliegt ein Kind bei Auseinandersetzungen, dann gibt es oft schmollend auf. Kinder schmollen auch, wenn ihnen Unrecht geschieht. Das Kind läßt den Kopf hängen, wendet seinen Blick ab und macht den typischen Schmollmund (Abb. 102 A – !ko-Buschkind aus der Kalahari). Es wird dann meist nicht länger belästigt, außerdem löst der Schmollmund Mitleid bzw. Zuneigung aus. Wahrscheinlich handelt es sich um eine beschwichtigende Verhaltensweise, die vermutlich angeboren ist, denn wir finden sie bei allen Menschen, auch bei Erwachsenen (Abb. 102 B – Waika-Indianer aus Venezuela), als → *Universalie.*

pouting behaviour □ Human behaviour: If a child has lost an argument or if he is done an injustice he may begin pouting. His head hangs, he turns his eyes away and makes a distinctive long face (Fig. 102 A – Bushman child from the Kalahari Desert). This expression tends to induce sympathy, and the child is then usually no longer perturbed. Pouting behaviour is probably an innate apeasement behaviour since it belongs to the → *universals.* It is found even in adults (Fig. 102 B – Waika Indian from Venezuela).

bouderie □ Comportement humain: Un enfant qui, lors de controverses ou de disputes, est perdant, abandonne souvent la partie en boudant. Les enfants boudent aussi lorsqu'ils sont victimes d'une injustice. L'enfant laisse tomber la tête, détourne les yeux et fait une moue typique (Fig. 102 A – enfant Boshiman du Kalahari). A partir de ce moment, on le laisse généralement tranquille; d'autre part, la moue déclenche pitié et sympathie. Il s'agit d'un comportement d'apaisement vraisemblablement inné étant donné que nous le trouvons chez tous les Hommes, même chez les adultes (Fig. 102 B – Indien Waika du Vénézuéla), en tant que comportement universel. → *universaux.*

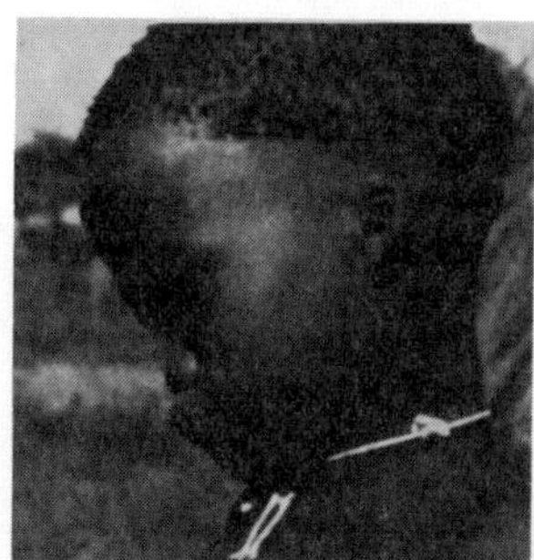

102 A

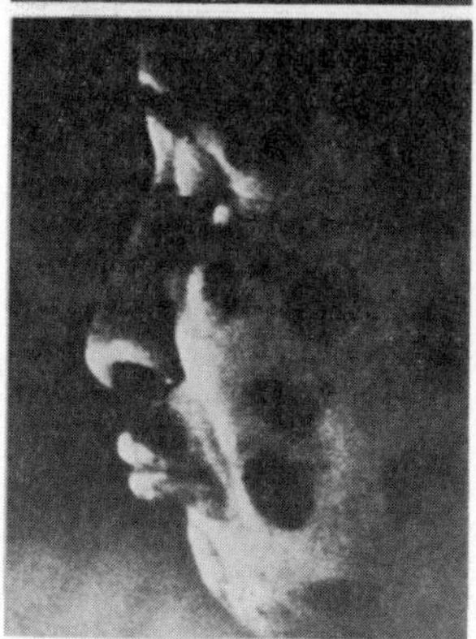

102 B

Schnabelflirt → Schnäbeln
E. bill flirtation □ F. flirt de becquetage

Schnabelgruß → Schnäbeln

Schnäbeln □ Bei Vogelarten, die in Dauerehe leben, kommt häufig Schnabelkontakt zwischen den Partnern vor. Die Schnäbel sind dabei ineinander verschränkt oder der Schnabel des ankommenden Partners wird zur Begrüßung umfaßt wie beim → *Futterbetteln* der Jungen (Abb. 103 – Mönchssittichpärchen, *Myiopsitta monachus*). Oft wird bei der Begrüßung Futter übergeben, → *Begrüßungsfüttern.* Es kann sich aber auch um einen einfachen Schnabelgruß handeln. Bei der Anpaarung spricht man, wenn Futter übergeben wird, von → *Balzfüttern,* ohne Futterübergabe wird auch der Ausdruck *Schnabelflirt* verwendet. Die Schnabelhaltung ist stets die gleiche wie beim → *Jungefüttern,* aus dem sich die Verhaltensweisen des Balzfütterns, Begrüßungsfütterns, Paarfütterns und das Schnäbeln entwickelt haben. Man kann auch von einem »ritualisierten Jungefüttern« sprechen (GWINNER 1964, WICKLER 1969).

103

billing □ In birds which maintain permanent pair bonds, mutual beak contact of the pair partners is common. The bills are either crossed or the bill of the arriving partner is grasped by the other as in → *food-begging* by the young (Fig. 103 – couple of *Myiopsitta monachus*). Often during this greeting food is transferred (→ *greeting feeding*), however this is not always the case. Transfer of food during pair formation is called → *courtship feeding* whereas in the absence of food exchange one speaks of *bill flirtation*. The bill posture is always the same as in → *feeding of young* from which the billing ceremony is derived, and this courtship feeding or food exchange ceremony may be referred to as »ritualized feeding of the young« (GWINNER 1964, WICKLER 1969).

becquetage □ Chez les espèces d'Oiseaux vivant en monogamie permanente, il existe souvent un contact de bec à bec. Lors de ce comportement, les becs sont entrecroisés ou le bec du partenaire qui arrive est saisi comme lors de la → *sollicitation alimentaire* (Fig. 103 – couple de *Myiopsitta monachus*). Souvent, lors de ce comportement de salut, il y a effectivement transmission de nourriture, → *nourrissage de salut,* mais il peut ne s'agir que d'un becquetage simple (*flirt de becquetage*). S'il y a transmission de nourriture lors de l'appariement, on parle de → *nourrissage de parade nuptiale.* La position du bec est toujours la même que lors du → *nourrissage des jeunes* à partir duquel les différents comportements tels que nourrissage de parade nuptiale, nourrissage de salut, nourrissage du partenaire et becquetage se sont développés. Tous ces comportements peuvent être considérés comme des formes ritualisées du nourrissage des jeunes (GWINNER 1964, WICKLER 1969).

Schnappreflex → Reflexverspätung
E. snapping reflex □ F. réflexe de happement

Schnauzentriller □ Eine epigame Verhaltensweise beim Stichling (PELKWIJK und TINBERGEN 1937). Ist das Stichlings-♀ nach dem Zickzacktanz dem ♂ gefolgt und ins Nest geschlüpft, hämmert das ♂ mit raschen Schnauzenschlägen gegen den Schwanzstiel des ♀ und löst damit die Eiablage aus (Abb. 104 – *Gasterosteus aculeatus*). → *Balzkette.*

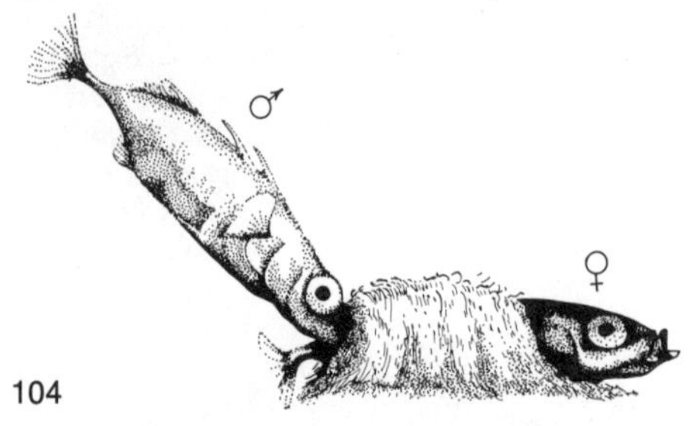

104

quivering □ An epigamic behaviour pattern in the stickleback (PELKWIJK and TINBERGEN 1937). After the ♀ stickleback has been led into the nest following the zig-zag dance, the ♂ strikes her with rapid quivering movements of the snout. This quivering induces spawning in the ♀ (Fig. 104 – *Gasterosteus aculeatus*). → *courtship chain.*

frémissement du museau □ Chez l'Epinoche, comportement épigame déclenchant la ponte (PELKWIJK et TINBERGEN 1937). Si après la danse zigzagante, la ♀ a suivi le ♂ et est entrée dans le nid, le ♂ effectue avec sa bouche des frémissements rapides contre le corps de la ♀ et déclenche ainsi la ponte (Fig. 104 – *Gasterosteus aculeatus*). → »*Balzkette*«.

Schnauzenzärtlichkeit □ Begrüßungsgebärden bei vielen Säugern, die sich Schnauze an Schnauze begrüßen. Der Ausdruck stammt von SCHENKEL (1947), der dieses Verhalten bei Wölfen beschrieb. Auch Schimpansen begrüßen einander mit einem Kuß, und schließlich ist unser menschlicher Kuß nichts anderes als eine »Schnauzenzärtlichkeit«. Beispiele für dieses Verhalten gibt EIBL-EIBESFELDT (1957).

nose caressing □ Greeting gestures in many mammals consist of rubbing noses. SCHENKEL (1947) coined the term when describing nose caressing in wolves. Chimpanzees greet each other with a kiss, and the human kiss may be considered as »nose caressing«. Further examples of this behaviour are given by EIBL-EIBESFELDT (1957).

caresse du museau □ Comportement de salut chez de nombreux Mammifères dont le contact est surtout naso-nasal ou de bouche à bouche. L'expression a été créée par SCHENKEL (1957) lors de ses observations sur le comportement social des Loups. Un comportement semblable existe également chez les Chimpanzés où le salut ressemble déjà beaucoup à un baiser et finalement, notre baiser humain n'est rien d'autre qu'une »caresse du museau«. Nous trouvons de nombreux exemples à ce sujet chez EIBL-EIBESFELDT (1957).

Schnüren → Gangart
E. loping □ F. filer

Schreckmauser □ Manche Vögel werfen bei starker Bedrängnis (Angst? Schreck?) und Gefiederkontakt einen Teil des Großgefieders, speziell die Schwanzfedern ab. In seltenen Fällen ist auch totaler Gefiederverlust beobachtet worden. Bisher kennt man dieses Verhalten von 51 Vogelarten, darunter 24 Passeriformes (DATHE 1955, DEMENTIEV 1958).

fright moult □ Several bird species lose part of their feathers during severe distress and restraint. The tail feathers are usually lost, but in rare cases the loss of all feathers has been reported. This fright-induced moult may occur in 51 species, 24 of which are Passeriformes (DATHE 1955, DEMENTIEV 1958).

mue phobique □ De nombreux Oiseaux perdent, lors d'un choc »émotionnel« très sévère (peur? effroi?), une grande partie de leur plumage, particulièrement les plumes de la queue. Dans quelques rares cas, la perte totale des plumes a été observée. Ce comportement a été constaté jusqu'à présent chez 51 espèces d'Oiseaux, dont 24 espèces de Passeriformes (DATHE 1955, DEMENTIEV 1958).

Schreckstellung → Akinesis
E. phobic posture □ F. posture phobique

Schreiweinen □ Ein Ausdrucksverhalten des neugeborenen Menschenkindes; es ist als eine Art → *Ruf des Verlassenseins* zu deuten.

intense crying □ An expressive behaviour of the new born infant which is a form of → *distress call.*

pleurer en criant □ Elément du comportement expressif chez le nouveau-né humain; il s'agit d'une sorte de → *vocalisation de détresse.*

Schulterbetonung □ Beim Menschen, insbesondere beim Mann, die Neigung, die Schultern deutlich hervorzuheben. LEYHAUSEN (1969) wies darauf hin, daß der Haarstrich bei Menschen – anders als bei Menschenaffen – an Schultern und Oberarm aufwärts gerichtet ist, so daß sich noch heute bei stark behaarten Menschen auf den Schultern Haarbüschel bilden. Beim → *Haarsträuben* würde dann vor allem der Schulterriß vergrößert. Bei unseren Vorfahren war dies sicher der Fall (Abb. 105 A – Rekonstruktion). Nach Abbau des Haarkleides blieb jedoch die Neigung erhalten, die Schultern zu betonen, – eine Neigung, die in den verschiedensten Kulturen unterschiedlich gelöst werden kann (Abb. 105 B – Waika-Indianer vom Orinoko; C – Kabuki-Schauspieler, Japan; D – Alexander II. von Rußland).

105

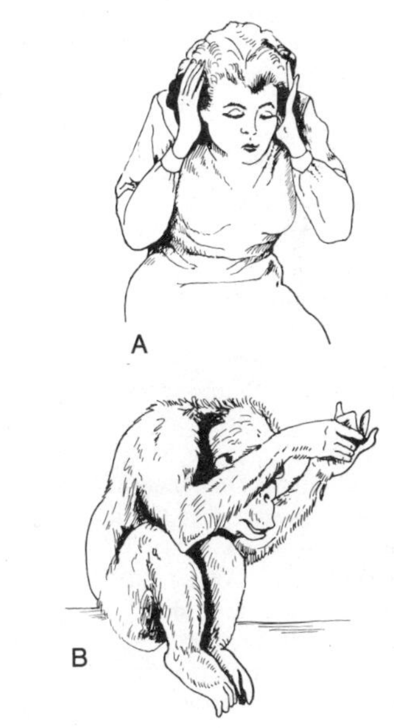

106

shoulder enhancement □ The tendency in humans to enlarge, especially in males, the shoulders. LEYHAUSEN (1969) pointed out that in men, unlike in anthropoid apes, the hair along shoulders and upper arms points upwards. Thus in hairy men regular hair tufts can be seen. During → *piloerection* the shoulder outline is especially enlarged. In our ancestors (Fig. 105 A – reconstruction) this must have certainly been the case. After loss of body hair the tendency remained to enhance the shoulders. This is accomplished on various ways in various peoples (Fig. 105 B – Waika Indian from the Orinoco; C – Kabuki actor, Japan; D – Alexander II of Russia).

accentuation de la carrure □ Chez les humains, particulièrement dans le sexe masculin, tendance à mettre en évidence ses épaules. LEYHAUSEN (1969) a signalé que chez les humains, contrairement à ce qu'on trouve chez les Anthropoides, le sens du poil au niveau des épaules et des bras est orienté vers le haut. Nous trouvons aujourd'hui encore, chez les hommes particulièrement poilus, de véritables touffes sur les épaules. Lors d'une → *horripilation,* les contours des épaules seraient alors agrandis; chez nos ancêtres préhistoriques, il en était certainement ainsi (Fig. 105 A – reconstitution). Après la perte des poils sur le corps, nous avons conservé la tendance à mettre en évidence nos épaules, tendance qui, dans les différentes cultures, s'exprime de différentes manières (Fig. 105 B – Indien Waika de l'Orinoque; C – acteur Kabuki du Japon; D – Alexandre II de Russie).

Schulterreaktion □ Eine Schreckreaktion und ein Versuch des Sich-Schützenwollens (SPINDLER 1958). Hört ein Mensch über sich plötzlich ein Geräusch, gleich ob von einer sichtbaren oder nicht sichtbaren Quelle, so zieht er die Schultern hoch, neigt sich etwas nach vorn unten und zieht den Kopf ein und verbirgt ihn in den Händen (Abb. 106 A). Oft werden die Hände auch schützend über den Kopf gehalten. Bei Menschenaffen kennen wir in der gleichen Situation eine ebensolche Reaktion (Abb. 106 B). Von anderen Säugern kennen wir ebenfalls ähnliche Reaktionen des Sich-Duckens bei plötzlichem Lärm und dem daraus resultierenden Erschrecken. Wahrscheinlich handelt es sich um ein abgeleitetes Versteckverhalten.

shoulder-reaction (pulling-in-head) □ A fright reaction and an attempt to protect one's head and neck (SPINDLER 1958). If a human suddenly hears a noise above himself, whether its source is visible or not, he will pull up his shoulders, pull in his head, bend slightly forward (Fig. 106 A). Often, the hands are raised protectively. In anthropoid apes we know the same reaction (Fig. 106 B). We know that other mammals will duck in response to sudden noises. It is possible that this behaviour is ritualized hiding.

réaction des épaules □ Réaction de crainte tendant à l'autoprotection (SPINDLER 1958). L'Homme, à la perception d'un bruit soudain venant d'une source visible ou invisible, hausse les épaules, se penche légèrement en avant et rentre la tête, puis essaie de la mettre à l'abri à l'aide de ses mains (Fig. 106 A). Souvent, les mains sont même élevées au-dessus de la tête pour bien la protéger. Chez les Anthropoïdes, nous observons la même réaction dans des situations comparables (Fig. 106 B). Chez d'autres Mammifères, nous connaissons également des réactions de blotissement en réponse à un bruit soudain et à l'effroi qui en résulte. Il s'agit probablement d'une ritualisation du comportement de mise à l'abri.

Schürzenwippen = Schamschürzenwippen → Schamweisen

Schüttelstrecken = Streckschütteln → Sichschütteln

E. stretch-shaking □ F. s'ébrouer en s'étirant

Schütteltanz → Putztanz

Schutzanpassung □ Überbegriff für alle in der Natur vorkommenden Möglichkeiten, einem Freßfeind zu entkommen. Weit verbreitet sind Tarntrachten, die das Tier in seiner Umgebung verbergen oder schwer sichtbar machen, wie → *Untergrundangleichung* in Form oder Farbe (Mimese); z. B. beim Schneehasen und dem Hermelin im Winter; die Wüstenfärbung vieler Vögel und Säuger; die Rindenfärbung bei an Baumstämmen ruhenden Lepidopteren; Gegenschattierung von Schwärmerraupen (Sphingidae) und Ästchenähnlichkeit von Spannerraupen (Geometridae). Als Ergänzung zu solchen Sichtschutztrachten haben sich außerdem dem Feindschutz dienende Verhaltensweisen entwickelt, wie z. B. → *Akinese,* Autotomie, Sichern als vorbeugende Aktivität, Sich-Fallenlassen, Sich-Lahmstellen, → *Verleiten,* das aktive Aufsuchen homochromer Orte, das Eingegrabensein von Ameisenlöwe, Libellenlarven und manchen Fischen, die sich nur zeitweise vergraben. Schützende Stellungen wie Somatolyse und → *Konturverfremdung,* Kollektivmimikry sowie der → *Konfusionseffekt* beim Fischschwarm. Neben diesen Tarnungen gibt es auch Warntrachten mit auffälligen, weithin sichtbaren Farben, insbesondere die Beispiele der Bates'schen Mimikry sowie die Entwicklung von → *Augenflekke(n),* vor allem bei Lepidopteren, in Verbindung mit Schreckstellungen. Vielleicht ist auch die → *Schreckmauser* der Vögel eine Schutzanpassung (BRUNS 1958, WICKLER 1968). → *Mimikry.*

protective adaptation □ Generalized term for all naturally occurring mechanisms for escaping a predator. Protective colouration is a widespread form of protective adaptation in which the animal resembles his environment, e. g. the colouration of snowshoe hare or ermine which changes with the season, or the colouration of desert birds or mammals. Substrate mimicry occurs in certain Lepidopterans which rest on the bark of trees, countershadowing in the caterpillars of Sphingid moths and dead twig resemblance of Geometrid moth caterpillars. In addition to camouflage, behavioural adaptations for predator protection have also occurred, as for example → *akinesis,* autotomy, collapsing or feigning death, → *distraction display,* active search for a camouflaging substrate, temporary self burial of ant lions, dragonfly larvae and several fish species, etc. Protective postures include somatolysis and → *contour alteration* as well as collective mimicry and the → *confusion effect* in fish. In addition to camouflaging colouration one also finds warning colouration where the animal exhibits conspicuous colours visible from a distance, e.g., Batesian mimicry, evolution of → *eye spots* in conjunction with fright postures in Lepidopterans, etc. Probably → *fright moult* in birds is also a form of protective adaptation (BRUNS 1958, WICKLER 1968). → *mimicry.*

adaptation protectrice □ Notion générale englobant tous les moyens existant dans la nature pour échapper à un prédateur potentiel. Les moyens de camouflage permettant à un animal de se rendre difficilement visible dans son environnement sont très répandus, comme par exemple → *l'homochromie* chez le Lièvre variable ou l'Hermine, la coloration désertique chez de nombreux Oiseaux et Mammifères, la coloration d'écorce chez des Lépidoptères se reposant sur les troncs d'arbres, l'ombre inversée chez les chenilles des Sphingidae et la ressemblance avec des brindilles chez les chenilles des Geometridae. Outre ces moyens de camouflage, se sont développés des comportements au service de la protection vis-à-vis du prédateur, comme par exemple → *l'acinèse,* l'autotomie, la vigilance en tant qu'activité préventive, se laisser tomber, faire le boiteux, les → *manoeuvre(s) de diversion,* la recherche active de substrats homochromes, l'enfouissement chez les larves des Fourmilions et certaines larves de Libellules ou encore chez de nombreux Poissons qui ne le font que temporairement. On connaît d'autre part, des postures protectrices comme la somatolyse, → *l'effacement des contours,* le mimétisme collectif et aussi → *l'effet de confusion* dans les bancs de Poissons. A côté de tous ces moyens de camouflage, il y a aussi des colorations dissuasives très vives et visibles de loin, comme les innombrables exemples de mimétisme Batésien ou le développement de → *taches ocellaires,* surtout chez des Lépidoptères en relation avec des positions d'effroi. Probablement, la → *mue phobique* chez les Oiseaux est également une forme d'adaptation protectrice (BRUNS 1958, WICKLER 1968). → *mimétisme.*

Schutzfärbung → Mimikry → Schutzanpassung
E. concealing colouration □ F. couleur protectrice

Schutzverhalten → Schutzanpassung
E. cryptic behaviour □ F. comportement d'autoprotection

Schwanzeinziehen → Schwanzhaltung
E. tail retraction □ F. la queue entre les jambes

Schwänzeltanz □ Bienensprache: – wenn die Futterquelle mehr als 100 m vom Stock entfernt ist, so tritt bei den Sammelbienen an die Stelle des → *Rundtanz(es)* der Schwänzeltanz. Außer den Daten, die auch der Rundtanz vermittelt, verkündet er den Stockgenossen die Entfernung und Richtung des Zieles (→ *Richtungsweisung*). Beim Schwänzeltanz läuft die Biene eine kurze Strecke geradlinig und dann – abwechselnd rechts herum und links herum – in einem Halbbogen zum Ausgangspunkt zurück. Dabei entspricht der Winkel der geradlinigen Strecke zur Senkrechten dem Winkel zwischen Futterquelle und Sonnenstand. Die geradlinige Laufstrecke wird durch seitliche Schwänzelbewegungen des Körpers und durch ein schnarrendes Geräusch betont (Abb. 107 – Schwänzeltanzschema der Honigbiene). Es kommen 13–15 Schwänzelbewegungen (nach rechts und links) auf 1 Sekunde. Das Geräusch entsteht durch wiederholte Vibrationsstöße (etwa 30 pro Sekunde). Die Vibrationen haben eine Frequenz von etwa 250 Hz. Der Schall wird wahrscheinlich durch die Brustmuskulatur erzeugt. Die → *Nachläuferinnen* können durch einen Piepton die Tänzerin zum Stillhalten und zur Abgabe von Futter veranlassen. Das Tanztempo dient der Entfernungsweisung. Mit zunehmender Entfernung des Zieles wird das Tanztempo langsamer. Die Gesetzmäßigkeit ließ sich bis zu einer Flugweite von 11 km verfolgen (VON FRISCH 1965).

waggle dance □ Language of bees: – if the food source is more than 100 meters away from the hive the forager bee performs a waggle dance instead of a → *round dance.* In addition to the information that is normally transferred by the round dance, the waggle dance also expresses the distance and an → *indication of direction* to the goal. In the waggle dance the bee runs a short straight distance and makes a curve alternately to the right and then to the left in a half arc returning each time to the starting point. The angle of the straight line to the plumb line corresponds to the angle of the food source in relation to sun position. The straight run part is accompanied by sidewise wagging movements and a rasping noise

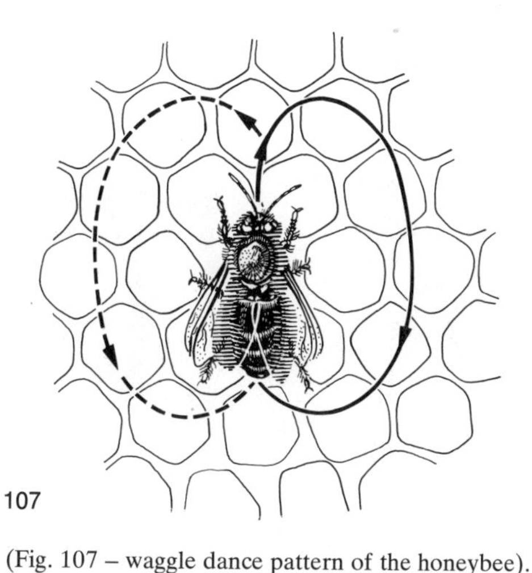

107

(Fig. 107 – waggle dance pattern of the honeybee). The wagging frequency is 13–15 right and left swings per second. The raspy noise is produced from repeated stamping vibrations (30/second) giving a sound of approximately 250 Hz. The sound production is probably controlled by the breast musculature. With certain peep tones, the → *follower bees* are able to stop the dancer and to induce it to release a food sample. The speed of the dance expresses the distance to the source with greater distances indicated by slower dances. The limit of predictability is about 11 kilometers flight distance (VON FRISCH 1965).

danse frétillante □ Langage des Abeilles: – lorsque la source de nourriture se trouve au-delà d'une distance de 100 m de la ruche, les butineuses exécutent à la place de la danse en rond une danse frétillante. Outre les données que contient la → *danse circulaire,* la danse frétillante donne aux ouvrières de la ruche des informations sur la distance et la direction de la source de nourriture *(→ indication de direction).* Lors de la danse frétillante, l'Abeille effectue un court trajet rectiligne, puis des demi-cercles alternativement en sens horaire et anti-horaire pour revenir à son point de départ. Pendant cette danse, l'Abeille transpose l'angle de la direction de la source de nourriture par rapport au soleil en un angle par rapport à la verticale (Fig. 107 – schéma de la danse frétillante de l'Abeille mellifère). Pendant le parcours rectiligne, l'Abeille effectue 13 à 15 frétillements latéraux de l'abdomen par seconde, accompagnés d'un bourdonnement composé de 30 émissions/sec, à la fréquence de 250 Hz environ; le bruit est probablement produit par la musculature pectorale. Les Abeilles qui suivent la danseuse parviennent à la stopper par une sorte de chuintement et à solliciter de la nourriture. La vitesse de la danse sert à indiquer la distance. Au fur et à mesure que la distance augmente, le rythme de la danse décroit. La validité de cette relation a pu être vérifiée jusqu'à une distance de 11 km (VON FRISCH 1965).

Schwanzhaltung □ Ein Ausdrucksverhalten bei

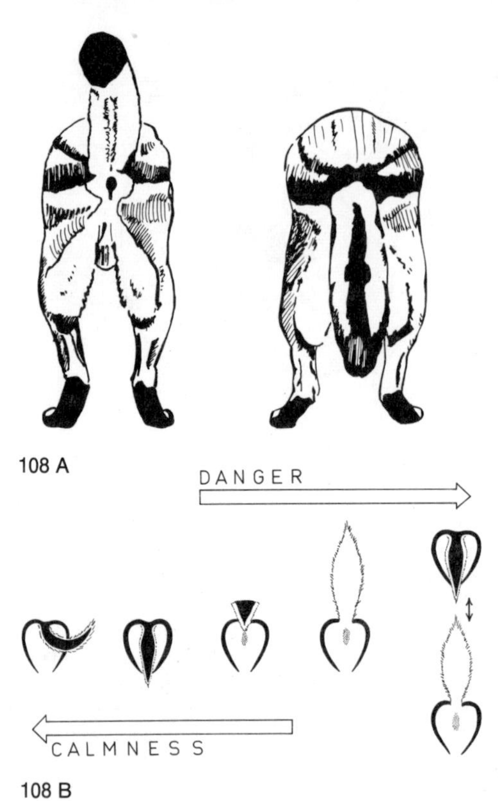

108 A

108 B

Säugetieren und Vögeln, welches als Signal bestimmte Motivationen anzeigt. Vor allem bei Caniden und dort insbesondere beim Wolf gut untersucht (SCHENKEL 1947). Wird der Schwanz steil nach oben getragen, zeigt dies Selbstsicherheit und ranghohe Stellung an, während bei Unterlegenheit verbunden mit Flucht der Schwanz zwischen die Hinterbeine geklemmt wird (Abb. 108 A – *Canis lupus*). Im Extremfall wird bei starker Hemmung der Schwanz nach vorn auf den Bauch geschlagen (ZIMEN 1976). ALVAREZ *et al.* (1976) haben beim Damhirsch das Ausdrucksverhalten untersucht und bei diesen Tieren ebenfalls den Aussagewert der Schwanzhaltung als Signal für bestimmte Stimmungen demonstrieren können (Abb. 108 B).

tail posture □ An expressive behaviour in mammals and birds which indicates specific motivations. Especially studied in canids, and there especially in wolves (SCHENKEL 1947). When the tail is raised up high, self-assurance and high rank is indicated. In lower-ranking individuals and during flight the tail is tucked between the hind legs (Fig. 108 A – *Canis lupus*). In a case of strong inhibition the tail is pressed against the belly (ZIMEN 1976). ALVAREZ *et al.* (1976) have studied the expressive behaviour of the fallow deer, *Dama dama,* and have demonstrated in these animals as well the significance of the tail posture as a signal for different motivations (Fig. 108 B).

posture de la queue □ Chez les Mammifères et les Oiseaux, comportement expressif signalant une motivation déterminée. Ce phénomène, particulièrement apparent chez les Canidae, a été surtout bien étudié chez le Loup (SCHENKEL 1947). La queue dressée en position verticale indique une situation dominante de l'animal et une confiance en soi. Par contre, en posture de soumission liée à la fuite, la queue est abaissée et coincée entre les pattes postérieures (Fig. 108 A – *Canis lupus*). Dans le cas extrême d'une très forte inhibition, la queue est même repliée en avant sous le ventre (ZIMEN 1976). ALVAREZ *et al.* (1976) ont étudié le comportement expressif chez le Daim, *Dama dama,* et ont pu démontrer que, chez ces animaux, la posture de la queue est aussi un signal pour différentes motivations (Fig. 108 B).

Schwanzsträuben □ Das Schwanzsträuben (SST) gibt bei Spitzhörnchen (Tupaiidae) Auskunft über bestimmte Erregungssituationen. Normalerweise haben sie die Schwanzhaare fast völlig angelegt (Abb. 109 A – *Tupaia glis*). Bei jeder Störung (plötzlicher Lärm, Geruch einer fremden Substanz, Kampf mit Artgenossen) kontrahieren sich die vom sympathischen Nervensystem innervierten Haarbalgmuskeln am Schwanz und richten die Schwanzhaare auf (Abb. 109 B). Der Schwanz bekommt dadurch ein buschiges Aussehen (VON HOLST 1969, 1973).

109 A 109 B

pilo-erection of the tail □ In tree shrews (Tupaiidae), raising the tail hair indicates arousal. Normally the tail hair is laid flat (Fig. 109 A – *Tupaia glis*). Following any disturbance (e. g. sudden noise, strange smell, fight with conspecifics), the sympathetically innervated hair follicle musculature in the tail contracts, thus raising the tail hair (Fig. 109 B). The tail then takes on a bushy appearance (VON HOLST 1969, 1973).

horripilation caudale □ Le hérissement des poils

de la queue chez les Tupaiidae donne des informations sur le degré d'excitation de l'animal. Normalement, ces poils sont appliqués contre l'axe de la queue (Fig. 109 A – *Tupaia glis*). Toute perturbation (bruit soudain, odeur d'une substance étrangère, lutte avec les congénères) déclenche une contraction des muscles capillaires de la queue innervés par le système nerveux sympathique, ce qui entraîne la pilo-érection (Fig. 109 B). La queue prend ainsi une apparence touffue (VON HOLST 1969, 1973).

Schwärmen → Hochzeitsflug → Volksteilung

E. swarming □ F. essaimage

Schwellenwert → Reizschwelle

E. threshold value, liminal value □ F. seuil, valeur liminaire

Seihen □ Eine Form der Nahrungsaufnahme bei Gänsen und Entenvögeln. Flamingos suchen ihre Nahrung im flachen Wasser und durchseihen dieses dabei mit dem Schnabel (Abb. 110 – *Phoenicopterus ruber*). Dabei erst erkennt man die Zweckmäßigkeit der sonderbaren Schnabelform.

110

sifting □ A method of feeding in geese and ducks. Flamingos seek their food in shallow water and strain it through their bill (Fig. 110 – *Phoenicopterus ruber*). Through observation of this behaviour the function of the unusual beak form was first determined.

passer, filtrer □ Une forme de recherche de la nourriture chez les Oies et les Canards. Les Flamants cherchent leur nourriture dans l'eau peu profonde et la filtrent avec leur bec recourbé (Fig. 110 – *Phoenicopterus ruber*). Cet acte révèle l'utilité de la forme inusuelle de leur bec.

Seitabstrecken → Räkelsyndrom

E. lateral stretching □ F. étirement latéral

Selbstmarkierung → Harnmarkierung

E. self-marking □ F. auto-marquage

Semantisierung □ WICKLER (1967) hat vorgeschlagen, in der Ethologie alle Vorgänge, die der Verbesserung der → *Verständigung* dienen, unter den Begriff Semantisierung zusammenzufassen. Diese kann einseitig vom Sender her erfolgen (senderseitige Semantisierung), und man spricht dann weiterhin von → *Ritualisierung*. Oft entwickeln sich jedoch Sender und Empfänger in wechselseitiger Anpassung, und manchmal kann auch eine empfangsseitige Semantisierung erfolgen. Das betrifft z. B. jede Entwicklung von Auslösemechanismen (→ *AAM*).

semantization □ WICKLER (1967) has proposed that all processes which lead to an improvement of → *communication* be collectively termed semantization. This can take place unilaterally from the sender (semantization from the sender) and is then called → *ritualization*. Often sender and receiver develop together by mutual adaptation, and sometimes a semantization on the part of the receiver only can take place. This concerns all development of releasing mechanisms (→ *IRM*).

sémantisation □ WICKLER (1967) a proposé de regrouper tous les systèmes qui contribuent à améliorer la → *communication* sous le terme de sémantisation. Celle-ci peut être unilatérale, venant de l'émetteur (sémantisation de l'émetteur), et est alors appelée → *ritualisation*. Souvent, émetteur et récepteur s'adaptent réciproquement; enfin, il s'établit parfois une sémantisation unilatérale du récepteur. C'est ce qui se passe dans toute évolution des mécanismes déclencheurs (→ *MDI*).

sensible Phase → Prägung

E. sensitive period □ F. période sensible

Sicheltanz □ Dieser Tanz tritt bei allen bisher untersuchten Bienenrassen mit Ausnahme der Krainer Bienen auf, und zwar als Übergang vom → *Rundtanz* zum → *Schwänzeltanz* (VON FRISCH 1965). Die Öffnung der Sichel weist in Richtung zum Ziel (Abb. 111 – Sicheltanzschema der Honigbiene).

sickle dance □ This bee dance occurs in all races of bees so far examined with the exception of Krainer bees; it represents a transition from the → *round dance* to the → *waggle dance* (VON FRISCH 1965). The opening of the sickle points toward the goal (Fig. 111 – sickle dance pattern of the honeybee).

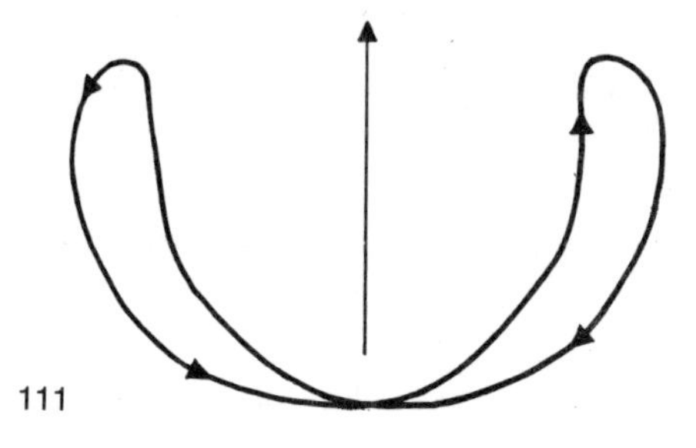

111

danse en croissant □ Cette forme de »danse« a été retrouvée chez toutes les races d'Abeilles mellifères, à l'exception de celle vivant dans la région de Krain en Yougoslavie (cf. *carnica*). Cette danse constitue une transition entre la → *danse circulaire* et la → *danse frétillante* (VON FRISCH 1965). La concavité du croissant est tournée vers la source de nourriture (Fig. 111 – schéma de la danse en croissant de l'Abeille mellifère).

Sichern → Schutzanpassung
E. vigilance □ F. vigilance

Sich-Lahmstellen → Verleiten
E. to be lame □ F. faire le boiteux

Sich-Schütteln □ Ein Komfortverhalten bei Vögeln und Säugern, welches je nach Art verschieden ausgeprägt ist. Es tritt fast regelmäßig auf, wenn ein Tier das Wasser verläßt, aber auch nach Staubbaden und speziell bei Säugern nach dem Auf-dem-Boden-Wälzen. Die Fähigkeit des willentlich aktiven Sich-Schüttelns scheint allen Primaten zu fehlen, obwohl beim Menschen eine unter bestimmten physiologischen Bedingungen, wie z. B. manchmal beim Urinieren, zu beobachtende Schüttelreaktion vielleicht als Verhaltensrelikt vorkommt. Bei Gänse- und Entenvögeln hat sich das Sich-Schütteln zum Streckschütteln abgewandelt; es besitzt bei Anatiden seinen festen Platz im Ablauf der Balzbewegungen des Erpels und geht dem → *Grunzpfiff* und dem → *Kurzhochwerden* voraus (LORENZ 1941). Die Intention zum Streckschütteln ist ein schwaches Gefiedersträuben. Es beginnt stets mit Schwanzwedeln, die Flügel werden leicht angehoben und der Hals nach vorn gestreckt (Abb. 112 – Schwimmente). Die Schüttelbewegung läuft vom Schwanz über den Körper zum Kopf. Der Schüttelvorgang ist artkonstant. Nach HEINROTH (1911) und PETZOLD (1964) soll der Auslöser ein *Unlustreiz* sein.

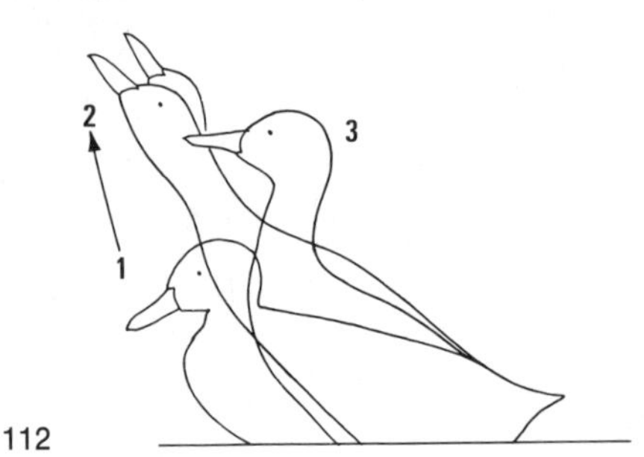

112

shaking □ A comfort movement in birds and mammals which is more or less pronounced depending on the species. It occurs frequently when the animal leaves the water but also occurs after dust bathing and specially in mammals after »rolling on the ground«. The ability for active intentional shaking is absent in all primates; however, the occasional occurrence of shaking in humans under certain physiological conditions such as urination may represent a vestigial behaviour. In geese and ducks shaking has become incorporated as »introductory shaking« into a sequence of epigamic displays preceding the → *grunt whistle* and → *head-up-tail-up* (LORENZ 1941). The intention movements preceding stretch-shaking are ruffling of the feathers and shaking of the tail. The wings are then lifted slightly and the neck stretched forward (Fig. 112 – Anatid duck). The shaking movement runs from the tail through the body and then to the head. The shaking movement itself is species-specific and its releaser, according to HEINROTH (1911) and PETZOLD (1964), is an *aversive stimulus*.

s'ébrouer □ Chez certains Oiseaux et Mammifères, comportement de confort plus ou moins prononcé selon l'espèce. Il est effectué d'une manière régulière après la sortie de l'eau, mais on peut l'observer aussi après le bain de poussière ou de sable et particulièrement chez les Mammifères, après s'être roulé au sol. La capacité et le besoin de s'ébrouer activement semblent manquer à tous les Primates. Chez l'Homme, cependant, on observe dans des conditions physiologiques particulières, par exemple lors de l'évacuation de l'urine, une réaction de secouement; il s'agit là probablement d'une relique comportementale. Chez les Oies et les Canards, l'action de s'ébrouer est devenu un ébrouement-étirement et constitue une étape du déroulement de la parade nuptiale du malard, précédant le → *grognement sifflé* et le → *»queue haute-tête haute«* (LORENZ 1941). L'intention de s'ébrouer en s'étirant est reconnaissable à un leger ébouriffement du plumage qui commence toujours par le secouement latéral de la queue, les ailes étant légèrement haussées et le cou étiré en avant (Fig. 112 – Canard Anatidae). Le mouvement d'ébrouement se déroule, en partant de la queue vers la tête. Le processus de cette action est caractéristique de l'espèce. Selon HEINROTH (1911) et PETZOLD (1964), le déclencheur de cette action serait un *stimulus aversif* ou *nociceptif*.

Sich-Strecken → Räkelsyndrom
E. self-stretching □ F. s'étirer

Sich-Totstellen → Akinesis
E. to feign death □ F. faire le mort

Sich-Verstellen → Verleiten
E. to feign □ F. feindre

Signal □ In der Ethologie und Soziobiologie jegliches Verhalten, welches dazu dient, von einem Individuum zu einem anderen oder von Gruppe zu Gruppe Informationen zu übermitteln, unabhängig davon, ob es auch noch andere Funktionen erfüllt. → *AM*, → *Auslöser*, → *Reiz*, → *Schlüsselreiz*.

signal □ In ethology and sociobiology, any behaviour that conveys information from one individual or from one group to another, regardless of whether it serves other functions as well. → *RM*, → *releaser*, → *stimulus*, → *key stimulus*.

signal □ En éthologie et sociobiologie, tout comportement servant à transmettre des informations d'un individu ou d'un groupe à un autre, indépendamment du fait que ce comportement peut avoir d'autres fonctions. → *MD,* → *déclencheur,* → *stimulus,* → *stimulus-clé.*

Signalfälschung → Mimikry
E. signal falsification □ F. falsification des signaux

Signalreiz → Schlüsselreiz
E. sign stimulus □ F. stimulus signal

Signalsprung □ Ein bei Fischen (Pomacentridae) sehr auffälliges, im Dienste der Werbung stehendes Verhalten (ABEL 1961, FRICKE 1973, HOLZBERG 1973). Der Fisch schwimmt dabei etwas nach oben, wobei sich die Körperlängsachse leicht nach vorne unten neigt. Dann kippt der ganze Körper nach unten, und der Fisch schwimmt schnell abwärts, um dann im leichten Bogen wieder die normale Körperlage einzunehmen. Der Signalsprung ist von Art zu Art sehr verschieden und wird vor allem vom ♂ ausgeführt (Abb. 113 A – *Dascyllus marginatus,* B – *Dascyllus trimaculatus*). Signalsprung kommt auch bei einigen Blenniiden und Tripterygiiden vor.

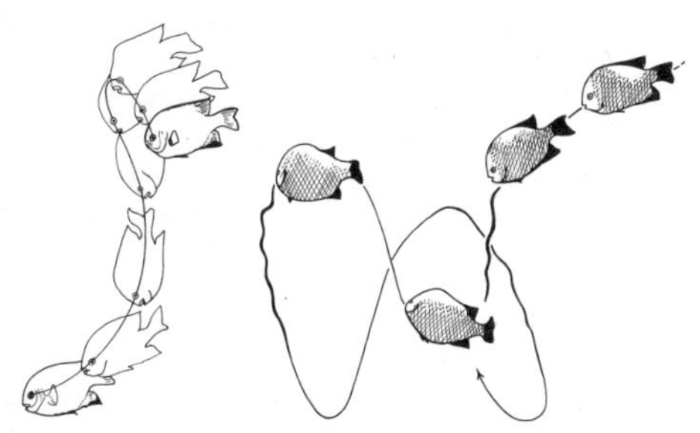

113 A 113 B

signal jump □ It is a conspicuous and common courtship behaviour of fish, especially of Pomacentridae (ABEL 1961, FRICKE 1973, HOLZBERG 1973). The fish swims upward, the rostral end pointing slightly downward. Then the fish drops swimming rapidly downwards and thus inscribing an arc. Signal jump differs from species to species, and is primarily executed by the ♂ (Fig. 113 A – *Dascyllus marginatus,* B – *Dascyllus trimaculatus*). It also occurs in certain Blenniids and Tripterygiids.
English speaking authors have suggested different terms corresponding to the German »Signalsprung«: *signal jump* (MYRBERG *et al.* 1967), *diving display* (SPANIER 1970), *loops with an exaggerated anguiliform motion* (STEVENSON 1963), *looping behaviour, along with zig-zag-swimming* (HELFREICH 1958).

saut-signal □ Chez les Poissons (Pomacentridae), élément très visible de la parade nuptiale (ABEL 1961, FRICKE 1973, HOLZBERG 1973). Le Poisson nage légèrement vers le haut tout en inclinant l'axe longitudinal de son corps vers le bas. Puis, tout le corps bascule vers le bas et le Poisson nage rapidement dans cette direction pour finalement rétablir sa position normale en exécutant une sorte de semi-looping. Ce saut de signalisation varie d'une espèce à l'autre et il est surtout effectué par le ♂ (Fig. 113 A – *Dascyllus marginatus,* B – *Dascyllus trimaculatus*). Ce comportement a également été observé chez d'autres Poissons, certains Blenniidae et Tripterygiidae par exemple.

Sippenduft → Gruppenduft

Skinner-Box → Futterspender

Sklavenhalterei □ Eine zwischenartliche Beziehung, bei welcher die Arbeiterinnen einer parasitischen (dulotischen) Ameisenart die Nester anderer Arten überfallen und dort vorwiegend die Brut, vor allem Puppen, in ihr eigenes Nest bringen, sie dort aufziehen und sie dann als Sklavenarbeiterinnen halten. Ein gutes Beispiel hierfür ist z. B. die Amazonenameise der Gattung *Polyergus* mit säbelförmigen Mandibeln, unfähig zu eigener Ernährung, welche darauf angewiesen ist, die Pflege auch der Brut durch die Sklaven von *Formica*-Arten vornehmen zu lassen. *Dulosis* wird auch von manchen Arten der Gattung *Harpagoxenus* praktiziert, welche sich ihre Sklaven bei *Leptothorax* besorgen (WILSON 1971, 1975).

slavery □ The relation in which workers of a parasitic (dulotic) ant species raid the nest of another species, capture brood, usually in the form of pupae, and rear them as enslaved nest mates. A good example are some species of the genus *Polyergus* with saber-shaped mandibles, unable to feed by themselves and consequently depending on slave workers of *Formica* species to rear and to care for the brood. Dulotic relations *(dulosis)* are frequent in some *Harpagoxenus* species which take their slaves among *Leptothorax* species (WILSON 1971, 1975).

esclavage □ Relation interspécifique lors de laquelle les ouvrières d'une espèce de Fourmis parasites (ou dulotiques) envahissent les nids d'autres espèces pour y enlever la progéniture, souvent sous forme de puppes, pour les emporter dans leur propre nid où elles sont élevées et ensuite utilisées comme ouvrières esclaves. Un exemple démonstratif est donné par les Fourmis Amazonéennes du genre *Polyergus* possédant des mandibules en forme de sabre, incapables de se nourrir elles-mêmes et qui par conséquent sont dépendantes des esclaves, surtout des espèces du genre *Formica,* pour effectuer les soins et l'élevage de leur propre progéniture. Ce comportement, aussi appelé *dulose,* est également pratiqué par quelques espèces du genre *Harpagoxenus* qui se procurent des esclaves chez *Leptothorax* (WILSON 1971, 1975).

Sonagramm = Sonogramm → Klangspektrogramm

Sozialduft → Gruppenduft
E. social odour □ F. odeur sociale

soziale Balz → Gruppenbalz
E. social courtship □ F. parade nuptiale sociale

Sozialhormone → Pheromone

Sozialverhalten □ Bei Menschen und Tieren

jene Verhaltensweisen, deren Frequenz durch die Anwesenheit eines Artgenossen geändert wird. Soziales Verhalten beginnt, sobald mindestens zwei Individuen miteinander in freundliche Beziehung treten.

social behaviour □ In animals and humans those behaviour patterns the frequency of which is changed by the presence of a conspecific. Social behaviour begins when at least two individuals interact with each other in a friendly fashion.

comportement social □ Chez les animaux et chez l'Homme, un certain nombre de comportements déterminés dont la fréquence est modifiée par la présence d'un congénère. Le comportement social est une notion très large et existe dès que deux individus entrent en relation »amicale«.

Soziogramm □ Die vollständige Beschreibung in Form eines Kataloges aller sozialen Verhaltensweisen einer Art. → *Ethogramm.*

sociogram □ The full description taking the form of a catalogue of all social behaviours of a species. → *ethogram.*

sociogramme □ La description complète sous forme d'un catalogue de tous les comportements sociaux d'une espèce. → *éthogramme.*

Soziohormone → Pheromone

Soziotomie → Volksteilung

E. sociotomy □ F. sociotomie

Sperren □ Singvogeljunge reagieren, wenn sie noch blind sind, auf taktile Reize, z. B. die Erschütterung des Nestes beim Ankommen des Altvogels, mit Aufreißen der Schnäbel (Sperreflex). Diese angeborene Sperreaktion ist bei den noch blinden Jungvögeln schweregerichtet. Die Vögel sperren mit gestrecktem Hals senkrecht nach oben (Abb. 114 – *Turdus merula*). Sehende Jungvögel sperren gezielt in Richtung des Altvogels bzw. auch auf eine Attrappe, mit der man die Sperrbewegungen ebenfalls auslösen kann (TINBERGEN und KUENEN 1939).

114

gaping □ Young song birds react to tactile stimuli, e. g. shaking the nest upon arrival of parent, with gaping of the beak. This innate reflex is even well developed in the blind young. They stretch their neck and open beak vertically upward (Fig. 114 – *Turdus merula*). Older young that can already see direct their gaping at the parent. Also a model can elicit gaping (TINBERGEN and KUENEN 1939).

bâillement de quête □ Parmi les Oiseaux, les jeunes Passereaux encore aveugles réagissent aux stimulations tactiles, par exemple à l'ébranlement du nid lors de l'arrivée de leurs parents, par une réaction qui consiste à ouvrir largement le bec. Cette réaction innée est orientée chez les jeunes Oiseaux encore aveugles selon l'axe de la pesanteur: le cou étant verticalement étiré, leur bec est dirigé vers le haut (Fig. 114 – *Turdus merula*). Par contre, les jeunes Oiseaux qui voient effectuent ce mouvement en direction du parent ou d'un leurre efficace (TINBERGEN et KUENEN 1939).

spezifisches Aktionspotential → SAP

E. specific action potential □ F. potentiel d'action spécifique

Spielverhalten □ Verhalten der Tiere und Menschen im Spiel mit folgenden Kriterien: Dem Spiel fehlt der spezifische Ernstbezug, und die gesetzmäßige Reihenfolge der Appetenzen und Instinktbewegungen ist aufgelöst, Spiele sind Verhaltensabläufe, deren Funktion weder aus den Handlungen noch aus ihren Effekten unmittelbar abgelesen werden kann. Es wird vermutet, daß es eine echte Spieltendenz (Spieltrieb) gibt, also eine eigene → *Handlungsbereitschaft* (Motivation) für Spielverhalten. Die Disposition zu fakultativem Lernen ist erhöht. In diesem Zusammenhang sind Bewegungsspiele von besonderem Interesse, denn dabei experimentiert das Tier mit seinem eigenen Bewegungskönnen. Es springt umher, wälzt sich am Boden und erfindet dabei auch neue Bewegungskoordinationen. Hierzu gehören auch die Kampfspiele als spielerische, nicht ernsthafte Auseinandersetzungen zwischen Jungtieren, in denen zwar → *Erbkoordinationen* von Kampfsituationen auftreten, aber ohne Ernstbezug ablaufen. Ein wesentliches Element im Verhalten des Menschen ist das konstruktive Spielen (Konstruktionsspiele), das oft von Vorbildern geleitet wird; es fällt aber auf, daß solche Handlungen, wie Hüttenbauen, bei unseren Stadtkindern spontan und ohne Vorbild auftreten.

play behaviour □ Play behaviour in animals and humans is determined according to the following criteria: play lacks a specific serious context and the regular sequence of appetitive and instinctive movement is absent. Play is a sequence of behavioural elements whose function is not apparent from the immediate actions themselves or their effects. Speculation suggests the existence of a true play tendency or drive with its own motivational system. The disposition for facultative learning is increased. In this context motoric play is particularly interesting: the animal experiments with its own motor capacities. It jumps about, frequently changes direction, rolls on the floor and in this way also »invents« new motor coordinations. In play fighting among young animals → *fixed action patterns* common to fighting behaviour occur, but are performed without serious intentions. An essential element in human play behaviour is play involving building of huts, dens, etc. This kind of play is usually imitative; however, it is noteworthy that such behaviour as hut building occurs in urban children spontaneously without any apparent model.

comportement ludique □ Chez l'animal comme chez l'Homme, le comportement ludique doit satisfaire aux critères suivants: le jeu ne peut être mis en relation avec une adaptation immédiate; l'enchaînement régulier des phases appétitives et des activités instinctives consommatoires fait défaut au cours du jeu. Les jeux sont des séquences comportementales dont la fonction n'est reconnaissable ni par les actions accomplies ni par leurs effets. On suppose qu'il existe une véritable pulsion ludique. Par le jeu, les possibilités d'apprentissage facultatif sont accrues. Dans ce contexte, les jeux moteurs sont d'un intérêt particulier étant donné que l'animal expérimente ainsi toutes ses possibilités d'activité propre. Il saute et court en changeant souvent de direction, se roule à terre et découvre ainsi de nouvelles coordinations motrices. Le comportement de jeu comprend aussi des combats ludiques entre jeunes animaux pendant lesquels de véritables → *coordinations héréditaires* caractéristiques des situations de combat apparaissent, mais se déroulent sans intention ni issue sérieuses. Un élément essentiel du comportement ludique humain réside dans les jeux de construction, souvent inspirés par l'imitation. Cependant, de telles activités, par exemple la construction de huttes, apparaissent d'une manière spontanée, même chez les enfants des grandes villes et en absence de tout modèle adéquat.

Spotten □ Man benutzt dieses Wort im Deutschen zweideutig, nämlich für das Nachahmen artfremder Laute bei Vögeln ebenso wie für das gegenseitige Sich-Verspotten beim Menschen. Im ersteren Fall sollte man vielleicht doch eher von *Spottgesang* sprechen. Das menschliche Spotten und Verspotten könnte man unter → *Spottverhalten* zusammenfassen.

parrotting □ Behaviour referring to the mimicking of species-foreign sounds by birds.

ridiculing or mocking □ This term as used in human ethology is defined under → *mocking behaviour.*

moquerie □ L'expression allemande *Spotten* est un mot à double sens. Il désigne à la fois l'imitation par les Oiseaux des chants et vocalisations d'autres espèces et les différentes manières de se moquer chez l'Homme. Dans le premier cas, on devrait plutôt parler *d'imitation* vocale et dans le deuxième cas, de → *comportement de moquerie.*

Spottgesang → Spotten

E. parrotting song □ F. imitation vocale

Spottverhalten □ Verhaltensweisen des Verspottens werden beim Menschen durch von der Gruppennorm abweichendes Verhalten ausgelöst, was beim Necken und Scherzen (→ *Scherzpartnerbeziehung*) nicht der Fall ist. Der Verspottete steht unter dem Druck der Gruppe und ist im allgemeinen bemüht, sich wieder an diese anzugleichen. Nach EIBL-EIBESFELDT (1973) hat das Spotten eine erzieherische Funktion. Man äfft nach und macht deshalb Verhaltensweisen lächerlich, die normalerweise Anstoß erregen. Bei Buschleuten kommen im Spottverhalten Zungezeigen und zwei Formen des weiblichen Sexualpräsentierens vor (→ *Schampräsentieren* und → *Schamweisen*).

mocking behaviour □ Mocking is elicited by the behaviour of individual group members who deviate from the group norm and in this manner is distinguished from → *joking relationship.* The recipient or object of mocking behaviour is under social pressure to readopt group-appropriate behaviour. According to EIBL-EIBESFELDT (1973), mocking behaviour serves an educational function. Behaviour is mocked and ridiculed that would otherwise induce confrontation. In bushmen society sticking out the tongue, as well as two forms of female genital presentation (→ *vulva presentation* and → *pubic presentation*) are interpreted as mocking behaviour.

comportement de moquerie □ Chez l'Homme, le comportement de moquerie est déclenché par des réactions déviantes par rapport à la norme du groupe, ce qui n'est pas le cas pour la taquinerie. La personne dont on se moque est ainsi sous la pression du groupe et s'efforce généralement de s'adapter et de se réinsérer dans le cadre social. Selon EIBL-EIBESFELDT (1973), la moquerie a une fonction éducative: on ridiculise ainsi des comportements qui normalement sont choquants pour la société. Chez les Boshimans, il existe dans le contexte de la moquerie l'acte de tirer la langue et deux formes de présentation sexuelle féminine (→ *présentation vulvaire* et → *présentation pubienne*).

Sprödigkeitsverhalten □ In den Paarungsvorspielen vieler Säuger ist das Weglaufen des ♀ vor dem ♂ keine echte, sondern eine ritualisierte Flucht. Es lädt als sog. Sprödigkeitsverhalten das ♂ zum Nachfolgen ein. In hoch differenzierter Form findet man ein solches Sprödigkeitsverhalten im menschlichen Flirtverhalten wieder. Beim → *Blickkontakt* beschränkt sich die ritualisierte Flucht auf Augenbewegungen, Wegsehen und Augenschließen.

coyness □ In the precopulatory play of many mammals the ♀ flees from the ♂ in a ritualized flight response. This is an invitation to the ♂ to follow and is present in a highly differentiated form in human flirting. For example, in humans this behaviour may be reduced to eye movements after initial → *eye contact.*

pruderie □ Dans les préliminaires de l'accouplement de nombreux Mammifères, le recul de la ♀ devant le ♂ n'est pas une fuite véritable, mais seulement un refus ritualisé: la ♀ invite ainsi le ♂ à la suivre.

Nous retrouvons la pruderie sous une forme hautement différenciée dans le flirt humain; lors du → *contact visuel,* la fuite ritualisée se limite à détourner la tête et le regard ou à fermer les yeux.

Spurbienen → Kundschafterinnen
E. scouting bees □ F. abeilles exploratrices

Stimmung → Handlungsbereitschaft
E. motivation □ F. motivation

Stigmergie → Stimmungsübertragung

Stimmungsübertragung □ Ansteckendes Verhalten von Mitgliedern einer Tiergruppe, welche zu annähernd gleicher Zeit das gleiche tun; es ist ein typisch soziales Verhalten und von vielen sozial lebenden Tieren bekannt. Hierher gehört die Sozialimitation, d. h. das Nachahmen einer Handlung oder einer Reihe von Handlungen durch Angehörige einer Sozietät, wobei der Anstoß vom Einzeltier ausgeht, auch *allomimetisches Verhalten* genannt (SCOTT 1946). Es dient der Synchronisation von Verhaltensweisen innerhalb einer Gruppe. So entleeren sich Lamas lokalisiert, und sobald ein Tier beginnt, stellen sich die anderen Herdenmitglieder in Reihen an der Harn- und Kotstelle auf (ALTMANN 1969). Die Sozialstimulation dagegen ist eine Wechselwirkung, d. h. gegenseitige Anregung zwischen Angehörigen einer Gruppe oder eines sozialen Verbandes; bei Tieren und Menschen gleich wirksam. Man spricht in diesem Zusammenhang auch von *Partnereffekt* oder *Gruppeneffekt* (GRASSE 1946, 1958, SOULAIRAC 1952, CHAUVIN 1958). Der Einfluß eines Artgenossen auf einen anderen kann die verschiedensten Konsequenzen haben. Oft wird durch die Gegenwart eines ♂ die Gonadenreife beim ♀ angeregt. In der Gruppe gehaltene Hühner fressen mehr als einzeln gehaltene Tiere, was auch für viele andere Arten gilt. Die Stimmungsübertragung bei sozialen Insekten und den Korrelationsmechanismus der individuellen Aufgaben und deren Regulierung, untermauert vor allem mit Untersuchungen an Termiten, nannte GRASSE (1959) *Stigmergie.* Das zum → *Räkelsyndrom* gehörende menschliche Gähnen kann auch der Stimmungsübertragung unterliegen und auf andere Mitmenschen ansteckend wirken.

mood induction, sympathetic induction □ Contagious behaviour in members of a group in which the individuals increasingly synchronize their actions. It is a typical pattern of social interaction known from a variety of animals living in social groups. It implies social imitation, i. e. mimicry of actions or sequences of actions of a single individual by the other members of the group; also called *allomimetic behaviour* (SCOTT 1946). It serves to synchronize behaviours within the group. Lamas, e. g., who perform their excretory function in a localized manner, line up at the excretory area as soon as one member of the herd urinates or defecates (ALTMANN 1969). In contrast, social stimulation is an interaction, i. e. a mutual stimulation between members of a group or a social association which is as effective in animals as in man. It can also be referred to as *partner effect* or *group effect* (GRASSE 1946, 1958, SOULAIRAC 1952, CHAUVIN 1958). The influence of the presence of conspecifics may have various effects. For example, the mere presence of a ♂ is often sufficient to stimulate gonadal development in the ♀. Group housed chicks and other animals eat more than individually isolated ones. GRASSE (1959) referred to sympathetic induction in social insects, when associated with the division of labour and its regulation, as *stigmergy*. Yawning in humans which is part of the → *stretching syndrome* may also be easily transferred to others and hence is considered under sympathetic induction.

induction allomimétique □ Comportement contagieux entre membres d'un groupe animal qui effectuent en même temps à peu près les mêmes activités; c'est un phénomène typiquement social, connu chez de nombreux animaux sociaux. Dans ce contexte, nous connaissons l'imitation sociale, c'est-à-dire le fait que tous les membres d'une société imitent une action ou une série d'actions dont l'impulsion vient généralement d'un seul individu, phénomène qu'on appelle aussi *comportement allomimétique* (SCOTT 1946). Ce comportement sert à la synchronisation des actions à l'intérieur d'un groupe. Il est particulièrement apparent chez les Lamas dont la défécation et la miction se font en un endroit déterminé; dès qu'un membre du groupe défèque ou urine, les autres congénères se mettent en file indienne et attendent leur tour pour exécuter le même comportement (ALTMANN 1969). La stimulation sociale, par contre, constitue une interaction réciproque entre les membres d'un groupe ou d'une société, opérant aussi bien chez les animaux que chez l'Homme, qu'il s'agisse de l'effet sur un partenaire ou de *l'effet de groupe* (GRASSE 1946, 1958, SOULAIRAC 1952, CHAUVIN 1958). L'influence d'un congénère sur un autre peut avoir les conséquences les plus variées. Souvent, la présence d'un ♂ stimule la maturation des gonades chez la ♀; les Poules élevées en groupe mangent davantage qu'un individu isolé, phénomène valable pour de nombreuses autres espèces. Chez les Insectes sociaux, l'induction allomimétique et les mécanismes de corrélation et de régulation des tâches individuelles, surtout mis en évidence chez les Termites, ont été appelés par GRASSE (1959) *stigmergie.* Le bâillement humain, phénomène physiologique appartenant au contexte du → *syndrome d'étirement,* peut être sujet à une induction sympathique et par conséquent, provoquer un effet contagieux envers d'autres Hommes.

Stockgeruch = Volksduft → Gruppenduft
E. hive odour □ F. odeur de la ruche

Stoffaustausch → Trophallaxie
E. stomodeal or proctodeal liquid exchange □ F. échange du liquide stomodéal ou proctodéal

Stößeln □ Ausdrucksverhalten bei Möwen

115

(Abb. 115 A – *Larus argentatus,* B – *Larus ridibundus,* C – *Rissa tridactyla*). Wird ein Silbermöwenpaar im Revier gestört, eilen beide Partner den Eindringling(en) entgegen, senken die Brust, knicken in den Fersengelenken ein, neigen den Kopf, senken das Zungenbein, was den Gesichtsausdruck stark verändert, und beginnen, rhythmisch gegen den Boden zu picken, ohne ihn zu berühren; manchmal kratzen sie dabei auch wie beim Muldescharren. Beim Stößeln (oder auch Gockern genannt) hört man im gleichen Rhythmus ein tiefes no-no-no-no. Während sie rufen, bewegt sich auffällig ihre Brust. Die Eindringlinge erwidern meist ebenfalls mit Stößeln. TINBERGEN (1958) deutet es als ein Drohverhalten und spricht auch von Drohstößeln oder Stößelruf. Man kennt es auch von anderen Möwen (NOBLE und WURM 1943). GOETHE (1956) nennt dieses Verhalten »Würgeln« und deutet es als Verlobungsanzeige gegenüber Reviernachbarn.

choking □ Expressive behaviour in gulls (Fig. 115 A – *Larus argentatus,* B – *Larus ridibundus,* C – *Rissa tridactyla*). If a herring gull pair is disturbed on its nesting territory, both animals rush up to the intruder, lower their breasts, bend the legs, nod their heads and lower their tongue bone hence sharply altering the facial expression. Then they begin a rhythmical jerking of the head as if to peck at the ground. Although they do not reach the ground, they may sometimes scrape it as in hollowing-out behaviour. During choking the gulls emit a deep rhythmic huoh-huoh-huoh-huoh sound. While they are calling, their breast moves in a conspicuous manner. Usually the intruder also responds with choking. TINBERGEN (1958) considered this behaviour as a threat display, sometimes referring to it as threat choking. It is also known in other gull species (NOBLE and WURM 1943). GOETHE (1956) termed this behaviour *Würgeln* and interpreted it as an »announcement of an engagement« to the territorial neighbours.

trépignement □ Comportement expressif des Goélands (Fig. 115 A – *Larus argentatus,* B – *Larus ridibundus,* C – *Rissa tridactyla*). Chez le Goéland argenté par exemple, lors de l'irruption du ou des intrus dans leur territoire, les deux membres du couple se précipitent à leur rencontre. Ils abaissent leur poitrine, plient leurs pattes au niveau de l'articulation du talon, penchent la tête en avant, abaissent leur os hyoïde, ce qui leur confère une expression faciale inhabituelle, et entament un picorage rythmique, mais sans toucher le sol. Quelquefois, ils effectuent des mouvements de grattage comme au cours du creusement du nid. Pendant cette activité, les deux partenaires émettent des vocalisations particulières et rythmiques comme no-no-no-no. Tout en criant, ils effectuent des mouvements bien apparents avec leur poitrine. Les intrus répondent généralement de même. TINBERGEN (1958) considère cette activité comme un comportement de menace. Il est également connu chez d'autres expèces de Goélands ou de Mouettes (NOBLE et WURM 1943). GOETHE (1956) qui appela ce comportement *Würgeln,* le considère comme l'annonce de la formation d'un couple aux occupants des territoires voisins.

Streckbewegung → Räkelsyndrom

E. stretching movement □ F. mouvement d'étirement

Streckschütteln → Sich-Schütteln

E. stretch-shaking □ F. s'ébrouer en s'étirant

Stridulation → Instrumentalverständigung

Suchautomatismus □ Bei Säuglingen von Tieren und Menschen ein seitliches Hin- und Herdrehen des Kopfes, das spontan oder nach Berührungsreizen in der Mundregion auftritt (Abb. 116). Es handelt sich um ein Suchverhalten, das endet, wenn der Säugling die Brustwarze in den Mund bekommt und

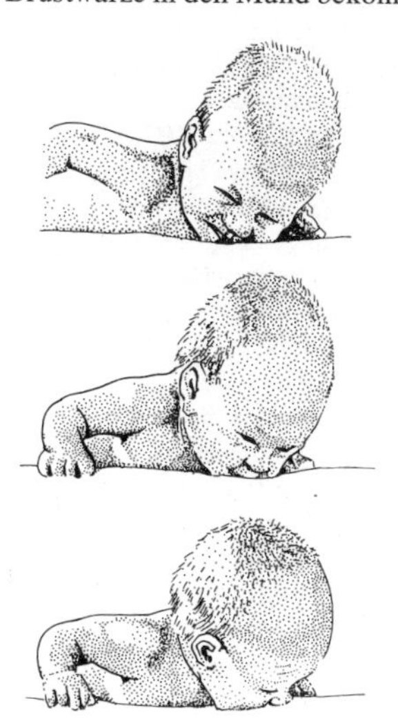

116

seine Lippen sich fest um den Saugzapfen schließen. Dieses ungerichtete Verhalten beobachtet man nur in den ersten Tagen nach der Geburt. Es wird schnell vom gerichteten Brustsuchen abgelöst (PRECHTL und SCHLEIDT 1950, 1951)

searching automatism □ In mammals and human infants a repeated movement of the head from one side to the other, occurring spontaneously or upon receiving a tactile stimulus in the mouth region (Fig. 116). The behaviour ceases when the neonate receives the nipple in its mouth and closes its lips around the opening. After its first few days of extrauterine life, the infant begins to display a goal-directed search for the breast (PRECHTL and SCHLEIDT 1950, 1951).

automatisme appétitif □ Recherche automatique du sein maternel. Chez les nouveaux-nés des animaux et de l'Homme, mouvement pendulaire de la tête qui peut apparaître spontanément ou survenir après des stimulations tactiles de la région buccale (Fig. 116). Il s'agit d'un comportement d'appétence qui s'achève dès que le nourrisson a le mamelon dans la bouche. Ce comportement non dirigé ne s'observe que dans les premiers jours après la naissance et il est rapidement remplacé par une recherche dirigée du mamelon (PRECHTL et SCHLEIDT 1950, 1951).

Suchbild □ Auf der Suche nach Nahrung bzw. Beute entwickelt ein Tier ein spezifisch auf eine bestimmte Beute bezogenes Schema, welches sich durch Erfolg und Mißerfolg entwickelt. Daraus ergibt sich ein Suchbild, nach welchem der Räuber selektiv vorgeht. Dies führt zu einer Änderung des Suchverhaltens und zur Bevorzugung eines bestimmten Nahrungstyps, der größeren Erfolg verspricht (CROZE 1970).

searching image □ In the search for food or prey an animal develops through trial and error a certain image of the goal object. From this behaviour results a searching image which produces a selective adaptation of the predator. This leads to a change in the animal's appetitive behaviour and to preference given to a determinate type of food which assures greater success (CROZE 1970).

image d'appétence □ Par son expérience lors de la recherche d'un aliment ou d'une proie, il se développe, chez un animal, une représentation particulière de l'objet recherché qui se constitue au gré des réussites et des échecs. Ceci produit une image d'appétence à laquelle le prédateur s'adapte sélectivement. Elle remanie le comportement appétitif et amène l'animal à préférer un type déterminé de nourriture qui augmente le rendement de son activité de prédation (CROZE 1970).

Suchverhalten → Appetenzverhalten
E. searching behaviour, seeking behaviour □ F. comportement de recherche

Sukzessivstrecken → Räkelsyndrom
E. successive stretching □ F. étirement successif

Symbiose □ Eine zeitweilige oder dauernde Verbindung bzw. Zusammenleben zwischen artverschiedenen Organismen (Symbionten) mit stark ausgeprägter gegenseitiger Abhängigkeit und mehr oder weniger gleichwertigem Nutzen beider Partner (FÜLLER 1958 – dort ausführliche Literaturangaben). Zwei Beispiele in der Ethologie: Anemonenfisch *Amphiprion* und Aktinie sowie die → *Putzsymbiose*.

mutualism □ A temporary or permanent bond or relationship between individuals of two different species in which strong mutual dependency occurs characterized by approximately equal benefits to both partners (FÜLLER 1958 – see also for detailed bibliography). Two ethological examples are the anemone fish, *Amphiprion* with the sea-anemone and the symbiots in → *cleaning symbiosis*.

symbiose □ Relation ou vie en communauté permanente ou temporaire entre des organismes d'espèces différentes. Entre ces symbiontes existe une interdépendance basée sur un avantage mutuel (FÜLLER 1958 – voir aussi pour bibliographie plus détaillée). Nous ne donnerons que deux exemples en éthologie: les Poissons du genre *Amphiprion* et l'Anémone de mer ainsi que la → *symbiose de nettoyage*.

Symbolhandlung □ Eine Handlung, die der ursprünglichen Sache nicht mehr dient, sondern nur noch »symbolisch« ausgeführt wird, wie z. B. das → *Futterlocken* bzw. → *Bodenpicken* bei Hühnervögeln, welches zu einem Signal im Dienste der Werbung verändert (ritualisiert) wurde.

symbolic action □ An action which no longer serves the original function, rather it has taken on an alternative or »symbolic« function. For example, → *food-calling* or → *ground pecking* in gallinaceous birds has evolved (or become ritualized) as a courtship signal.

activité symbolique □ Action qui n'est plus au service de la motivation initiale et qui, par conséquent, n'est exécutée que d'une manière symbolique; ainsi, → *l'attraction alimentaire* ou le → *picorage au sol* chez les Galliformes est devenu un signal d'attraction sexuelle dans le contexte de la parade nuptiale et apparaît donc sous une forme ritualisée.

symbolische Unterwerfung → Grußformeln → Vertrauensbeweis
E. symbolic submission □ F. soumission symbolique

Sympaedium → Kinderfamilie

Symphagium → Freßgemeinschaft

Syngenophagie → Verwandtenfresserei

T

Tagesperiodik → Tagesrhythmik

Tagesrhythmik □ Ein 24-Stunden-Rhythmus oder *Tag-Nacht-Rhythmus* im Verhalten und anderen physiologischen Körperfunktionen. In vielen Fällen in Form einer endogenen Periodik im Lebewesen selbst vorhanden. Über → *Zeitgeber,* wie Sonnenauf- und Sonnenuntergang (Hell-Dunkel-Rhythmus) wird die endogene Rhythmik in einen zeitlichen Einklang gebracht, d. h. synchronisiert (BÜNNING 1963 – dort ausführliche Literatur). HUTELAND (1798) schrieb dazu: »Die 24-stündige Periode, welche durch die regelmäßige Umdrehung unseres Erdkörpers auch allen seinen Bewohnern mitgeteilt wird, . . . ist gleichsam die Einheit unserer natürlichen Chronologie«.

circadian rhythm □ A 24-hour rhythm or *day-night rhythm* in the behaviour and other physiological body functions. In many cases it represents an endogenous rhythm present in the organism itself. Through external signals or → *time synchronizer* such as the sun rise or sunset, the endogenous rhythm is regulated or synchronized with the environment (see BÜNNING 1963 for extensive references). HUTELAND (1798) wrote: »The 24-hour cycle, arising from the regular revolution of the earth, is impressed on all inhabitants of the earth, and at the same time constitutes the unit of our natural chronology«.

rythme circadien □ Rythme de 24 heures ou *rythme jour-nuit* modulant le comportement et les fonctions physiologiques du corps. Dans de nombreux cas, un tel rythme existe sous forme d'une périodicité endogène dans l'être vivant lui-même. Par ailleurs, le rythme endogène est déterminé par des → *synchroniseurs temporels* venant de l'environnement tels que le lever et le coucher du soleil(rythme lumière-obscurité; – BÜNNING 1963, voir pour bibliographie exhaustive). HUTELAND (1798) écrivait déjà: »La période de 24 heures, transmise à tous les habitants de la terre par la rotation régulière de notre planète, est en quelque sorte l'unité de notre chronologie naturelle«.

Tag-Nacht-Rhythmus → Tagesrhythmik
E. day-night rhythm □ F. rythme jour-nuit

Tandemlauf → Rekrutierungsverhalten
E. tandem running □ F. marche en tandem

Tänzerinnenattrappe → Nachläuferinnen
E. bee dancer model □ F. leurre imitant une abeille danseuse

Tanzsprache □ Allgemeiner, von KARL VON FRISCH geprägter Begriff für die Verständigung bei Bienen für die → *Entfernungsweisung* und → *Richtungsweisung,* welche durch den → *Rundtanz* und den → *Schwänzeltanz* den Stockgenossen übermittelt werden. Außer diesen beiden wichtigen Tänzen gibt es noch den → *Putztanz,* → *Rucktanz,* → *Rumpellauf,* → *Sicheltanz, Schütteltanz* und → *Zittertanz* für andere soziale Interaktionen.

dance language □ A general term for the mode of communication among bees with regard to the distance (→ *distance indication*) and direction (→ *indication of direction*) of a food source or other goal. These two properties are elaborated by the → *round dance* and → *waggle dance.* In addition to these important dances, there are also → *grooming dance,* → *jerk dance,* → *bumping run,* → *sickle dance, shaking dance* and → *quivering dance* for other social interactions.

langage dansé □ Notion générale qui désigne les différents moyens de communication servant, chez les Abeilles, à → *l'indication de distance* et à → *l'indication de direction* d'une source de nourriture, signalée aux congénères de la même ruche, respectivement, à l'aide de la → *danse circulaire* et de la → *danse frétillante.* En dehors de ces deux danses principales, nous connaissons la → *danse de nettoyage,* la → *danse saccadée,* la → *marche tapageuse,* la → *danse en croissant,* la *danse secouée* et la → *danse frémissante* valables pour d'autres interactions sociales.

Tarnfärbung → Schutzanpassung
E. cryptic colouration □ F. coloration de camouflage

Tarnkleid → Schutzanpassung
E. concealing pattern □ F. livrée de camouflage

Tarnstellung → Konturverfremdung
E. concealing posture □ F. attitude de camouflage

Tarntracht → Schutzanpassung
E. concealing pattern □ F. livrée de camouflage

täuschendes Verhalten → Verleiten
E. distraction display □ F. comportement de diversion

Täuschung → Mimikry → Verleiten
E. deception □ F. feinte

Taxis □ Orientierungsbewegung oder Orientierungsreaktion frei im Raum lebender Organismen im Gegensatz zum → *Tropismus.*

Eine Orientierungsbewegung, welche auf einen Reiz bezogen ist, bezeichnet man oft auch als Reizantwort. KÜHN (1919) hat ein Begriffssystem der Taxien aufgestellt, das heute noch vielfach verwendet wird. FRAENKEL und GUNN (1961) haben dieses System ausgebaut. Die darin enthaltenen Begriffe erlauben, die äußeren Merkmale einfacher Orientierungsbewegungen in Kurzform zu kennzeichnen (z. B. phototaktisch = auf Licht reagierend). Die zusätzlichen Bezeichnungen wie »positiv« = Hinwenden und »negativ« = Wegwenden geben uns Auskunft über die Ausrichtung zum Reiz. Eine positiv phototaktische Raupe z. B. kriecht der Lichtquelle entgegen. Neben einer Reihe solcher Bezeichnungen gibt es drei Kernbegriffe, wie Phobotaxis, Tropotaxis, Telotaxis, die auch eine Unterscheidung der physiologischen Mechanismen erlauben (KOEHLER 1949, 1950, LINDAUER 1973, SCHÖNE 1973, 1974). Nachstehend wollen wir eine Reihe von Bezeichnungen aufführen, die in der Verhaltensforschung häufig Anwendung finden.

Anemotaxis. Orientierungsreaktion und Lokomotion zur Windrichtung. Handelt es sich um eine transversale Orientierung, so spricht man von Anemomenotaxis.

Astrotaxis. Orientierung nach Himmelskörpern.
Chemotaxis. Eine durch chemische Reize, Geruch und Geschmack induzierte Orientierungsbewegung bei frei beweglichen Tieren.
Geotaxis. Durch die Schwerkraft bedingte Orientierung; bekannt auch von freibeweglichen niederen Pflanzen, z. B. der Algengattung *Volvox*.
Klinotaxis. Eine einfache Form direkter Orientierung, bei welcher die seitlichen Bewegungen des Vorderkörpers die fortlaufende Prüfung des Orientierungsreizes in verschiedenen Richtungen im Raum gestatten.
Menotaxis. Eine Winkeleinstellung zur Reizquelle. Das Tier hält an einem bestimmten »Ungleichgewicht« der Erregung beider Seiten fest. Man könnte sich die Schräg- oder Querorientierung aber auch telotaktisch vorstellen: Das Tier fixiert den Reiz mit einer entsprechenden seitlichen Fixierstelle des Sinnesorgans.
Mnemotaxis. Orientierung nach der Erinnerung; der Ausdruck wird heute aber nicht mehr gebraucht.
Phobotaxis. Orientierung in einem Reizgefälle, die auf Schreckreaktionen beruht. Pantoffeltierchen, die z. B. in stark kohlendioxydhaltiges Wasser geraten, stoppen, machen eine Wendung und schwimmen dann weiter. Gelangen sie erneut in höhere Kohlendioxydkonzentration, dann wiederholt sich der Vorgang. Die Wendungen sind bezüglich des Reizes ungerichtet; das Pantoffeltierchen gelangt nur durch »Versuch und Irrtum« aus der Gefahrenzone. Ähnliche Mechanismen beschreiben FRAENKEL und GUNN unter dem Namen → *Kinese*.
Phototaxis. Orientierungsbewegung nach dem Licht; positiv = zum Licht hin; negativ = vom Licht weg. Die Lichtrückenorientierung z. B. ist eine positive Phototaxis.
Telotaxis. Das geradlinige Ansteuern einer Reizquelle bzw. eines Zieles. Bei ihr kommt es nicht auf die Unterschiede der Erregung beider Seiten an, sondern auf den Ort der Reizung. Die Erregung der gereizten Rezeptorstelle löst eine Wendung bestimmter Größe aus, die das Tier in die Reizrichtung bringt, so daß nach vorn gerichtete Stellen der Sinnesorgane vom Reiz getroffen werden.
Topotaxis. Im Reizfeld gerichtete Wendung.
Tropotaxis. Eine symmetrische Einstellung zum Reiz. Die Erregung der Sinnesorgane der rechten und linken Körperseite fließen in entsprechend rechte und linke Zentren. Die zentrale Ungleichheit der Erregung löst Wendungen aus, bis die Erregungen beider Seiten gleich groß sind und sich dadurch die Waage halten.

taxis □ Orientating reaction or movement in freely moving organisms as opposed to → *tropism* in sessile organisms. An orientating movement directed toward or in relation to a stimulus may also be designated as a response. KÜHN (1919) developed a system of terms that is still in widespread use. FRAENKEL and GUNN (1961) have extended this concept. The terms represent an abbreviated means of describing simple orientation movements (for example, phototactic = reacting to light). Additional reference, such as »negative« = away from and »positive« = toward, give information about the organism's mode of reaction to the stimulus. A positive phototactic moth will move toward a light source. In addition to the above, there are three terms which permit some designation of the physiological mechanisms; these are phobotaxis, tropotaxis and telotaxis (KOEHLER 1949, 1950, LINDAUER 1973, SCHÖNE 1973, 1974). These terms and others frequently used in ethology are listed below.
anemotaxis. Orientation and locomotion in relation to the wind direction. If it is a transverse orientation, it is called anemomenotaxis.
astrotaxis. Orientation to heavenly bodies (star orientation).
chemotaxis. An orientational reaction in free moving animals induced by a chemical stimulus, smell or taste.
geotaxis. A reaction in reference to gravity, also known to occur in free-moving lower plants such as in the algae genus *Volvox*.
klinotaxis. A simple form of directed movement in which the sidewise movement of the anterior part of the body results in a continuous testing of the position of the orientational stimulus in space.
menotaxis. An angular orientation in relation to the stimulus source. The animal maintains a certain imbalance in the response of each side of the body to the stimulus. One can also describe this diagonal or cross orientation as a telotaxis: the animal simply fixates the stimulus with a lateral fixation receptor of the sense organ.
mnemotaxis. Orientation according to memory; this term is no longer in common usage.
phobotaxis. Orientation in a stimulus situation in which aversion occurs. *Paramecium* that enter water with a high carbon dioxide content stop, turn, and continue their movement. Should they again encounter the carbon dioxide, they repeat this manoeuvre. The turns are not directed in relation to the stimulus; only through »trial and error« does the *Paramecium* succeed in avoiding the danger zone. FRAENKEL and GUNN describe similar mechanisms as → *kinesis*.
phototaxis. Orientation movements to the light; positive = toward the light; negative = away from the light. The dorsal light orientation for example, is a positive phototaxis.
telotaxis. The direct approach to a stimulus source or goal. Here it is not a matter of a lateral differential in excitation, but rather of the position or location of the stimulus source. The excitation of the receptor site elicits a turn of a given degree. This turn brings the organism in the vicinity of the stimulus so that the anteriorly directed sense receptor sites can encounter the stimulus.

topotaxis. A turn in the stimulus field.
tropotaxis. A symmetric position in relation to the stimulus. The excitations of the bilateral sense organs flow into corresponding right and left centers. Any unequal central representation of the stimuli elicits turning until the excitation of both sides is identical, thus restoring the balance.

taxie □ Mouvement ou réaction d'orientation des organismes se déplaçant librement dans l'espace, par opposition au → *tropisme.* Un mouvement d'orientation relatif à une stimulation peut aussi être appelé réponse ou riposte. KÜHN (1919) a élaboré une classification des différentes taxies qui est encore souvent utilisée et que FRAENKEL et GUNN (1961) ont élargie. Ces notions permettent de désigner les caractères externes des mouvements d'orientation simples (par exemple phototactique = dirigé vers la lumière). Des désignations supplémentaires telles que »positif« ou »négatif« nous renseignent sur l'orientation par rapport au stimulus. Une chenille phototactiquement positive, par exemple, évolue vers une source de lumière. A côté de toute une série de désignations de ce type, nous connaissons trois notions principales: phobotaxie, tropotaxie et télotaxie, qui permettent une définition des mécanismes physiologiques en jeu (KOEHLER 1949, 1950, LINDAUER 1973, SCHÖNE 1973, 1974). Nous donnons ci-après une liste des désignations fréquemment utilisées en éthologie:
anémotaxie. Réaction d'orientation et locomotion par rapport à la direction du vent. S'il s'agit d'une orientation transversale par rapport au vent, on parle d'anémoménotaxie.
astrotaxie. Orientation d'après les corps célestes.
chimiotaxie. Mouvement d'orientation induit chez les animaux libres par les stimulations chimiques telles que les odeurs et les saveurs.
clinotaxie. Forme élémentaire d'orientation, au cours de laquelle des mouvements d'oscillation latérale de la partie antérieure du corps permettent l'examen permanent du stimulus dans différentes directions spatiales.
géotaxie. Réponse d'orientation dans le champ de la pesanteur. Ce phénomène est également connu chez les végétaux inférieurs non fixés, par exemple chez les Algues du genre *Volvox.*
ménotaxie. Orientation transverse selon un angle défini par rapport à la source de stimulation. L'animal maintient une certaine inégalité entre les excitations naissant au niveau des récepteurs sensoriels bilatéraux. Une telle orientation oblique ou transversale pourrait aussi être conçue sur le modèle télotactique: l'animal fixerait le stimulus par un point de fixation latéral bien déterminé de son organe sensoriel.
mnémotaxie. Orientation d'après des repères mis en mémoire. Cette expression n'est plus guère utilisée en éthologie.
phobotaxie. Orientation dans un gradient d'intensités stimulatrices, basée sur une réaction phobique. Des paramécies par exemple, arrivant dans une eau riche en gaz carbonique, s'arrêtent, changent de direction et continuent à nager. Dans le cas où elles arrivent à nouveau dans une zone où la concentration en gaz carbonique est élevée, le même phénomène se répète. Ces changements de direction ne sont pas orientés par rapport au stimulus, mais les paramécies arrivent néanmoins »par essais et erreurs« à se dégager de la zone dangereuse. Des mécanismes semblables sont décrits par FRAENKEL et GUNN sous le nom de → *cinèse.*
phototaxie. Mouvement d'orientation par rapport à la lumière; positif = vers la lumière; négatif = à l'opposé de la lumière. L'orientation photo-dorsale, par exemple, est une variante de la phototaxie positive.
télotaxie. Trajet rectiligne visant une source de stimulation ou un but précis. Dans la télotaxie, c'est l'emplacement de la source de stimulation qui est important, et non pas la différence d'excitation entre les deux côtés. L'excitation d'un récepteur stimulé déclenche un virage d'un angle déterminé qui amène l'animal dans la direction du stimulus, de manière à ce que les zones des organes sensoriels dirigées vers l'avant soient atteintes par le stimulus.
topotaxie. Rotation orientée dans un champ de stimulation.
tropotaxie. Orientation symétrique par rapport au stimulus. L'excitation des organes sensoriels des côtés droit et gauche du corps est acheminée vers des centres bilatéraux correspondants. L'inégalité des excitations au niveau central déclenche des pivotements jusqu'à ce que les excitations des deux côtés soient équilibrées.

Telotaxis → Taxis

Temperaturkundschafterin □ Einzelindividuen bei Ameisen, die während der Überwinterung in gewissen Zeitabständen an die Nestoberfläche laufen, um dort die Außentemperatur zu prüfen. Je wärmer es wird, umso länger bleiben sie und umso mehr Wärme geben sie bei der Rückkehr an ihre Nestgenossen ab und umso mehr Ameisen nehmen am Temperaturprüfen teil, bis schließlich die Wärme in der Temperaturkammer so ansteigt, daß das ganze Volk erwacht und sich an die Nestoberfläche begibt (WILSON 1971, 1973).

temperature messenger □ Individual ants that run along the nest surface of the overwintering ant colony testing the external temperature. The warmer it becomes the longer the ant remains near the surface and the more heat is transferred to the nest mates when he returns. This induces more individuals to participate until finally the temperature in the nest quarters rises to the point that the entire colony awakes and moves to the surface (WILSON 1971, 1973).

messagère thermique □ Chez les Fourmis, individus qui, au cours de l'hibernation de la colonie, se dirigent de temps à autre vers la surface du nid pour sonder la température extérieure. Plus il fait chaud, plus ils restent longtemps au dehors et davantage de chaleur est transmise au retour à leurs congénères; par la suite, un nombre croissant de Fourmis participe à ce sondage thermique. Ce processus continue jusqu'à ce que la température intérieure du nid atteigne un niveau tel que l'ensemble de la société se réveille et se dirige alors vers la surface (WILSON 1971, 1973).

Termitenangeln → Werkzeuggebrauch
E. fishing for termites □ F. pêche de termites

territoriale Hierarchie □ In streng linearen → *Rangordnung(en)* haben meist nur die ranghöchsten Tiere räumliche Reviere, und den rangniedrigen stehen nur noch Standorte oder bestimmte Verstecke zu. Man spricht bei einem solchen Verhalten von einer raumgegliederten territorialen Hierarchie.

territorial hierarchy □ In strictly linear → *rank order(s)* only highest ranking animals usually obtain a spacious territory while the lower ranking animals are pushed to the fringe. One refers in such instances to a spacially separated territorial hierarchy.

hiérarchie territoriale □ Dans les → *hiérarchie(s) sociale(s)* strictement linéaires, souvent seuls les animaux du rang le plus élevé possèdent des territoires étendus. Les individus de rang inférieur ne possèdent que des emplacements ponctuels ou des cachettes. En pareil cas, on parle d'une hiérarchie par structuration territoriale de l'espace.

Territorialität → Revierverhalten
E. territoriality □ F. territorialité

Territorium → Revier

Territoriumsmarkierung → Reviermarkierung

Territoriumssicherung → Revierverteidigung
E. patrolling of territory □ F. mesures de sécurité en vue de la défense du territoire

Thanatose → Akinesis

Tischgenossenschaft → Freßgemeinschaft

Topotaxis → Taxis

Totschütteln □ Viele Carnivora, vor allem aber die Canidae, nehmen Beutetiere an beliebiger Stelle rasch und oft nur lose mit den Zähnen auf und schleudern sie schnell wieder weg. Aus diesem Wegschleudern wird dann ein kräftiges und mehrmaliges Abschütteln, welches beim Beutetier zeitweilige Atemlähmung verursachen kann (LEYHAUSEN 1965). Noch stärkeres Abschütteln führt schließlich den Tod durch Genickbruch herbei (Totschütteln). Bei fast allen Feliden findet man dagegen den perfektionierten Tötungsbiß *(→ Nackenbiß)*.

death shake □ Many carnivores, especially the canids, pick up the prey loosely with the teeth, shake it, and flip it aside. This shaking can result in respiratory collapse in the prey (LEYHAUSEN 1965). Additional more intense shaking leads to death by dislocation of the neck (death shake). In feline species, however, the prey is usually killed by a highly perfected → *neck bite*.

secouement fatal ou léthal □ Chez de nombreux Carnivores, en particulier chez les Canidae, la proie est souvent saisie avec les dents n'importe comment, puis rapidement rejetée. La reprise de la proie se transforme par la suite peu à peu en un mouvement de secouement qui peut provoquer chez la proie une paralysie respiratoire partielle (LEYHAUSEN 1965). Un secouement de plus en plus fort peut alors entraîner la mort par fracture de la nuque. Chez presque tous les Felidae, par contre, la proie est tuée par une morsure très précise dans la nuque *(→ morsure à la nuque)*.

Tötungsbiß → Nackenbiß
E. killing bite □ F. morsure fatale ou léthale

Tötungshemmung □ Im innerartlichen Kampfverhalten verhindern Verhaltenselemente, wie Beschwichtigungsgebärden (→ *Beschwichtigungsverhalten*) und Demutstellung (→ *Demutsverhalten*), eine ernsthafte Verletzung oder gar die Tötung des Rivalen und sind in diesem Sinne als Hemm-Mechanismen der Tötung von Artgenossen anzusehen. Bei Arten ohne Tötungshemmung gegenüber fremden Artgenossen gibt es sie aber gegenüber Rudel- bzw. Gruppenmitgliedern.

killing inhibition □ In intra-specific fights serious injury or killing the rival are inhibited by behavioural elements such as appeasement gestures (→ *appeasement behaviour*) and submissive postures (→ *submissive behaviour*). These behaviours may be regarded in this sense as inhibitory mechanisms for killing conspecifics. In forms in which there is no inhibitory mechanism in regard to strange conspecifics, there is a killing inhibition toward other group members.

inhibition de la mise à mort □ Lors des combats intraspécifiques, des éléments comportementaux tels que gestes d'apaisement *(→ comportement d'apaisement)* et attitudes de soumission *(→ comportement de soumission)* évitent généralement des blessures graves ou même la mise à mort des combattants; en ce sens, on peut les considérer comme des mécanismes inhibiteurs de la mise à mort des congénères. Chez les espèces dépourvues de toute inhibition à l'égard des congénères étrangers, ces mécanismes inhibiteurs existent cependant envers les individus appartenant au même groupe.

T-Phänomen → Reflexverspätung
E. T-phenomenon □ F. phénomène T

Tradierung → Tradition

Tradition □ Man kennt bei hochentwickelten und sozial lebenden Tieren eine Anzahl von Fällen, in denen sich individuell erworbenes Wissen über die Lebensdauer des erwerbenden Individuums hinaus in der Sozietät erhält. Diesen Vorgang, durch den erlerntes Wissen und Können von einem Individuum an ein anderes und von einer Generation an eine andere weitergegeben wird, nennt man in der Ethologie *Tradierung* oder *Traditionsbildung*. Das wohl bekannteste Beispiel einer solchen Tradierung kennen wir von japanischen Makaken, bei denen ein ♀ das Waschen von Süßkartoffeln erfand, das im Laufe der Zeit zu einem gruppenspezifischen Merkmal wurde. Würde man experimentell keine Süßkartoffeln mehr füttern, bis alle Gruppenmitglieder, die diese Fähigkeit besitzen, gestorben sind, so käme es wahrscheinlich zu einem Traditionsabriß (KAWAI 1965, LORENZ 1973).

tradition □ There are several known examples in highly-developed social species in which knowledge (behaviour) learned by an individual is retained (practiced) among the group members beyond the life span of the individual who introduced it. This transfer of knowledge or information from one individual to another and one generation to the next is referred to in ethology as a tradition. The best known example is that of the Japanese Macaques in which a female began to wash sweet potatoes. This behaviour eventually became a characteristic of the group. If one would remove the sweet potatoes until all individuals possessing this behaviour had died, it would probably result in a collapse of tradition (KAWAI 1965, LORENZ 1973).

tradition □ On connaît chez les animaux supérieurs vivant en société, un certain nombre de cas où un savoir individuellement acquis peut être transmis, au-delà de la durée de vie de l'individu qui a appris, et conservé dans la société. Un tel processus par lequel des connaissances et des capacités acquises peuvent être transmises d'un individu à un autre, de génération en génération, est appelé en éthologie tradition. L'exemple le plus connu est sans doute celui d'une jeune femelle de Macaque Japonais qui a inventé le lavage des patates douces avant de les manger. Ce comportement est devenu avec le temps un caractère propre au groupe. Pour provoquer une rupture de tradition de ce comportement, on devrait, d'une manière expérimentale, cesser de nourrir ces animaux avec des patates douces jusqu'à ce que tous les membres du groupe exerçant cette pratique soient décédés (KAWAI 1965, LORENZ 1973).

Traditionsabriß → Tradition
E. collapse of tradition □ F. rupture de tradition

Traditionsbildung → Tradition
E. formation of tradition □ F. formation de tradition

Tragling □ Von HASSENSTEIN (1973) geprägter Ausdruck für Säugerjunge, die in den ersten Lebenstagen oder -wochen von ihren Müttern stets mit herumgetragen werden. Die enge Bindung zwischen Kind und Mutter schließt ein Ablegen des Jungen aus, denn der verlorene Kontakt mit der Mutter bedeutet höchste Gefahr. Die Jungtiere halten sich hauptsächlich mit Händen und Füßen am Fell der Mutter fest, wie z. B. bei Halbaffen und Affen, beim Faultier und beim Koala (einem baumkletternden Beutler). Junge Fledermäuse, die ebenfalls zu den Traglingen gehören, halten sich noch zusätzlich mit ihren Lippen an den Zitzen der Mutter fest. Junge Menschenaffen halten sich zwar an der Mutter fest, bedürfen aber in der Regel der Unterstützung seitens der Mutter, die einen Arm um das Junge legt. Der menschliche Säugling ist nach seinen Verhaltenselementen ein ehemaliger Tragling, was man bei Neugeborenen am → *Handgreifreflex* erkennen kann. Der Verlust des Haarkleides zwingt uns, das Kind ständig aktiv herumzutragen; wenigstens tun Naturvölker das noch. WICKLER (1969) bezeichnete solche Jungtiere einschließlich des menschlichen Säuglings als *Elternhocker*.

parent-clinger □ A term coined by HASSENSTEIN (1973) for suckling young mammals which are continuously carried by the mother during the first days or weeks of life. The close tie between the mother and offspring precludes leaving the new born animal alone. The loss of contact to the mother represents extreme danger to the young. Young animals hold to the mother's fur with hands and feet as, e. g., in lemurs and monkeys, in sloths and in koalas. Young bats, which are also parent clingers, also grasp the mother's teats with their lips. Young anthropoid apes also hold on to the mother but normally have to be supported by the mother as well. The human infant was perhaps once a parent clinger as is suggested by the → *grasping reflex*. The loss of hair cover demands that the child be continuously actively transported; at least, such is often the case in primitive societies. WICKLER (1969) referred to such young animals including infants as parent huggers.

»agrippeur« □ Cette expression a été créée par HASSENSTEIN (1973) pour désigner tous les jeunes Mammifères qui, pendant les premiers jours ou les premières semaines de leur vie, sont constamment portés par leur mère. Ce lien étroit entre enfant et mère exclut le dépôt du jeune, car la perte de contact avec la mère signifie le plus grand danger. Ces jeunes animaux s'accrochent surtout avec les mains et les pieds à la fourrure de leur mère; c'est par exemple le cas chez les Prosimiens et Simiens, chez le Paresseux et le Koala (un Marsupial grimpeur). Les jeunes Chauves-Souris qui peuvent également être considérées comme des »agrippeurs«, se fixent en outre avec les lèvres aux mamelles de leur mère. Les jeunes Anthropoïdes s'agrippent également dans la fourrure de leur mère, mais d'une manière générale, ils ont besoin d'un soutien de la part de la mère qui les entoure de son bras. Dans l'espèce humaine, le nourrisson a conservé à titre de relique évolutive, certains éléments du comportement des »agrippeurs«, comme en témoigne le → *réflexe de préhension* du nouveau-né. La perte des poils nous oblige à porter constamment l'enfant d'une manière active; du moins, les peuples primitifs le font encore. WICKLER (1969) a désigné ces jeunes, y compris le nourrisson humain, par le terme de *Elternhocker* (cramponneurs aux parents).

Tragstarre □ Carnivoren und Nager ergreifen

ihre Jungen mit dem Maul im Nacken (→ *Nackenbiß)* und tragen sie auf diese Weise von einem Ort zum anderen oder ins Nest zurück. Die Jungen nehmen dabei eine passive Haltung ein und lassen mit leicht gekrümmtem Rücken Kopf und Beine einfach hängen; sie verfallen in eine Tragstarre (Abb. 117 – Fuchswelpe). Da der Nackenbiß zum Töten oder auch der Tötungsbiß diesem Tragegriff gleicht, löst er beim nicht lebensgefährlich verletzten Beutetier oft die Tragstarre aus. Die bewegungslose Beute wird, vor allem bei Katzen, oft »für tot« abgelegt und hat dann eine Fluchtmöglichkeit. Die Tragstarre dient demnach der Selbst- und Arterhaltung, bietet einen Selektionsvorteil und bleibt als ursprünglich juvenile Verhaltensweise über die Jugendentwicklung hinaus erhalten (LEYHAUSEN 1965, 1975).

117

limp posture □ Carnivores and rodents grasp their young with the mouth at the back of the neck (= *neck bite*). In this manner they are carried from one place to another or returned to the nest. The young animal assumes a passive posture or »transport catalepsy« and with a slightly bent back, allows the head and legs to hang (Fig. 117 – fox whelp). Since the neck bite or killing bite is similar in form to this grip for carrying the young, often prey which are not seriously wounded assume a cataleptic posture. The motionless prey is often dropped by the predator, especially in cats, and hence has an additional opportunity to escape. The limp posture has a self- and species preservation value, offers a selective advantage and hence persists as an originally juvenile behaviour pattern into adulthood (LEYHAUSEN 1965, 1975).

immobilisation de transport □ Carnivores et Rongeurs saisissent leurs jeunes par la nuque et les portent ainsi d'un endroit à un autre ou au nid. Les jeunes adoptent une position passive et, le dos légèrement courbé, laissent pendre tête et pattes; ils se trouvent ainsi dans un réflexe de roideur (Fig. 117 – renardeau). Etant donné que la → *morsure à la nuque* pour tuer une proie ressemble à cette manière de porter les jeunes, elle déclenche souvent chez une proie qui n'est pas dangereusement blessée, cet état cataleptique. La proie immobilisée est alors souvent considérée comme morte et, surtout chez les Chats, déposée par terre ce qui lui donne une possibilité de fuite. La catalepsie réflexe est donc un phénomène au service du maintien de l'individu et de l'espèce; elle confère un avantage sélectif et persiste en tant que comportement initialement juvénile jusqu'au stade adulte (LEYHAUSEN 1965, 1975).

Tragwirt → Phoresieverhalten
E. carrier host □ F. hôte transporteur, hôte phorétique

Transmittersubstanzen → Aktionssubstanzen

Transponiervermögen → Richtungsweisung
E. information transfer □ F. capacité de transposition

Transportwirt → Phoresieverhalten
E. carrier host □ F. hôte transporteur, hôte phorétique

Treiben □ Bei jenen Wiederkäuern, die kein → *Paarungskreisen* ausführen, kommt ein Paarungslauf vor, bei welchem der Bock das Weibchen über eine gerade Strecke treibt und dabei auch den → *Laufschlag* ausführt (Abb. 118 – Treiben bei *Muntiacus muntjak*). Treiben während des Paarungsverhaltens ist auch von Tauben bekannt (KLINGHAMMER *in litt.*).

precopulatory chasing, driving □ In those Bovidae in which → *display circling* is absent, a display chasing occurs in which the ♂ chases the ♀ for some distance in a straight line and then strikes her with the outstretched forelegs (→ *foreleg-kick;* Fig. 118 – *Muntiacus muntjak*). Driving may also be seen during courtship behaviour in doves and pigeons (KLINGHAMMER *in litt.*).

118

poursuite précopulatoire □ Chez un certain nombre de Ruminants où il n'y a pas de → *parade nuptiale circulaire,* il existe une poursuite sexuelle au cours de laquelle le ♂ pourchasse la ♀ en ligne droite et peut lui administrer de temps à autre un coup de patte (→ »*Laufschlag*«; Fig. 118 – *Muntiacus muntjak*). La poursuite précopulatoire a également été observée lors de la formation du couple chez les Tourterelles et les Pigeons (KLINGHAMMER *in litt.*).

Treteln □ Ein Auf-der-Stelle-Treten; bei Eidechsen beobachtet man dieses Verhalten bei Rivalenkämpfen. Der Verlierer legt sich flach auf den Boden und beginnt zu treteln und wird dann nicht mehr angegriffen; erst dann läuft er davon. Man deutet dieses Treteln als ritualisierte Flucht (KITZLER 1942).

»running on the spot«, »treadling« □ Running on the spot occurs in lizards during dominance fighting. The loser lays flat on the ground and begins to tread without touching the ground. The attacker then desists and the loser is able to flee. »Running on the spot« is interpreted as ritualized flight (KITZLER 1942).

piétinement □ Mouvement de marche sur place connu chez les Lézards au cours des combats intraspécifiques. Le perdant se couche à plat sur le sol et commence à piétiner; il n'est plus alors attaqué et ce n'est qu'à ce moment-là qu'il s'enfuit. Ce piétinement est interprété comme une fuite ritualisée (KITZLER 1942).

Trieb □ Der Trieb basiert auf spezifischen Antriebskräften, deren Produktion rhythmisch-automatisch erfolgt und zu inneren Spannungszuständen führt, welche von sich aus ohne äußeren Anlaß zur Entladung drängen. Triebe sind daher die Quelle des Spontanverhaltens. In ihrer Entwicklung folgen sie eigenen, endogenen Gesetzen (LEYHAUSEN 1952). → *Drang*, → *Handlungsbereitschaft*.

drive □ A drive has a specific internal source of energy which is produced in an automatic rhythmic manner and which leads to a state of internal tension. This tension is the source of spontaneous behaviour since it presses for its own release even in the absence of external stimuli. The build up of drives occurs in a characteristically defined manner (LEYHAUSEN 1952). → *urge*, → *motivation*.

pulsion □ La pulsion est conçue comme un ensemble de forces d'incitation à l'action, s'accumulant de façon automatique et rythmique et conditionnant la constitution d'un état de tension interne spécifique. Celui-ci peut conduire à une décharge, en l'absence de toute intervention extérieure. De ce fait, les pulsions sont la source de tout comportement spontané. Dans leur développement, elles obéissent à des lois endogènes qui leur sont propres (LEYHAUSEN 1952). → *Drang*, → *motivation*.
En principe, *Trieb* correspondrait en français à *tendance* (PIERON 1968), état latent de *Drang* (pulsion), mais la priorité au sens éthologique du terme devra être donnée à l'usage qui est d'employer pulsion pour traduire le mot Trieb; on parle par exemple de la → *satisfaction d'une pulsion*, de → *l'interpénétration de la pulsion et de l'apprentissage* ou encore de → *l'hypothèse de la pulsion agressive*.

Triebbefriedigung □ Die → *Endhandlung* beendet das → *Appetenzverhalten* und »befriedigt« somit auch den der Appetenz zugrunde liegenden Trieb. Man könnte somit annehmen, daß Triebbefriedigung stellvertretend für Endhandlung stehen könnte. Zahlreiche Arbeiten zum Phänomen des »Durstes« haben aber gezeigt, daß die Komponenten der Endhandlung des Trinkens, Durststillung und Ablauf der Trinkbewegungen bzw. die Anzahl der Saugbewegungen, voneinander abhängig sind. Wenn Säuglinge eine bestimmte Menge in 20 min saugend aufgenommen haben, schlafen sie befriedigt ein. Haben die Sauger jedoch eine zu große Öffnung, so daß die gleiche Menge oder 50 % mehr in nur 5 min ersaugt wird, dann bleibt der Saugtrieb unbefriedigt, und die Säuglinge saugen im Leerlauf weiter (→ *Trinken im Leerlauf*). Ähnliches beobachtete man bei Kälbern, die sich aus Eimern erst satt tranken und dann an → *Ersatzobjekt(en)* lutschten (SPITZ 1957, PLOOG 1964). → *Zungenschlagen*.

drive satisfaction □ The → *consummatory act* ends the → *appetitive behaviour* and »satisfies« the drive underlying the appetence. One could assume that drive satisfaction may be represented by the consummatory act. However, it is well documented in studies of thirst, for example, that the components of the consummatory act of drinking, the relieving of thirst and the sequence of drinking movements, e. g., the number of sucking movements, are dependant of each other. When an infant consumes a certain quantity in 20 minutes of sucking it may fall asleep. However, if due to a larger opening in the tit, it consumes the same quantity or even 50 % more in 5 minutes, the infant will continue to nurse (→ *vacuum drinking*), since the sucking drive has not been satisfied. A similar observation can be made in calves which have drunk their fill from a bucket and then continue to suck on a → *substitute object* (SPITZ 1957, PLOOG 1964). → *substitute nursing*.

»satisfaction« d'une pulsion □ L' → *acte consommatoire* termine le → *comportement appétitif* et satisfait ainsi la pulsion qui sous-tendait ce comportement. On pourrait donc supposer que la satisfaction d'une pulsion est un équivalent strict de l'acte consommatoire. Cependant, de nombreux travaux, par exemple au sujet de la »soif«, ont montré que les différentes composantes de l'acte consommatoire de la désaltération, c'est-à-dire l'absorption de liquide et le déroulement des mouvements de déglutition ou le nombre de mouvements de tétée, sont inter-dépendantes. Si des nourrissons, par exemple, ont ingurgité une certaine quantité de liquide en tétant pendant 20 minutes, ils s'endormiront satisfaits. Si la tétine, par contre, possède une ouverture trop large permettant d'absorber la même quantité de liquide ou même jusqu'à 50 % en plus en 5 minutes, la pulsion de tétée reste insatisfaite et la tétée continue à vide. Nous connaissons des faits semblables chez les

Veaux qui ont bu dans des seaux jusqu'à satiété et qui sucent ensuite des → *objet(s) de remplacement* (SPITZ 1957, PLOOG 1964). → *claquement de langue,* → *boire à vide.*

Trieb-Dressur-Verschränkung □ Verhaltensweisen, welche aus einer → *Erbkoordination* und aus einer Komponente, die erlernt werden kann oder muß (→ *Erwerbskoordination),* bestehen, nennt man Trieb-Dressur-Verschränkung, da beide Komponenten synchron ineinander »verschränkt« auftreten (LORENZ 1932). Der jeweilige angeborene und erlernte bzw. erworbene Anteil kann meist nur durch saubere Experimentierarbeit eruiert werden. Ein gutes Beispiel einer angeborenen Bewegungskoordination, verschränkt mit einer erlernten Orientierung, finden wir im → *Aufspießen* der Würger (LORENZ und ST. PAUL 1968). Trieb-Dressur-Verschränkungen erhöhen die Plastizität des Verhaltens und sind besonders häufig bei höheren Wirbeltieren.

instinct-learning intercalation □ Behaviour patterns that consist of an innate component (→ *fixed action pattern*) and a component that can or must be learned (→ *acquired behaviour pattern*). This type of pattern is termed an »intercalation« because both innate and learned components appear together already interwoven (LORENZ 1932). The respective innate and acquired aspects can only be determined through careful experimentation. A good example of an innate motor pattern interacting with an acquired orientation is the → *impaling* behaviour of the shrikes (LORENZ and ST. PAUL 1968). Instinct-learning intercalations increase the plasticity of behaviour and are particularly common in higher vertebrates.

interpénétration de la pulsion et de l'apprentissage □ Comportement composé d'une → *coordination héréditaire* et d'une composante dont l'apprentissage est possible ou nécessaire (→ *coordination acquise*). On parle dans ce cas d'interpénétration étant donné que les deux composantes apparaissent d'une manière synchronisée (LORENZ 1932). La part de l'inné et la part apprise ou acquise ne peuvent être élucidées qu'à l'aide d'un travail d'expérimentation minutieux. Un bon exemple d'interpénétration entre une coordination héréditaire et un mouvement d'orientation qui doit être appris est constitué par le comportement de → *l'embrochage* des Pies-Grièches (LORENZ et ST. PAUL 1968). Les interpénétrations de l'inné et de l'acquis augmentent la plasticité du comportement et apparaissent fréquemment chez les Vertébrés supérieurs.

Triebhandlung □ Ein auf ererbten (angeborenen) Bahnen des Zentralnervensystems beruhender Handlungsablauf, der als solcher ebenso wenig veränderlich ist wie seine histologische Grundlage oder irgendein morphologisches Merkmal (LORENZ 1932, 1973). Ein durch das Ansprechen aus einem → *AAM* und aus einer durch ihn in Gang gebrachten → *Erbkoordination* zusammengesetzter Ablauf ist eine Funktionsganzheit, wie wir sie aus dem Tierreich gut kennen. Sie wurde von HEINROTH (1911) als »arteigene Triebhandlung« bezeichnet.

instinctive behaviour pattern □ A behaviour sequence involving inherited (innate) pathways (aspects) of the central nervous system. Such sequences are no more subject to modification than are histological or morphological traits of the individual (LORENZ 1932, 1973). The instinctive behaviour pattern, which is initiated by an → *IRM* eliciting a sequence of → *fixed action pattern(s)*, represents a functional unit. This unit was first referred to by HEINROTH (1911) as »species specific behaviour«.

activité instinctive □ Déroulement d'actes qui dépend d'un circuit nerveux héréditairement programmé (inné) et qui, à ce titre, est aussi peu variable que sa base histologique ou qu'un caractère morphologique quelconque (LORENZ 1932, 1973). Ce déroulement fait appel pour son déclenchement à un → *MDI* qui met en route une → *coordination héréditaire;* il peut être considéré comme une entité fonctionnelle comme nous en connaissons de bons exemples dans le comportement animal. Cette unité comportementale a été désignée par HEINROTH (1911) comme une »activité instinctive spécifique«.

Triebhandlungskette □ Der starre Ablauf einer aus → *Triebhandlung(en)* zusammengesetzten → *Erbkoordination.*

innate response chain □ The sequence of → *fixed action pattern(s)* composing an → *instinctive behaviour pattern.*

chaîne d'activités instinctives □ Le déroulement d'une → *coordination héréditaire* composée → *d'activité(s) instinctive(s).*

Trieb-Hypothese □ Die hypothetische Annahme eines Aggressionstriebes, welche sich hauptsächlich auf Befunde am Cichliden *Etroplus maculatus* stützt. Nach dieser Hypothese (LORENZ 1950, 1963) produziert eine Instanz im Zentralnervensystem unabhängig von Außenreizen ständig aggressive Triebenergie. Sie läßt die Bereitschaft für Kampfhandlungen (= Aggressivität) ansteigen, bis passende Reize die → *Endhandlung* auslösen. Dadurch wird die Energie wieder verbraucht (→ *Triebbefriedigung*). Das geschieht normalerweise im Rahmen von Grenzkämpfen täglich viele Male, so daß die Aggressivität nie sehr hoch ansteigt. Wenn aber, wie in der Isolation, die Endhandlung länger ausbleibt, sammelt sich mehr und mehr Energie an, und es kommt zu einem sog. Triebstau. Als Folge davon zeigt das Tier erstens ein → *Appetenzverhalten,* das auf eine Triebbefriedigung gerichtet ist, und reagiert zweitens auf immer weniger spezifische Reize; es entsteht also eine Schwellensenkung, bis es schließlich die Handlung im Leerlauf ausführt. Daß isolierte Individuen in der Isolation aggressiver werden, konnte für Einsiedlerkrebse (COURCHESNE und BARLOW 1971), Schwertträger (FRANCK und WIL-

HELMI 1973, WILHELMI 1974), Korallenbarsche (RASA 1971), Kampfhähne (KRUIJT 1964) und Mäuse (GARATTINI und SIGG 1969) positiv beantwortet werden. Bei Anabantiden (LAUDIEN 1965, WARD 1967) und Cichliden (HEILIGENBERG 1964, HEILIGENBERG und KRAMER 1972) aber wurde nach Isolierung eine Schwellenerhöhung für aggressive Handlungen festgestellt. In einer rezenten, ausgezeichneten Untersuchung (REYER 1975) konnte die Annahme einer endogenen Produktion aggressiver Triebenergie für *Etroplus maculatus* weder kausal noch funktionell gestützt werden. PLACK (1973) stellt die Existenz eines Aggressionstriebes generell in Frage.

drive model □ The postulation of an endogenous aggressive drive supported primarily from work with the Cichlid *Etroplus maculatus*. According to this concept (LORENZ 1950, 1963) a process in the central nervous system continuously produces energy of an aggressive nature independently of external stimuli. The drive for aggressive interaction increases until an appropriate stimulus releases the → *consummatory act*. In this fashion the energy is dissipated (→ *drive satisfaction*). This normally occurs in context of territorial fights several times daily such that the aggression never increases to a high level. If the consummatory act is blocked, such as may occur when the animal is isolated, the action specific energy accumulates. The animal then displays an → *appetitive behaviour* for release of the energy by reacting to less and less specific stimuli. A reduction of the threshold occurs until the behaviour is finally executed *in vacuo*. Several species including hermit crabs (COURCHESNE and BARLOW 1971), swordtails (FRANCK and WILHELMI 1973, WILHELMI 1974), coral fish (RASA 1971), fighting cocks (KRUIJT 1964), and mice (GARATTINI and SIGG 1969) show isolation-induced aggression. However, Anabantids (LAUDIEN 1965, WARD 1967) and Cichlids (HEILIGENBERG 1964, HEILIGENBERG and KRAMER 1972) appear to show an increased threshold for aggression following isolation. A recent elegant study by REYER (1975) failed to support causally or functionally an endogenous aggressive drive in *Etroplus maculatus*. PLACK (1973) suggests that the existence itself of an aggressive drive is questionable.

hypothèse de la pulsion agressive □ Cette hypothèse est basée en particulier sur les résultats des recherches effectuées sur le Cichlidé *Etroplus maculatus*. Selon cette hypothèse (LORENZ 1950, 1963), une instance située dans le système nerveux central produirait, indépendamment des stimulations exogènes et d'une manière permanente, une énergie pulsionnelle agressive. Cette énergie augmente la disposition à des actions agonistiques (= agressivité) jusqu'à ce que des stimuli adéquats déclenchent → *l'acte consommatoire* ce qui consommerait cette énergie (→ *satisfaction d'une pulsion*). Ce phénomène se réalise normalement et de nombreuses fois par jour à l'aide des disputes aux limites des territoires de sorte que l'agressivité n'atteint jamais un degré très élevé. Par contre, en isolement, où l'acte consommatoire ne peut se produire, l'énergie agressive serait de plus en plus accumulée et il se produirait donc une accumulation pulsionnelle. Par la suite, ce phénomène se traduit par le fait que l'animal montre en premier lieu un → *comportement appétitif* qui est dirigé vers une satisfaction de la pulsion et il réagit par ailleurs de plus en plus envers des stimuli moins adéquats. Il se produit un abaissement du seuil jusqu'à ce que l'action soit effectuée à vide. Le fait que des individus isolés deviennent plus agressifs a pu être démontré chez les Pagures (COURCHESNE et BARLOW 1971), chez les Poissons du genre *Xiphophorus* (FRANCK et WILHELMI 1973, WILHELMI 1974), chez *Etroplus maculatus* (RASA 1971), chez les Coqs combattants (KRUIJT 1964) et chez le Souris (GARATTINI et SIGG 1969). – Chez les Anabantidae (LAUDIEN 1965, WARD 1967) et Cichlidae (HEILIGENBERG 1964, HEILIGENBERG et KRAMER 1972), par contre, il a été démontré que ces animaux subissent en isolement une augmentation du seuil pour les activités agressives. Dans une analyse récente et excellente (REYER 1975), la supposition d'une production endogène d'énergie pulsionnelle agressive chez *Etroplus maculatus* n'a pu être confirmée, et ceci ni de façon causale ni de façon fonctionnelle. PLACK (1973) met en doute d'une manière générale l'existence d'une pulsion agressive.

Triebleben = Instinktverhalten → Instinkt
E. instinctive behaviour □ F. vie instinctive

Trillern □ Regenpfeifer vollführen mit einem schräg nach vorn gehaltenen Fuß Zitterbewegungen gegen den Boden, die der Nahrungssuche dienen und besonders im Watt Erfolg haben. Das Stichling-♂ betrillert sein ♀ im Nest zum Ablaichen *(→ Schnauzentriller)*. Triller ist weiter eine Bezeichnung für verschiedene Vogelrufe, z. B. den Ruf des Ziegenmelkers.

quivering □ The golden plover executes a quivering movement against the ground with his foot held forward at an oblique angle. This serves to facilitate his search for food and is particularly effective in tidal shoals. The stickleback male quivers for the female when inducing her to spawn in the nest (→ *Schnauzentriller*). Quivering is also a designation for various bird calls, for example the call of the goatsuckers.

tremblement □ Comportement alimentaire chez les Pluviers (Charadridae). Ces Oiseaux, une patte posée en avant, effectuent au contact du sol des tremblements ayant pour effet de faire sortir de petites proies qui y vivent. Ce comportement est particulièrement efficace dans le sol meuble et humide de la mer de Wadden dégagée par la marée basse. Une autre forme de tremblement existe chez les Epinoches où le ♂ effectue ces mouvements avec la bouche sur le corps de la ♀, déclenchant le comportement de ponte chez cette dernière (→ *frémissement* *du museau*). L'expression allemande »Triller« signifie également sifflet ou roulade chez un certain nombre d'Oiseaux, par exemple la trille de l'Engoulevent.

Trinken im Leerlauf □ Das Trinken bzw. Trinkbewegungen ohne Aufnahme von Flüssigkeit. → *Triebbefriedigung.*

vacuum drinking □ Drinking movements without accompanying uptake of fluid. → *drive satisfaction.*

»boire à vide« □ Le fait de boire ou d'effectuer des mouvements de boire sans absorber de liquide. → *satisfaction d'une pulsion.*

Triumphgeschrei □ Epigames Ausdrucksverhalten während der Paarbildung bei Gänsen. Das ♂ führt zunächst Scheinangriffe auf Objekte aus, die es normalerweise meidet. Es kehrt danach »triumphierend« zu seiner Erwählten zurück und droht an ihr vorbei (Abb. 119). Stimmt das ♀ in das Triumphgeschrei ein, dann ist eine Verteidigungsgemeinschaft gegründet, die Voraussetzung für eine erfolgreiche Aufzucht der Jungen. Dieses Verhalten dient als Grußzeremonie auch weiterhin der Paarbindung (LORENZ 1943, FISCHER 1965, RADESÄTER 1975).

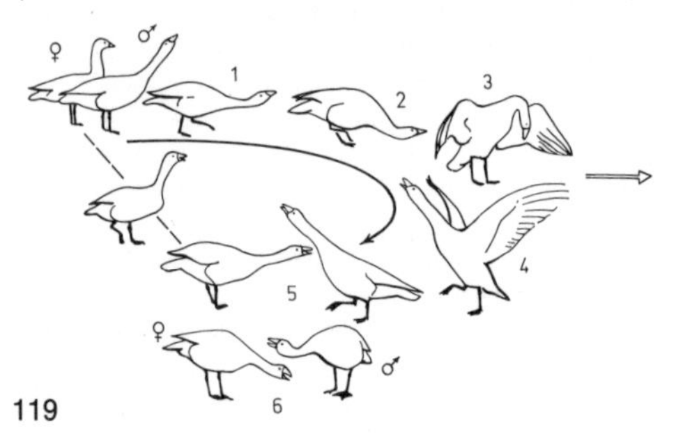

119

triumph ceremony □ Epigamous behaviour pattern during pair formation in geese. The ♂ at first makes sham attacks toward objects that are normally avoided. Following such an attack he »triumphantly« returns to his intended mate and threatens beyond her (Fig. 119). If she joins the »triumph ceremony«, a defensive alliance has been formed, which is a prerequisite for the successful rearing of a brood. This behaviour continues to maintain the pair-bond by functioning as a greeting ceremony (LORENZ 1943, FISCHER 1965, RADESÄTER 1975).

cérémonial de triomphe □ Comportement épigame lors de la formation du couple chez les Oies. Le jars dirige tout d'abord ses pseudo-agressions vers des objets qui le laissent ordinairement indifférent. Il se tourne ensuite »triomphalement« vers sa ♀ qu'il a choisie et la menace (Fig. 119). Mêle-t-elle sa voix aux cris de triomphe du jars, il se forme alors une communauté de défense qui est nécessaire pour la réussite de l'élevage de la couvée. Ce comportement continuera de servir à la cohésion du couple et fonctionnera comme cérémonie d'accueil (LORENZ 1943, FISCHER 1965, RADESÄTER 1975).

Trophallaxie □ Mund-zu-Mund-Austausch von Nahrungsflüssigkeit bei sozialen Insekten (WHEELER 1918, 1928); – von ROUBAUD (1916) *Ökotrophobiose* genannt. Die Verhaltensmechanismen der taktilen Kontaktaufnahme mit Hilfe der Antennen, die den Nahrungsaustausch zustande bringen, hat MONTAGNER (1966) eingehend bei sozialen Faltenwespen der Gattung *Vespa* untersucht. Die ♂♂ der Wespen sind anatomisch und morphologisch nicht auf einen Sinneskontakt zur Auslösung von Trophallaxie angepaßt und müssen sich, um zur begehrten Nahrung zu kommen, zwischen zwei austauschende Arbeiterinnen schieben. Bei Ameisen werden durch den Austausch von Nahrungstropfen aus dem Sozialmagen Informationen über die Lokalisierung von Beute oder Nistmaterial übermittelt (Abb. 120 A – Nahrungsaustausch zwischen zwei Arbeiterinnen von *Formica rufa*). Neben der klassischen *stomodealen* Trophallaxie konnte TOROSSIAN (1958) einen *proctodealen* Nahrungsaustausch nachweisen, der besonders bei der Larvenaufzucht zu beobachten ist. In Ameisenstaaten leben häufig myrmecophile Käfer, wie die Brenthidae, die regelmäßig als Nehmende, aber auch als Gebende mit ihren Wirtsameisen Nahrungsaustausch praktizieren (Abb. 120 B – Trophallaxie zwischen *Amorphocephalus coronatus,* Coleoptera Brenthidae, und einer Arbeiterin von *Camponotus cruentatus;* TOROSSIAN 1966). Vor allem die Staphylinidae haben den jeweiligen spezifischen Trophallaxis-Kode ihrer Wirtsarten gebrochen und haben als »nehmende« Sozialparasiten an der Nahrungsübergabe Anteil und ernähren auf diese Weise sich und ihre Brut (Fig. 120 C – Fütterungskontakt zwischen einer *Formica*-Arbeiterin und einer Larve von *Atemeles pubicollis,* Coleoptera Staphylinidae; HÖLLDOBLER 1967 a + b). Das Phänomen der Trophallaxie gilt als ein grundlegend verbindender Faktor für die Angehörigen eines Insektenstaates (Übersicht bei WILSON, 1971 und 1975).

trophallaxis □ Mouth-to-mouth exchange of food in social insects (WHEELER 1918, 1928); – referred to as *ecotrophobiosis* by ROUBAUD (1916). MONTAGNER (1966) studied behavioural mechanisms of the genus *Vespa* involving tactile stimulation of the antennae which leads to trophallaxis. The male wasps are not morphologically adapted for eliciting trophallaxis, and in order to obtain their preferred food must interpose themselves between two workers. In ants the transfer of food droplets also conveys information about the location of prey or nest material (Fig. 120 A – food exchange between two workers of *Formica rufa*). In addition to the classical transfer of stomach contents, TOROSSIAN (1958) described a transfer of *proctodeal* material commonly observed during larval feeding. In ant colonies, myrmecophile beetles, e. g., Brenthidae, practice trophallaxis, both giving to and taking from the ants (Fig. 120 B – food exchange between *Amorphocephalus coronatus,* Coleoptera Brenthidae, and

120 A

120 B

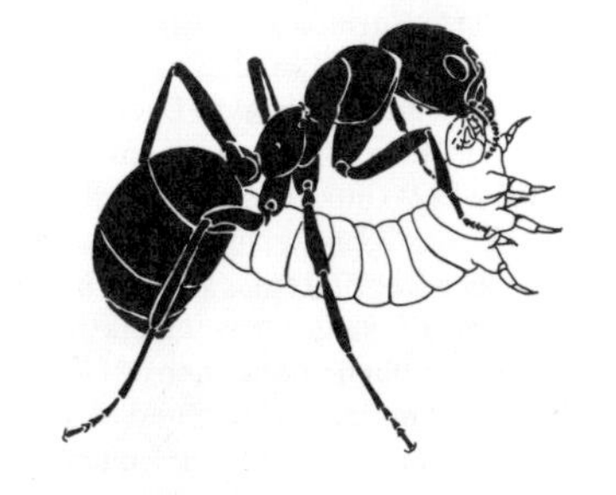

120 C

a worker of *Camponotus cruentatus;* TOROSSIAN 1966). The Staphylinidae have actually broken the specific trophallaxis code of their host species and have become social parasites supporting themselves and their young in this fashion (Fig. 120 C – feeding of a larva of *Atemeles pubicollis* by an ant worker of the genus *Formica;* HÖLLDOBLER 1967 a + b). The phenomenon of trophallaxis may be regarded as a basic social bonding mechanism for members of an insect colony (summarized by WILSON 1971 and 1975).

trophallaxie □ Echange d'un liquide nutritif de bouche à bouche chez les Insectes sociaux (WHEELER 1918, 1928); – comportement que ROUBAUD (1916) appela *écotrophobiose*. Les mécanismes comportementaux du contact tactile à l'aide des antennes rendant possible un tel échange d'éléments nutritifs ont été étudiés d'une manière approfondie chez les Guêpes du genre *Vespa* par MONTAGNER (1966). Les ♂♂ de ces Guêpes sont anatomiquement et morphologiquement inaptes à un tel contact sensoriel déclenchant la trophallaxie et doivent, pour obtenir la nourriture convoitée, s'intercaler entre deux ouvrières en contact trophallactique. Chez les Fourmis, l'échange du liquide nutritif du jabot social contient des informations sur la localisation des proies ou du matériel de nidification (Fig. 120 A – échange de nourriture entre deux Fourmis de l'espèce *Formica rufa*). A côté de la trophallaxie *stomodéale* classique, TOROSSIAN (1958) a pu démontrer un échange *proctodéal* d'un liquide nutritif, phénomène qui s'observe surtout lors de l'élevage des larves. Par ailleurs, dans des fourmilières, on trouve souvent des Coléoptères myrmécophiles tels que les Brenthidae qui non seulement profitent des échanges trophallactiques en tant que receveurs, mais aussi comme donneurs, avec leurs Fourmis-hôtes (Fig. 120B–échange de nourriture entre *Amorphocephalus coronatus,* Coleoptera Brenthidae, et une Fourmi de l'espèce *Camponotus cruentatus;* TOROSSIAN 1966). Les Staphylinidae sont les Coléoptères qui ont le mieux réussi à déchiffrer le code trophallactique spécifique de différentes espèces-hôtes; ils se conduisent exclusivement comme receveurs et ne se nourrissent, eux-mêmes et leur progéniture, qu'à l'aide des échanges trophallactiques avec des Fourmis (Fig. 120 C – nourrissage d'une larve d'*Atemeles pubicollis* par une ouvrière du genre *Formica*). Ils sont ainsi devenus de véritables parasites sociaux (HÖLLDOBLER 1967 a + b). Le phénomène de la trophallaxie chez les Insectes sociaux peut être considéré comme un facteur fondamental de cohésion (résumé chez WILSON 1971 et 1975).

Trophobionten → Trophobiose

Trophobiose □ Eine Beziehung zwischen Ameisen und anderen Insekten, von welchen die Ameisen Honigtau und sonstige Futterstoffe erhalten, insbesondere von Blattläusen und anderen Homopteren oder auch von manchen Schmetterlingsraupen aus den Familien der Lycaenidae und Riodinidae. Als Gegenleistung bieten die Ameisen ihren Futterlieferanten Schutz und Pflege. Die Honigtau liefernden Insekten werden generell als *Trophobionten* bezeichnet (WILSON 1971, 1975).

trophobiosis □ The relationship in which ants receive honey dew from aphids and other Homopterans, or the caterpillars of certain butterflies of the families Lycaenidae and Riodinidae, and provide these insects with protection in return. The insects providing the honey dew are referred to as *trophobionts* (WILSON 1971, 1975).

trophobiose □ La relation entre Fourmis et un certain nombre d'Aphidae et d'autres Homoptères ou encore les chenilles d'un certain nombre de Lépidoptères des familles des Lycaenidae et des Riodinidae, desquels les Fourmis obtiennent la miellée, excréments liquides et sucrés très recherchés. Ces Insectes, appelés généralement *trophobiontes,* sont de leur côté protégés et soignés par les Fourmis (WILSON 1971, 1975).

Tropismus □ Im Raum gerichtetes Wachstum von Pflanzen und festsitzenden Tieren. Die beiden bekanntesten Arten tropischer Bewegungen sind der Phototropismus und der Geotropismus: Hält man z. B. eine Keimpflanze von *Sinapsis* in der Nähe eines Fensters in Wasserkultur in einem durchsichtigen Behälter, so krümmt sich nach kurzer Zeit der Stengel mit den Blättern dem Licht zu, die Wurzel dagegen von der Seite des Lichteinfalles ab (Abb. 121). Ähnliche Erscheinungen finden wir im Tierreich bei Bryozoen, Octokorallen und Ascidien (HARTMANN 1953).

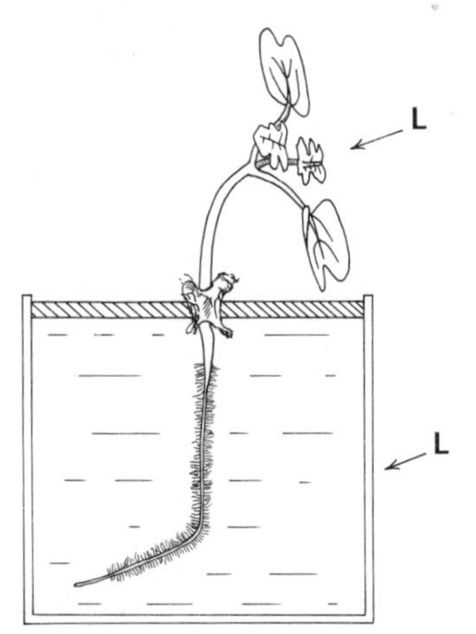

121

tropism □ The spatial orientation in growth of plants and sessile animals. The two best known examples are phototropism and geotropism. As an example, if one holds a newly germinated *Sinapsis* plant in a water culture near a window in a transparent container, the stem which sprouts will grow sideways toward the light side and the root in the opposite direction (Fig. 121). This pattern may also be seen in Bryocytes, Ascidia and other sessile animal organisms (HARTMANN 1953).

tropisme □ Croissance orientée dans l'espace chez les Végétaux et les animaux fixés. Les deux formes les plus connues de mouvements tropiques sont le phototropisme et le géotropisme. Voici un exemple de phototropisme: Si l'on maintient une plantule du genre *Sinapsis* en culture aquatique dans un récipient transparent près d'une fenêtre, le pédoncule et les feuilles de cette plantule s'orienteront après peu de temps vers la lumière ce qui correspond à une courbure phototropique positive. La racine, par contre, montre une réaction phototropique négative et se courbe à l'opposé de la lumière (Fig. 121). Nous connaissons des phénomènes semblables dans le monde animal chez les Bryozoaires, les Octocoralliaires et les Ascidies (HARTMANN 1953).

Tropotaxis → Taxis

Turnierkampf → Kommentkampf

U

Überfließhandlung = Überflußhandlung → Leerlaufhandlung

Überflußhandlung → Leerlaufhandlung

E. overflow activity □ F. débordement d'activité

Überlaufhandlung = Überflußhandlung → Leerlaufhandlung

übernormaler Auslöser □ Ein Reiz, der eine bestimmte Verhaltensweise besser auslöst, als es unter normalen natürlichen Bedingungen vom »zuständigen« Auslöser erzielt wird. Der angeborene Auslösemechanismus (→ *AAM*) paßt im Normalfall zu den Merkmalen des Objektes oder der Situation, auf welche die Reaktion gemünzt ist. Wäre es anders, so könnte die Reaktion bei unpassender Gelegenheit ausgelöst werden, und es entstünde Unordnung. Das genaue experimentelle Studium angeborener Auslösemechanismen hat Reizsituationen erkennen lassen, die die natürliche an Wirksamkeit übertreffen. Die erste übernormale Attrappe entdeckten KOEHLER und ZAGARUS (1937) bei ihren Untersuchungen am Sandregenpfeifer, *Charadrius hiaticula,* der gleichgroße weiße Eiattrappen mit schwarzen Flecken, also kontrastreicher als die normalen, gegenüber seinen eigenen bevorzugt. Beim Austernfischer, *Haematopus ostralegus,* wurde experimentell nachgewiesen, daß er ein übernormales Fünfergelege seinem normalen Dreiergelege bevorzugt. Bei der Wahl zwischen einem Ei der Silbermöwe (Abb. 122 – links) und seinem eigenen (vorn) rollt er lieber das Riesenei in sein Nest ein (TINBERGEN 1948). Ein Beispiel eines solchen übernormalen Schlüsselreizes (TINBERGEN – Instinktlehre) aus der Natur finden wir beim Sperrachen des Kuckucks, der an Auffälligkeit in der Färbung und in der Größe im Auslösewert den der Nestlinge seiner Wirtsarten übertrifft.

122

supra-normal releaser, super-releaser □ A stimulus that elicits a particular behaviour pattern more easily than the natural releaser which normally elicits the behaviour. The innate releasing mechanism (→ *IRM*) responds specifically to the characteristics of the object or situation for which the reaction has evolved. Otherwise the response might be released by inappropriate stimuli and thus lead to chaos. Precise experimental studies of innate releasing mechanisms have defined stimulus configurations which are more effective than the natural releasers. KOEHLER and ZAGARUS (1937) discovered the first super-releaser while studying the ringed plover *Charadrius hiaticula;* these birds preferred white eggs with black spots to their own eggs of similar size due to the spotted model's greater contrast. The oystercatcher, *Haematopus ostralegus,* prefers a clutch of five eggs to its own normal clutch of three. In a choice between an egg of the herring gull (Fig. 122 – left) and its own one (in front), it prefers to roll the giant egg into its nest (TINBERGEN 1948). An example of a super-releaser found in nature is the cuckoo hatchling which is more conspicuous in colouration and larger than the nestlings of its host species (TINBERGEN – Study of Instinct).

déclencheur supra-normal □ Stimulus qui déclenche un comportement donné mieux que le déclencheur adéquat dans les conditions naturelles. Le mécanisme déclencheur inné (→ *MDI*) est, dans les cas normaux, adapté aux caractères de l'objet ou de la situation à laquelle correspond la réaction; s'il en était autrement, une réaction pourrait être déclenchée à un moment inadéquat et il se produirait alors un désordre. L'étude expérimentale et précise des mécanismes déclencheurs a révélé des situations stimulatrices qui dépassent l'efficacité naturelle. Le premier leurre supra-normal a été découvert par KOEHLER et ZAGARUS (1937) lors de recherches sur le Grand Gravelot, *Charadrius hiaticula,* qui préférait des œufs artificiels de couleur blanche avec des taches noires, donc plus contrastés que les siens qui sont à considérer comme normaux. Chez l'Huîtrier-Pie, *Haematopus ostralegus,* il a été démontré expérimentalement que l'Oiseau préfère une ponte composée de 5 œufs à une ponte normale constituée de 3 œufs. Le même Oiseau ayant le choix entre un œuf de Goéland argenté (Fig. 122 – à gauche) et son propre œuf (au premier plan), préfère de loin l'œuf géant qu'il essaie de rouler vers son nid (TINBERGEN 1948). Un exemple d'un tel stimulus-clé supra-normal (TINBERGEN – Etude de l'Instinct) se retrouve dans la nature chez le jeune Coucou dont la coloration, la taille et l'apparence du bec largement ouvert constitue un effet déclencheur bien supérieur que celui des Oisillons des espèces parasitées.

Übersprungbewegung □ KORTLANDT und TINBERGEN haben (1940) unabhängig voneinander diese Erscheinung gesehen und analysiert. Es kommt zu Übersprunghandlungen bei überstarker Erregung, die sich auf dem üblichen Wege nicht zu entladen vermag. Die häufigste Ursache ist der Konflikt zwischen zwei antagonistischen Trieben, wie z. B. während einer Kampfhandlung der Konflikt zwischen Angriff und Flucht (→ *Konfliktverhalten*), auch die Reizung einer »erschöpften Reaktion« oder die vorzeitige Beendigung einer angelaufenen Handlung oder auch das Ausbleiben einer notwendigen äußeren Reizung bei angelaufener Handlung. Meist treten in diesem Zusammenhang solche Verhaltenselemente auf, die im Normalverhalten sehr häufig aktiviert werden, also Verhaltensweisen der Nahrungsaufnahme, Komfortbewegungen, Ruhestellungen. Kämpfende Silbermöwen können in höchster Kampferregung im Übersprung plötzlich Nistmaterial abreißen (→ *Grasrupfen*); bei Limicolen beobachtet man häufig Schlafstellungen (Übersprungschlaf); Enten zeigen Gefiederputzen (Übersprungputzen) – dieses Verhalten wurde zusätzlich noch ritualisiert (→ *Scheinputzen*). Bei Hühnervögeln beobachtet man ein »Übersprungfressen«, wobei zwar auf den Boden gepickt wird wie beim Fressen, jedoch keine Nahrung aufgenommen wird. Auch dieses Verhalten tritt in ritualisierter Form als → *Bodenpicken* in der Werbung auf. – Der Ausdruck »Übersprungbewegung« bzw. »Übersprunghandlung« ist bislang in verschiedener Weise gebraucht worden. Bereits MAKKINK (1936) verglich die Erscheinung mit dem elektrischen Funken, der auf eine andere Leitungsbahn überspringt, und sprach von *sparking over.* KIRKMAN (1937) prägte die Bezeichnung *substitute activity,* die auch TINBERGEN (1939) übernahm, wogegen jedoch von psychologischer Seite Bedenken geäußert wurden. RAND (1943) sprach von *irrelevant movements.* TINBERGEN und VAN IERSEL (1947) gebrauchten dann *displacement activity,* ein Ausdruck, der sich dann bis jetzt in der englischen Literatur eingebürgert hat. – LEYHAUSEN (1952) und SEVENSTER (1961) haben neue Hypothesen zum Problem der Übersprungbewegung aufgestellt.

displacement activity □ KORTLANDT (1940) and TINBERGEN (1940) observed and analysed this phenomenon independently of each other. Excess excitation which cannot be released through normal channels may lead to displacement activity. The most frequent causes are conflicts between opposing drives, as for example, the conflict during a fight between attack and flight (→ *conflict behaviour*). Other causes are continued stimulation of an »exhausted response«, premature termination of an ongoing action or absence of necessary external signal during an ongoing action. Usually the resulting behavioural responses are elements which normally occur very frequently, e. g., feeding behaviour, comfort movements, resting postures. Fighting herring gulls may suddenly rip grass from the ground (→ *grass pulling*); Limicolans may sleep in place (displacement sleeping); ducks preen (displacement preening) and this behaviour, moreover, has also become ritualized (→ *sham preening*). In fowl one encounters »displacement feeding« in which the ground is pecked but no food consumed. This pecking behaviour also occurs during courtship in a ritualized form (→ *ground pecking*). – The term displacement activity has been used in a number of ways. MAKKINK (1936) referred to this subject as »sparking over« in an analogy to electrical energy that jumps from one conductor to another. KIRKMAN (1937) coined the term »substitute activity« which TINBERGEN (1939) adopted but which was subsequently opposed by psychologists. RAND (1943) then introduced »irrelevant movements« and in 1947 TINBERGEN and VAN IERSEL suggested »displacement activity« which is the now generally accepted term. LEYHAUSEN (1952) and SEVENSTER (1961) have developed new hypotheses on the phenomenon of displacement activity.

activité substitutive □ KORTLANDT (1940) et TINBERGEN (1940) ont découvert ce phénomène indépendamment et l'ont analysé par la suite. Les activités substitutives se produisent lors d'une surexcitation qui ne peut se décharger par la voie usuelle. Les causes les plus courantes sont: le conflit entre deux tendances antagonistes, comme par exemple pendant l'engagement d'un combat le conflit entre l'attaque et la fuite (→ *comportement conflictuel*), l'activation d'une réaction qui a subi une extinction temporaire, l'interruption prématurée d'une activité en voie d'accomplissement, l'absence d'une stimulation exogène nécessaire à la poursuite d'une activité amorcée. Dans ce contexte, apparaissent à titre substitutif les éléments comportementaux qui, dans les conditions usuelles, sont les plus fréquemment activés, c'est-à-dire les comportements d' alimentation et de confort et les positions de repos. Ainsi, les Goélands argentés peuvent, lors d'un combat où ils déploient une excitation extrême, effectuer soudain le → *désherbage* comme activité substitutive. Chez de nombreux Limicoles, on observe souvent des positions de repos telles que le sommeil substitutif. Chez les Canards, il apparaît des mouvements de nettoiement du plumage (nettoiement substitutif); de plus, ce comportement a subi une ritualisation et apparaît, dans le contexte de la parade nuptiale, sous la forme d'un → *pseudo-nettoyage*. Chez de nombreux Gallinacés, on observe un comportement alimentaire substitutif au cours duquel les animaux picorent effectivement le sol, sans pour autant prendre de la nourriture. Ce comportement apparaît également sous une forme ritualisée dans la parade nuptiale (→ *picorage au sol*). La notion d'activité substitutive, définie à l'origine par KORTLANDT et par TINBERGEN, a cependant été utilisée de différentes manières. MAKKINK (1936) tenta une interprétation analogique de ce phénomène, en le désignant comme un *sparking over*, allusion au phénomène de l'étincelle électrique qui saute d'une ligne à une autre. KIRKMAN (1937) créa le terme de *substitute activity* qui a été également utilisé par TINBERGEN (1939), mais sur lequel les psychologues ont exprimé des réserves, certes, car il désigne une activité dirigée vers un substitut. RAND (1943) parla de *irrelevant movements*. TINBERGEN et VAN IERSEL (1947) ont finalement utilisé *displacement activity*, expression qui d'une manière générale est maintenant utilisée dans la littérature anglaise. LEYHAUSEN (1952) et SEVENSTER (1961) ont formulé de nouvelles hypothèses à propos des activités substitutives.

Übersprunghandlung → Übersprungbewegung

Übersprungputzen → Scheinputzen

E. displacement preening □ F. nettoyage substitutif

Übersprungverhalten → Übersprungbewegung

E. displacement behaviour □ F. comportement substitutif

Übervölkerungseffekt → Massensiedlungseffekt

E. overpopulation effect □ F. effet de surpopulation

Ultraschall-Orientierung → Echo-Orientierung

E. ultrasonic orientation □ F. orientation par ultra-sons

Ultraschallpeilung → Echo-Orientierung

Umgehungsmanöver → Verleiten

Umkehrdressur □ Bei der Umkehrdressur wird nach gefestigter Dressur die Bedeutung eines Dressurmerkmals vom Negativen zum Positiven geändert und nach neuerlicher Festigung wieder umgekehrt (BITTERMANN 1965, BUCHHOLTZ 1973, RENSCH 1973).

reversal learning □ In reversal learning a conditioned response to a set of stimuli is reversed such that the previously incorrect choice is now correct. Following acquisition of the new set of contingencies, the rewarded signal is again reversed, etc. (BITTERMANN 1965, BUCHHOLTZ 1973, RENSCH 1973).

apprentissage avec inversions □ Il s'agit d'une situation où, après une discrimination bien établie, le signal à valence jusqu'alors négative est ensuite renforcé positivement. Une fois cette nouvelle situation bien apprise, le dressage est de nouveau inversé, etc. (BITTERMANN 1965, BUCHHOLTZ 1973, RENSCH 1973).

umorientiertes Verhalten → Übersprungbewegung

E. re-oriented behaviour □ F. comportement redirigé

Umprägung → Prägung

E. reversal imprinting □ F. renversement d'empreinte

Umstimmung □ Ein Wechsel von einer Stimmung auf eine andere; von VON HOLST und VON ST. PAUL (1960) experimentell näher untersucht. Oft wiederholte Reizung vermag die Grundstimmung eines Tieres zu ändern. Ein Huhn, das spontan anhaltend schimpfte, konnte durch künstliche Aktivierung des Sitzdranges so »umgestimmt« werden, daß es auch nach Aufhören des Hirnreizes sitzen blieb. Weiter läßt sich eine Verhaltensweise durch wiederholte Reizung von einem Punkt aus ermüden; zur Auslösung bedarf es dann immer stärkerer Reize. Zeigt der betreffende Punkt I keine Reaktion mehr und reizt man danach einen Punkt II, so ist die Reaktion von dort oft noch voll intensiv auslösbar. In diesem Falle hat man also von Reizpunkt I nur eine leitende Struktur, die allein vom ersten Reizfeld bestimmt wird und nicht das motorische Zentrum ermüdet. Solche »Ermüdungserscheinungen« an einer leitenden Struktur nannten VON HOLST und VON ST. PAUL eine »zentrale lokale Adaptation« und unterschieden diese von dem als »Umstimmung« bezeichneten Vorgang.

change of mood □ A change from one motivational state or mood to another, studied experimentally by VON HOLST and VON ST. PAUL (1960). Repetitive stimulation of an animal often induces change in the basic mood. A hen, for example, that was spontaneously and persistently aggressive was induced to sit by artificial activation of a sitting urge via electrical stimulation of the brain. The mood was altered such that the hen remained sitting even after the brain stimulation was removed. However, a behaviour pattern may also wane following repeated stimulation of a given site (Point I). It then requires increasingly stronger stimulation for its elicitation. When Point I shows no further responsiveness, the behaviour pattern may be elicited in its full-blown form from stimulation of an adjacent site (Point II). Point I may be interpreted as located along a conducting pathway which may be exhausted without attendant fatigue of the motor center. Such a fatigue in a conducting structure was referred to as central local adaptation by VON HOLST and VON ST. PAUL in order to distinguish it from motivational switch processes.

commutation de la motivation □ Phénomène expérimentalement étudié par VON HOLST et VON ST. PAUL (1960): Des stimulations électriques répétées sont en mesure de changer la motivation de base d'un animal. Une Poule, par example, émettant spontanément des cris de mécontentement et dont la pulsion de repos a été artificiellement activée, peut subir un changement de motivation, en sorte qu'elle reste couchée, même après l'arrêt de la stimulation cérébrale. D'autre part, un comportement peut être épuisé par des stimulations répétées d'un même point; pour le déclenchement de ce comportement, il est alors nécessaire d'effectuer des stimulations de plus en plus intenses. Si le point I ne déclenche plus aucune réaction et qu'on stimule alors un point II, la réaction peut être parfaitement déclenchable à partir de ce second point. Dans un pareil cas, à partir du point I, seule une voie nerveuse particulière a été soumise à l'épuisement, mais nullement le centre moteur intéressé. De tels phénomènes d'épuisement d'une voie de conduction ont été appélés par VON HOLST et VON ST. PAUL »adaptation centrale localisée«, distincte du phénomène d'épuisement total de la réactivité spécifique qui est la source du changement de motivation.

Umwegversuch □ Eine Versuchsanordnung zur Prüfung des spontan einsichtigen Verhaltens, welches gleichzeitig zur Lösung des Problems und zur Erlangung des begehrten Objektes (meist Nahrung) führt. Bei solchen Versuchen zeigen Tiere oft Leistungen, die man als spezifisch menschliche »höhere Hirnleistungen« einzustufen pflegt (FISCHEL 1961, O. VON FRISCH 1962, BUCHHOLTZ 1973, RENSCH 1973). Die beiden beigefügten Abbildungen zeigen Umweglösungen bei einem Hund (Abb. 123 A) und einem Zwergchamäleon (Abb. 123 B – *Microsaurus punilus*).

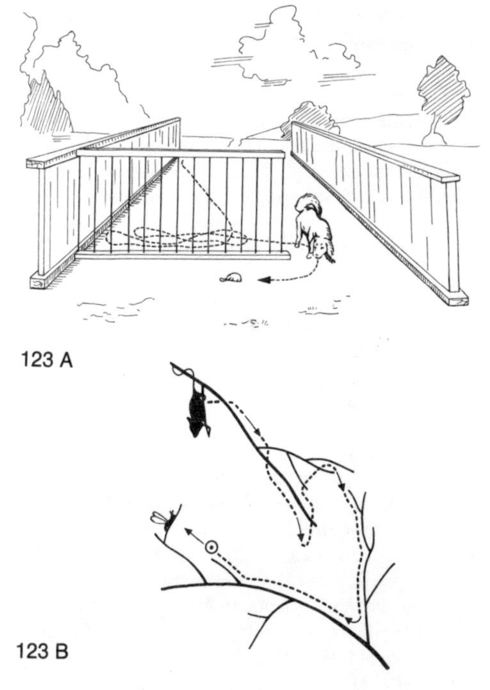

123 A

123 B

detour experiment □ An experiment to evaluate spontaneous insight behaviour. Solution of the detour problem is usually rewarded with food. Frequently in such experiments the animal may demonstrate capabilities which we are accustomed to associate with humans (FISCHEL 1961, O. VON FRISCH 1962, BUCHHOLTZ 1973, RENSCH 1973). The two accompanying illustrations show detour learning in a dog (Fig. 123 A) and in a dwarf chameleon (Fig. 123 B – *Microsaurus punilus*).

expérience de détour □ Expérience permettant d'étudier des comportements mettant en jeu la compréhension subite d'une situation-problème (»intuition concrète« ou »Einsicht«) et permettant à la fois la résolution du problème et l'obtention de l'objet recherché, en général un appât alimentaire. Dans ces expériences, de nombreux animaux montrent des performances comparables aux »activités mentales supérieures« spécifiquement humaines (FISCHEL 1961, O. VON FRISCH 1962, BUCHHOLTZ 1973, RENSCH 1973). Les deux figures ci-jointes montrent la résolution du problème de détour chez un Chien (Fig. 123 A) et chez un Caméléon nain (Fig. 123 B – *Microsaurus punilus*).

Universalien □ Bei allen Menschen, ganz gleich welcher Rasse oder Kultur, im gleichen Kontext vorkommende Verhaltensweisen, welche auch die gleichen Reaktionen auslösen, die gleiche Bedeutung haben und gleich verstanden werden, bezeichnet man in der Ethologie als Universalien. Das Weinen, → *Lachen*, → *Lächeln*, Ausdruck des Mißtrauens, Ausdruck der Trauer, Schmerzgrimasse und Drohgesicht sind Beispiele hierfür und mit Sicherheit angeboren und nicht kulturbedingt (EIBL-EIBESFELDT 1973, MORIN und PIATTELLI-PALMARINI 1974).

universals □ In ethology universals are behaviour patterns which occur in the same context in all humans regardless of their culture or race and which elicit the same response, have the same meaning and are interpreted in the same way. Crying, → *laughing*, → *smiling*, expressions of mistrust, expressions of sadness, pain and threat are examples of cultural universals which are certainly innate and unrelated to culture (EIBL-EIBESFELDT 1973, MORIN and PIATTELLI-PALMARINI 1974).

universaux □ Comportements existant chez tous les Hommes quelles que soient leur race et leur appartenance culturelle. Ces conduites apparaissent dans le même contexte, déclenchent les mêmes réactions, sont chargées des mêmes significations et aboutissent à la même compréhension. Exemples: les pleurs, le → *rire*, le → *sourire*, l'expression de la méfiance ou du deuil, les grimaces de douleur et les expressions de menace. Ces éléments comportementaux sont incontestablement innés et nullement dépendants d'une structure culturelle particulière (EIBL-EIBESFELDT 1973, MORIN et PIATTELLI-PALMARINI 1974).

Untergrundangleichung □ Bei vielen Lebewesen kennen wir eine Angleichung der Körperfarbe an den Untergrund, wie z. B. bei Wüstenmäusen, Wüstenlerchen, oder die jahreszeitliche Anpassung bei Schneehühnern (*Lagopus mutus*), Schneehasen (*Lepus timidus*) und Hermelin (*Mustela erminea*), die im Winter weiß werden (BRUNS 1958). Es scheint sich hier um genetisch fixierte Merkmale zu handeln. Versuche mit *Lepus timidus*, die 1854/55 von Nordnorwegen, wo sie im Winter weiß sind, nach den Färöer-Inseln, 350 km nördlich von Schottland, transportiert wurden, zeigten auf diesen Inseln mit durch den Golfstrom bedingtem temperiertem Winterklima im Winter 1890, also nach 35 Jahren, nur 5–6 weiße Tiere unter 100 Abschüssen (COUTURIER 1955). Auf mehrere Generationen erweist sich diese Anpassung als ein umweltlabiles Merkmal. Von vielen Cephalopoden, Crustaceen, Fischen und Chamaeleoniden kennen wir eine spontane Untergrundangleichung, bei Trichopteren-Larven erkennen wir diese im Köcherbau, und mittelbraun eingebrachte Libellenlarven (*Aeschna*), die in zwei Gruppen aufgeteilt auf schwarzem und weißem Untergrund gehalten werden, passen sich bei der nächsten Häutung an ihre neue Umwelt adäquat an (KRIEGER 1954).

camouflage □ A number of organisms exhibit a camouflage of the body colour with the substrate as for example, desert mice or desert larks; the seasonal adaptation of the ptarmigans *(Lagopus mutus)*, the snow-shoe hare *(Lepus timidus)*, and the ermine *(Mustela erminea)* which turn white in the winter (BRUNS 1958). Camouflage is under genetic control as demonstrated by an experiment in which snowshoe hares were transported in 1854–1855 from northern Norway to Faeroe Islands, 350 km north of Scotland. These islands have a mild winter climate due to the influence of the Gulf Stream. In 1890, 35 years later, only 5 or 6 white animals appeared among 100 pelts taken that winter (COUTURIER 1955). Over several generations this adaptation proved to be an environmentally labile trait. In many cephalopods, crustaceans, fish and chamaeleonids a spontaneous camouflaging occurs; in Trichopteran larvae camouflaging occurs in the cocoon construction. Brown dragonfly larvae (*Aeschna*) placed on a black or white substrate will adapt their colour to the particular substrate during the next moult (KRIEGER 1954).

homochromie □ Chez de nombreux animaux, adaptation de la coloration corporelle à l'environnement, comme par exemple chez les Souris du désert ou les Alouettes du désert, ou adaptation saisonnière, comme chez *Lagopus mutus*, *Lepus timidus* et *Mustela erminea*, dont le plumage ou le pelage deviennent blancs en hiver (BRUNS 1958). Il semble qu'il s'agisse de caractères génétiquement fixés. Par ailleurs, des observations ont été effectuées sur des *Lepus timidus* qui ont été transportés en 1854/55 du Nord de la Norvège où ils deviennent blancs en hiver, sur les Iles Feroë à 350 km au nord de l'Ecosse, îles qui possèdent grâce à la proximité du Gulf Stream un climat hivernal tempéré. En hiver 1890, c'est-à-dire 35 ans après, seuls 5 à 6 animaux blancs ont été rencontrés parmi 100 animaux tués (COUTURIER 1955). Nous voyons donc que sur plusieurs générations, cette adaptation s'avère comme un caractère labile sous l'influence de l'environnement. Par ailleurs, nous connaissons des exemples d'adaptation spontanée à l'environnement chez les Céphalopodes, les Crustacées, les Poissons et les Chamaeleonidae. Les larves des Trichoptères adaptent parfois la construction de leur fourreau au milieu dans lequel elles vivent. Les expériences effectuées sur des larves de Libellules du genre *Aeschna,* capturées dans la nature avec une coloration brun moyen, divisées en deux groupes et élevées sur un fond noir ou blanc, ont montré une adaptation adéquate à leur environnement lors de la mue suivante (KRIEGER 1954).

Unterlegenheitsfärbung → Inferioritätsfärbung

Unterlegenheitsgeste = Demutsgebärde → Demutsverhalten

Unterwerfung → Demutsverhalten
E. submission □ F. soumission

Unterwerfungsgebärde = Demutsgebärde → Demutsverhalten

Unterwerfungszeremonie → Demutsverhalten
E. submission ceremonial □ F. cérémonie de soumission

Urinmarkieren → Harnmarkierung

Urinspritzen → Harnspritzen

Urinwaschen → Harnmarkierung
E. urine washing □ F. lavage à l'urine

V

Ventilsitten □ Verhaltensweisen beim Menschen zur unblutigen Ableitung aufgestauter Aggression. Verschiedene australische Stämme kommen zu bestimmten Zeiten zusammen, um sich zu beschimpfen und um nach bestimmten Regeln miteinander zu raufen. Von Eskimos wissen wir, daß sie viele ihrer Dispute in Gesangsduellen erledigen. In Bayern und Österreich mit Hilfe von Verbalinjurien in Form von Wechselgesängen. Auch Kampfsportarten dienen als mögliche Ventile der Aggression. Weitere Beispiele von Ventilsitten finden wir bei BOHANNAN (1966).

safety valve customs □ Behaviour which leads to non-violent release of pent-up aggression in human beings. A number of Australian tribes meet regularly to argue and scuffle with each other. Eskimos are known to solve disputes through singing duels. Verbal insults in Bavaria and Austria may also take the form of singing contests. Sports may serve as valves for aggression. BOHANNAN (1966) gives other examples.

soupape de sûreté □ Comportement humain tendant à liquider de manière non sanglante une pulsion agressive qui s'est accumulée. Nous connaissons ces phénomènes chez diverses tribus d'Aborigènes australiens qui, à des périodes déterminées, se réunissent pour s'insulter mutuellement et pour lutter selon des règles bien établies. Les Esquimaux également règlent la plupart de leurs disputes à l'aide de duels chantés. En Bavière et en Autriche, on échange des injures verbales présentées sous forme de duos chantés. Des sports de combat peuvent également servir d'exécutoire à l'agression. D'autres exemples du même ordre sont décrits dans BOHANNAN (1966).

Verbalinjurien → Ventilsitten
E. verbal insults □ F. injures verbales

Verblümungssitten □ Im menschlichen Verhalten ritualisierte Umgangsformen der indirekten Aussage oder des indirekten Ausdrucks. Entlehnt von »etwas durch die Blume sagen«, d. h. verschleiert bzw. »verblümt«. Vor allem bei Abweisungen werden Entschuldigungen angebracht wie »Wir bedauern sehr, Ihnen mitteilen zu müssen«, um diese zu entschärfen und um das soziale Band zu erhalten. Ähnlich verhalten sich die Menschen in der Liebeswerbung. Hierzu nur ein Beispiel: In der Eifel bringt der Bursche eine Flasche Wein, wenn er die Erwählte besucht. Er stellt sie auf den Tisch, und wenn das Mädchen ein Glas holt, gibt es damit sein Einverständnis kund.

figurative customs □ In human behaviour a ritualized form of figurative speech or expression. Especially used during negation or rejection as for example, »we regret to have to inform you . . .«. Figurative customs serve to preserve the social bond. Another example of figurative customs comes from courtship behaviour in the Eifel region of Germany. The young man usually brings a bottle of wine when he visits a young lady. If she fetches a glass, this signifies her interest in continuing the interaction.

déclarations voilées □ Dans le comportement humain, formes ritualisées de déclarations ou d'expressions indirectes. L'expression allemande »etwas durch die Blume sagen« (litt. »dire quelque chose avec des fleurs«) veut dire »voiler« une déclaration. Particulièrement lors d'un refus, on emploie souvent des excuses comme »nous regrettons beaucoup d'être obligés de vous dire . . .«. Ces formules servent à l'apaisement et au maintien du lien social. Nous connaissons des modes d'expression semblables dans le comportement de cour chez l'Homme, dont nous ne citerons qu'un exemple: dans la région de l'Eifel en Allemagne, un garçon amoureux d'une jeune fille se rend chez elle avec une bouteille de vin et la pose sur la table; si la jeune fille cherche un verre, elle exprime son accord.

Vereitelung → Frustration

Verhalten → Ethologie
E. behaviour □ F. comportement

Verhaltensablauf → Erbkoordination → Erwerbkoordination
E. behavioural sequence □ F. déroulement d'un comportement, d'une conduite

Verhaltensanalogie → Konvergenz
E. behavioural analogy □ F. analogie comportementale, convergence éthologique

Verhaltensdifferenzierung = Ethospezies → Ethosystematik
E. behavioural differentiation □ F. différenciation comportementale

Verhaltensforschung → Ethologie
E. objective study of behaviour □ F. étude objective du comportement

Verhaltensinventar → Ethogramm
E. behavioural inventory □ F. inventaire comportemental

Verhaltensnachahmung = Ethomimikry→ Mimikry
E. behavioural mimicry □ F. mimétisme comportemental

Verhaltensontogenese → Ethogenese
E. behavioural ontogeny □ F. ontogenèse du comportement

Verhaltensrepertoire → Ethogramm
E. behavioural repertoire □ F. répertoire comportemental

Verhaltensumschlag □ Bei zunehmender Reizintensität kann ein plötzlicher »Umschlag« von einem Verhalten auf ein anderes vorkommen, wie z. B. bei hirngereizten Hühnern bei Erhöhung der Reizstärke ein Verhaltensumschlag von Angriff zu Flucht beobachtet werden konnte (VON HOLST und VON ST. PAUL 1960).

behavioural change □ During increased stimulus intensity a sudden transition from one behavioural state to another may occur, e. g. during brain stimulation in chickens a sudden switch from attack to escape may occur (VON HOLST and VON ST. PAUL 1960).

commutation comportementale □ Dans les expériences de stimulation centrale une augmentation de la tension stimulatrice peut entrainer un passage soudain d'un comportement à un autre. Cette commutation a été constatée expérimentalement chez les Poules, avec passage d'un comportement agressif à un comportement de fuite (VON HOLST et VON ST. PAUL 1960).

Verhaltensumstimmung → Umstimmung
E. motivational switch □ F. commutation motivationnelle

Verlassenheitssyndrom → Hospitalismus
E. infant isolation syndrome □ F. syndrome d'abandon

Verlegen □ Eine echte Instinkthandlung, auf welcher das Nestbauverhalten der Schwäne aufbaut (PETZOLD 1964). Den die → *Erbkoordination* »Verlegen« bedingenden Affekt nennt PORTIELJE (1936) *Aufrafftrieb*. Beim Verlegen von Nistmaterial steht der Schwan unbeweglich, nur der Hals wird ausgiebig gedreht und langgestreckt. Im Sitzen auf dem Nest wird in gleicher Weise verfahren. Wenn der Schwan in allernächster Nähe des Nestes steht, legt er das mit dem Schnabel herangeholte Material aufs Nest, aber nur, wenn er es, ohne einen Schritt zu tun, erreichen kann. Das Verlegen kann als Einzelhandlung oft ganz unsinnig wirken, ist letzten Endes aber doch nützlich, da der Schwan beim Verlegen meist mit dem Kopf vom Nest abgewandt steht und stets nach hinten legt (Abb. 124 – *Cygnus atratus*).

sideways building □ A true instinctive component in the nest building behaviour of swans (PETZOLD 1964). The motivational foundation of the → *fixed action pattern* of sideways building is termed *gathering drive* by PORTIELJE (1963). During sideways building the swan moves only its neck to gather material. If it is standing next to the nest or sitting on it, it lays all nest material in reach on the nest with the bill. Sometimes single movements appear to be purposeless; however, in the long run they are useful, since the swan stands with the head away from the nest and always places the material behind him (Fig. 124 – *Cygnus atratus*).

transfert de matériaux □ Action instinctive véritable sur laquelle est basée la construction du nid chez les Cygnes (PETZOLD 1964). La → *coordination héréditaire* correspondante dépend de la *pulsion à ramasser* (PORTIELJE 1936). Pendant le transfert des matériaux dc nidification, le Cygne reste immobile; seul le cou est étiré et effectue des mouvements latéraux de grande ampleur. Si l'animal est assis sur son nid, il agit de même. S'il se trouve à proximité immédiate du nid, il pose les matériaux qu'il a ramassés avec son bec directement sur le nid, s'il peut le faire sans bouger d'un seul pas. Le transfert de matériaux, surtout en tant qu'action isolée, peut parfois paraître aberrant; en fin de compte, ce comportement est malgré tout utile étant donné que pendant son accomplissement, l'animal détourne toujours sa tête du nid et que les matériaux récoltés sont toujours déposés derrière lui et arrivent ainsi auprès du nid (Fig. 124 – *Cygnus atratus*).

Verlegenheitsgebärde □ Bei leichter Verlegenheit verdecken Menschen in aller Welt, aller Ras-

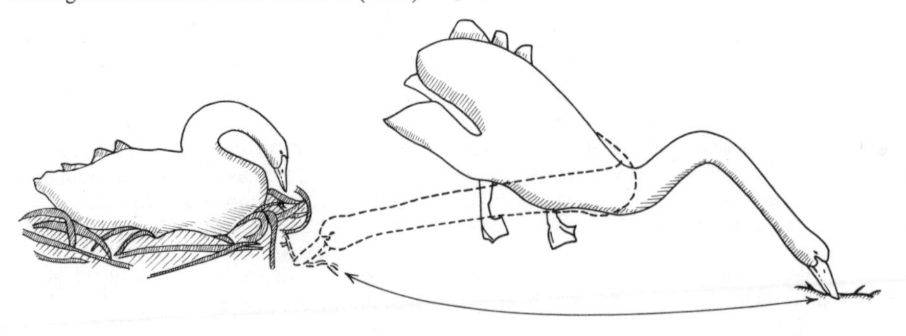

124

sen und Kulturen, ganz oder teilweise ihr Gesicht (Abb. 125 A – !ko Buschmädchen aus der Kalahari; B – Bayaka-Pygmäe aus Zentralafrika). Es handelt sich hierbei wohl um eine ritualisierte Versteckbewegung. Man beobachtet dieses Verhalten auch bei flirtenden Mädchen und sogar bei Blindgeborenen kann man sehen, wie sie bei Verlegenheit entweder den Blick senken oder das Gesicht hinter den Händen verbergen (*Gesichtverbergen*).

125 A

125 B

gesture of embarrassment □ In all cultures and races, a person which is mildly embarrassed partially covers the face (Fig. 125 A – !ko Bushman girl from the Kalahari Desert; B – Pygmy Bayaka from Central Africa). This may represent a ritualized hiding behaviour. A similar behaviour occurs in flirting girls. Even congenitally blind individuals either drop their glance or hide the face with the hands when embarrassed (*hiding the face*).

geste d'embarras □ Dans des situations d'embarras, les Hommes du monde entier, quelle que soit leur appartenance raciale ou culturelle, se recouvrent souvent la figure, soit complètement, soit partiellement (Fig. 125 A – jeune fille Boshiman du Kalahari; B – Pygmée Bayaka d'Afrique Centrale). Il s'agit probablement d'un mouvement ritualisé permettant de se cacher. On observe ce comportement également chez les jeunes filles pendant le flirt. Chez les enfants nés aveugles, on peut observer que, dans des situations d'embarras, ils baissent les yeux ou/et abritent leur visage derrière leurs mains (*se cacher le visage*).

Verlegetrieb = Aufrafftrieb → Verlegen

Verleiten □ Verhaltensweisen, insbesondere bei bodenbrütenden Vögeln, die der Irreführung eines Feindes dienen. Es sind Ablenkmanöver, die bei Gefahr die Aufmerksamkeit des Feindes auf den Altvogel lenken, indem meist Lokomotionsbehinderungen wie Lahmstellen, Scheinhinken oder Flügelverletzungen vorgetäuscht werden, um den Raubfeind vom Nest oder von den Jungen fortzulocken. Weit genug vom Nest oder von den Jungen entfernt, stellt der Altvogel das Ablenkungsmanöver plötzlich ein und fliegt auf Umwegen zu seinem Nest (in sein Revier) zurück. Von Kiebitzen und Regenpfeifern ist dieses Verhalten wohlbekannt.

distraction display □ Behaviour pattern serving to divert a potential predator from the nest; common in ground nesting birds (broken wing ruse). The predator is distracted by the adult bird which feigns some locomotor impairment such as limping or wing injury. The predator is led some distance from the nest or young, then the adult stops feigning and flies a circuitous route back to its nest (or territory). This behaviour is well known from plovers.

manoeuvre de diversion □ Comportement existant surtout chez les Oiseaux nidificateurs terrestres, destiné à tromper un prédateur potentiel. Ces manœuvres de diversion servent à attirer l'attention du prédateur sur l'Oiseau adulte qui simule des défaillances locomotrices, comme faire semblant de boiter ou feindre d'avoir une aile blessée, pour détourner le danger du nid ou des jeunes. Quand la distance est suffisante, l'Oiseau adulte arrête brusquement sa manœuvre et s'envole pour regagner son territoire par des chemins détournés. Ce comportement est très bien connu chez les Vanneaux, les Pluviers et les Gravelots.

Verneinung □ Verhaltensweisen des Ablehnens beim Menschen. In Griechenland und in der Türkei sowie in vielen arabischen Ländern des Nahen Ostens verneint man durch ruckartiges Hoch- und Zurückwerfen des Kopfes, wobei die Augen fest geschlossen werden; bekannt auch als »ANANOUEIN« der Griechen des klassischen Altertums. Waikas tun dies auch bei Ablehnung eines Blickkontaktes als Kontaktabbruch (Abb. 126 A – Waika-Indianer vom Orinoko; B – Grieche). Bei betonter Verneinung werden oft eine oder auch beide Hände mit der Handfläche zum Gesprächspartner angehoben, ähnlich wie wir es bei entrüsteter Ablehnung tun. Wahrscheinlich läßt sich diese spezielle Form der Vernei-

126 A

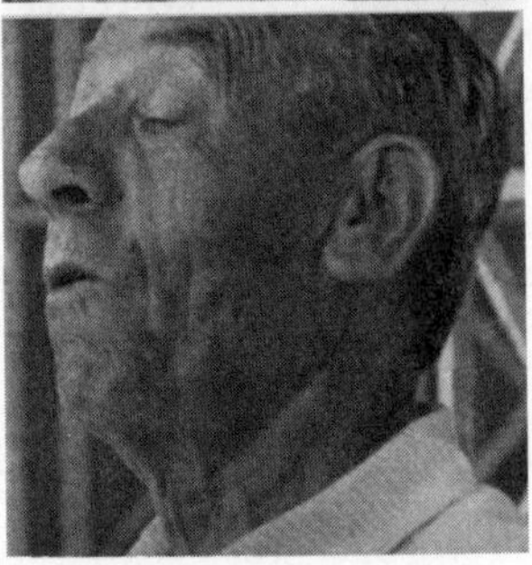

126 B

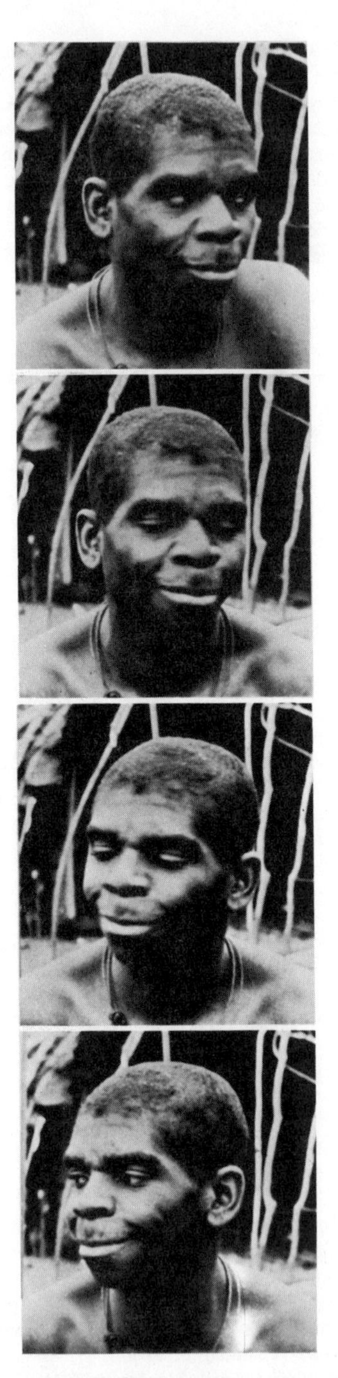

126 C

nung aus einer allgemein menschlichen Geste der sozialen Ablehnung ableiten. Wir Mitteleuropäer haben mit vielen anderen Völkern das seitliche, abweisende Kopfschütteln zum Nein ritualisiert (Abb. 126 C – Bayaka-Pygmäe aus Zentralafrika; von oben nach unten Bild 14, 34, 43, 57 aus einem 16 mm-Film, 25 B/sec). Das meiste in unseren Gesten ist traditionell gestaltet, oft jedoch ist eine angeborene Basis nachweisbar. Es ist zu vermuten, daß sich unser Kopfschüttel-Nein stammesgeschichtlich aus der ablehnenden Schüttelbewegung bei anderen Säugern entwickelt hat.

negation (or rejection) □ Human behaviour patterns which reject or negate. In Greece, Turkey and many Arab countries of the Near East one declines by throwing the head up and back with the eyes closed; this is also known from the ancient Greek civilization (»ANANOUEIN«). In the Waika tribe a similar gesture occurs when mutual eye contact is not desired (Fig. 126 A – Waika Indian from the Orinoco; B – Greek). During more emphatic rejection the hands may be raised with their palms toward the other speaker. This gesture of rejection is probably derived from a more general form of social rejection. Western Europeans have a ritualized lateral head shake in common with many other cultures as a rejection gesture (Fig. 126 C – Bayaka Pygmy from Central Africa; picture 14, 34, 43, 57 of a 16 mm film, 25 pictures/sec). Although most such gestures are culturally determined often an innate basis is detectable. It is possible that our head shaking is phylogenetically related to rejection shaking movements in other mammals.

négation □ Comportement de refus dans l'espèce humaine. En Grèce, en Turquie et dans de nombreux pays arabes du Proche Orient, on nie en hochant la tête vers le haut, les yeux bien fermés, et en effectuant en même temps un mouvement en arrière. Ce geste est déjà connu des Grecs de l'antiquité classique (»ANANOUEIN«). Les Waikas effectuent aussi le même mouvement pour éviter un contact visuel ou en général pour rompre un contact (Fig. 126 A – Indien Waika de l'Orinoque; B – Grec). Dans le cas d'un refus plus insistant, les deux mains sont souvent soulevées et les paumes dirigées vers le partenaire, parfois en signe d'indignation. Il est probable que cette forme tout à fait spéciale du refus se soit développée à partir d'un geste humain plus général signifiant le rejet social. Les peuples d'Europe Occidentale, avec de nombreux autres, ont ritualisé en négation le secouement latéral de la tête (Fig. 126 C – Pygmée Bayaka de l'Afrique Centrale; de haut en bas image 14, 34, 43, 57 d'un film 16 mm, 25 images/sec). La plupart de nos gestes se sont formés par voie de tradition; pourtant on peut souvent y déceler une base innée. Il est vraisemblable que le »non« indiqué en secouant la tête latéralement se soit phylogénétiquement développé à partir d'un secouement de la tête existant chez de nombreux Mammifères en signe de refus.

Verrechnungsmechanismus □ Zentraler zerebraler Mechanismus, über den man bisher wenig präzise Daten besitzt. Hierzu zwei Beispiele: Würden die → *Kundschafterinnen* bei Bienen bei ihrer → *Richtungsweisung* stets den zu Beginn des → *Schwänzeltanz(es)* angezeigten Winkel zur Sonne beibehalten, dann würde das die Nestgenossen in die Irre führen, da ja die Sonne scheinbar wandert und sich der Winkel damit in gesetzmäßiger Weise ändert. Versuche von VON FRISCH (1965) haben nun aber gezeigt, daß die Bienen diese scheinbare Sonnenwanderung mit Hilfe eines noch nicht näher bekannten »Verrechnungs«-Mechanismus kompensieren und immer den derzeit tatsächlichen Sonnenwinkel anzeigen. Ist die Futterquelle nur über einen Umweg zu erreichen, so tanzen die Bienen die Luftlinie als Richtung, geben aber in der Entfernung die Zeit bzw. die Länge des Umweges an. Auch hier wird »verrechnet«. → *Dauertanz.*

compensatory mechanism □ The existence of a central (cerebral) mechanism has been postulated which can compensate for interim changes in the external environment in directing behaviour responses. For example, if → *scouting bees* continued to give the same information concerning the sun's azimuth (→ *indication of direction*) throughout their dance their nestmates would be misdirected because the relative position of the sun changes in a defined manner. Experiments of VON FRISCH (1965) show that the bee compensates for this change by some, as yet undefined, mechanism. In all cases the bee transmits information utilizing the correct position of the sun. In another example, if the food source can be reached only by a detour, then the bee's dance portrays the directional component as a straight line but the distance component as the time or length of the necessary detour. → *persistent dance.*

mécanisme de compensation □ On suppose l'existence de mécanismes nerveux centraux de compensation sur lesquels nous avons cependant jusqu'à présent peu de données précises. Voici deux exemples: Si les → *abeilles exploratrices* conservaient, dans leurs → *danse(s) frétillante(s)* prolongées, un angle invariable par rapport au soleil, → *l'indication de direction* résultante induirait en erreur les autres ouvrières de la ruche, étant donné que la terre tourne et que l'angle que fait la direction du but avec celle du soleil se modifie en conséquence. Les expériences de VON FRISCH (1965) ont démontré que les Abeilles tiennent compte du déplacement apparent du soleil à l'aide d'un mécanisme de compensation encore mal connu et qu'elles indiquent toujours l'angle effectif par rapport au soleil. Si la source de nourriture n'est accessible que par un détour, les Abeilles dansent alors en indiquant la direction en ligne droite à partir de la ruche, mais signalent pour la distance, la longueur du détour ou le temps nécessaire pour parvenir à la source de nourriture. Il y a donc, là aussi, un mécanisme de compensation. → *danse persistante.*

Verschleppung → Phoresieverhalten
E. carrying off □ F. trainage

Verständigung □ Signalübertragung und Informationsaustausch als Voraussetzung eines sozialen Verhaltens. Verständigungs- oder Kommunikationsmittel sind taktile, chemische, optische, akustische und elektrische Reize. Zu den taktilen Mitteln der Verständigung gehören die Handlungen der sozialen Körperpflege, zu den chemischen die → *Pheromone* und Sekrete aus Duftdrüsen, bei Säugern auch Urin und Kot, wie z. B. bei der → *Reviermarkierung.* Optische Kommunikationsmittel sind bestimmte Bewegungen einschließlich der Mimik, aber auch bestimmte Formen und Farben. Akustische Mittel sind Lautäußerungen und instrumentale Lauterzeugungen (→ *Instrumentalverständigung*). Elektrische Verständigung schließlich kennen wir von schwach elektrischen Fischen.

Generell lassen sich zwei Gruppen unterscheiden:

1. Signale, die mit summativer Wirkung einen → *AAM* ansprechen und die Art oder das Geschlecht betreffen.

2. Die gestaltenden Reizmuster, die der Individualkennzeichnung dienen und somit über angeborene gestaltbildende Mechanismen auf obligatorische Lernvorgänge bezogen sind.

Am häufigsten erfolgt intraspezifischer Informationsaustausch. Wir kennen aber auch Verständigung zwischen verschiedenen Arten, wie z. B. im Bereich der → *Symbiose* oder zwischen Honiganzeigern (Indicatoridae) und dem Honigdachs (*Mellivora*); dem Honiganzeiger folgen auch Pygmäen zum Auffinden des Honigs. Die Warnrufe vieler Vögel gehören ebenfalls zur interspezifischen Kommunikation, da sie von vielen anderen Arten richtig verstanden werden. → *Warnverhalten.*

Die Sprache des Menschen ist ebenfalls ein Verständigungsmittel, jedoch gleichzeitig durch die vielen verschiedenen Sprachen eine kulturell bedingte

Kommunikationsbarriere erster Ordnung.

communication, understanding □ Signal transmission and information exchange as a prerequisite of social behaviour. Modes of understanding or communication include tactile, chemical, optical, acoustical or electrical stimuli. Among tactile stimuli we include such epimeletic stimuli as grooming; among the chemical stimuli belong → *pheromones* and scent signals, e. g., urine, feces in the → *marking of territory*. Optical signals include certain movement patterns as well as mimics and certain forms or colours. Acoustic stimuli are vocalizations and instrumentally produced sounds (→ *mechanical communication*). Electrical communication is known from the weakly electric fishes.

In general we distinguish two groups:

1. Signals that address an innate releasing mechanism (→ *IRM*) with additive properties; these deal with the species or the sex.
2. Patterns of stimuli which serve individual recognition and which must be relayed via inherited pattern receptor channels to learning mechanisms in higher centers.

Most frequent exchange of information is intraspecific; however, interspecific communication is also common. For example, in → *mutualism* between the honey guide (Indicatoridae) and the honey badger (*Mellivora*). Even pygmies may follow the honey guide to a honey source. The alarm calls of many bird species are widely understood. → *warning behaviour*.

Language is also a mode of communication, but due to the variety of tongues it may also constitute a *cultural barrier* of the first order.

communication □ Transmission de signaux et échange d'informations comme condition préalable au comportement social. Les moyens de communication sont des stimuli tactiles, chimiques, optiques, acoustiques et électriques. Parmi les modalités tactiles de communication, on peut citer les soins corporels mutuels; parmi les moyens chimiques, il faut mentionner les → *phéromones* et les sécrétions des glandes odorantes, sans oublier chez les Mammifères l'urine et les matières fécales lors du → *marquage du territoire*. Les moyens de communication optiques sont représentés par certains types de mouvement, y compris la mimique faciale, mais aussi par certaines formes et colorations corporelles. Les moyens acoustiques sont les vocalisations ou les émissions sonores instrumentales (→ *communication instrumentale*). La communication électrique est connue chez les Poissons faiblement électriques.

D'une manière générale, on peut distinguer deux groupes de moyens de communication:

1. Signaux dont les effets sommatifs élicitent un mécanisme déclencheur inné (→ *MDI*) et concernent la reconnaissance du congénère ou du partenaire sexuel.
2. La configuration des signaux servant à la reconnaissance individuelle et qui, à l'aide de mécanismes innés de structuration perceptive, sont dépendants du processus d'apprentissage obligatoire.

La plupart des échanges d'information sont intraspécifiques, mais nous connaissons également des communications entre espèces différentes, comme dans le domaine de la → *symbiose* ou comme par exemple entre les Oiseaux Indicatoridae et un Mammifère, le Ratel *(Mellivora)*. Les Pygmées suivent également les Indicatoridae à la recherche du miel. Les cris d'alarme de nombreux Oiseaux font également partie de la communication interspécifique, étant donné qu'ils sont correctement interprétés par de nombreuses autres espèces. → *comportement d'avertissement*.

Le langage humain est également un moyen de communication, mais en même temps, compte tenu de la diversité extraordinaire des différentes langues, une *barrière culturelle* de premier ordre à la communication.

Versuch und Irrtum □ Ein Mechanismus des »Lernens am Erfolg« durch Erfahrung positiver oder negativer Folgen einer Handlung. Es handelt sich um eine Selbstdressur durch Eigentätigkeit. Vögel meiden z. B. die stechenden und giftigen, schlecht schmeckenden Vespidae erst, wenn sie mit ihnen schlechte Erfahrungen gemacht haben. Das Gegenteil hierzu sind Verhaltensweisen der spontanen → *Einsicht*.

trial and error □ A mode of learning by experience. For example, birds avoid prickly, poisonous or bad-tasting vespid species only after an unpleasant initial encounter with them. → *Insight* is conceptually opposite to »trial and error« learning.

essai et erreur □ Un mécanisme d'apprentissage par réussite dépendant de l'expérience de l'essai et de l'erreur. Il s'agit d'un auto-dressage basé sur l'activité propre du sujet. Certains Oiseaux, par exemple, n'évitent les Insectes piquants, venimeux et de mauvais goût tels que les Vespidae qu'après avoir fait de mauvaises expériences. L'apprentissage par essai et erreur s'oppose à la → *compréhension soudaine*.

Versuchs-Erfolgs-Verhalten □ Selbstdressur bedingt durch den Erfolg, z. B. bei der Rabenkrähe, die durch den Erfolg ein bestimmtes → *Suchbild* der Nahrung, d. h. des Freßbaren entwickelt hat, das im folgenden beibehalten wird.

trial and success behaviour □ Self-training in which a successful act is repeated in similar contexts. For example, the carrion crow develops a → *searching image* of its food by early success with that type of food.

comportement par essai et réussite □ Autodressage conditionné par la réussite, comme par exemple chez la Corneille qui a développé une certaine → *image d'appétence* ou de recherche concernant les aliments comestibles, image qui est retenue et maintenue par la suite.

Verteidigungsgemeinschaft □ Bei vielen Tierarten finden wir ein soziales Zusammenscharen zur gemeinsamen Verteidigung gegen den Feind. So greifen Dohlen einen Hund an, der eine Dohle im Maul trägt. Vom Moschusochsen ist ebenfalls gemeinsame Verteidigung bekannt.

mutual defense group □ Various animal species defend themselves as a group against predators. Jackdaws will attack in mass a dog carrying another jackdaw in its mouth. Mutual defense is also seen in musk oxen.

association de défense mutuelle □ Coopération sociale pour la défense en commun contre un ennemi ou un prédateur potentiel existant chez diverses espèces animales. Les Choucas, par exemple, attaquent en commun un Chien si ce dernier porte un de leurs congénères dans sa gueule. Nous connaissons également un alignement de défense en commun chez le Bœuf musqué.

Vertrauensbeweis □ Im menschlichen Verhalten, vor allem beim Grüßen bekundetes Vertrauen, indem man sich »wehrlos« zeigt, um keine Aggressionen auszulösen. Germanen z. B. legten bei Audienzen außerhalb des Burgtores die Waffen ab; grüßende Massais stoßen den Speer vor sich in den Boden. Diese Vertrauensbekundungen sind kulturell abgewandelt.

demonstration of trust □ In human behaviour demonstration of trust occurs primarily during the greeting gesture as a display of defenselessness. Examples: in the presence of their hosts Teutons would leave their weapons outside the castle gates; Massai greet each other by throwing their spears into the ground in front ot them. Demonstration of trust differs between cultures.

témoignage de confiance □ Dans les salutations, chez l'Homme, manifestation tendant à mettre en évidence l'absence de moyens de défense susceptibles de déclencher une agression. Ainsi, les anciens Germains déposaient leurs armes hors de l'enceinte du château-fort avant d'être reçus par leur hôte. Actuellement, quand les Massaï se saluent, ils piquent leur sagaie devant eux dans le sol. Ces manifestations sont modifiés, au gré des multiples courants du développement culturel.

Verwandtenfresserei □ Das Sich-Einander-Auffressen innerhalb einer Familie. Dieser Überbegriff umfaßt auch die beiden Möglichkeiten, wie Jungtiere innerhalb der Familie zu Beute werden können. Fressen z. B. Nestgeschwister einander auf, wie das bei Eulen und Greifvögeln vorkommt, spricht man von *Kainismus* oder *Adelphophagie*. Es wird meist das kleinere Nestgeschwister Opfer der größeren. Werden die Jungen aber von den Eltern aufgefressen, so spricht man von *Kronismus*. Bekannt ist dieses Verhalten vom Weißstorch, vor allem aber von Säugern; es wird wohl durch dichteabhängige Streß-Erscheinungen ausgelöst (VON HOLST 1969 – Versuche an *Tupaia*). DEEGENER (1918) faßte beide oben genannte Begriffe unter dem Ausdruck *Syngenophagie* zusammen.

syngenophagy □ Eating one another within a family. This super-category includes the cases whereby young of one family become prey. Thus, when owls and other raptors eat their siblings one speaks of *cainism* or *adelphophagy*. Usually, the larger eat the smaller. When parents eat their young, then it is called *Kronismus* in German. This behaviour is known in the white stork, and especially in some mammals. It is probably triggered by density-dependent stress-conditions (VON HOLST 1969 – experiments on *Tupaia*). DEEGENER (1918) combined these two terms in one: *syngenophagy*.

syngénophagie □ Le fait de s'entre-dévorer au sein d'une même famille. Cette notion englobe les deux possibilités de prédation des jeunes au sein d'une famille. Si les frères et sœurs occupant un même nid se dévorent entre eux, comme chez les Chouettes et les Rapaces, on parle de *caïnisme* ou *adélphophagie*. Dans les cas les plus fréquents, c'est le plus jeune qui devient la proie de plus âgés. Par contre, si les jeunes deviennent la proie de leurs propres parents, on parle de *cronisme*. Ce dernier comportement est connu chez les Cigognes et surtout chez de nombreux Mammifères; il est probablement déclenché par des phénomènes de stress dépendants d'une densité de population excessive (VON HOLST 1969 – recherches sur *Tupaia*). DEEGENER (1918) avait proposé le terme de syngénophagie pour désigner les deux phénomènes décrits ci-dessus.

Vielfachwahl □ Eine von HUNTER (1913) entwickelte und bis heute gebräuchliche Methode zur Erforschung der Merkfähigkeit und der Wahrnehmung. Ein Tier z. B. wird in einem Vielfachwahlapparat darauf dressiert, eine von mehreren Türen zu wählen, die durch eine bestimmte Markierung (z. B. ein aufleuchtendes Lämpchen) gekennzeichnet ist. Beherrscht es diese Aufgabe, dann hindert man das Tier daran zu handeln, solange das positive Dressursignal sichtbar ist. Erst eine Weile danach darf es wählen, und die Höchstzeit eines solchen Handlungsaufschubs gilt als Maß für die Erinnerungsfähigkeit. Die Zuverlässigkeit dieser Methode wird von MAIER und SCHNEIRLA (1935) und TINBERGEN (1951) bezweifelt.

multiple choice □ A method developed by HUNTER (1913) and still in use for studying memory and perception. An animal is trained in a choice chamber to choose, e. g., one of several doors marked with some signal, e. g., with a light bulb. After it has mastered this task, it is prevented from exercising its choice as long as the signal is on. After a time lapse the animal again is allowed to choose, and the longest period which elapses with a subsequently correct response serves as the measure of performance. The reliability of this method has been questioned by MAIER and SCHNEIRLA (1935) and TINBERGEN (1951).

choix multiple □ Méthode de recherche pour déceler la performance de mémoire et de perception, développée par HUNTER (1913). Un animal, dans un appareil ou un dispositif à choix multiple, est conditionné à choisir parmi plusieurs portes celle qui présente un caractère particulier (par exemple allumage d'une petite lampe). A partir du moment où l'animal maîtrise le problème, on l'empêche d'agir tant que le signal positif de dressage est visible. Ce n'est qu'après un certain temps que l'animal est à nouveau autorisé à choisir et le temps maximum d'une telle réaction différée est considéré comme la mesure de la capacité de mémoire. MAIER et SCHNEIRLA (1935) et TINBERGEN (1951) contestent la validité de cette méthode.

Vitaminkot → Coecotrophie
E. caecum contents □ F. matière caecotrophe

Volksduft → Gruppenduft

Volksteilung □ Die Vermehrung von Insektenstaaten durch Abwanderung einer oder mehrerer fortpflanzungsfähiger Formen unter Begleitung einer Arbeiterinnengruppe aus dem elterlichen Nest, wodurch untereinander verwandte Einheiten entstehen, die zum »Fortbestand« oder zur Perpetuation der Elternkolonie beitragen. Diese Art der Staatenvermehrung wird in der Ameisenliteratur als *Hesmose* und in der Termitenliteratur als *Soziotomie* bezeichnet. Das *Schwärmen* der Honigbienen kann als eine Sonderform von Volksteilung angesehen werden (WILSON 1971, 1975).

colony fission □ The multiplication of colonies in social insects by the departure of one or more reproductive forms, accompanied by a group of workers from the parental nest, leaving behind comparable units to perpetuate the »parental« colony. This mode of colony multiplication is referred to occasionally as *hesmosis* in ant literature and *sociotomy* in termite literature. *Swarming* in honey bees can be regarded as a special form of colony fission (WILSON 1971, 1975).

fission de colonie □ La multiplication des colonies chez les Insectes sociaux par le départ d'une ou de plusieurs formes sexuées, accompagnées d'un groupe d'ouvrières, du nid parental. De telles unités semblables entre elles garantissent la perpétuation de la colonie parentale. Ce mode de multiplication des colonies est appelé *hesmose* pour les Fourmis et *sociotomie* pour les Termites. *L'essaimage* des Abeilles mellifères peut être considéré comme une forme spéciale de la fission de colonie (WILSON 1971, 1975).

Vorangepaßtheit = stammesgeschichtliche Anpassung → Vorprogrammierung
E. phylogenetic adaptation □ F. adaptation phylogénétique

Vornherumkratzen □ Kratzweise von Vögeln, bei der der Flügel am Körper angelegt bleibt und das Bein zum Kratzen direkt nach vorn zum Kopf angehoben wird, im Gegensatz zum → *Hintenherumkratzen.*

direct scratching under the wing □ When birds scratch their heads in this fashion, they bring the foot up without moving the wing; in contrast to → *scratching over the wing.*

grattage direct de la tête □ Mouvement de grattage de la tête chez les Oiseaux pendant lequel l'aile reste accolée au corps, en position normale, et la patte est directement orientée vers l'avant, donc vers la tête. → *grattage par-dessus l'aile.*

Vorprogrammierung □ Die stammesgeschichtliche Vorangepaßtheit von Verhaltensweisen. Während man bei Tieren im allgemeinen von angeboren spricht, hat sich in der → *Humanethologie* der Ausdruck Vorprogrammierung durchgesetzt. Wie weit die Vorprogrammierung im menschlichen Sozialverhalten im einzelnen geht, wissen wir heute noch nicht. Vieles weist darauf hin, daß Verhaltensweisen wie Rangstreben, Bereitschaft zur Unterordnung und zum Gehorsam, Intoleranz gegen Außenseiter, Aggression, aber auch unsere altruistischen Neigungen und der Drang, ein freundliches Band zu stiften, im umfassenden Sinne durch stammesgeschichtliche Anpassungen vorgezeichnet (vorprogrammiert) sind (EIBL-EIBESFELDT 1973). Dies steht im Gegensatz zur → *Milieutheorie,* die davon ausgeht, daß der Mensch als *Tabula rasa* (unbeschriebenes Blatt) geboren würde und nach Belieben manipuliert werden könne.

pre-programmed behaviour □ Phylogenetically pre-programmed behaviour in humans; in animals one generally speaks of innate behaviour. How extensive pre-programmed behaviour is in human social behaviour is presently unknown. Considerable evidence suggests that dominance, subordinance, intolerance of outsiders, aggression as well as altruistic tendencies and the need to form amicable groups may have evolved as a phylogenetic adaptation (EIBL-EIBESFELDT 1973). This contrasts with the extreme position of → *learning theory* that man is born as a *tabula rasa* (unwritten page) and hence can be manipulated at will.

programmation □ Préadaptation phylogénétique du comportement. Alors que nous parlons en éthologie animale d'une manière générale de comportements innés, en → *éthologie humaine* l'expression programmation est plus couramment utilisée. A l'heure actuelle, nous ne savons pas encore dans quelle mesure la programmation joue un rôle dans le comportement social chez l'Homme. Tout laisse à penser cependant que les comportements manifestant l'ambition, la subordination et l'obéissance, l'intolérance envers les rivaux, les pulsions agressives ainsi que nos tendances altruistes et la disposition à nouer des liens amicaux sont, d'une manière globale, ébauchés (programmés) par des adaptations phylogénétiques (EIBL-EIBESFELDT 1973). Ceci est évidemment en opposition avec la → *théorie du milieu* qui, elle, part de l'idée que le nouveau-né humain est une *tabula rasa* et peut être manipulé à volonté.

W

Wachesitzen □ Bei vielen sozialen Affenarten hat man beobachtet, daß einige ♂♂ an der Peripherie ihrer Gruppe »Wache« sitzen. Dieses Verhalten ist gegen Nachbargruppen der eigenen Art gerichtet. Die Schildwachen sitzen immer mit dem Rücken zur eigenen Gruppe und stellen dabei ihre Geschlechtsorgane zur Schau, die bei diesen Tieren besonders bunt gefärbt sind (WICKLER 1966). → *Genitalpräsentieren.*

sentinel behaviour □ In many social apes several ♂♂ may sit on the periphery as if to stand watch. The sentinels usually sit with their back turned to their own group and direct their attention to neighbouring conspecific groups. In this sitting position their brightly coloured sex organs are exposed (WICKLER 1966). → *genital presentation.*

comportement de sentinelle □ Chez de nombreuses espèces de Primates sociaux, on observe plusieurs animaux ♂♂ en position de sentinelle à la périphérie du groupe. Ce comportement est dirigé vers les groupes voisins appartenant à la même espèce. Les sentinelles sont toujours assises, le dos tourné vers leur propre groupe, et exposent vers l'extérieur leurs organes génitaux, souvent vivement colorés chez ces animaux (WICKLER 1966). → *présentation des organes génitaux.*

Wahlhandlung → Wahlversuch

E. discrimination operation □ F. opération de choix

Wahlvermögen → Wahlversuch

E. discrimination capacity □ F. capacité de discrimination

Wahlversuch □ Eine Versuchsanordnung zur Prüfung der Leistungs- und Lernfähigkeit von Tieren bei der Unterscheidung von bestimmten Farben, Formen, Mustern und Zahlen (Abb. 127 – Kleine Zibetkatze, *Viverricula malaccensis,* bei der Wahl zwischen ungleichen und gleichen Streifen). Die vom Versuchsleiter gewünschte richtige Wahl wird belohnt, um das zu prüfende Individuum darauf zu dressieren. Ein Fehler wird natürlich nicht belohnt, aber oft durch einen Strafreiz negativ verstärkt. Das Wahlvermögen wird in späteren Versuchen genau geprüft, indem man nicht nur eine Zweifachwahl (Präferenzversuch, Simultanmethode) durchführt, sondern in einer → *Vielfachwahl* den Schwierigkeitsgrad für das zu wählende »Objekt« erhöht. Bereits Libellenlarven der Gattung *Aeschna* lernten zwischen Gelb-genießbar und Violett-ungenießbar zu unterscheiden (KOEHLER 1924). Besonders ausführlich abgehandelt findet sich die komplexe Problematik des Wahlvermögens bei Tieren in BUCHHOLTZ (1973) und RENSCH (1973); beide Abhandlungen mit umfangreicher weiterführender Literatur.

discrimination test □ A type of test for determi-

127

ning the learning capacity of animals. Certain colours, forms, patterns or numbers of objects are used and a choice of the correct one as designated by the experimenter is rewarded (Fig. 127 – *Viverricula malaccensis* choosing between unequal and equal stripes). A wrong choice, rather than being rewarded, often results in a punishment (negative reinforcement). The animal's performance is subsequently tested in a choice situation (preference test) where a variety of objects is presented hence increasing the difficulty of the discrimination test *(→ multiple choice).* Even dragonfly larvae of the genus *Aeschna* learn to distinguish a yellow pleasant stimulus from an unpleasant violet one. The complex problems of discrimination capacity in animals are extensively treated by BUCHHOLTZ (1973) and RENSCH (1973); both books also contain more detailed references.

épreuve de discrimination □ Technique expérimentale permettant d'examiner la capacité des animaux à reconnaître spontanément ou après apprentissage certains nombres, couleurs, formes et structures. Le choix considéré comme »correct« par l'expérimentateur est récompensé ce qui permet d'en augmenter la probabilité (Fig. 127 – *Viverricula malaccensis* choisissant entre des rayures inégales et égales). Les erreurs ne sont évidemment pas récompensées, mais font parfois l'objet d'un renforcement négatif par un stimulus nociceptif. La capacité de choix est minutieusement testée dans des expériences ultérieures en soumettant l'animal non seulement à une épreuve de choix entre deux possibilités, mais en augmentant le degré de difficulté pour l'objet à choisir. On présentera alors un dispositif de → *choix multiple.* Même des larves de Libellules du genre *Aeschna* apprennent à faire la discrimination entre le jaune-consommable et le violet-inconsommable (KOEHLER 1924). Les problèmes complexes de la capacité de discrimination chez les animaux sont particulièrement bien traités dans BUCHHOLTZ (1973) et RENSCH (1973); ces deux livres contiennent également une bibliographie exhaustive et détaillée.

Warntracht → Schutzanpassung

E. warning colouration □ F. coloration de mise en garde, coloration d'avertissement

Warnverhalten □ Warnverhalten ist ein Überbegriff für zwei unterschiedliche Verhaltensweisen: es umfaßt einmal das *den Feind Warnen* sowie *vor dem Feind Warnen.* Letzteres kann auch als Alarmverhalten bezeichnet werden, auf welches Artgenossen wie andere Arten gleichermaßen ansprechen können; meist handelt es sich um interspezifisch verständliche Warnlaute (MARLER 1956; Abb. 128 – Warnlaute 5 verschiedener Singvogelarten). Das den *Feind Warnen* ist eine aktive Schutzanpassung, die ein Abschrecken des Angreifers bewirken soll, wie z. B. das plötzliche Vorzeigen der → *Augenflecke* bei vielen Schmetterlingen. Droh- und Schrecktrachten oder das plötzliche Annehmen einer Warnfärbung sind ebenfalls eine aktive Schutzanpassung. Man spricht oft auch von *aposematischem Verhalten.* Warntrachten treten auch bei ungenießbaren oder übelschmeckenden Tieren auf und werden von genießbaren nachgeahmt. → *Mimikry.*

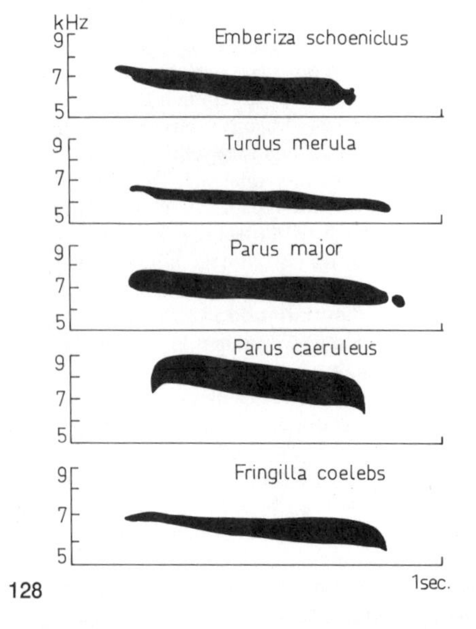

128

warning behaviour □ Warning behaviour is a general term for two different behaviour patterns: it includes *warning the predator* and *warning against the predator.* The latter one may also be termed alarm behaviour, and conspecifics as well as other species may equally react to it; a typical example are interspecifically understandable alarm calls (MARLER 1956; Fig. 128 – warning calls of five different Passerine species). Warning the predator is an active adaptation for defense which is supposed to frighten the predator, e. g., the sudden revealing of → *eye spots* in many lepidopterans. Threat colouration or a sudden shift of colouration may also represent an active defensive adaptation. Such adaptations are sometimes called *aposematic behaviour.* Display of threat colouration also occurs in unpalatable or bad-tasting animals and is often then mimicked by less repulsive species. → *mimicry.*

comportement d'avertissement □ Le comportement d'avertissement constitue une notion générale pour deux comportements différents: cette notion inclut à la fois »l'avertissement adressé aux ennemis ou prédateurs potentiels de l'espèce *(mise en garde)*)« et »l'avertissement destiné aux congénères« et éventuellement aux autres espèces risquant le même danger *(manoeuvre ou comportement d'alarme);* généralement, il s'agit de cris d'alarme dont la compréhension est interspécifique au sens le plus large (MARLER 1956; Fig. 128 – cris d'avertissement chez cinq différentes espèces de Passereaux). L'avertissement envers le prédateur potentiel est une forme de protection active, entraînant une réaction d'effroi chez le prédateur (exposition brusque des → *taches ocellaires* des ailes chez les Papillons en danger). Les colorations phobiques et l'apparition spontanée de colorations d'avertissement sont également une adaptation active de protection, phénomène qu'on appelle aussi *comportement aposématique.* Ces colorations d'avertissement apparaissent aussi chez les animaux incomestibles ou de mauvais goût et sont souvent imitées par des animaux comestibles. → *mimétisme.*

Wasserspucken → Beutespucken

Weben □ Ein offenbar vom → *Bodenforkeln* abgeleitetes und ritualisiertes Verhalten bei Gazellen und Antilopen. Der Bock stößt nur noch zu Beginn mit den Hörnern ins Gras oder in die Stauden. Danach bewegt er die gesenkten Hörner parallel zum Erdboden oft lange und anhaltend hin und her (Abb. 129 – *Antilope cervicapra* ♂). Nach WALTHER (1968) handelt es sich um eine dynamisch-optische Markierung (besser: Signalisierung) der augenblicklichen Position, die auch von nicht-territorialen ♂♂ ausgeführt wird.

weaving □ A ritualized behaviour in gazelles and antilopes apparently derived from → *ground rutting.* The buck thrusts his horns into the grass or bushes and then holding them parallel to the ground moves them side to side (Fig. 129 – *Antilope cervicapra* ♂).

129

According to WALTHER (1968) this constitutes a dynamic optic marking (or better signalling) of the animal's position. Weaving may also occur in non-territorial ♂♂.

tissage □ Comportement ritualisé chez les Gazelles et les Antilopes, probablement dérivé du → *râclement du sol.* Mais c'est seulement au début de cette activité que le mâle enfonce légèrement ses cornes dans l'herbe ou les buissons; par la suite, il effectue, souvent avec persistance, des mouvements latéraux de va-et-vient avec ses cornes abaissées et maintenues parallèles au sol (Fig. 129 – *Antilope cervicapra* ♂). Selon WALTHER (1968), il s'agit d'un marquage ou, mieux, d'une signalisation visuelle dynamique de la position instantanée de l'animal. Cette manœuvre est également exécutée par des ♂♂ non territoriaux.

Wechselgesang → Duettgesang
E. antiphonal duet □ F. duo antiphonal

Wechselgespräch □ Beim Menschen gibt es Gespräche als bandstiftende Rituale, bei welchen kaum sachliche Informationen ausgetauscht werden. Wohl aber enthält ein solches Gespräch die soziale Information, daß man am Partner interessiert und bereit ist, ihm zuzuhören und zu antworten. MORRIS (1968) nannte diese Gespräche *grooming talks* (Putzgespräche), da sie, wie beim gegenseitigen Lausen, eine Funktion der freundlichen Kontaktherstellung haben (JOLLY 1972).

grooming talk, small talk □ In humans certain sorts of conversations in which substantive exchange of information does not occur may serve group-bonding functions. Such conversation does contain social information – that the partner is interested in a person and ready to listen and answer. MORRIS (1968) called such conversations »grooming talks« since they function to maintain friendly contact just as »lousing« does (JOLLY 1972).

dialogue intime □ Chez l'Homme, forme de communication verbale non-linguistique: c'est un rituel de maintien du contact social au cours duquel il n'y a guère d'échange d'informations objectives. Par contre, un tel dialogue contient des informations sociales et signifie qu'on s'intéresse au partenaire et qu'on est prêt à l'écouter et à lui répondre. MORRIS (1968) a appelé ce genre de conversations *grooming talks* parce qu'elles permettent l'établissement de contacts amicaux, comme lors de l'épouillage réciproque (JOLLY 1972).

Wechselterritorien → Revier
E. rotating territories □ F. territoires rotatives

Wegsehen → Hinterkopfzudrehen
E. facing away, looking away □ F. regarder ailleurs, détourner la tête

Weibchenbewachung □ Eine epigame Verhaltensweise bei Libellen, die nach der Begattung in verschiedenen Formen vorkommt. Bei den meisten ursprünglichen Zygoptera und einigen Anisoptera werden nach der Radstellung nur die Kopulationsorgane getrennt, und das ♂ bleibt auch während der

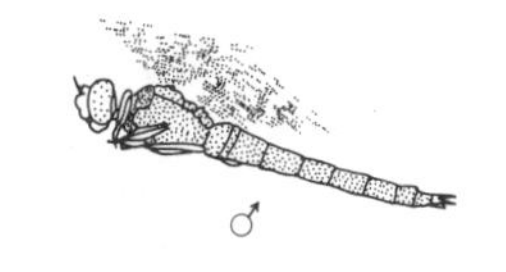

130 A

130 B

Eiablage mit seinen Abdominalzangen am Prothorax des ♀ verankert. Bei höher entwickelten Familien, wie z. B. den Libellulidae und den Calopterygidae, trennen sich die Partner völlig voneinander und das ♀ wird zur Sicherung der störungsfreien Eiablage von seinem ♂ bewacht (Abb. 130 A – *Orthetrum brunneum,* Libellulidae; B – *Calopteryx virgo,* Calopterygidae). Das ♂ achtet darauf, daß das ♀ nicht vor Beendigung der Eiablage aus dem Revier fliegt; gleichzeitig wehrt das ♂ Angreifer ab und übt so auch eine schützende Funktion aus (HEYMER 1966, 1969, 1973, 1974).

guarding of the female □ An epigamic behaviour pattern in dragonflies which occurs in various forms following mating. In the most primitive Zygoptera and certain Anisoptera, following copulation, only the copulatory organs are separated, and the ♂ remains fastened to the female's prothorax during egg-laying. In less primitive forms, e. g., Libellulidae and Calopterygidae, the partners completely separate from each other, and the ♀ is guarded during egg-laying by the ♂ (Fig. 130 A – *Orthetrum brunneum,* Libellulidae; B – *Calopteryx virgo,* Calopterygidae). The ♂ assures that the ♀ does not fly out of his territory before completing egg-laying. At the same time he wards off conspecific intruders, thus serving a protective function (HEYMER 1966, 1969, 1973, 1974).

surveillance de la femelle □ Comportement épigame chez certaines Libellules apparaissant sous des formes différentes après l'accouplement. Chez la plupart des Zygoptères primitifs et chez quelques Anisoptères, seuls les organes copulateurs se séparent après l'accouplement, mais les pinces abdominales du ♂ restent insérées dans les creux prothoraciques de la ♀. C'est ainsi qu'il accompagne la ♀ sur le lieu de ponte. Chez les familles plus évoluées, comme par exemple les Libellulidae et les Calopterygidae, les partenaires, après avoir terminé l'accouplement, se séparent complètement et la ♀ est surveillée pendant la ponte par son ♂ (Fig. 130 A – *Orthetrum brunneum,* Libellulidae; B – *Calopteryx virgo,* Calopterygidae). Le ♂ veille à ce que la ♀ ne quitte pas le lieu de ponte ou le territoire avant la fin de la ponte; en même temps, il chasse les congénères intrus et exerce ainsi également une fonction protectrice (HEYMER 1966, 1969, 1973, 1974).

Weibchenrudel = Brunftharem → Herden

Weibchenschema □ Merkmale und Eigenschaften, die in ihrer summativen Reizwirkung das Werbeverhalten des ♂ auslösen. VOGEL (1954, 1957, 1958) untersuchte das Weibchenschema bei der Fliege *Sarcophaga carnaria* und konnte die Größenordnungen bestimmter Auslösereize feststellen. Eine schwarze Kugel an einer Angel löst Anflüge des ♂ aus, wenn sie in 36 cm/sec Geschwindigkeit im Abstand von 30 cm vom ♂ bewegt wird. Mit zunehmendem Durchmesser steigt der Auslösewert und erreicht sein Maximum bei zwei- bis dreifacher Größe. Siebenfache Größe kann Fluchtverhalten auslösen. Bei manchen Libellen (Calpterygidae) löst ein einzelner ♀-Flügel an der Angel bereits Werbeverhalten aus (BUCHHOLTZ 1951, HEYMER 1973).

female schema, female pattern □ Characteristics and properties which taken together elicit the courtship behaviour of the male. VOGEL (1954, 1957, 1958) studied the female pattern in the fly, *Sarcophaga carnaria* and distinguished certain releasing stimuli. A black ball attached to a line attracts the ♂ when it moves at 36 cm/sec at a distance of 30 cm from the ♂. With increasing diameter the attractivity increases reaching a maximum at 2–3 x the original size. An increase of 7 x the diameter elicits withdrawal. In many dragonflies (Calopterygidae) a single wing of a ♀ attached to a line will elicit courtship (BUCHHOLTZ 1951, HEYMER 1973).

schème de la femelle □ Caractères perceptifs déclenchant, selon un effet stimulateur sommatif, la parade nuptiale chez le ♂. VOGEL (1954, 1957, 1958) a effectué des recherches sur le schème de la ♀ chez la Mouche *Sarcophaga carnaria* et a pu préciser l'ordre de grandeur des stimuli déclencheurs décisifs. Par exemple, une boule noire suspendue au bout d'un fil à pêche déclenche la parade nuptiale chez le ♂ à condition que cette boule soit animée d'une vitesse de 36 cm/sec et située à 30 cm du ♂. Avec un diamètre croissant de la boule, la valeur de déclenchement augmente et atteint son maximum avec un leurre 2 à 3 fois plus grand que la Mouche elle-même. Par contre, un leurre 7 fois plus grand déclenche un comportement de fuite. Chez un certain nombre de Libellules, comme par exemple les Calopterygidae, une seule aile d'une ♀ adulte présentée au bout d'un fil peut déclencher la parade nuptiale chez le ♂ (BUCHHOLTZ 1951, HEYMER 1973).

Werbeverhalten □ Verhaltensweisen und Signale zur Anlockung des Geschlechtspartners und Abbau von Kontaktscheu. Abstimmung der Partner aufeinander, damit die Begattung und Befruchtung ermöglicht wird. Oft werben ♂♂ und ♀♀, in der Regel, aber nicht immer ist das ♂ der aktivere Werbepartner. Bei Seenadeln z. B. wirbt das ♀ um die ♂♂. HUXLEY (1923, 1938) bezeichnete Balzhandlungen während der Anpaarung, bei der beide Partner über annähernd den gleichen Verhaltensschatz verfügen, als *mutual courtship* (gemeinsame Balz). Gehen die Balzhandlungen vorwiegend von einem Partner aus, nannte er sie *unilateral display.* Bei Arten mit unterschiedlichem Balzverhalten der sich anpaarenden Individuen spricht ARONSON (1949) von *behavioural dichotomy.* → *Balz* oder *Balzverhalten* wird hauptsächlich für Vögel und Fische verwandt. Bei Säugern spricht man auch von *Brunft* bzw. → *Brunst; – Östrus* ist nur für ♀♀ gebräuchlich.

courtship behaviour □ Behaviour patterns and signals, that attract the sex partner, that reduce shyness, and that synchronize the partners with each other permiting mating and conception. Often both ♂♂ and ♀♀ court; however, more frequently the ♂ is the active partner. In the pipefish it is the ♀ that courts the ♂♂. HUXLEY (1923, 1938) termed courtship during mating in which both partners participate equally as »mutual courtship«. If one partner is the active one, he referred to this as »unilateral display«. When the sexes distinctly differ in courtship behaviour, ARONSON (1949) termed this »behavioural dichotomy«. The German terms → *Balz* and *Balzverhalten* especially refer to birds and fish. In mammals, one speaks of *Brunft* or → *Brunst;* oestrus only applies to ♀♀. → *courtship,* → *heat* and *rut.*

comportement de parade nuptiale □ Comportements et signaux assurant l'attraction du partenaire sexuel et la diminution de la crainte mutuelle. Il s'agit d'une coordination réciproque entre partenaires, permettant l'accouplement et la fécondation. Souvent, ♂♂ et ♀♀ interviennent dans la parade nuptiale; en règle générale, mais pas toujours, le ♂ est le partenaire plus actif. Chez les Aiguilles de mer, par exemple, c'est la ♀ qui courtise le ♂. HUXLEY (1923, 1938) parle de *mutual courtship* (parade nuptiale mutuelle) lorsque les deux partenaires, lors de l'appariement, possèdent approximativement le même répertoire comportemental. Si les activités de parade nuptiale sont exécutées surtout par l'un des partenaires, on parle de manœuvres unilatérales (*unilateral display*). Chez les espèces où chacun des deux partenaires dispose d'un répertoire comportemental distinct, ARONSON (1949) parle de dichotomie comportementale (*behavioural dichotomy*). Les termes allemands → *Balz* et *Balzverhalten* sont utilisés, d'une manière générale, pour les Oiseaux et les Poissons. Chez les Mammifères, on parle en allemand de *Brunft* ou de → *Brunst;* oestrus ne s'applique qu'à l'état physiologique des ♀♀. → *parade nuptiale,* → *rut.*

Werkzeuggebrauch □ Aus dem Tierreich sind verschiedene Beispiele bekannt, daß Tiere einen Gegenstand aus ihrer Umwelt als »Werkzeug« benutzen, um besser an Nahrung zu kommen. Von Cephalopoden ist bekannt, daß sie beim Öffnen von Lamellibranchiern mit einem ihrer Fangarme einen Stein zwischen die Schalen klemmen, um zu verhindern, daß die Muschel sich wieder schließt (TETRY 1948). Der Spechtfink *Cactospiza pallida* (Abb. 131 A) benutzt einen Kaktusstachel als Werkzeug, um damit Insekten aus dem Holz zu stochern (EIBL-EIBESFELDT und SIELMANN 1962, 1964). Der Seeotter holt Steine vom Meeresgrund, um damit auf dem Rücken liegend Muscheln auf seinem Bauch aufzuklopfen. Schimpansen in der freien Natur benutzen einen Halm oder ein dünnes Ästchen (Abb. 131 B) zum Termiten- und Ameisenangeln (GOODALL 1965, HLADIK 1973). Sie benutzen aber auch Stöcke zum Schlagen und Werfen (KORTLANDT 1968).
Bereits niedere Affen, wie z. B. *Cebus hypoleucos,* können im Experiment 3 Kisten übereinandertürmen, um an Nahrung zu kommen (BIERENS DE HAAN 1931). Weitere Beispiele von Werkzeugbenutzung in Gefangenschaft und im Versuch siehe vor allem RENSCH (1973), der dieses Thema ausführlich abhandelt.
Werkzeuggebrauch ist aber nicht auf Nahrungsbeschaffung beschränkt. Von einigen Grabwespen der Gattung *Ammophila* ist bekannt, daß sie einen kleinen Kiesel benutzen, um bei der definitiven Verschließung des Nestes die Erde damit festzuklopfen (PECKHAM 1904).

tool using □ Several examples are known in which animals use some object in their environment to facilitate food acquisition. Certain cephalopods use,

131 A

131 B

when opening large bivalves (Lamellibranchiata), a stone as a block to prevent the shells from closing (TETRY 1948). The woodpecker finch, *Cactospiza pallida* (Fig. 131 A), uses cactus spines as tools for piercing insects and removing them from bark (EIBL-EIBESFELDT and SIELMANN 1962, 1964). The sea otter uses stones from the tidal shoals to crack mussels open against his belly as he swims on his back. Chimpanzees in nature use a straw or small twig (Fig. 131 B) to fish for termites or ants (GOODALL 1965, HLADIK 1973). They also use sticks for beating and throwing (KORTLANDT 1968).
Even among the lower Simiens, as for example *Cebus hypoleucos,* it has been observed in experimental conditions that these animals are capable of

stocking three boxes one on top of the other in order to obtain food (BIERENS DE HAAN 1931). Other examples of tool using in captivity and in experiments are cited in RENSCH (1973) who made a thorough study of these problems.
Tool using is not limited to obtaining food: certain dipper wasps of the genus *Ammophila* use a small stone to press down the earth when definitively closing the nest (PECKHAM 1904).

usage d'instruments □ Nous connaissons plusieurs exemples où les animaux utilisent des objets provenant de leur environnement comme instruments, leur permettant de se procurer plus facilement leur nourriture. Certains Céphalopodes, pour ouvrir des Lamellibranches, utilisent un caillou comme cale entre les deux valves pour les empêcher de se refermer (TETRY 1948). Parmi les Oiseaux, le Pinson *Cactospiza pallida* des Iles Galapagos utilise une épine de Cactus comme instrument (Fig. 131 A) pour extraire des Insectes vivant dans des branches creuses (EIBL-EIBESFELDT et SIELMANN 1962, 1964). La Loutre de mer plonge au fond de la mer pour ramener des cailloux dont elle se sert pour casser des Moules sur son ventre alors qu'elle se tient allongée sur le dos. Les Chimpanzés, dans la nature, utilisent une tige ou une petite branche (Fig. 131 B) pour pêcher des Termites ou des Fourmis (GOODALL 1965, HLADIK 1973), mais ils utilisent aussi des bâtons pour taper sur un adversaire ou pour les lancer à distance (KORTLANDT 1968).
Même chez les Simiens inférieurs, comme par exemple chez *Cebus hypoleucos,* nous observons qu'ils sont capables, en expérience, de superposer trois caisses l'une sur l'autre, pour atteindre la nourriture recherché (BIERENS DE HAAN 1931). D'autres exemples d'usage d'instruments en captivité et en expérience se trouvent surtout dans RENSCH (1973) qui a traité ces problèmes d'une manière approfondie.
L'usage d'instruments ne se limite pas à l'appropriation d'aliments: Certains *Ammophiles* utilisent un petit gravier pour tasser ou damer la terre, lors de la fermeture définitive d'un nid (PECKHAM 1904).
Werkzeug et *tool = outil* sont des appellations courantes en allemand et en anglais. En français, cependant, les spécialistes de la psychologie comparée font une distinction intéressante entre la notion *d'outil,* réservée au cas de l'espèce humaine, et celle, plus générale, *d'instrument.* Selon cette distinction, un outil est un instrument privilégié, conçu pour un usage précis, transformé à cet usage, et surtout employé de façon permanente à cet usage. – L'épine de *Cactospiza* et même le bâton des Chimpanzés sont, selon ces critères, des instruments temporaires et/ou polyvalents, mais non des outils au sens strict du terme (MEDIONI *in litt.*).

»Werkzeughandlungen« □ Von LORENZ (1939) geprägter Ausdruck für im Dienste vieler Dränge stehende Verhaltensweisen. Als Beispiel sei angeführt, daß Sitzen bei Hühnern einmal vom Schlafdrang und ein andermal vom Brutdrang her aktiviert wird. Laufen und Fliegen können ebenfalls im Dienste verschiedener Dränge stehen. HEINZ (1949) sprach bei Putzhandlungen ohne sicheren Nachweis von Putzstimmung von »Gebrauchshandlungen«.

»behavioural tools« □ In 1939 LORENZ coined the term behavioural tools for identical behaviour patterns that are expressions of several drives. For example, sitting in the hen is activated from an urge to sleep as well as from broodiness. HEINZ (1949) spoke about grooming movements as »useful actions« since he had no evidence of a true »grooming drive«.

»activités instrumentales« □ Expression créée par LORENZ (1939) pour désigner certains comportements au service de pulsions multiples. Citons à titre d'exemple la position assise des Poules, activée d'une part par la pulsion de sommeil et d'autre part par la pulsion de couvaison. La marche et le vol peuvent également être activés par des pulsions différentes. HEINZ (1949) parle d'activités instrumentales dans le cas de mouvements de nettoyage sans preuve certaine d'une motivation de ce comportement.

Werkzeugherstellung □ Schimpansen können ein Ästchen entsprechend herrichten, um damit Termiten und Ameisen zu angeln. Ist eine Wasserstelle nicht mit dem Maul erreichbar, sammeln sie eine Handvoll Blätter, zerknüllen und zerkauen diese zu einem saugfähigen Schwamm, den sie ins Wasser tauchen und aussaugen. Eine solche adäquate Werkzeugherstellung ist im Tierreich wohl nur von Schimpansen bekannt, während → *Werkzeuggebrauch* weiter verbreitet ist (GOODALL 1965).

tool making □ Chimpanzees can fashion a twig for use in catching termites. If a water source is situated out of reach of their mouth, they collect a handful of leaves, chew them into a spongy consistency and dip them into the water. Then they suck the water from the leaves. Only chimpanzees among animals are known to construct tools (GOODALL 1965) although several animal species use tools. → *tool using.*

fabrication d'instruments □ Les Chimpanzés sont capables d'arranger une petite branche de façon à ce qu'elle soit bien adaptée à la pêche aux Termites ou aux Fourmis. S'il leur est impossible d'atteindre une réserve d'eau avec leur bouche, ils cueillent une poignée de feuilles, les froissent et les mâchent pour fabriquer ainsi une »éponge« qui ensuite est trempée dans l'eau et sucée. Une telle fabrication d'instruments adéquats n'est connue dans le monde animal que chez les Chimpanzés bien que l' → *usage d'instruments* soit plus répandu (GOODALL 1965).

Wertbegriff □ Ein gewöhnlich typisch menschliches Denksymbol, jedoch ist eine solche Auffassung nur zutreffend, soweit es sich um ideelle, etwa moralische, ethische, wissenschaftliche, künstlerische oder religiöse Wertvorstellungen handelt. Auf Materielles bezogene, averbale Wertbegriffe, etwa Münzwerten entsprechend, können auch bei höheren Tieren vorausgesetzt werden. Eine solche Assoziation eines Objektes zu einem bestimmten Wertgegenstand gibt es z. B. bei Schimpansen, die den unterschiedlichen Symbolwert von Marken verschiedener Größe und Farbe erlernten. Sie konnten z. B. blaue, weiße und Messingmarken in einen Automaten werfen und dafür 2, 1 oder 0 Weinbeeren eintauschen. Bald benutzten sie nur noch die blauen Marken, die für sie den »Wert« von zwei Beeren hatten. Dieses »Wertbegriff«-Vermögen konnte auch bei einigen niederen Affen nachgewiesen werden (WOLFE 1936, COWLES 1937, KAPUNE 1966).

value concept □ A typical human thought concept which is applicable only to idealistic, moral, ethical, scientific, artistic or religious issues. In animals nonverbal value concepts may occur as for example, the value associated with a coin. Such associations of objects with values are seen in chimpanzees which learn the significance of coins of different size and colour. Among blue, white and brass coins which produce respectively, 2, 1 or 0 grapes when dropped into a slotmachine, chimpanzees will learn to use only the blue ones. This »value concept« capability is also known from a few lower primates (WOLFE 1936, COWLES 1937, KAPUNE 1966).

notion de valeur □ Concept typiquement humain, qui, au sens propre, ne concerne que des valeurs idéales d'ordre moral, éthique, scientifique, artistique ou religieux. Les concepts de valeurs matérielles et averbales, correspondant par exemple à des valeurs monétaires, peuvent également être envisagées chez certains animaux supérieurs. L'association d'un objet à une valeur déterminée existe par exemple chez les Chimpanzés qui avaient appris la valeur symbolique de jetons de taille et de couleur différentes. Ils pouvaient utiliser des jetons bleus, blancs et en laiton en les introduisant dans un distributeur automatique pour les échanger contre 2, 1 ou 0 raisins. Très vite, ils n'utilisaient plus que les jetons bleus qui possédaient pour eux la valeur de deux raisins. Une telle capacité d'appréhender la »notion de valeur« a également été démontrée chez quelques Primates inférieurs (WOLFE 1936, COWLES 1937, KAPUNE 1966).

Willkomm-Saugen □ Von Papuas aus West-Irian bekanntes Begrüßungsverhalten, wobei die Besucher eines Dorfes zum Empfang erst einmal an der Brust der Frau des Dorfhäuptlings kurz saugen mußten (EIBL-EIBESFELDT 1970). HEYMER hat das selbst vor vielen Jahren in einem Farbfilm gesehen, wo auch die Weißen am Hütteneingang an der Brust der Frau saugen mußten.

welcome sucking, nursing □ A type of greeting ceremony found among the Papuans of West Irian in which the visitor to the village must first briefly suck on the breast of the village chief's wife (EIBL-EIBESFELDT 1970). HEYMER has seen this in a colour film several years ago where visitors did this on entering a dwelling.

tétée de bienvenue □ Comportement de salut et de bienvenue chez les Papous de la Nouvelle Guinée occidentale. Les visiteurs arrivant dans un village doivent, lors de la réception, effectuer d'abord une brève tétée à la poitrine de la femme du chef du village (EIBL-EIBESFELDT 1970). HEYMER a vu, il y a bien des années, un film en couleurs où même les Européens devaient effectuer cette tétée à l'entrée de la case de réception.

Winkmodus □ Krabben drohen mit der Schere, ihrer Waffe, gegen Artgenossen und gegen Artfremde. Bei manchen Gattungen und Arten wurde dieses Drohen ritualisiert und zu einem Signal beim Werbeverhalten (HEDIGER 1933, CRANE 1943, 1957, 1966, ALTEVOGT 1955, 1957, SCHÖNE und SCHÖNE 1963). *Grapsus grapsus* hebt die Scheren rhythmisch und droht nur. *Goniopsis cruentata* droht ähnlich wie *Grapsus*, und in etwas abgeleiteter Form wird die Scherenbewegung als »Winken« bei der Werbung verwendet. Bei den »Winkerkrabben« der Gattung *Uca* hat sich eine Schere des ♂ zu einer sog. »Winkschere« vergrößert, und jede Art entwickelte einen spezifischen »Winkmodus« (Abb. 132 A – *Uca rhizophorae,* vertikales Winken; B – *Uca annulipes,* seitliches Winken; C – *Uca pugilator,* kreisendes Winken mit seitwärts gehaltener Schere).

fiddling display, claw waving modes □ In crabs the claw of the cheliped is used in threatening conspecifics as well as other crab species. In several genera and species this threatening has been ritualized to a courtship signal (HEDIGER 1933, CRANE 1943, 1957, 1966, ALTEVOGT 1955, 1957, SCHÖNE and SCHÖNE 1963). *Grapsus grapsus* raises and lowers the claw rhythmically only as a threat. *Goniopsis cruentata* threatens in a similar fashion and in a derived form of this, »waves« in courtship. In the ge-

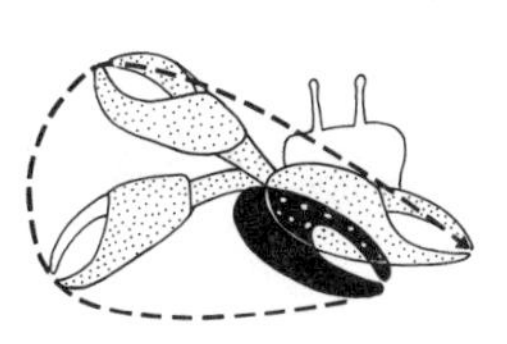

132 A B C

nus *Uca* the claw of one cheliped has become modified into a »waving claw« and each species has developed a special waving form (Fig. 132 A – *Uca rhizophorae,* vertical waving; B – *Uca annulipes,* lateral waving; C – *Uca pugilator,* circular waving with the claw in lateral position).

agitation chélipédale □ Normalement, les Crabes menacent leurs congénères et même leurs adversaires d'autres espèces à l'aide de leurs pinces. Dans quelques genres et espèces, ce comportement a été ritualisé et est devenu un signal intégré à la parade nuptiale (HEDIGER 1933, CRANE 1943, 1957, 1966, ALTEVOGT 1955, 1957, SCHÖNE et SCHÖNE 1963). *Grapsus grapsus,* par exemple, lève tout simplement les pinces d'une manière rythmique et se contente de menacer. *Goniopsis cruentata* a un comportement de menace semblable à celui de *Grapsus* et, sous une forme légèrement modifiée, utilise le mouvement de ses pinces comme signal de la parade nuptiale. Chez les Crabes du genre *Uca,* une pince seulement est hypertrophiée chez le ♂ et dans ce genre, chaque espèce a développé une manière spécifique de faire signe (Fig. 129 A – *Uca rhizophorae,* mouvement vertical; B – *Uca annulipes,* mouvement latéral; C – *Uca pugilator,* mouvement circulaire, pince en position latérale).

Wippschwimmen → Putzertanz

Wirkmalträger → Funktionskreis
E. functional cue bearer □ F. support d'information fonctionnelle

Wirkungsgefüge → Funktionskreis
E. functional cycle □ F. cycle fonctionnel

Wirkwelt → Funktionskreis
E. motor field □ F. champ d'action

Wirkorgan → Funktionskreis
E. central effector □ F. structure effectrice centrale

Wischbewegung □ Haben Frösche und Kröten eine zu große oder ungenießbare Nahrung aufgenommen, öffnen sie weit ihr Maul und versuchen, diese wieder hervorzuwürgen und auszuspucken, wobei eine Vorderextremität eine von der Seite nach der Mitte und nach vorn führende Wischbewegung ausführt, um das abgelehnte Objekt zu entfernen. Es handelt sich wohl eher um einen Wischreflex, dessen Koordination auch dann noch erhalten bleibt, wenn man bei einem spinalen Frosch das ausführende Bein desafferentiert (HERING 1896, EIBL-EIBESFELDT 1951, LESCURE 1965).

wiping reflex, expulsion reflex □ If frogs or toads eat some oversized or unpalatable object, they try to expel it by opening the mouth wide and moving the forelegs forward from the side in a wiping movement. This is a type of wiping reflex since the coordination remains intact when the spinal frog's forelegs are deafferentiated (HERING 1896, EIBL-EIBESFELDT 1951, LESCURE 1965).

mouvement d'essuyage □ Lorsqu'une Grenouille ou un Crapaud a avalé une proie trop grande ou non comestible, il ouvre largement sa bouche et essaie de la recracher. En même temps, il effectue avec une de ses pattes antérieures un mouvement d'essuyage latéral de l'extérieur vers l'intérieur pour se débarrasser de la proie rejetée. Il s'agit probablement là d'un réflexe d'essuyage dont la coordination persiste même chez les Grenouilles spinales dont le membre effecteur a été désafférenté (HERING 1896, EIBL-EIBESFELDT 1951, LESCURE 1965).

Wischreflex → Wischbewegung

Witterungsmarken = Duftmarken → Duftmarkierung

Wrickschwimmen □ Bezeichnung für die Schwimmbewegungen bei Tieren (Amphibien, Reptilien, echte Wassersäuger), die beim Schwimmen die Beine noch an den Körper anlegen und den Antrieb durch schlängelnde Bewegungen von Rumpf und Schwanz gewinnen. Eine solche Schwimmweise kennen wir auch von manchen am Wasser lebenden Säugern, wie z. B. *Potamogale velox,* Insectivora (DUBOST 1965). Die meisten echten Landsäuger haben diese Eigenschaft verloren und können nur noch mit den Extremitäten schwimmen.

undulation swimming □ A term for the swimming movements of animals (Amphibians, Reptilians, true water mammals) which hold their legs against the body and propel themselves by snake-like movements. The same swimming behaviour is known from some terrestrial mammals such as *Potamogale velox,* Insectivora (DUBOST 1965). Most true land mammals have lost this characteristic and can swim only by using the limbs.

nage ondulante □ Terme désignant la nage particulière d'un certain nombre d'animaux (Amphibiens, Reptiles, véritables Mammifères aquatiques) chez lesquels, lors de la locomotion subaquatique, les 4 membres sont repliés le long du corps, la propulsion se faisant uniquement par des mouvements ondulatoires du corps et de la queue. Une telle forme de nage se trouve aussi chez certains Mammifères vivant près le l'eau, comme par exemple *Potamogale velox,* Insectivora (DUBOST 1965). La plupart des Mammifères terrestres ont perdu cette possibilité comportementale et ne peuvent nager qu'à l'aide de leurs membres.

Würgeln → Stößeln

Wutkopulation □ Das Aufreiten auf Artgenossen gleichen Geschlechtes dient bei vielen Säugern als Rangdemonstration aggressiven Charakters. SCHENKEL (1948) beschreibt es von Wölfen, EIBL-EIBESFELDT (1950) von Hausmäusen, ZUCKERMANN (1932) von Pavianen, CARPENTER (1942) von Makaken und KOFORD (1963) von Rhesus-Affen, bei welchen man solche »Wutkopulationen« auch im Verlaufe aggressiver Auseinandersetzungen beobachtete, welche die aggressiv Erregten gegen unbeteiligte Dritte ausführten. Solche Verhaltensweisen kommen auch bei Menschen vor (KOSINSKI 1966).

rage copulation □ Mounting a same-sex conspecific among many mammals represents an aggressive-dominance demonstration. SCHENKEL (1948) described it in wolves, EIBL-EIBESFELDT (1950) from the house mouse, ZUCKERMANN (1932) from baboons, and CARPENTER (1942) from macaco monkeys. KOFORD (1963) observed dominance copulation in rhesus monkeys during aggressive encounters in which the antagonists mounted other conspecifics. Such behaviour patterns also occur in humans (KOSINSKI 1966).

copulation de colère □ Le chevauchemant de congénères du même sexe constitue, chez de nombreux Mammifères, une démonstration de dominance à caractère agressif. SCHENKEL (1948) le décrit chez les Loups, EIBL-EIBESFELDT (1950) chez les Souris domestiques, ZUCKERMANN (1932) chez les Babouins, CARPENTER (1942) chez les Macaques et KOFORD (1963) chez les Singes Rhésus chez lesquels la monte homosexuelle se produit aussi lors de disputes à caractère agressif. Les animaux excités et agressifs exécutent cette manœuvre sur des tiers non concernés. De tels comportements existent également chez l'Homme (KOSINSKI 1966).

Wutmimik → Drohmimik
E. rage mimic □ F. mimique de colère

Wutverhalten → Drohverhalten
E. rage behaviour □ F. comportement de colère

X

Xenobiosis □ Die Beziehung zwischen zwei verschiedenen Ameisenarten, bei welchen die eine, wie z. B. *Megalomyrmex symmetochus,* in der Kolonie der anderen, *Sericomyrmex amabilis,* lebt und im Nest ihrer Wirtsart völlig frei herumläuft und von den Wirtstieren durch → *Trophallaxis* oder andere Mechanismen Nahrung erhält, aber ihre Brut selbst und getrennt aufzieht (WILSON 1971, 1975).

xenobiosis □ The relation in which colonies of one ant species, for example *Megalomyrmex symmetochus,* live in the nests of another, *Sericomyrmex amabilis,* and move freely among the hosts, obtaining food from them by regurgitation *(→ trophallaxis)* or other means, but still keeping their brood separate (WILSON 1971, 1975).

xénobiose □ La relation entre deux espèces de Fourmis lors de laquelle une espèce, comme par exemple *Megalomyrmex symmetochus,* vit à l'intérieur de la colonie d'une autre espèce telle que *Sericomyrmex amabilis,* circule librement parmi ses hôtes et obtient d'eux la nourriture par régurgitation (→ *trophallaxie*) ou d'autres mécanismes. Cependant, sa progéniture est élevée séparément et par elle-même (WILSON 1971, 1975).

Z

Zählvermögen □ Bei Vögeln und Säugetieren die Fähigkeit für ein weitgehendes Generalisationsvermögen, im Rahmen von Versuchsanordnungen averbale Zahlenbegriffe zu bilden. Entsprechende jahrelange Experimente hat vor allem KOEHLER (1937–1955) durchgeführt. Das Zählvermögen gipfelte in der Leistung eines Kolkraben (1943), der nach einer langen Reihe von Versuchen in der Lage war, zunächst auf einer Anweistafel eine Anzahl von 2–6 Punkten abzulesen und in einer Gruppe von Futterschälchen den Deckel von jenem abzuwerfen, das die gleiche Anzahl Punkte trug. Wir kennen solche Leistungen auch von anderen Vögeln (Abb. 133 – Dohle, die die Zahl 3 wie vorgegeben wählt, obwohl die Anordnung der 3 Punkte anders ist als auf der Anzeigetafel). Bei Delphinen (*Tursiops truncatus*) konnte ein spontanes Zählvermögen ohne Dressur festgestellt werden. Die Tiere antworteten auf eine beliebige Anzahl von menschlichen Zurufen mit der gleichen Anzahl von Lauten zurück (BUCHHOLTZ 1973, RENSCH 1973 – bei beiden Autoren ausführliche Literaturangaben).

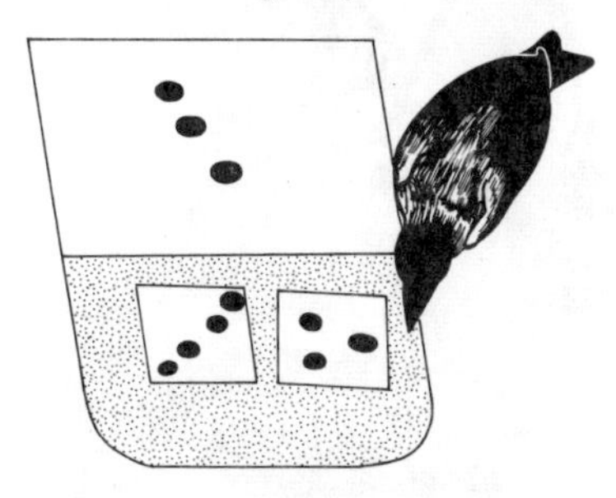

133

counting ability □ In birds and mammals the capacity for forming generalizations in non verbal counting or number recognition. KOEHLER (1937–1955) conducted numerous experiments on this subject. He found the zenith of performance in the raven, *Corvus corax* (1943). This bird was able to observe two to six dots on a board and recognize tops of food containers bearing the same number of dots. Such ability is also present in other birds (Fig. 133 – jackdaw recognizing the relation between three dots on a board and three dots on food containers even if they are in a different pattern). The dolphin (*Tursiops truncatus*) shows a counting ability without training. These animals answer a particular number of human calls with the same number of calls (BUCHHOLTZ 1973, RENSCH 1973 – see also for more detailed references).

aptitude à compter □ Chez les Oiseaux et les Mammifères, capacité de généralisation permettant d'acquérir à la suite d'un apprentissage institué par l'Homme, des concepts informulés de nombres. Des recherches à ce sujet ont été principalement effectués pendant de nombreuses années par KOEHLER (1937–1955). La performance la plus remarquable a été obtenue chez un Corbeau (1943) qui, après une longue série d'expériences, était capable de »lire« d'abord un nombre donné de 2 à 6 points sur un tableau et de choisir ensuite une cupule de nourriture dont le couvercle portait le même nombre de points. Nous connaissons de telles performances également chez d'autres Oiseaux (Fig. 133 – Choucas choisissant, comme indiqué sur le tableau, le nombre 3 bien que la disposition des 3 points soit différente). Chez les Dauphins (*Tursiops truncatus*), on a observé une capacité de compter sans dressage préalable. Ces animaux répondent à un nombre déterminé d'appels humains par le même nombre de vocalisations spécifiques (BUCHHOLTZ 1973, RENSCH 1973 – voir aussi ces deux auteurs pour une bibliographie exhaustive).

Zähnefletschen □ Ein Entblößen der Zähne als Beißdrohung, besonders bei Raubtieren (Abb. 134 – *Canis lupus*). Das Zähneentblößen bei Affen und Menschen bei der Wutmimik bezeichnet man als Zähnezeigen oder → *Drohmimik*.

134

bared teeth display □ A threat gesture occurring especially in carnivores (Fig. 134 – *Canis lupus*). In apes and man referred to as → *threat mimic*.

montrer les dents, retrousser les babines □ Menace de morsure, surtout chez les Carnivores (Fig. 134 – *Canis lupus*). Chez les Singes et chez l'Homme, le même geste qui survient pendant la colère est aussi une → *mimique de menace*.

Zähnewetzen □ Bei Nagern und Wiederkäuern, die die Zähne als Waffe benutzen, ein geräuschvolles Aufeinanderreiben der Zähne durch laterale Kieferbewegungen als Drohgeste (Zähneknirschen). Bei Nagern mag es auch dem Schärfen der Zähne dienen; außerdem ist es eine notwenige Aktion der Abnützung, da Nagerzähne ständig wachsen.

teeth grinding □ A threat gesture in rodents and ruminants which use their teeth as weapons; the teeth are noisily rubbed together by sideways movements of the mandible. In rodents this movement may also serve to sharpen the teeth; moreover, it helps to wear off the teeth which in rodents are continuously growing.

grincement des dents □ Chez les Rongeurs et les Ruminants qui utilisent les dents comme armes, le grincement des dents résulte d'une friction bruyante par mouvements latéraux des mâchoires et constitue un geste de menace. Chez les Rongeurs, l'affutage, et donc l'usure des dents, est une nécessité, étant donné leur croissance permanente.

Zähnezeigen → Zähnefletschen

Zärtlichkeitsfüttern → Mund-zu-Mund-Füttern

Zeitgeber □ Bestimmte Außenreize, die eine Aktivitätsperiodik, z. B. die → *Tagesrhythmik*, regulieren. Es gibt aktuelle Zeitgeber, wie z. B. der Hell-Dunkel-Wechsel, der eine exogene Periodik vorgibt. Latente Zeitgeber treten in Aktion, wenn der erstere ausfällt (*Innere Uhr*, BÜNNING 1963). Schirmt man z. B. einen Menschen in einem unterirdischen Bunker sorgfältig von allen Umwelteinflüssen ab, dann wird auch bei ihm eine Spontanfrequenz sichtbar. Auch seine Periodik ist circadian, d. h. sie weicht nur wenig von der normalen 24-Stunden-Periodik ab, was ihren endogenen Ursprung beweist (ASCHOFF und WEVER 1962, ASCHOFF 1966).

time synchronizer, Zeitgeber □ Certain external stimuli that control an activity rhythm, e. g. → *circadian rhythm*. Recurrent cues such as the light and dark change impose an external periodicity. Latent cues come into play when the primary *Zeitgeber* is absent (*internal clock*, BÜNNING 1963). If a human being is carefully kept isolated from all environmental influences in a subterranian bunker, one can also observe a spontaneous frequency. Man's periodicity is also circadian, that is, it deviates slightly from the normal 24-hour periodicity, which proves its endogenous origin (ASCHOFF and WEVER 1962, ASCHOFF 1966).

synchroniseur temporel □ Certains stimuli exogènes réglant un rythme d'activité, comme par exemple le → *rythme circadien*. On distingue des synchroniseurs actuels, telle que l'alternance jour-nuit, et des synchroniseurs latents, capables de se substituer aux synchroniseurs actuels quand ils sont défaillants (*horloge interne*, BÜNNING 1963). Si un humain est isolé avec soin de toutes les influences de l'environnement, dans un emplacement souterrain (bunker), on peut également observer une fréquence spontanée. La périodicité humaine est aussi circadienne, c'est-à-dire qu'elle dévie légèrement de la périodicité normale de 24 heures, ce qui prouve son origine endogène (ASCHOFF et WEVER 1962, ASCHOFF 1966).

Zeitsinn → Tagesrhythmik
E. time sense □ F. sens du temps

Zentrenhierarchie → Hierarchie
E. hierarchical organization of neural centers □ F. hiérarchie des centres nerveux

Zielvorstellung □ Bei höheren Tieren eine gewisse Fähigkeit, ein erwartetes oder erreichbares Ziel zu erstreben. Versteckt man vor den Blicken eines Rhesus-Affen eine Banane und vertauscht diese dann unsichtbar gegen ein weniger begehrtes Salatblatt, so sucht der Affe dann in sichtlicher Erregung nach der von ihm »erwarteten« Banane. Man kann also annehmen, daß er eine gewisse »Vorstellung« von seinem »Ziel« hat.

goal expectation □ In higher animals an ability to strive for an imagined or expected goal. If one puts a banana in the view of a Rhesus monkey and then secretly exchanges it against a less appealing object like a lettuce leaf, the monkey will refuse this object and will excitedly search for something, i. e. for the »expected« banana. One can hence postulate a certain goal expectation.

représentation d'un but, notion d'un but □ Chez les animaux supérieurs, capacité de tendre vers un but attendu. Par exemple lorsqu'on cache, sous les yeux d'un Singe Rhésus, une banane et qu'on l'échange ensuite à l'insu de l'animal contre une feuille de salade, moins convoitée, il cherchera avec beaucoup d'insistance et avec une excitation visible la banane »attendue«. On peut donc supposer que cet animal a une certaine »représentation« de son »but«.

zirkadianer Rhythmus → Tagesrhythmik

Zirkeln □ Eine Bewegungsweise der Nahrungssuche (HEINROTH 1928), die darin besteht, daß der Vogel den geschlossenen Schnabel in den Boden, zwischen Graswurzeln und unter Baumrinde steckt und die Schnabelhälften dann mit großer Kraft auseinanderspreizt (Abb. 135 – *Sturnus vulgaris*); so erweitert der Vogel Ritzen und Löcher und gelangt in die Bohrgänge von Würmern und Larven. Nach LORENZ (1949) gehört die Zirkelbewegung bestimmter Vogelarten zu den Instinktbewegungen, die eine eigene Spontaneität zeigen (WICKLER 1970, NEWEKLOWSKI 1972).

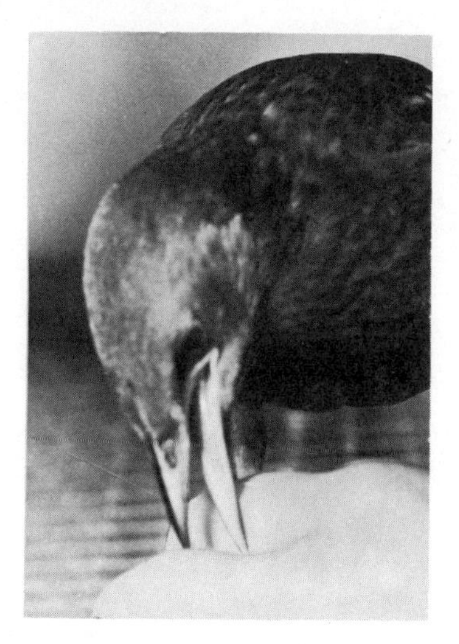

135

»Zirkeln« (prying movements) □ A feeding behaviour in birds (HEINROTH 1928) in which the bird inserts his beak in a crevice in the ground, between grass roots or under the bark of trees and then forcefully opens the beak (Fig. 135 – *Sturnus vulgaris*), thus expanding the hole and increasing his access to the worms and larvae. According to LORENZ (1949) this behaviour is instinctive in certain birds and has its own underlying drive (WICKLER 1970, NEWEKLOWSKI 1972).

forage rotatif, Zirkeln □ Notion créée par HEINROTH (1928) pour désigner un mode de recherche de la nourriture chez un certain nombre de Passereaux. L'Oiseau enfonce son bec fermé dans le sol, entre les racines des herbes ou dans une fente de l'écorce des arbres, l'écarte ensuite avec une grande force et effectue des mouvements circulaires (Fig.135 – *Sturnus vulgaris*). Ainsi l'Oiseau élargit des fentes et des trous pour atteindre des Vers et des Larves dans leurs galeries. Selon LORENZ (1949), le forage rotatif est, chez un certain nombre d'Oiseaux, un mouvement instinctif vrai, sous-tendu par une pulsion autonome (WICKLER 1970, NEWEKLOWSKI 1972).

Zittertanz □ Nicht selten verfallen die Bienen, vor allem bei Beunruhigung und widrigen Umständen, in einen Zittertanz. Sie laufen langsam auf vier Beinen auf den Waben herum. Das vorderste Beinpaar wird erhoben getragen. Infolge von zuckenden Bewegungen der Beine macht der Körper zitternde Ausschläge nach vorn und hinten, nach rechts und links. Den Stockgenossen gibt der Zittertanz keine Information (VON FRISCH 1965).

quivering dance □ During disturbance or noxious stimulation bees adopt a quivering dance. They wander slowly about the comb on four legs. The bee holds his most anterior appendages raised and moves in a jerking fashion rocking back and forth, left and right. This dance does not impart specific information to the hive mates (VON FRISCH 1965).

danse frémissante □ En cas de dérangement et de circonstances adverses dans la ruche, il est fréquent que les Abeilles adoptent une danse tremblante. Elles évoluent alors lentement sur le rayon en se soutenant avec quatre pattes et en relevant leurs pattes antérieures. Par suite des mouvements saccadés de leurs pattes, le corps effectue des balancements tremblants en avant et en arrière, vers la droite et vers la gauche. Ce curieux comportement ne donne aucune information aux compagnes de la ruche (VON FRISCH 1965).

Züngeln □ Beim Menschen das kurze Vorstrecken der Zunge und manchmal eine Leckbewegung in der Luft (Abb. 136 A – Mädchen aus Norddeutschland, B – Waika-Indianer vom Orinoko). Im menschlichen Flirt- und Werbeverhalten ist es wahrscheinlich ein ritualisiertes Lecken als heterosexuelle Aufforderungsgeste.

Bei vielen Schlangen dient das schnelle Vorstrecken der Zunge, das man ebenfalls als Züngeln bezeich-

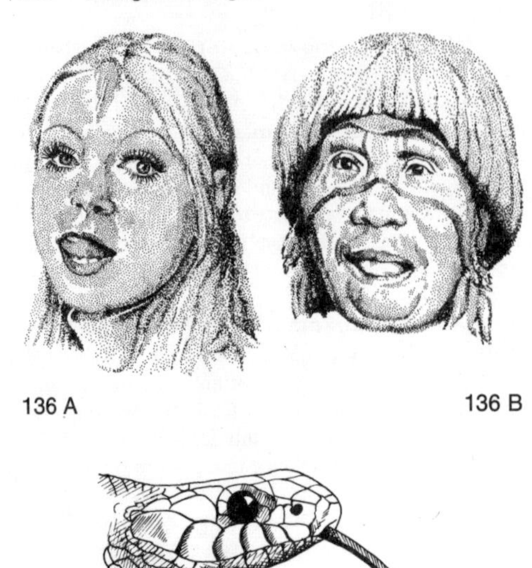

136 A 136 B

136 C

net, der olfaktorischen Orientierung und Spurenkontrolle mit Hilfe des Jacobson'schen Organs (Abb. 136 C – *Natrix natrix* ♀).

tongue-flicking □ In humans the brief extension of the tongue often accompanied by licking movements. Tongue-flicking or tonguing is found in human flirtation and courtship and probably represents ritualized licking (Fig. 136 A – girl from Northern Germany, B – Waika Indian from the Orinoco River).

In many snakes tongue-flicking serves orientation and olfactory control of trails involving Jacobson's organ (Fig. 136 C – *Natrix natrix* ♀).

darder la langue □ Chez l'Homme, mouvement rapide en avant de la langue avec parfois un léchage »à vide« (Fig. 136 A – jeune fille d'Allemagne du Nord, B – Indien Waika de l'Orinoque). Il s'agit là probablement d'un léchage ritualisé en tant que geste de sollicitation hétérosexuel dans le comportement de flirt et de cour chez l'Homme.

Chez les Serpents, darder la langue est un comportement d'orientation et de contrôle olfactif des traces à l'aide de l'organe de Jacobson (Fig. 136 C – *Natrix natrix* ♀).

Zungenschlagen □ Eine Leerlauf-Saugstereotypie, die man bei retardierten Kälbern beobachtet, wenn man sie aus Eimern tränkt. Sie trinken dabei zu schnell und können den Saugtrieb nicht befriedigen und entwickeln die Gewohnheit, an Ringen und Stallketten oder an anderen Kälbern zu lutschen. Die dabei in der Entwicklung Zurückgebliebenen zeigen dieses Syndrom. → *Triebbefriedigung*, → *Trinken im Leerlauf.*

substitute nursing □ A stereotyped vacuum sucking in some calves which have been fed from buckets rather than allowed to be nursed. Since the sucking drive is not satisfied after rapidly consuming milk from a bucket, these calves often suck on metal rings or chains or even other calves and sometimes this syndrome persists beyond the normal nursing stage. → *drive satisfaction*, → *vacuum drinking.*

claquement de la langue □ Stéréotypie de tétée à vide qu'on observe chez les Veaux retardés et qui n'ont jamais véritablement tété. Leur nourriture a toujours été présentée dans des seaux. Les Veaux boivent alors trop rapidement et ne peuvent satisfaire leur pulsion de tétée; ils développent l'habitude de sucer des objets métalliques, tels que bagues ou chaînes, ou même d'autres Veaux. Ce symptôme peut persister après le sevrage. Ceux qui, par cette manière d'élevage, sont retardés dans leur développement montrent le syndrome de claquement de la langue. → *satisfaction d'une pulsion,* → *boire à vide.*

Zungenschnappen □ Das blitzschnelle Vorschnellen und Wiedereinziehen oder -einrollen der Zunge beim Beuteerwerb von Fröschen, Kröten und vom Chamäleon (Abb. 137 – *Bufo bufo*).

tongue snapping □ The quick extension and retraction of the tongue for catching prey in frogs, toads or chameleons (Fig. 137 – *Bufo bufo*).

137

happement avec la langue □ Projection rapide de la langue et rétraction ou enroulement lors de la capture de proie chez les Grenouilles, Crapauds et Caméléons (Fig. 137 – *Bufo bufo*).

Zungezeigen □ Das Herausstrecken der Zunge ist bei Menschen eine weitverbreitete Geste verächtlicher Ablehnung. Es scheint sich um ein umweltstabiles Verhalten zu handeln, da man es in vielen Kulturen antrifft (Abb. 138 – Negerjunge aus Tanzania).

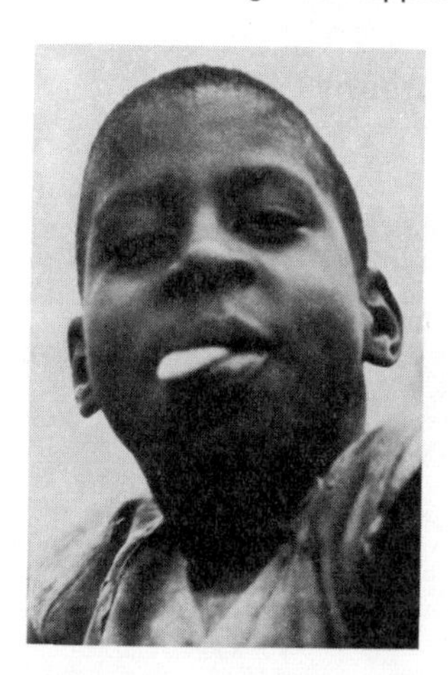

138

sticking out the tongue □ Widespread gesture of contempt and rejection among humans. It appears to be an environmentally stable behaviour, since it is found in many cultures (Fig. 138 – black boy from Tanzania).

tirer la langue □ Mouvement dédaigneux et exprimant le mépris et le rejet, très répandu dans l'espèce humaine. Ce comportement semble être stable par rapport à l'environnement et aux phénomènes d'acculturation, étant donné qu'on le trouve dans de nombreuses cultures avec la même signification (Fig. 138 – jeune garçon noir de Tanzania).

Bibliography

ABEL, E. F., 1961: Freiwasserstudien über das Fortpflanzungsverhalten des Mönchsfisches, *Chromis chromis* L., einem Vertreter der Pomacentriden im Mittelmeer. Z. Tierpsychol. **18,** 441–449

ABEL, E. F., 1971: Zur Ethologie von Putzsymbiosen einheimischer Süßwasserfische im natürlichen Biotop. Oecologia **6,** 133–151

ABELOOS, M., 1964: Comportement Animal et Evolution. Bull. Psychol. **17,** 11–14

ADRIAN, E. D., 1950: The Control of Nerve-Cell Activity. Symp. Soc. Exp. Biol. **4,** Cambridge

ALLAND, A., 1972: The Human Imperative. Columbia Univ. Press, New York
F. La dimension humaine – Réponse à Konrad Lorenz. Edit. du Seuil Paris, 189 pp.

ALLEN, M. D., 1959 a: The Occurrence and Possible Significance of the »Shaking« of Honeybee Queens by the Workers. Anim. Behaviour **7**, 66–69

ALLEN, M. D., 1959 b: The "Shaking" of Worker Honeybee by Other Workers. Anim. Behaviour **7,** 233–240

ALTEVOGT, R., 1955: Beobachtungen und Untersuchungen an indischen Winkerkrabben. Z. Morph. Ökol. Tiere **43,** 255–277

ALTEVOGT, R., 1957: Untersuchungen zur Biologie, Ökologie und Physiologie indischer Winkerkrabben. Z. Morph. Ökol. Tiere **46,** 1–110

ALTMANN, D., 1969: Harnen und Koten bei Säugetieren. Die Neue Brehm-Bücherei H. **404,** A. Ziemsen Verlag Wittenberg-Lutherstadt, 104 pp.

ALTUM, B., 1868: Der Vogel und sein Leben. Wilhelm Riemann Verlag Münster, 196 pp.

ALVAREZ, F., BRAZA, F., and NORZAGARY, A., 1976: The Use of the Rump Patch in Fallow Deer, *Dama dama.* Behaviour **56,** 298–308

ANDREE, R., 1889: Ethnographische Parallelen und Vergleiche. Stuttgart

ANDREW, R. J. and KLOPMAN, R. B., 1974: Urine Washing – Comparative Notes. in MARTIN, R. D., DOYLE, G. A. and WALKER, A. C. [Eds.]: Prosimian Biology, Duckworth London, 302–312

ANGERMEIER, W. F., 1976: Kontrolle des Verhaltens – Das Lernen am Erfolg. Springer Verlag Berlin/Heidelberg/New York,195 pp.

APFELBACH, R., 1967: Kampfverhalten und Brutpflegeform bei *Tilapia.* Naturwiss. **54,** 72

APFELBACH, R., und DÖHL, J., 1976: Verhaltensforschung – Eine Einführung. G. Fischer Verlag Stuttgart, 210 pp.

ARONSON, L. R., 1949: An Analysis of Reproductive Behaviour in the Mouth-Breeding Cichlid Fish *Tilapia macrocephala* BLEEKER. Zoologica N. Y. **34**, 133–157

ASCHOFF, J., 1966: Tagesrhythmus des Menschen bei völliger Isolation. Umschau **12,** 378–383

ASCHOFF, J. und WEVER, R., 1962: Spontanperiodik des Menschen bei Ausschluß aller Zeitgeber. Naturwiss. **49,** 337–342

BACKHAUS, D., 1958: Beitrag zur Ethologie der Paarung einiger Antilopen. Zuchthygiene **2,** 281–294

BAERENDS, G. P., 1941: Fortpflanzungsverhalten und Orientierung der Grabwespe *Ammophila campestris.* Tijdschr. Entomol. **84,** 68–275

BAERENDS, G. P., 1958: Comparative Methods and the Concept of Homology in the Study of Behaviour. Arch. Neerl. Zool. **13,** 401–417

BAERENDS, G. P. and BAERENDS-VAN ROON, J. M., 1950: An Introduction to the Study of the Ethology of Cichlid Fishes. Behaviour, Suppl. **1,** 1–243

BANDURA, A. and WALTHERS, R. H., 1963: Social Learning and Personality Development. Holt, Rinehart and Winston, New York

BASTOCK, M., MORRIS, D. and MOYNIHAN, M., 1953: Some Comments on Conflict and Thwarting in Animals. Behaviour **6,** 66–84

BECHTEREW, W., 1913: Objektive Psychologie oder Psychoreflexologie. Leipzig

BECKER, G., 1971: Magnetfeld-Einfluß auf die Galeriebau-Richtung bei Termiten. Naturwiss. **58,** 60

BECKER-CARUS, Ch., SCHÖNE, H. *et al.,* 1972: Zusammenfassung der Diskussionsbeiträge auf dem Ethologentreffen 1970 in Bern über Motivation, Handlungsbereitschaft und Trieb. Z. Tierpsychol. **30,** 321–326

BENEDIKT, R., 1934: Patterns of Culture. Houghton and Mifflin, Boston/New York

BIERENS DE HAAN, J. A., 1931: Werkzeuggebrauch und Werkzeugherstellung bei einem niederen Affen, *Cebus hypoleucos.* Z. vgl. Physiol. **13,** 639–695

BITTERMANN, E. F., 1965: The Evolution of Intelligence. Scient. American. **212,** 92–100

BLOOMFIELD, L., 1933: Language. New York

BLUME, D., 1967: Ausdrucksformen unserer Vögel. Neue Brehm-Bücherei H. **342,** A. Ziemsen Verlag Wittenberg-Lutherstadt, 184 pp.

BOHANNAN, P., 1966: Law and Warfare. Nat. Hist. Press, New York

BOWDITCH, H., 1871: The All-or-None Law. Philadelphia

BOWLBY, J., 1958: The Nature of the Child's Tie to His Mother. Int. Journ. Psychoanalysis **39,** 350–373

BRESTOWSKY, M., 1972: Zur Motivation der Grenzflächenappetenz junger Tilapien (Cichlidae). Vortrag – **III.** Deutsches Ethologentreffen Radolfzell

BROSSET, A., 1974: Structure sociale des populations de Chauves-Souris. J. Psychol. **71**, 85–102

BRUNS, H., 1958: Schutztrachten im Tierreich. Die Neue Brehm-Bücherei H. **207,** A. Ziemsen Verlag Wittenberg-Lutherstadt, 107 pp.

BUCHHOLTZ, CH., 1951: Untersuchungen an der Libellengattung *Calopteryx* LEACH, unter besonderer Berücksichtigung ethologischer Fragen. Z. Tierpsychol. **8,** 273–293

BUCHHOLTZ, CH., 1973: Das Lernen bei Tieren. Grundbegriffe der Modernen Biologie Bd. **11,** Gustav Fischer Verlag Stuttgart, 160 pp.

BUECHNER, H. K. and SCHLOETH, R., 1965: Ceremonial Mating Behaviour in Uganda Kob (*Adenota kob thomasi* NEUMANN). Z. Tierpsychol. **22,** 209–225

BÜHLER, K., 1922 a: Die geistige Entwicklung des Kindes. Jena

BÜHLER, K., 1922 b: Handbuch der Psychologie, **1.** Die Struktur der Wahrnehmung

BÜNNING, E., 1963: Die physiologische Uhr. Springer Verlag Berlin, 153 pp.

BÜRGER, M., 1959: Einige vergleichende Untersuchungen über Putzbewegungen bei Lagomorpha und Rodentia.

Der Zool. Garten N. F. **24,** 434–506

BURKHARDT, D. *et al.,* 1966: Signale in der Tierwelt – vom Vorsprung der Natur. Moos Verlag München, 150 pp.

BUSNEL, R. G., 1964: Acoustic Behaviour in Animals. Elsevier London/Amsterdam

BÜTZLER, W., 1974: Kampf- und Paarungsverhalten, soziale Rangordnung und Aktivitätsperiodik beim Rothirsch. Fortschr. d.Verhaltensforsch. H. **16,** Verlag Paul Parey Berlin und Hamburg, 80 pp.

CARPENTER, C. R., 1942: Societies of Monkeys and Apes. Biol. Symp. **8,** 177–204

CHANCE, M. R. A., 1962: An Interpretation of Some Agonistic Postures; the Role of "Cut-off" Acts and Postures. Symp. Zool. Soc. London **8,** 71–89

CHANCE, M. R. A. and RUSSELL, W. M. S., 1959: Protean Display: a Form of Allaesthetic Behaviour. Proc. Zool. Soc. London **132,** 65–70

CHARLES-DOMINIQUE, P. et HLADIK, C. M., 1971: Le Lepilemur du Sud de Madagascar: Ecologie, alimentation et vie sociale. La Terre et la Vie **25,** 3–66

CHAUVIN, R., 1968 a: Ethologie et Comportement. Le Comportement, Symposium de l'Association de Psychologie de la langue française, Paris, P. U. F., 51–62

CHAUVIN, R. [Ed.], 1968 b: L'effet de groupe chez les animaux. Coll. Int. C. N. R. S. Paris, **173,** 390 pp.

CHAUVIN, R., 1969: Psychophysiologie **II** – Le comportement animal. Masson et Cie. Paris, 418 pp.

CHAUVIN, R. [Ed.], 1972: Modèles animaux du comportement humain. Coll. Int. C. N. R. S. Paris, **198,** 378 pp.

CHAUVIN, R., 1975: L'Ethologie – Etude biologique du comportement animal. P. U. F. Paris, 236 pp.

CHRISTEN, A., 1974: Fortpflanzungsbiologie und Verhalten bei *Cebuella pygmaea* und *Tamarin tamarin.* Fortschr. d. Verhaltensforsch. H. **14,** Verlag Paul Parey Berlin und Hamburg, 78 pp.

COGHILL, G. E., 1929: Anatomy and the Problem of Behaviour. Cambridge Univ. Press

COURCHESNE, E. and BARLOW, G. W., 1971: Effect of Isolation on Components of Aggressive and Other Behaviour in the Hermit Crab, *Pagurus samuelis.* Z. vgl. Physiol. **75,** 32–48

COUTURIER, M., 1955: Le pelage du lièvre variable des Alpes, *Lepus timidus varius* MÜLLER, 1901. La Terre et la Vie, 101–116

COWLES, J. T., 1937: Food-token as Incentives for Learning by Chimpanzees. Comp. Psychol. Monogr. **14,** 1–96

CRAIG, W., 1918: Appetites and Aversions as Constituents of Instincts. Biol. Bull. Woods Hole **34,** 91–107

CRANE, J., 1943: Display, Breeding and Relationship of the Fiddler Crabs (Brachyura, genus *Uca*). Zoologica **28,** 217–223

CRANE, J., 1957: Basic Patterns of Display in Fiddler Crabs. Zoologica **42,** 69–82

CRANE, J., 1966: Combat, Display and Ritualization of Fiddler Crabs. Philos. Trans. Roy. Soc. London B **251,** 459–472

CROOK, J. H., 1970: Social Behaviour in Birds and Mammals. – Essays on the Social Ethology of Animals and Man. Academic Press London and New York, 492 pp.

CROZE, H., 1970: Searching Image in Carrion Crows. Fortschr. d. Verhaltensforsch. H. **5,** Verlag Paul Parey Berlin und Hamburg, 86 pp.

CURIO, E., 1963: Probleme des Feinderkennens bei Vögeln. Proc. XIII. Intern. Ornith. Congr., 206–239

CURIO, E., 1975: Die innerartliche Variabilität der Beutewahl beuteerfahrungsloser *Anolis.* Experimentia **31,** 45

DAANJE, A., 1951: On the Locomotory Movements in Birds, and the Intention Movements Derived from them. Behaviour **3,** 48–98

DARLING, F. F., 1937: A Herd of Red Deer. Oxford University Press

DATHE, H., 1955: Über die Schreckmauser. J. Ornith. **96**

DAVID, J. H. M., 1973: The Behaviour of the Bontebok, *Damaliscus dorcas* (PALLAS 1766), with Special Reference to the Territorial Behavior. Z. Tierpsychol. **33,** 38–107

DEEGENER, P., 1918: Die Formen der Vergesellschaftung im Tierreiche. Ein systematisch-soziologischer Versuch. Verlag von Veit u. Co. Leipzig, 420 pp.

DECKERT, G., 1968: Der Feldsperling. Die Neue Brehm-Bücherei H. **398,** A. Ziemsen Verlag Wittenberg-Lutherstadt, 90 pp.

DECROLY, O., 1934: Comment l'enfant arrive à parler? Centrale du P. E. S. de Belgique, 2 Vol.

DEMENTIEV, G. P., 1958: On the Autotomy in Birds (russ.). Zool. Journal **37,** 251–256

DIAMOND, J. M. and TERBORGH, J. W., 1968: Dual Singing by New Guinea Birds. The Auk **85,** 62–82

DIJKGRAAF, S., 1960: Spallanzani's Unpublished Experiments on the Sensory Basis of Object Perception in Bats. Isis **51,** 9–20

DOLLARD, J., DOOB, L., MILLER, N., and SEARS, R., 1939: Frustration and Aggression. New Haven, Yale Univ. Press

DOLLO, L., 1893: Les lois de l'évolution. Bull. Soc. Belg. Geol., Tome **7,** Bruxelles

DOLLO, L., 1895: Sur la Phylogénie des Dipneustes. Bull. Soc. Belg. Geol., Tome **9,** Bruxelles

DOLLO, L., 1903: *Eochelone brabantica.* Bull. Acad. Belg., 792–801

DUBOST, G., 1965: Quelques renseignements biologiques sur *Potamogale velox.* Biologia Gabonica **1,** 257–272

DUBOST, G., 1975: Le comportement du Chevrotain africain, *Hyemoschus aquaticus* OGILBY (Artiodactyla; Ruminantia). Z. Tierpsychol. **37,** 403–501

DUBOST, G. et GENEST, H., 1974: Le comportement social d'une colonie de Maras, *Dolichotis patagonum,* dans le Parc de Branféré. Z. Tierpsychol. **35,** 225–302

EIBL-EIBESFELDT, I., 1950: Beiträge zur Biologie der Haus- und der Ährenmaus nebst einigen Beobachtungen an anderen Nagern. Z. Tierpsychol. **7,** 558–587

EIBL-EIBESFELDT, I., 1951 a: Zur Fortpflanzungsbiologie und Jugendentwicklung des Eichhörnchens. Z. Tierpsychol. **8,** 370–400

EIBL-EIBESFELDT, I., 1951 b: Nahrungserwerb und Beuteschema der Erdkröte. Behaviour **4,** 1–35

EIBL-EIBESFELDT, I., 1953: Eine besondere Form des Duftmarkierens beim Riesengalago, *Galago crassicaudatus.* Säugetierkdl. Mitt. **1,** 171–173

EIBL-EIBESFELDT, I., 1955: Über Symbiosen, Parasitismus und andere besondere zwischenartliche Beziehungen tropischer Meeresfische. Z. Tierpsychol. **12,** 203–219

EIBL-EIBESFELDT, I., 1956: Angeborenes und Erworbenes in der Technik des Beutetötens – Versuche am Iltis, *Putorius putorius.* Z. Säugetierk. **21,** 135–137

EIBL-EIBESFELDT, I., 1957: Die Ausdrucksformen der Säugetiere. in KÜKENTHAL: Handbuch der Zoologie **8,** 1–88

EIBL-EIBESFELDT, I., 1959: Der Fisch *Aspidontus taenia-*

tus als Nachahmer des Putzers *Labroides dimidiatus*. Z. Tierpsychol. **16,** 19–25

EIBL-EIBESFELDT, I., 1965: *Nannopterum harrisi* (Phalacrocoracidae): Brutablösung. Film **E 596,** I. W. F. Göttingen

EIBL-EIBESFELDT, I., 1967: Grundriß der vergleichenden Verhaltensforschung – Ethologie. Piper Verlag München, 528 pp. 4. Aufl., 1974, 629 pp.
E. 1970: Ethology – The Biology of Behaviour. Holt, Rinehart and Winston New York, 530 pp.
F. 1972: Ethologie – Biologie du Comportement. Naturalia et Biologia, Jouy-en-Josas, 576 pp.

EIBL-EIBESFELDT, I., 1968: Zur Ethologie des menschlichen Grußverhaltens. Beobachtungen an Balinesen, Papuas und Samoanern nebst vergleichenden Bemerkungen. Z. Tierpsychol. **25,** 727–744

EIBL-EIBESFELDT, I., 1970: Liebe und Haß – zur Naturgeschichte elementarer Verhaltensweisen. Piper Verlag München, 293 pp.
E. 1972: Love and Hate. Holt, Rinehart and Winston New York
F. 1972: Contre l'agression – Contribution à l'histoire naturelle des comportements élémentaires. Edit. Stock Paris

EIBL-EIBESFELDT, I., 1971a: Eine ethologische Interpretation des Palmfruchtfestes der Waika (Venezuela) nebst einigen Bemerkungen über die bindende Funktion von Zwiegesprächen. Anthropos **66,** 767–778

EIBL-EIBESFELDT, I., 1971 b: Vorprogrammierung im menschlichen Sozialverhalten. Mitt. Max-Planck-Ges. **5,** 307–338

EIBL-EIBESFELDT, I., 1972 a: The Cross-Cultural Documentation of Social Behaviour. Modèles animaux du comportement humain, Coll. Int. C. N. R. S. Paris, **198,** 227–239

EIBL-EIBESFELDT, I., 1972 b: Die !ko-Buschmann-Gesellschaft: Gruppenbindung und Aggressionskontrolle. Piper Verlag München, 226 pp.

EIBL-EIBESFELDT, I., 1973 a: Der vorprogrammierte Mensch – Das Ererbte als bestimmender Faktor im menschlichen Verhalten. Verlag Fritz Molden Wien/München, 288 pp.
F. 1976: L'Homme programmé. L'inné, facteur déterminant du comportement humain. Flammarion Paris

EIBL-EIBESFELDT, I., 1973 b: The Expressive Behaviour of the Deaf and Blindborn. in M. V. CRANACH and I. VINE [Eds.]: Non-Verbal Behaviour and Expressive Movements, Academic Press London, 163–194

EIBL-EIBESFELDT, I., 1975: Krieg und Frieden – aus der Sicht der Verhaltensforschung. Piper Verlag München, 316 pp.

EIBL-EIBESFELDT, I. und HASS, H., 1966: Zum Projekt einer ethologisch orientierten Untersuchung menschlichen Verhaltens. Mitt. Max-Planck-Ges. H. **6,** 383–396

EIBL-EIBESFELDT, I. und LORENZ, K., 1974: Die stammesgeschichtlichen Grundlagen menschlichen Verhaltens, in HEBERER, G.: Evolution der Organismen, 3. Aufl., **III,** 572–624

EIBL-EIBESFELDT, I. und SIELMANN, H., 1962: Beobachtungen am Spechtfinken *Cactospiza pallida* (Fringillidae). J. Ornith. **103,** 92–101

EIBL-EIBESFELDT, I. und SIELMANN, H., 1965: *Cactospiza pallida* (Fringillidae): Werkzeuggebrauch beim Nahrungserwerb. Film **E 597,** Publ. zu Wiss. Filmen, I. W. F. Göttingen, **1 A,** 385–390

EISENBERG, J., 1967: Englisch-Deutsche Fachausdrücke, in: A Comparative Study in Rodent Ethology with Emphasis on the Evolution of Social Behaviour. Proc. Nat. Mus. **122,** 1–51

EMERSON, A. E., 1956: Ethospecies, Ethotypes and the Evolution of *Apicotermes* and *Altognathotermes* (Isoptera; Termitidae). Am. Mus. Nov. **1771,** 1–31

ERIKSON, E. H., 1966: Ontogeny of Ritualization in Man. Philos. Trans. Roy. Soc. London **251,** 337–349

ESCH, H., 1964: Beiträge zum Problem der Entfernungsweisung in den Schwänzeltänzen der Honigbienen. Z. vgl. Physiol. **48,** 534–546

FALLET, M., 1958: Zum Sozialverhalten des Haussperlings *Passer domesticus* L. Zool. Anz. **161,** 178–187

FEEKES, F., 1972: »Irrevelant« Ground Pecking in Agonistic Situations in Burmese Red Junglefowl *Gallas g. spadiceus.* Behaviour **43,** 186–326

FISCHER, H., 1965: Das Triumphgeschrei der Graugans. Z. Tierpsychol. **22,** 247–304

FISHER, J. and HINDE, R., 1949: The Opening of Milk Bottles by Birds. Brit. Birds **42,** 347–358

FOPPA, K., 1968: Lernen, Gedächtnis, Verhalten – Ergebnisse und Probleme der Lernpsychologie. Kiepenheuer u. Witsch Köln, 466 pp.

FOX, M. W. [Ed.], 1968: Abnormal Behaviour in Animals. W. B. Saunders Company Philadelphia/London/Toronto, 563 pp.

FRAENKEL, G. S. and GUNN, D. L., 1961: The Orientation of Animals – Kineses, Taxes and Compass Reactions. Dover Publications Inc. New York/London, 376 pp.

FRANK, D. und WILHELMI, U., 1973: Veränderungen der aggressiven Handlungsbereitschaft männlicher Schwertträger, *Xiphophorus helleri,* nach sozialer Isolation (Pisces, Poeciliidae). Experientia **29,** 896–897

FRANK, F., 1956: Das Duftmarkieren der Großen Wühlmaus, *Arvicola terrestris* (L.). Z. Säugetierk. **21,** 172–175

FRANK, I., HADELER, U. und HARDER, W., 1951: Zur Ernährungsphysiologie der Nagetiere – über Bedeutung der Coecotrophie und die Zusammensetzung der Coecotrophe. Pflügers Archiv **253,** 173–180

FRICKE, H. W., 1973: Ökologie und Sozialverhalten des Korallenbarsches *Dascyllus trimaculatus* (Pisces, Pomacentridae). Z. Tierpsychol. **32,** 225–256

FRISCH, K. V., 1965: Tanzsprache und Orientierung der Bienen. Springer Verlag Berlin/Heidelberg/New York, 578 pp.

FRISCH, K. V. und LINDAUER, M., 1954: Himmel und Erde in Konkurrenz bei der Orientierung der Bienen. Naturwiss. **41,** 245–253

FÜLLER, H., 1958: Symbiose im Tierreich. Die Neue Brehm-Bücherei H. **227,** A. Ziemsen Verlag Wittenberg-Lutherstadt, 227 pp.

GARATTINI, S. and SIGG, E. B. [Eds.], 1969: Aggressive Behaviour. Exerpta Medica Foundation, Amsterdam

GAUTIER-HION, A., 1971: Répertoire comportemental du Talapoin, *Miopithecus talapoin.* Biologia Gabonica **7,** 295–361

GENEST, H. and DUBOST, G., 1974: Pair-Living in the Mara, *Dolichotis patagonum.* Mammalia **38,** 155–162

GOETHE, F., 1938: Beobachtungen über das Absetzen von Witterungsmarken beim Baummarder. Deutscher Jäger **13**

GOETHE, F., 1956: Die Silbermöwe. Die Neue Brehm-Bücherei H. **182,** A. Ziemsen Verlag Wittenberg-Lutherstadt, 95 pp.

GOODALL, J., 1965: Chimpanzees of the Gombe Stream

Reserve; in DE VORE, I.: Primate Behaviour. New York, 425–473
GRAF, W., 1956: Territorialism in Deer. J. of Mammalogy **37,** 156–170
GRASSÉ, P.-P., 1946: Sociétés Animales et effet de groupe. Experientia **2,** 77–82
GRASSÉ, P.-P., 1958: L'effet de groupe sur l'animal et sur l'Homme. J. Psych. norm. path. 129–148
GRASSÉ, P.-P., 1959: La construction du nid et les coordinations inter-individuelles chez *Bellicositermes natalensis* et *Cubitermes sp.* – La théorie de la stigmergie. – Essai d'interprétation du comportement des termites constructeurs. Insectes Sociaux **6,** 41–84
GRZIMEK, B., 1944: Die Radfahrer-Reaktion. Z. Tierpsychol. **6,** 41
GUHL, A. M. and SCHEIN, W. M., 1968: A Glossary of Terms Used in Animal Behaviour. Anim. Behav. in Laboratory and Field, No. **838,** 10 pp.
GWINNER, E., 1964 a: Untersuchungen über das Ausdrucks- und Sozialverhalten der Kolkraben. Z. Tierpsychol. **21,** 657–748
GWINNER, E., 1964 b: Beobachtungen über Nestbau und Brutpflege des Kolkraben, *Corvus corax,* in Gefangenschaft. J. Ornith. **105**

HACKER, F., 1971: Aggression – Die Brutalisierung der modernen Welt. Verlag Fritz Molden Wien, 464 pp.
HAECKEL, E., 1866: Generelle Morphologie. Berlin
HALDER, U., 1976: Ökologie und Verhalten des Banteng, *Bos javanicus,* in Java – Eine Feldstudie. Mammalia depicta, H. **10,** Verlag Paul Parey Hamburg/Berlin, 124 pp.
HALDER, U. und SCHENKEL, R., 1972: Das Riechgähnen bei Rindern (Bovidae). Z. Säugetierk. **37,** 232–245
HAMMANN, E., 1957: Wer hat die Initiative bei den Ausflügen der Jungkönigin, – die Königin oder die Arbeiterinnen? Insectes Sociaux **4,** 91–106
HARDER, W., 1949: Zur Morphologie und Physiologie des Blinddarmes der Nagetiere. Verh. Deutsch. Zoologen Mainz, 95–109
HARLOW, H. F. and HARLOW, M. K., 1962 a: The Effect of Rearing Conditions on Behavior. Bull. Menninger Clin. **26,** 213–224
HARLOW, H. F. and HARLOW, M. K., 1962 b: Social Deprivation in Monkeys. Scient. Americ. **207,** 137–146
HARTMANN, M., 1953: Allgemeine Biologie. Gustav Fischer Verlag Stuttgart, 940 pp.
HASSENBERG, L., 1965: Ruhe und Schlaf bei Säugetieren. Die Neue Brehm-Bücherei H. **338,** A. Ziemsen Verlag Wittenberg-Lutherstadt, 160 pp.
HASSENSTEIN, B., 1966: Kybernetik und biologische Forschung. Handb. d. Biol., Athenaion Verlag Frankfurt, **1,** 631–719
HASSENSTEIN, B., 1972: Das spezifisch Menschliche nach den Resultaten der Verhaltensforschung; in Neue Anthropologie **II,** 60–97
HASSENSTEIN, B., 1973: Verhaltensbiologie des Kindes. Piper Verlag München, 459 pp.
HAYDAK, M. H., 1929: Some New Observations of the Bee Life. Cesky Vcelar **63,** 133–135
HAYDAK, M. H., 1945: The Language of the Honey-Bee. Americ. Bee Journal **85,** 316–317
HEDIGER, H., 1933: Beobachtungen an der marokkanischen Winkerkrabbe *(Uca tangeri).* Verh. Schweiz. Naturforsch. Ges. **114,** 388–389
HEDIGER, H., 1942: Wildtiere in Gefangenschaft. Basel
E. 1950: Wild Animals in Capativity. Dover Publishing
F. 1953: Les animaux sauvages en captivité. Edit. Payot Paris, 236 pp.
HEDIGER, H., 1943: Zur Biologie und Psychologie der Flucht bei Tieren. Biol. Zbl. **54,** 21–40
HEDIGER, H., 1953: Ein symbioseartiges Verhältnis zwischen Flußpferd und Fisch. Säugetierkundl. Mitt. **1,** 75–76
HEDIGER, H., 1957: Beobachtungen zum Markierungsverhalten einiger Säugetiere. Z. Säugetierk. **22,** 57–76
HEDIGER, H., 1959: Die Angst des Tieres; in Die Angst. Studien aus dem C. G. Jung-Institut Zürich **10,** 7–33
HEDIGER, H., 1964: Man as a Social Partner of Animals and Vice-Versa. Symp. Zool. Soc. London **14,** 291–300
HEDIGER, H., 1967: Die Straßen der Tiere. Zürich
HEIDEMANN, G., 1973: Zur Biologie des Damwildes. Mammalia depicta H. **9,** Verlag Paul Parey Berlin/Hamburg, 95 pp.
HEILIGENBERG, W., 1964: Versuch zur ganzheitsbezogenen Analyse des Instinktverhaltens eines Fisches, *Pelmatochromis subocellatus kribensis.* Z. Tierpsychol. **21,** 1–52
HEILIGENBERG, W. and KRAMER, U., 1972: Aggressiveness as a Function of External Stimulation. J. Comp. Physiol. **77,** 332–340
HEINROTH, O., 1911: Beiträge zur Biologie, namentlich Ethologie und Psychologie der Anatiden. Verh. **5.** Int. Ornith. Kongr. 1910 Berlin, 589–702
HEINROTH, O. und M., 1928: Die Vögel Mitteleuropas. 4 Bde., Verlag für Kunst und Wissenschaft Leipzig, Nachdruck 1966
HEINROTH, O., 1930: Über bestimmte Bewegungsweisen der Wirbeltiere. Sitz. Ber. Ges. Naturf. Freunde Berlin, 333–342.
HEINZ, H. J., 1949: Vergleichende Beobachtungen über die Putzhandlungen bei Dipteren im allgemeinen und *Sarcophaga carnaria* im Besonderen. Z. Tierpsychol. **6,** 330–371
HEINZ, H. J., 1967: Conflicts, Tensions and Release of Tensions in a Bushman Society. The Institute for the Study of Man in Africa, Isma Papers No. **23**
HELFEREICH, P., 1958: The Early Life History and Reproductive Behaviour of the Maomao *Abudefduf abdominalis.* Doct. Diss. Univ. Hawaii, 228 pp.
HENDRICHS, H. und HENDRICHS, U., 1971: Freilanduntersuchungen zur Ökologie und Ethologie der Zwergantilope *Madoqua kirki* (GÜNTHER 1880). Ethologische Studien, Piper Verlag München, 11–75
HERING, H. E., 1896: Über Bewegungsstereotypien nach centripetaler Lähmung. Arch. exp. pathol. Pharmakol. **38,** 266–283
HESS, E. H., 1956: Space Perception in the Chick. Scient. Americ. **195,** 71–80
HESS, E. H., 1965: Attitude and Pupil Size. Scient. Americ. **212,** 46–54
HESS, E. H. *et al.,* 1965: Pupil Response of Hetero- and Homosexual Males to Pictures of Men and Women: A Pilot Study. J. Abnorm. Psychol. **70,** 165–168
HESS, E. H., 1973: Imprinting – Early Experience and the Developmental Psychobiology of Attachment. Van Nostrand Reinhold Co., New York
D. 1975: Prägung – Die frühkindliche Endwicklung von Verhaltensmustern bei Mensch und Tier. Kindler Verlag München, 542 pp.
HESS, E. H., 1975: The Tell-Tale Eye. Van Nostrand Reinhold Co. New York
HEYMER, A., 1966: Etudes comparées du comportement inné de *Platycnemis acutipennis* SELYS 1841 et de *Platyc-*

nemis latipes RAMBUR 1842 (Odon. Zygoptera). Ann . Soc. Ent. France, N. S. **2,** 39–73

HEYMER, A., 1969: Fortpflanzungsverhalten und Territorialität bei *Orthetrum coerulescens* und *O. brunneum.* Rev. Comp. Animal **3,** 1–24

HEYMER, A., 1972a: Ethologische Freiwasserbeobachtungen an Putzsymbiosen im Mittelmeer. Rev. Comp. Animal **6,** 17–24

HEYMER, A., 1972b: Comportement social et territorial des Calopterygidae (Odon. Zygoptera). Ann. Soc. Ent. France **8**, 3–53

HEYMER, A., 1973 a: Verhaltensstudien an Prachtlibellen.Fortschr. d. Verhaltensforsch. H.**11,** Verlag Paul Parey Berlin/Hamburg, 100 pp.

HEYMER, A., 1973 b: Etude du comportement reproducteur et analyse des mécanismes déclencheurs innés (MDI) optiques chez les Calopterygidae (Odonata; Zygoptera). Ann. Soc. Ent. France N. S. **9,** 219–255

HEYMER, A., 1973c: Das hochspezialisierte Beutefangverhalten der Larve von *Cordulegaster annulatus* (LATR. 1805) – eine ökologische Einnischung (Odon. Anisoptera). Rev. Compt. Animal **7**, 183–189

HEYMER, A., 1974 a: Hochentwickelte Verhaltensweisen und Erhaltung ursprünglicher Körpermerkmale bei Libellen. Image Roche No. **58,** 11–19
E. The Ancient Dragonfly: Behavioural Evolution but Morphological Standstill. Roche Image No. **58,** 11–19
F. Comportements hautement évolués et survivance de caractères morphologiques archaïques chez les Libellules. Roche Image No. **58,** 11–19

HEYMER, A., 1974 b: Das Jäger- und Sammlervolk der Pygmäen. Roche Image No. **62,** 17–26
E. The Hunting and Food-gathering Pygmies. Roche Image No. **62,** 17–26
F. Les Pygmées, peuple de chasseurs-collecteurs. Roche Image No. **62,** 17–26

HILL, W. C. O., 1938: A curious Habit Common to Lorisids and Platyrrhine Monkeys. Ceyl. J. Sci. (B) **21,** 65

HINDE, R. A., 1954: Factors Governing the Changes in Strength of a Partially Inborn Response, as Shown by the Mobbing Behaviour of the Chaffinch *Fringilla coelebs.* Proc. Roy. Soc. London, S. B. **142,** 306–358

HINDE, R. A., 1966: Animal Behaviour – A Synthesis of Ethology and Comparative Psychology. McGraw-Hill Book Co., 534 pp.

HINGSTON, R. W. G., 1929: Instinct and Intelligence. MacMillan and Co., New York

HINSCHE, G., 1935: Der Schnappreflex nach „Nichts" bei Anuren. Zool. Anz. **111,** 113–122

HLADIK, C. M., 1973: Alimentation et activité d'un groupe de Chimpanzés réintroduits en forêt gabonaise. La Terre et la Vie **27,** 343–413

HLADIK, C. M. *et al.,* 1971: La caecotrophie chez un Primate phyllophage du genre *Lepilemur* et les correlations avec les particularités de son appareil digestif. C. R. Acad. Sc. Paris **272,** 3191–3194

HÖLLDOBLER, B., 1967 a: Verhaltensphysiologische Untersuchungen zur Myrmecophilie einiger Staphilinidenlarven. Verh. D. Z. G. Heidelberg, 428–434

HÖLLDOBLER, B., 1967 b: Zur Physiologie der Gast-Wirt-Beziehungen (Myrmecophilie) bei Ameisen: I. Das Gastverhältnis der *Atemeles-* und *Lomechusa-*Larven (Col. Staphilinidae) zu *Formica* (Hym. Formicidae). Z. vgl. Physiol. **56,** 1–21

HÖLLDOBLER, B., 1976: Recruitment Behaviour, Home Range Orientation and Territoriality in Harvester Ants, *Pogonomyrmex.* Behav. Ecol. Sociobiol. **1,** 3–44

HOLLOWAY, R. L. [Ed.], 1974: Primate Aggression, Territoriality and Xenophobia. Academic Press, New York and London, 513 pp.

HOLST, D. V., 1969: Sozialer Stress bei Tupajas *Tupaia belangeri.* Z. vgl. Physiol. **63,** 1–58

HOLST, D. V., 1973: Sozialverhalten und sozialer Streß bei Tupajas. Umschau **73,** 8–12

HOLST, E. V., 1934: Studien über Reflexe und Rhythmen beim Goldfisch. Z. vgl. Physiol. **20,** 582–599

HOLST, E. V., 1935: Erregungsbildung und Erregungsleitung im Fischrückenmark. Pflüg. Arch. **235,** 345–359

HOLST, E. V., 1937: Vom Wesen der Ordnung im Zentralnervensystem. Naturwiss. **25,** 625–631 und 641–647

HOLST, E. V., 1939: Die relative Koordination als Phänomen und als Methode zentralnervöser Funktionsanalyse. Erg. Physiol. **42,** 228–306

HOLST, E. V., 1950: Die Tätigkeit des Statolithenapparates im Wirbeltierlabyrinth. Naturwiss. **37,** 265–272

HOLST, E. V., 1969: Zur Verhaltensphysiologie bei Tieren und Menschen. Gesammelte Abhandlungen, Piper Verlag München, **I,** 294 pp.

HOLST, E. V., 1970: Zur Verhaltensphysiologie bei Tieren und Menschen. Gesammelte Abhandlungen, Piper Verlag München, **II,** 299 pp.

HOLST, E. V. und MITTELSTÄDT, H., 1950: Das Reafferenzprinzip. Naturwiss. **37,** 464–476

HOLST, E. V. und ST. PAUL, U. V., 1960: Vom Wirkungsgefüge der Triebe. Naturwiss. **18,** 409–422

HOLZBERG, S., 1973: Beobachtungen zur Ökologie und zum Sozialverhalten des Korallenbarsches *Dascyllus marginatus* RÜPPELL (Pisces, Pomacentridae). Z. Tierpsychol. **33,** 492–513

HOOFF, J. A. R. A. M. VAN, 1972: A Comparative Approach to the Phylogeny of Laughter and Smiling; in HINDE, R. A. [Ed.]: Non-verbal Communication. Cambridge University Press, 209–241

HUNTER, W. S., 1913: The Delayed Reaction in Animals and Children. Behav. Monogr. **2,** 21–30

HUTELAND, D. C. W., 1798: Die Kunst das menschliche Leben zu verlängern. Jena

HUXLEY, J. S., 1914: The Courtship Habits of the Great Crested Grebe, *Podiceps cristatus;* with an Addition to the Theory of Sexual Selection. Proc. Zool. Soc. London, 491–562

HUXLEY, J. S., 1923: Courtship Activities in the Red-throated Diver, *Gavia stellata;* together with a Discussion of the Evolution of Courtship in Birds. J. Linn. Soc. Zool. London **53,** 253–292

HUXLEY, J. S., 1938: Darwin's Theory of Sexual Selection and the Data Subsumed by it, in the Light of Recent Research. Americ. Natur. **72,** 416–433

HUXLEY, J. S., 1966: A Discussion on Ritualization of Behaviour in Animals and Man. Phil. Trans. Roy. Soc. London **251,** 247–526

IERSEL, J. V., 1953: An Analysis of the Parental Behaviour of the Three-Spined Stickleback *Gasterosteus aculeatus.* Behaviour, Suppl. **3**

IMMELMANN, K. [Ed.], 1974: Verhaltensforschung. Sonderband in Grzimeks Tierleben, Kindler Verlag München, 660 pp.

IMMELMANN, K., 1975: Wörterbuch der Verhaltensforschung. Taschenbuch – Kindler Verlag München, 133 pp.

JAISSON, P., 1974: Proposition d'un néologisme: Ethogenèse. Rev. Compt. Anim. **8,** 351

JANDER, R., 1968: Über die Ethometrie von Schlüsselreizen, die Theorie der telotaktischen Wahlhandlung und das Potenzprinzip der terminalen Cumulation bei Arthropoden. Z. vgl. Physiol. **59,** 319–356

JANDER, U., 1966: Untersuchungen zur Stammesgeschichte von Putzbewegungen von Tracheaten. Z. Tierpsychol. **23,** 799–844

JAVANOVIC, U. J. [Ed.], 1973: The Nature of Sleep. Gustav Fischer Verlag Stuttgart, 308 pp.

JENNINGS, H. S., 1906: Behaviour of Lower Organisms. New York

JEWELL, P. A. and LOIZOS, C., 1966: Play, Exploration and Territory in Mammals. Academic Press, London and New York, 280 pp.

JOLLY, A., 1972: The Evolution of Primate Behaviour. MacMillan Co. New York
D. 1975: Die Entwicklung des Primatenverhaltens. Gustav Fischer Verlag Stuttgart, 318 pp.

JUNGIUS, H., 1969: Beiträge zur Biologie des Großriedbocks. Diss. Universität Kiel

KAPUNE, TH., 1966: Untersuchungen zur Bildung eines »Wertbegriffes« bei niederen Primaten. Z. Tierpsychol. **23**, 324–363

KAWAI, M., 1958: On the Rank System in a Natural Group of Japanese Monkeys. Primates **1**, 84–98

KAWAI, M., 1965: Newly Acquired Precultural Behavior of Natural Troop of Japanese Monkeys on Koshima Island. Primates **6,** 1–30

KEITER, F. [Ed.]. 1969: Verhaltensforschung im Rahmen der Wissenschaften vom Menschen. Musterschmidt-Verlag Göttingen, 248 pp.

KIRCHHOFER, R., 1960: Über das Harnspritzen des Großen Mara. Z. Säugetierk. **25,** 112–127

KIRKMANN, F. B., 1937: Bird Behaviour. London and Edinburgh

KITZLER, G., 1942: Die Paarungsbiologie einiger Eidechsen. Z. Tierpsychol. **4,** 353–402

KLOPFER, P. H., 1969: Habitats and Territories, a Study of the Use of Space by Animals. Basis Books New York, 117 pp.

KLOPFER, P. H., 1974: An Introduction to Animal Behavior – Ethology's First Century. 2. Edition, Prentice-Hall Inc., Englewood Cliffs, New Jersey, 332 pp.

KOEHLER, O., 1924: Sinnesphysiologische Untersuchungen an Libellenlarven. Verh. Deutsch. Zool. Ges. **29,** 83–91

KOEHLER, O., 1937: Können Tauben »zählen«? Z. Tierpsychol. **1,** 39–48

KOEHLER, O., 1941: Vom Lernen unbekannter Zahlen bei Vögeln. Naturwiss. **29,** 201–218

KOEHLER, O., 1943: Zähl-Versuche an einem Kolkraben und Vergleichsversuche an Menschen. Z. Tierpsychol. **5,** 575–712

KOEHLER, O., 1949: Die Analyse der Taxisanteile instinktartigen Verhaltens. Symp. Soc. Exp. Biol. Cambridge **4,** 268–302

KOEHLER, O., 1950 a: Zählende Vögel und vorsprachliches Denken. Zool. Anz. Suppl. **13,** 219–238

KOEHLER, O., 1950 b: Physiological Mechanisms in Animal Behaviour. Academic Press New York

KOEHLER, O., 1954 a: Das Lächeln als angeborene Ausdrucksbewegung. Z. menschl. Vererb. u. Konstitut.-Lehre **32,** 390–398

KOEHLER, O., 1954 b: Das Lächeln des Säuglings. Umschau **54,** 321–324

KOEHLER, O., 1955: Zählende Vögel und vergleichende Verhaltensforschung. Acta **II,** 588–598

KOEHLER, O., 1964 a: Stellungnahme zum Ausdruck »Schlüsselreiz« in der deutschen Übersetzung Tinbergens Instinktlehre. **3**. Aufl., p. 28–36

KOEHLER, O., 1964 b: Verzeichnis der ethologischen Fachausdrücke: Englisch-Deutsch und Deutsch-Englisch; in TINBERGEN, N.: Instinktlehre, Verlag Paul Parey Hamburg/Berlin, 234–256

KOEHLER, O. und ZAGARUS, A., 1937: Beiträge zum Brutverhalten des Halsbandregenpfeifers, *Charadrius hiaticula.* Beitr. Fortpfl.-biol. Vögel **13,** 1–9

KOENIG, L., 1973: Das Aktionssystem der Zwergohreule, *Otus scops scops* (L. 1758). Fortschr. d. Verhaltensforsch. H. **13,** Verlag Paul Parey Hamburg/Berlin, 124 pp.

KOENIG, O., 1951: Das Aktionssystem der Bartmeise, *Panurus biarmicus.* Österr. Zool. Zeitschr. **3,** 1–82 und 247–325

KOENIG, O., 1970: Kultur und Verhaltensforschung – Einführung in die Kultur-Ethologie. dtv-Verlag, 290 pp.

KOENIG, O., 1975: Urmotiv Auge – Neuentdeckte Grundzüge menschlichen Verhaltens. Piper Verlag München, 556 pp.

KOFORD, C. B., 1963 a: Rank of Mothers and Sons in Bands of Rhesus Monkeys. Science **141,** 356–357

KOFORD, C. B., 1963 b: Group Relations in an Island Colony of Rhesus Monkeys; in SOUTHWICK, CH. [Ed.]: Primate Social Behaviour, Princeton N. Y., 136–152

KÖHLER, W., 1921: Intelligenzprüfungen an Menschenaffen. Springer Verlag Berlin/Heidelberg/New York, 3. Aufl. 1973, 234 pp.

KÖHLER, W., 1947: Gestalt Psychology. New American Library New York
F. 1964: Psychologie de la forme. NRF, Gallimard, 373 pp.

KÖHLER, W., 1971: Die Aufgabe der Gestaltpsychologie. Walter de Gruyter Berlin, 122 pp.

KÖHLER, W., 1973: L'intelligence des Singes supérieurs. P. U. F. Paris, 191 pp.

KORRINGA, P., 1957: Lunar Periodicity; in HEDGEPOTH, J. W. [Ed.]: Treatise on Marine Ecology and Palaeoecology, **1.** Ecology Mem. **67,** Geol. Soc. Am. Baltimore, 917–934

KORTLANDT, A., 1940 a: Eine Übersicht der angeborenen Verhaltensweisen des mitteleuropäischen Kormorans, *Phalacrocorax carbo sinensis,* ihre Funktion, ontogenetische Entwicklung und phylogenetische Herkunft. Arch. Neerl. Zool. **4,** 401–442

KORTLANDT, A., 1940 b: Wechselwirkung zwischen Instinkten. Arch. Neerl. Zool. **4,** 442–520

KORTLANDT, A., 1940 c: Methode van onderzoeken en interpreteren van doelstrevende gedragscoördinatie bij in het wild levende aalschovers, *Phalacrocorax carbo sinensis.* Neerl. Tijdschr. voor Psychol. **VIII**

KORTLANDT, A., 1968: Handgebrauch bei freilebenden Schimpansen; in RENSCH, B.: Handgebrauch und Verständigung bei Affen und Frühmenschen, Verlag Hans Huber Bern und Stuttgart, 54–102

KOSINSKI, J., 1966: The Painted Bird. Pocket Book New York

KRAMER, G., 1952: Die Sonnenorientierung der Vögel. Zool. Anz. Suppl. **16,** 72–84

KRAMER, G., 1957: Experiments on Birds Orientation and their Interpretation. Ibis **96,** 173–185

KRAMER, G., 1959: Recent Experiments on Bird Orientation. Ibis **101,** 399–416

KRIEGER, F., 1954: Untersuchungen über den Farbwechsel der Libellenlarven. Z. vgl. Physiol. **36,** 352–366

KRUIJT, J., 1964: Ontogeny of Social Behaviour in Burmese Red Jungle Fowl, *Gallus gallus spadiceus.* Behaviour, Suppl. **12**

KRUUG, H., 1966: Clan-system and Feeding Habits of Spotted Hyenas, *Crocuta crocuta* ERXLEBEN. Nature **209,** 1257–1258

KUENZER, E. und P., 1962: Untersuchungen zur Brutpflege der Zwergcichliden. Z. Tierpsychol. **19,** 56–83

KUENZER, P., 1968: Die Auslesung der Nachfolgereaktion bei erfahrungslosen Jungfischen von *Nannacara anomala* (Cichlidae). Z. Tierpsychol. **25,** 257–314

KÜHN, A., 1919: Die Orientierung der Tiere im Raum. Gustav Fischer Verlag Jena

KUNKEL, P. und KUNKEL, I., 1964: Beiträge zur Ethologie des Hausmeerschweinchens, *Cavia porcellus* L. 1758. Säugetierk. Mitt. **3,** 168–171

KURT, F., 1968: Das Sozialverhalten des Rehes, *Capreolus capreolus* L. Mammalia depicta, Verlag Paul Parey Hamburg und Berlin, 102 pp.

KURTH, G. und EIBL-EIBESFELDT, I. [Eds.], 1975: Hominisation und Verhalten. Gustav Fischer Verlag Stuttgart, 411 pp.

LAMPRECHT, J., 1970: Duettgesang beim Siamang, *Symphalangus syndactylus.* Z. Tierpsychol. **27,** 186–204

LAMPRECHT, J., 1971: Verhalten. In HERDER **8,** 591–653

LAUDIEN, H., 1965: Untersuchungen über das Kampfverhalten der Männchen von *Betta splendens* REGAN (Anabantidae, Pisces). Z. Wiss. Zool. **172,** 133–178

LAWICK-GOODALL, J. V., 1971: In the Shadow of Man. William Collins Sons + Co. London
D. 1971: Wilde Schimpansen – 10 Jahre Verhaltensforschung am Gombe-Strom. Rowohlt Hamburg
F. 1971: Les Chimpanzées et moi. Ed. Stock Paris

LAWICK, H. V. and LAWICK-GOODALL, J. V., 1970: Innocent Killers. William Collins Sons + Co. London
D. 1972: Unschuldige Mörder. Rowohlt Hamburg. 232 pp.

LENNEBERG, E., 1967: The Biological Foundation of Language. John Wiley and Sons New York
D. 1972: Biologische Grundlagen der Sprache. Suhrkamp Verlag

LESCURE, J., 1965: L'alimentation et le comportement de prédation chez *Bufo bufo.* Thèse Sorbonne, Paris No. **5504,** 164 pp.

LEUTHOLD, W. und LEUTHOLD, B. M., 1973: Notes on the Behaviour of Two Young Antelopes Reared in Captivity. Z. Tierpsychol. **32,** 418–424

LEVI, L. [Ed.], 1971: Society, Stress and Disease: The Psychosocial Environment and Psychosomatic Diseases. London/New York

LEYHAUSEN, P., 1952 a: Theoretische Überlegungen zur Kritik des Begriffes der Übersprungbewegung. Vortrag II. Intern. Etholog. Konferenz Buldern; in LORENZ, K. und LEYHAUSEN, P., 1969: Antriebe tierischen und menschlichen Verhaltens, Piper Verlag München, 77–88

LEYHAUSEN, P., 1952 b: Das Verhältnis von Trieb und Wille in seiner Bedeutung zur Pädagogik. Lebendige Schule (Schola) **7,** 521–542

LEYHAUSEN, P., 1954: Die Entdeckung der relativen Koordination. – Ein Beitrag zur Annäherung von Physiologie und Psychologie. Studium Generale **7,** 45–60

LEYHAUSEN, P., 1965: Über die Funktion der relativen Stimmungshierarchie, dargestellt am Beispiel der phylogenetischen und ontogenetischen Entwicklung des Beutefangs von Raubtieren. Z. Tierpsychol. **22,** 412–498

LEYHAUSEN, P., 1967 a: Zur Naturgeschichte der Angst. Politische Psychologie **6,** 94–112

LEYHAUSEN, P., 1967 b: Biologie von Ausdruck und Eindruck. Psychologische Forschung **31**

LEYHAUSEN, P., 1969: Experimentelle Untersuchung eines angeborenen Auslösemechanismus. 11. Ethol. Conf. Rennes, Vortrag

LEYHAUSEN, P., 1975: Verhaltensstudien an Katzen. Fortschr. d. Verhaltensforsch. H. **2,** Verlag Paul Parey Berlin/Hamburg, 4. Aufl., 232 pp.

LIMBAUGH, C., PEDERSON, H. and CHACE, F. A., 1961: Shrimps that Clean Fishes. Bull. Marine Science Gulf Caribbean **11,** 237–257

LINDAUER, M., 1948: Über die Einwirkung von Duft- und Geschmacksstoffen sowie anderer Faktoren auf die Tänze der Bienen. Z. vgl. Physiol. **31,** 348–412

LINDAUER, M., 1952: Ein Beitrag zur Frage der Arbeitsteilung im Bienenstaat. Z. vgl. Physiol. **34,** 299–345

LINDAUER, M., 1954: Dauertänze im Bienenstock und ihre Beziehung zur Sonnenbahn. Naturwiss. **41,** 506–507

LINDAUER, M., 1955: Schwarmbienen auf Wohnungssuche. Z. vgl. Physiol. **37,** 263–324

LINDAUER, M. [Ed.]. 1973: Orientierung der Tiere im Raum – Teil 1: Sinnes- und neurophysiologische Grundlagen. Fortschr. d. Zool. **21,** H. 2–3, 370 pp.
Teil 2: Intraspezifische Kommunikation. Fortschr. d. Zool. **22,** H. 1, 135 pp.

LISSMANN, H. W., 1946: The Neurological Basis of Locomotory Rhythm in the Spinal Dogfish *Scyllium canicula* and *Acanthias vulgaris.*
I. Reflex Behaviour, J. Exp. Biol. **23,** 143–161
II. The Effect of Deafferentation. J. Exp. Biol. **23,** 162–176

LOEB, J., 1905: Einleitung in die vergleichende Gehirnphysiologie und vergleichende Physiologie mit besonderer Berücksichtigung der wirbellosen Tiere. Leipzig

LÖHRL, H., 1965: Rabenkrähe, *Corvus corone* – Einemsen. Wiss. Film **E 948,** IWF Göttingen

LORENZ, K., 1931: Beiträge zur Ethologie sozialer Corviden. J. Ornith. **79,** 67–127

LORENZ, K., 1932: Beobachtungen über das Erkennen der arteigenen Triebhandlungen der Vögel. J. Ornith. **80,** 50–98

LORENZ, K., 1935: Der Kumpan in der Umwelt des Vogels. J. Ornith. **83,** 137–413

LORENZ, K., 1937: Über die Bildung des Instinktbegriffes. Naturwiss. **25,** 289–300; 307–318; 325–331

LORENZ, K., 1939: Vergleichende Verhaltensforschung. Verhandl. DZG Rostock, Zool. Anz. Suppl. **12,** 69–102

LORENZ, K., 1941: Vergleichende Bewegungsstudien an Anatinen. J. Ornith. **89,** 194–294

LORENZ, K., 1943: Die angeborenen Formen möglicher Erfahrung. Z. Tierpsychol. **5,** 235–409

LORENZ, K., 1949: Die Beziehung zwischen Kopfform und Zirkelbewegung bei Sturniden und Icteriden; in Ornithologie als biologische Wissenschaft, Stresemann-Festschrift, Heidelberg, 135–157

LORENZ, K., 1950: The Comparative Method in Studying Innate Behaviour Patterns. Symp. Soc. Exp. Biol. **IV,** 221–268

LORENZ, K., 1951: So kam der Mensch auf den Hund. Verlag Borotha-Schoeler Wien, 211 pp.

LORENZ, K., 1959 a: Psychologie und Stammesgeschichte; in HEBERER, G. [Ed.]: Evolution der Organismen, G. Fischer Verlag Stuttgart

LORENZ, K., 1959 b: Gestaltwahrnehmung als Quelle wissenschaftlicher Erkenntnis. Z. Exp. Angew. Psychol. **6,** 118–165

LORENZ, K., 1963: Das sogenannte Böse – Zur Naturgeschichte der Aggression. Verlag Borotha-Schoeler, 388 pp.
E. 1963: On Aggression. Harcourt, Brace and World Inc. N. Y.
F. 1969: L'agression – Une histoire naturelle du mal. Club français du livre, Paris

LORENZ, K., 1965 a: Evolution and Modification of Behaviour. The University of Chicago Press Chicago
F. 1967: Evolution et modification du comportement – L'inné et l'acquis. Payot Paris, 152 pp.

LORENZ, K., 1965 b: Über tierisches und menschliches Verhalten – Aus dem Werdegang der Verhaltenslehre. Piper Verlag München, Bd. **1,** 412 pp.
E. 1970: Studies in Animal and Human Behavior. Havard University Press Cambridge, Vol. **1**

LORENZ, K., 1966: Über tierisches und menschliches Verhalten – Aus dem Werdegang der Verhaltenslehre. Piper Verlag München, Bd. **2**, 398 pp.
E. 1971: Studies in Animal and Human Behavior. Harvard University Press Cambridge, Vol. **2**
F. 1970: Essais sur le comportement animal et humain. Ed. du Seuil, un seul volume, 484 pp.

LORENZ, K., 1973: Die Rückseite des Spiegels. Piper Verlag München, 338 pp.

LORENZ, K. und LEYHAUSEN, P., 1968: Antriebe tierischen und menschlichen Verhaltens. Piper Verlag München, 472 pp.
E. 1973: Motivation of Human and Animal Behavior. Van Nostrand Reinhold Co. New York

LORENZ, K. und ST. PAUL, U. V., 1968: Die Entwicklung des Spießens und Klemmens bei den drei Würgerarten *Lanius collurio, L. senator* und *L. excubitor.* J. Ornith. **109,** 137–156

LORENZ, K. und TINBERGEN, N., 1938: Taxis und Instinkthandlung in der Eirollbewegung der Graugans. Z. Tierpsychol. **2,** 1–29

LOUCH, CH. D., 1956: Adrenocortical Activity in Relation to the Density and Dynamics of Three Confined Populations of *Microtus pennsylvanicus.* Ecology **37,** 701–713

LÜLING, K. H., 1969: Das Beutespucken vom Schützenfisch *Toxotes jaculatrix* und Zwergfadenfisch *Colisa lalia.* Bonn. Zool. Beitr. **20,** 416–422

LÜLING, K. H., 1973: *Colisa lalia* (Anabantidae) – Beutespucken. Wiss. Film **E 1674,** IWF Göttingen

LUTTENBERGER, F., 1975: Zum Problem des Gähnens bei Reptilien. Z. Tierpsychol. **37,** 113–137

MACKINTOSH, N. J., 1974: The Psychology of Animal Learning. Academic Press London/New York, 730 pp.

MAIER, N. R. F. and SCHNEIRLA, T. C., 1935: Principles of Animal Psychology. New York

MAKKINK, G. F., 1936: An Attempt at an Ethogram of the European Avocet, *Recurvirostra avosetta,* with Ethological and Psychological Remarks. Ardea **25,** 1–60

MANLEY, G., 1960: The Agonistic Behaviour of the Black-Headed Gull. Diss. Oxford

MARLER, P., 1955: Study of Fighting in Chaffinches (1). Behaviour in Relation to the Social Hierarchy. Brit. J. Anim. Beh. **3,** 111–117

MARLER, P. and HAMILTON, W. J., 1966: Mechanism of Animal Behaviour. John Wiley and Sons New York, 711 pp.
D. 1972: Tierisches Verhalten. BLV München, 706 pp.

MARTIN, H. und LINDAUER, M., 1973: Orientierung im Erdmagnetfeld; in Orientierung der Tiere im Raum, G. Fischer Verlag Stuttgart, **I,** 211–228

MASCHWITZ, U., 1964: Gefahrenalarmstoffe und Gefahrenalarmierung bei sozialen Hymenopteren. Z. vgl. Physiol. **47,** 596–655

MASCHWITZ, U., HÖLLDOBLER, B. und MÖGLICH, M., 1974: Tandemlaufen als Rekrutierungsverhalten bei *Bothroponera tesserinoda* FOREL (Formicidae, Ponerinae). Z. Tierpsychol. **35,** 113–123

MAYR, E., 1963: Animal Species and Evolution. The Belknap Press, Harvard University Press, Cambridge, Mass.
D. 1967: Artbegriff und Evolution. Verlag Paul Parey, Hamburg und Berlin, 617 pp.

MEISCHNER, I., 1959: Verhaltensstudien an Pelikanen. Zool. Garten **25,** 104–126

MEISENHEIMER, J., 1921: Geschlecht und Geschlechter im Tierreich. G. Fischer Verlag Jena

MEISSNER, K., 1976: Homologieforschung in der Ethologie. VEB Gustav Fischer Verlag Jena, 184 pp.

MERKEL, F. W. und WILTSCHKO, W., 1965: Magnetismus und Richtungsfinden zugunruhiger Rotkehlchen, *Erithacus rubecula.* Die Vogelwarte **23,** 71–77

MERTENS, R., 1946: Die Warn- und Drohreaktion der Reptilien. Abh. Senkenberg. Naturf. Ges. Frankfurt

MEYER-HOLZAPFEL, M., 1940: Triebbedingte Ruhezustände als Ziel von Appetenzhandlungen. Naturwiss. **28,** 237

MILUM, V. G., 1947: Grooming Dance and Associated Activities of the Honeybee Colony, Ill. Acad. Sc. Trans. **40,** 194–196

MILUM, V. G. 1955: Honeybee Communication. Americ. Bee Journ. **95,** 97–104

MITTELSTAEDT, H., 1953: Über den Beutefangmechanismus der Mantiden. Zool. Anz. Suppl. **16,** 102–106

MITTELSTAEDT, H., 1954: Regelung und Steuerung bei der Orientierung der Lebewesen. Regelungstechnik **10,** 226–232

MITTELSTAEDT, H. und MITTELSTAEDT, M. L., 1973: Mechanismen der Orientierung ohne richtende Außenreize. Orientierung der Tiere im Raum I, Fortschr. d. Zool. **21,** 46–58

MONFORT-BRAHAM, N., 1975: Variations dans la structure sociale du Topi, *Damaliscus korrigum* OGILBY, au Parc National de l'Akagera, Rwanda. Z. Tierpsychol. **39,** 332–364

MONTAGNER, H., 1966: Le mécanisme et les conséquences des comportements trophallactiques chez les Guêpes du genre *Vespa.* Thèse Nancy, 143 pp.

MONTAGU, M. F. A., 1962: Culture and the Evolution of Man. Oxford Univ. Press New York

MORIN, E. et PIATTELLI-PALMARINI, M. [Eds.], 1974: L'Unité de l'Homme – Invariants biologiques et universaux culturels. Ed. du Seuil Paris, 830 pp.

MORRIS, CH. W., 1946: Signs, Language and Behavior. New York

MORRIS, D., 1958: The Reproductive Behaviour of the Ten-Spined Stickleback *Pygosteus pungitius.* Behaviour Suppl. **6**

MORRIS, D., 1967: The Naked Ape. J. Cape Ltd. London
D. 1968: Der nackte Affe. Droemer Verlag München, 391 pp.

F. 1968: Le singe nu. Edit. Grasset, 277 pp.
MOYER, K. E., 1968: Internal Impulses to Aggression. Trans. of the New York Academy of Sciences, Ser. II **31,** 104–114
MOYER, K. E., 1971 a: Experimentelle Grundlagen eines physiologischen Modells aggressiven Verhaltens; in SCHMIDT-MUMMENDY, A. und SCHMIDT, H. D. [Eds.]: Aggressives Verhalten, Juventa München
MOYER, K. E., 1971 b: The Physiology of Aggression. Markham Press Chicago
MOYNIHAN, M., 1955: Some Aspects of Reproductive Behavior in the Black-headed Gull, *Larus ridibundus,* and Related Species. Behavior Suppl. **4,** 1–201
MÜLLER, D., 1961: Quantitative Luftfeind-Attrappenversuche bei Auer- und Birkhühnern (*Tetrao urogallus* L. und *Lyrurus tetrix* L.). Naturforsch. **16** b, 551–553
MÜLLER, K., 1965: Handbuch der Psychologie. Vol. 1 u. 2, Allgem. Psych.
MURALT, A. V., 1939: Aktionssubstanzen. Basel
MURALT, A. V., 1946: Die Signalübermittlung im Nerven. Basel
MYRBERG, A. A., BRAHY, B. D. and EMERY, A. R., 1967: Field Observation on Reproduction of the Damselfish *Chromis multilineata* (Pomacentridae) with Additional Notes on General Behavior. Copeia **4,** 819–827

NEWEKLOWSKI, W., 1972: Untersuchungen über die biologische Bedeutung und die Motivation der Zirkelbewegung des Stars *Sturnus v. vulgaris* L. Z. Tierpsychol. **31,** 474–502
NICOLAI, J., 1959: Familientradition in der Gesangstradition des Gimpels, *Pyrrhula pyrrhula.* J. Ornith. **100,** 39–46
NICOLAI, J., 1969: Beobachtungen an Paradieswitwen *Steganura paradisea* L., *Steganura obtusa* CHAPIN und der Strohwitwe *Tetraenura fischeri* REICHENOW in Ostafrika. J. Ornith. **110,** 421–447
NOBLE, G. K. and WURM, M., 1943: The Social Behaviour of the Laughing Gull. Ann. N. Y. Acad. Sci. **45,** 179–220
NOLTE, A., 1958: Beobachtungen über das Instinktverhalten von Kapuzineraffen, *Cebus apella,* in der Gefangenschaft. Behaviour **12,** 182–207

OHM, D., 1958 a: Die ontogenetische Entwicklung des Kampfverhaltens bei *Aequideus portalegrensis* und *Ae. latifrons* (Cichlidae). Verh. Deutsch. Zool. Ges. Frankfurt/M.
OHM, D., 1958 b: Vergleichende Beobachtungen am Balzverhalten von *Aequideus* (Cichlidae). Wiss. Z. Humboldt Univ. Berlin, Math.-Nat. Reihe **8,** 357–404
OHM, D., 1964: Die Entwicklung des Kommentkampfverhaltens bei Jungcichliden. Z. Tierpsychol. **21,** 308–325
OZORIO DE ALMEIDA, M. et PIERON, H., 1924 a: Sur les effets de l'extirpation de la peau chez la Grenouille. C. R. Soc. de Biol. **90,** 420–422
OZORIO DE ALMEIDA, M., et PIERON, H., 1924 b: Action de la peau sur l'etat général du systeme nerveux chez la Grenouille. C. R. Soc. de Biol. **90,** 422–425
OZORIO DE ALMEIDA, M. et PIERON, H., 1924 c: Sur le rôle de la peau dans le maintien du tonus musculaire chez la Grenouille. C. R. Soc. de Biol **90,** 478–481

PAPI, F. und PARDI, L., 1959: Nuovi reperti sull'orientamento lunare di *Talitrus saltator.* Z. vgl. Physiol. **41,** 583–596
PAWLOW, I. P., 1927: Conditioned Reflexes. Oxford
PECKMAN, G. und PECKMAN, E., 1904: Instinkt und Gewohnheiten der solitären Wespen. Berlin
PEIPER, A., 1956: Die Eigenart der kindlichen Hirntätigkeit. Leipzig, 2. Aufl.
PELKWIJK, J. J. TER und TINBERGEN, N., 1937: Eine reizbiologische Analyse einiger Verhaltensweisen von *Gasterosteus aculeatus.* Z. Tierpsychol. **1,** 193–204
PETZOLD, H. G., 1964: Vergleichend-ethologische Beobachtungen. Beitr. Vogelk. **10,** 1–126
PIERON, H., 1968: Vocabulaire de la Psychologie. Presses Universitaires de France, Paris, 571 pp.
PILLERI, G. and KNUCKEY, J., 1969: Behaviour Patterns of Some Delphinidae Observed in Western Mediterranean. Z. Tierpsychol. **26,** 48–72
PILTERS, H., 1956: Das Verhalten der Tylopoden. Handb. d. Zool. **8,** 1–24
PLACK, A. [Hrsg.], 1973: Der Mythos vom Aggressionstrieb. List Verlag München, 399 pp.
PLOOG, D., 1964: Verhaltensforschung und Psychiatrie; in Psychiatrie der Gegenwart, Springer Verlag Berlin, 291–443
PLOOG, D., BLITZ, J. and PLOOG, F., 1963: Studies on Social Sexual Behavior of the Squirrel Monkey *Saimiri sciureus.* Fol. Prim., 29–66
PORTIELJE, A. E. J., 1936: Ein bemerkenswerter Grenzfall von Polygamie bzw. akzessorischer Promiskuität beim Höckerschwan, zugleich ein Beitrag zur Ethologie bzw. Psychologie von *Cygnus olor.* J. Ornith. **84,** 140–158
PORTMANN, A., 1953: Das Tier als soziales Wesen. Rhein Verlag Zürich, 381 pp.
PRECHTL, H. F. R., 1955: Die Entwicklung der frühkindlichen Motorik I–III. Wiss. Filme **C 651, C 652, C 653,** IWF Göttingen
PRECHTL, H. F. R. und SCHLEIDT, W., 1950: Auslösende und steuernde Mechanismen des Saugaktes I. Z. vgl. Physiol. **32,** 252–262
PRECHTL, H. F. R. und SCHLEIDT, W., 1951: Auslösende und steuernde Mechanismen des Saugaktes II. Z. vgl. Physiol. **33,** 53–62
PREVOST, J., 1961: Ecologie du Manchot Empereur. Exp. Pol. Franç. Publ. **222,** Edition Hermann Paris, 204 pp.
PRINZINGER, R., 1976: Temperatur- und Stoffwechselregulation der Dohle (*Coleus monedula* L.), Rabenkrähe (*Corvus corone corone* L.) und Elster (*Pica pica* L.). Anz. Ornith. Ges. Bayern **15,** 1–47

QUERENGÄSSER, A., 1973: Über das Einemsen von Singvögeln und die Reifung dieses Verhaltens. J. Ornith. **114,** 96–117

RÄBER, H., 1949: Das Verhalten gefangener Waldohreulen und Waldkäuze zur Beute. Behaviour **2,** 1–95
RADESÄTER, T., 1975: Interactions between ♂ and ♀ During the Triumph Ceremony in the Canada Goose, *Branta canadensis* L. Z. Tierpsychol. **39,** 189–205
RAND, A. L., 1943: Some Irrelevant Behaviour in Birds. The Auck **60,** 168–171
RASA, A., 1971: Appetence for Aggression in Juvenile Damsel Fish. Fortschr. d. Verhaltensforsch. H. **7,** Paul Parey Verlag Hamburg/Berlin, 68 pp.
REMANE, A., 1952: Die Grundlagen des natürlichen Systems der vergleichenden Anatomie und der Phylogenetik. Leipzig
REMANE, A., 1971: Sozialleben der Tiere. G. Fischer Verlag Stuttgart, 177 pp.
RENGGLI, F., 1976: Angst und Geborgenheit – Soziokultu-

relle Folgen der Mutter-Kind-Beziehung im ersten Lebensjahr. Rowohlt Taschenbuch, Hamburg, 284 pp.

RENSCH, B., 1973: Gedächtnis, Begriffsbildung und Planhandlung bei Tieren. Verlag Paul Parey Hamburg/Berlin, 274 pp.

REUTER, O. M., 1913: Lebensgewohnheiten und Instinkte bei Insekten. Berlin

REYER, H.-U., 1975: Ursachen und Konsequenzen von Aggressivität bei *Etroplus maculatus* (Cichlidae, Pisces). Z. Tierpsychol. **39,** 415–454

RICHELLE, M., 1970: Malentendus sur les apports du conditionnement. Rev. Compt. Animal **4,** 22–31

ROE, A. and SIMPSON, G. G. [Eds.], 1964: Behaviour and Evolution. Yale University Press New Haven and London, 557 pp.

ROTHGÄNGER, G. und H., 1973: Über spezielle Verhaltensweisen fliegender Mauersegler. Der Falke **20,** 124–130

ROUBAUD, E., 1916: Recherches biologiques sur les Guêpes solitaires et sociales d'Afrique. Ann. Sc. Nat. Zool. **1,** 1–160

RUSSELL, E. S., 1943: Perceptual and Sensory Signs in Instinct Behaviour. Proc. Linn. Soc. London **154,** 195–216

RYALL, R. W., 1958: Effect of Drugs on Emotional Behaviour in Rats. Nature **182,** 1606–1607

SAMBRAUS, H. H., 1971: Das Sexualverhalten des Hausrindes, speziell des Stieres. Fortschr. d. Verhaltensforsch. H. **6,** Verlag Paul Parey Berlin/Hamburg, 54 pp.

SAMBRAUS, H. H., 1973: Das Sexualverhalten der domestizierten einheimischen Wiederkäuer. Fortschr. d. Verhaltensforsch. H. **12,** Verlag Paul Parey Berlin/Hamburg, 100 pp.

SCHAEFER, H. H., 1960: Suggested German Translation of Expressions in the Field and Operant Conditioning. J. of Exper. Analysis of Behaviour **3,** 171–182

SCHALLER, F., 1960: Das Phoresie-Problem vergleichend ethologisch gesehen. Forsch. u. Fortschr. **34,** 1–7

SCHALLER, G. B., 1967: The Deer and the Tiger. Chicago

SCHALLER, G. B., 1973: On the Behaviour of Blue Sheep, *Pseudovis nayaus.* J. of the Bombay Nat. Hist. Soc. **69,** 523–537

SCHENKEL, R., 1947: Ausdrucksstudien an Wölfen. Behaviour **1,** 81–129

SCHENKEL, R., 1956: Zur Deutung der Phasianidenbalz. Ornith. Beob. **53,** 182–201

SCHENKEL, R., 1958: Zur Deutung der Balzleistungen einiger Phasianiden und Tetraoniden. Ornith. Beob. **55,** 65–95

SCHENKEL, R., 1967: Submission, its Features and Function in the Wolf and Dog. Americ. Zool. **7,** 319–329

SCHEYGROUND, A., HOOFF, J. A. R. A. M. VAN und ARONDS, H., 1973: Lachen apen om hun eigen streken? in Apenstreken, Blaricum, Bigot und van Rossum, 36–51

SCHJELDERUP-EBBE, TH., 1922 a: Beiträge zur Sozialpsychologie des Haushuhns. Z. Psychol. **88,** 225–252

SCHJELDERUP-EBBE, TH., 1922 b: Soziale Verhältnisse bei Vögeln. Z. Psych. **90,** 106–107

SCHLEIDT, M., 1954: Untersuchungen über die Auslösung des Kollerns beim Truthahn, *Meleagris gallopavo.* Z. Tierpsychol. **11,** 417–435

SCHLEIDT, W., 1961 a: Über die Auslösung der Flucht vor Raubvögeln bei Truthühnern. Naturwiss. **48,** 141–142

SCHLEIDT, W., 1961 b: Reaktionen von Truthühnern auf fliegende Raubvögel und Versuche zur Analyse ihrer AAMs. Z. Tierpsychol. **18,** 534–560

SCHLEIDT, W., 1962: Die historische Entwicklung der Begriffe »Angeborenes auslösendes Schema« und „Angeborener Auslösemechanismus" in der Ethologie. Z. Tierpsychol. **19,** 697–722

SCHLEIDT, W., 1974: How "Fixed" is the Fixed Action Pattern? Z. Tierpsychol. **36,** 184–211

SCHMID, H., 1964: Zur Frage der Störung des Bienengedächtnisses durch Narkosemittel, zugleich ein Beitrag zur Störung der sozialen Bindung durch Narkose. Z. vgl. Physiol. **47,** 559–595

SCHMIDT, H. D., 1960: Bigotry in School-Children. Commentary **29,** 253–257

SCHMIDT, R. S., 1972: Mecanism of Clasping and Releasing (Unclasping) in *Bufo americanus.* Behaviour **43,** 85–96

SCHNEIDER, F., 1961: Beeinflussung der Aktivität des Maikäfers durch Veränderung der gegenseitigen Lage magnetischer und elektrischer Reize. Mitt. Schweiz. Ent. Ges. **33,** 223–237

SCHNEIDER, K. M., 1930: Das Flehmen I. Der Zool. Garten **3,** 183–198

SCHNEIDER, K. M., 1931: Das Flehmen II. Der Zool. Garten **4,** 349–364

SCHNEIDER, K. M., 1932: Das Flehmen III. Der Zool. Garten **5,** 200–226

SCHNEIDER, K. M., 1933: Das Flehmen IV. Der Zool. Garten **6,** 287–292

SCHÖNE, H., 1973 a: Verhalten und Orientierung. Z. Tierpsychol. **33,** 287–294

SCHÖNE, H., 1973 b: Raumorientierung, Begriffe und Mechanismen. Fortschr. Zool. **21,** 1–19

SCHÖNE, H., 1974: Spatial Orientation in Animals; in KINNE, O. [Ed.] Marine Ecology, Vol. **2,** Wiley London

SCHÖNE, H. und SCHÖNE, H., 1963: Balz und andere Verhaltensweisen der Mangrovenkrabbe, *Goniopsis cruentata,* und das Winkverhalten der eulitoralen Brachyuren. Z. Tierpsychol. **20,** 641–656

SCHUETT, F., 1933: Studies in Mass Physiology: the Effect of Numbers upon Oxygen Consumption of Fishes. Ecology **14,** 106–122

SCHULER, W., 1974: Die Schutzwirkung künstlicher Bates'scher Mimikry abhängig von Modellähnlichkeit und Beuteangebot. Z. Tierpsychol. **36,** 71–127

SCOTT, J. P., 1946: The Analysis of Social Organisation in Animals. Ecology **37,** 213–221

SEIBT, U. und WICKLER, W., 1972: Individuen-Erkennen und Partnerbevorzugung bei der Garnele *Hymenocera picta* DANA. Naturwiss. **59,** 40–41

SEIFERT, A., 1950: Die Struktur der Erinnerung; in Philosophia naturalis

SEITZ, A., 1940: Die Paarbildung bei einigen Cichliden I. Z. Tierpsychol. **4,** 40–84

SEITZ, A., 1943: Die Paarbildung bei einigen Cichliden II. Z. Tierpsychol. **5,** 74–101

SELIGMANN, S., 1922: Die Zauberkraft des Auges und das Berufen. Hamburg

SELIGMANN, S., 1927: Die magischen Heil- und Schutzmittel aus der unbelebten Natur mit besonderer Berücksichtigung der Mittel gegen den bösen Blick. Stuttgart

SELOUS, E., 1907: Observations Tending to Throw Light on the Question of Sexual Selection in Birds, Including a Day to Day Diary on the Breeding Habits of the Ruff. Zoologist **11,** 60–65, 161–182, 367–381

SEVENSTER, P., 1961: A Causal Analysis of Displacement Activity: Fanning in *Gasterosteus aculeatus.* Behaviour Suppl. **9**

SHIOVITZ, K. A., 1975: The Process of Species-Specific

Song Recognition by the Indigo Bunting, *Passerina cyanea,* and its Relationship to the Organization of Avian Acoustical Behaviour. Behaviour **55,** 128–179

SIEBENALER, J. B. and CALDWELL, D. K., 1956: Cooperation among Adult Dolphins. J. of Mammalogy **37,** 126–128

SKINNER, B. F., 1966: The Phylogeny and Ontogeny of Behaviour. Science **153,** 1205–1213

SKINNER, B. F., 1971: Beyond Freedom and Dignity. A Knopf, New York

SORENSEN, E. R. and GAJDUSEK, D. C., 1966: The Study of Child Behavior and Development in Primitive Cultures. Journ. Americ. Acad. Pediatrics **37,** 149–243

SOULAIRAC, A., 1952: L'effet de groupe dans le comportement sexuel du Rat mâle. Coll. Int. C. N. R. S. Paris, **34,** 91–97

SPALLANZANI, L. (1729–1799) 1793: see DIJKGRAAF, S., 1960

SPANIER, E., 1970: Analysis of Sounds and Associated Behaviour of the Domino Damselfish *Chromis trimaculatis.* M. Sc. Thesis, Univ. Tel-Aviv

SPENCER, H., 1899: The Principles of Psychology. London

SPINDLER, P., 1954: Vererbung von Verhaltensweisen, I. Verhalten bei akustischem Schreckreiz. – Die »Hals-Schulter-Reaktion« bei Schimpanse und Mensch. Film **CT 1043** der Bundesstaatl. Hauptst. f. Lichtbild und Bildungsfilm Wien

SPITZ, R., 1945: Hospitalism. The Psychoanalytic Study of the Child I, Int. Univ. Press New York, 53–74

SPITZ, R., 1957: Die Entstehung der ersten Objektbeziehung. Klett-Verlag Stuttgart

SPITZ, R., 1965: The First Year of Life. Int. Univ. Press New York

STEVENSON, R. A., 1963: Life History and Behaviour of *Dascyllus albisella,* a Pomacentrid Reef Fish. Diss. Univ. Hawaii

STOKES, A. W. [Ed.], 1974: Territory. Benchmark Papers in Animal Behavior, Dowdon, Hutchinson and Ross, Inc., Stroudsburg, Penn., 398 pp.

STRESEMANN, E., 1935: Die Benutzung von Ameisen zur Gefiederpflege. Orn. Mber. **43,** 134–138

SZYMANSKI, J. S., 1913: Ein Versuch, die für das Liebesspiel charakteristischen Körperstellungen und Bewegungen bei der Weinbergschnecke künstlich hervorzurufen. Pflügers Archiv **149**, 471–482

TEMBROCK, G., 1954: Rotfuchs und Wolf, ein Verhaltensvergleich. Z. Säugetierk. **19,** 152–159

TEMBROCK, G., 1959: Tierstimmen. Die Neue Brehm-Bücherei H. **250,** A. Ziemsen Verlag Wittenberg-Lutherstadt, 285 pp.

TEMBROCK, G., 1961: Verhaltensforschung – eine Einführung in die Tier-Ethologie. VEB Gustav Fischer Verlag Jena, 371 pp.

TEMBROCK, G., 1967: Grundlagen der Tierpsychologie. Akademie Verlag Berlin, 207 pp.
F. 1967: Eléments de psychologie animale. Gauthier-Villars Paris, 155 pp.

TEMBROCK, G., 1968: Grundriß der Verhaltenswissenschaften; in Grundbegriffe der modernen Biologie, Gustav Fischer Verlag Stuttgart, Bd. **3,** 207 pp.

TEMBROCK, G., 1972: Tierpsychologie. Die Neue Brehm-Bücherei H. **455,** A. Ziemsen Verlag Wittenberg-Lutherstadt, 181 pp.

TETRY, A., 1948: Les outils chez les êtres vivants. Gallimard Paris, 345 pp.

TINBERGEN, E. A. and TINBERGEN, N., 1972: Early Childhood Autism – An Ethological Approach. Fortschr. d. Verhaltensforsch. H. **10,** Verlag Paul Parey Berlin/Hamburg, 53 pp.

TINBERGEN, N., 1932: Über die Orientierung des Bienenwolfes *Philanthus triangulum.* Z. vgl. Physiol. **16,** 305–335

TINBERGEN, N., 1939: On the Analysis of Social Organization among Vertebrates, with Special Reference to Birds. Americ. Midl. Natural. **21,** 210–234

TINBERGEN, N., 1940: Die Übersprungbewegung. Z. Tierpsychol. **4,** 1–40

TINBERGEN, N., 1942: An Objectivistic Study of the Innate Behaviour of Animals. Biblioth. biotheor. **1**, 39–98

TINBERGEN, N., 1948: Social Releasers and the Experimental Method Required for Their Study. Wilson Bull. **60,** 6–52

TINBERGEN, N., 1949: De functie van de rode vlek op de snavel van de zilvermeeuw. Bijdragen tot de Dierkunde **28,** 453–465

TINBERGEN, N., 1951: The Study of Instinct. Oxford University Press
D. 1972: Instinktlehre. 5. Aufl., Verlag Paul Parey Hamburg/Berlin, 256 pp.
F. 1971: L'étude de l'instinct (dernière édition). Payot Paris, 312 pp.

TINBERGEN, N., 1955: Social Behaviour in Animals. Methuen + Co. Ltd.
D. 1955: Tiere untereinander – Soziales Verhalten bei Tieren. Verlag Paul Parey Berlin/Hamburg, 150 pp.

TINBERGEN, N., 1958: The Herring Gull's World. Oxford
D. 1958: Die Welt der Silbermöwe. Musterschmidt Verlag Göttingen, 279 pp.

TINBERGEN, N., 1959 a: Comparative Studies of the Behaviour of Gulls (Laridae) – a Progress Report. Behaviour **15,** 1–70

TINBERGEN, N., 1959 b: Einige Gedanken über Beschwichtigungsgebärden. Z. Tierpsychol. **16,** 651–665

TINBERGEN, N., 1961: Curious Naturalists. Country Life Ltd. London
D. 1961: Wo die Bienenwölfe jagen. Verlag Paul Parey Berlin/Hamburg, 228 pp.

TINBERGEN, N. and IERSEL, J. J. A. VAN, 1947: Displacement Reaction in the Three-Spined Stickleback. Behaviour **1,** 56–63

TINBERGEN, N. und KUENEN, D. J., 1939: Über die auslösenden und richtungsgebenden Reizsituationen der Sperrbewegung bei jungen Drosseln. Z. Tierpsychol. **3,** 37–60

TINBERGEN, N. and PERDECK, A. C., 1950: On the Stimulus Situation Releasing the Begging Response in the Newly Hatched Herring Gull Chick *Larus argentatus.* Behaviour **3,** 1–38

TIRALA, H., 1923: Die Form als Reiz. Zool. Jahrb. Abtg. Allg. Zool. u. Physiol. **39,** 395–442

THORPE, W. H., 1951: The Definition of Some Terms Used. Animal Behaviour **9,** 34–40

THORPE, W. H., 1963: Antiphonal Singing in Birds as Evidence for Avian Auditory Reaction Time. Nature **197,** 774–776

TOROSSIAN, CL., 1958: L'aliment proctodéal chez la Fourmi *Dolichoderus quadripunctatus.* C. R. Ac. Sciences **246,** 3524–3526

TOROSSIAN, CL., 1966: Recherches sur la biologie et l'éthologie des Myrmécophiles. III. – Etude expérimentale de la spécificité du couple Myrmécophile-Fourmis, entre le Coléoptère Brenthidae, *Amorphocephalus coronatus* et diverses espèces de Fourmis. Insectes Sociaux **13,** 39–58

TSCHANZ, B., 1968: Trottellummen – Die Entstehung der persönlichen Beziehung zwischen Jungvogel und Eltern. Fortschr. der Verhaltensforsch. H. **4,** Verlag Paul Parey Berlin/Hamburg, 103 pp.

UEXKÜLL, J. V. 1921: Umwelt und Innenwelt der Tiere. 2. Aufl., Berlin

UEXKÜLL, J. V., 1937: Umweltforschung. Z. Tierpsychol. **1,** 33–34

UEXKÜLL, J. V. und KRISZAT, G., 1934: Streifzüge durch die Umwelten von Tieren und Menschen – Ein Bilderbuch unsichtbarer Welten, Verständl. Wissensch. Berlin 1956 (neue Aufl.)

VIERKE, J., 1973: Das Wasserspucken der Arten der Gattung *Colisa;* Pisces, Anabantidae. Bonn. Zool. Beitr. **24,** 62–104

VOGEL, G., 1954: Das optische Weibchenschema bei *Musca domestica.* Naturwiss. **41,** 482

VOGEL, G., 1957: Verhaltensphysiologische Untersuchungen über die den Weibchensprung des Stubenfliegenmännchens auslösenden optischen Faktoren. Z. Tierpsychol. **14,** 309

VOGEL, G., 1958: Supernormale Auslösereize bei *Sarcophaga carnaria.* Zool. Beitr. N. F. **4,** 69–76

VOLKELT, H., 1914: Die Vorstellungen der Tiere – Arbeiten zur Entwicklungsphysiologie.
Herausg. v. KRUEGER, F.

VOLKELT, H., 1937: Tierpsychologie als genetische Ganzheitspsychologie. Z. Tierpsychol. **1,** 49–65

WAHLERT, H. V., 1961: Le comportement de nettoyage de *Crenilabrus melanocercus* (Labridae, Pisces) en Méditerranée. Vie et Milieu **12,** 1–10

WALTHER, F., 1958 a: Zum Kampf- und Paarungsverhalten einiger Antilopen. Z. Tierpsychol. **15,** 340–380

WALTHER, F., 1958 b: Zum Kampf- und Paarungsverhalten einiger Antilopen. Z. Tierpsychol. **16,** 631–665

WALTHER, F., 1966: Mit Horn und Huf. Verlag Paul Parey Berlin/Hamburg, 171 pp.

WALTHER, F., 1968: Verhalten der Gazellen. Neue Brehm-Bücherei H. **373,** A. Ziemsen Verlag Wittenberg-Lutherstadt, 144 pp.

WARD, R. W., 1967: Ethology of Paradise Fish, *Macropodus opercularis,* I. Differences between Domestic and Wild Fish. Copeia 1967, 809–813

WATSON, J. B., 1930: Behaviorism. Norton and Comp. New York
D. 1968: Behaviorismus. Kiepenheuer und Witsch Köln, 295 pp.

WATSON, J. R., 1910: The Impaling Instinct in Shrikes. The Auk, 459

WIEPKEMA, P. R., 1961: An Ethological Analysis of the Reproductive Behaviour of the Bitterling *Rhodeus amarus.* Arch. Neerl. Zool. **14,** 103–199

WHEELER, W. M., 1918: A Study of Some Ant Larvae with a Consideration of the Origin and Meaning of Social Habits among Insects. Proc. Amer. Philos. Soc. **57,** 293–343

WHEELER, W. M., 1928: Social Life among Insects. New York

WHITMAN, CH. O., 1899: Animal Behavior. Biol. Lect. Mar. Biol. Lab., Woods Hole, 285–338

WHITMAN, CH. O., 1919: The Behavior of Pigeons. Publ. Carnegie Inst. **257,** 1–161

WICKLER, W., 1956: Eine Putzsymbiose zwischen *Corydoras* und *Trichogaster.* Z. Tierpsychol. **13,** 46–49

WICKLER, W., 1961 a: Über das Verhalten der Blenniiden *Runula* und *Aspidontus.* Z. Tierpsychol. **18,** 421–440

WICKLER, W., 1961 b: Über die Stammesgeschichte und den taxonomischen Wert einiger Verhaltensweisen bei Vögeln. Z. Tierpsychol. **18,** 320–342

WICKLER, W., 1961 c: Ökologie und Stammesgeschichte von Verhaltensweisen. Fortschr. Zool. **13,** 303–365

WICKLER, W., 1962: Zur Stammesgeschichte funktionell korrelierter Organ- und Verhaltensmerkmale: Ei-Attrappen und Maulbrüten bei afrikanischen Cichliden. Z. Tierpsychol. **19,** 129–164

WICKLER, W., 1963: Zum Problem der Signalbildung, am Beispiel der Verhaltensmimikry zwischen *Aspidontus* und *Labroides.* Z. Tierpsychol. **20,** 657–679

WICKLER, W., 1965 a: Die äußeren Genitalien als soziale Signale bei einigen Primaten. Naturwiss. **52,** 269–270

WICKLER, W., 1965 b: Die Evolution von Mustern der Zeichnung und des Verhaltens. Naturwiss. **52,** 335–341

WICKLER, W., 1965 c: Über den taxonomischen Wert homologer Verhaltensmerkmale. Naturwiss. **52,** 441–444

WICKLER, W., 1966: Ursprung und biologische Deutung des Genitalpräsentierens männlicher Primaten. Z. Tierpsychol. **23,** 422–437

WICKLER, W., 1967: Vergleichende Verhaltensforschung und Phylogenetik; in HEBERER, G. [Ed.]: Evolution der Organismen I, 3. Aufl., G. Fischer Verlag Stuttgart, 420–508

WICKLER, W., 1968: Mimikry. Kindler Verlag München, 255 pp.
E. 1968: Mimicry. McGraw-Hill Book Co., New York and Toronto
F. 1968: Mimétisme. Librairie Hachette Paris

WICKLER, W., 1969: Sind wir Sünder? – Naturgesetze der Ehe. Verlag Droemer Knaur München, 280 pp.

WICKLER, W., 1970: Stammesgeschichte und Ritualisierung. Zur Entstehung tierischer und menschlicher Verhaltensmuster. Piper Verlag München, 282 pp.

WICKLER, W., 1971: Die Biologie der Zehn Gebote. Piper Verlag München, 225 pp.

WICKLER, W., 1972: Duettieren zwischen artverschiedenen Vögeln im Freiland. Z. Tierpsychol. **31,** 98–103

WICKLER, W. und SEIBT, U., 1972: Über den Zusammenhang des Paarsitzens mit anderen Verhaltensweisen bei *Hymenocera picta* DANA. Z. Tierpsychol. **31,** 163–170

WILSON, E. O., 1971: The Insect Societies. Belknap Press of Harvard Univ. Press Cambridge, 548 pp.

WILSON, E. O., 1975: Sociobiology – The New Synthesis. Belknap Press of Harvard Univ. Press Cambridge, 697 pp.

WILTSCHKO, W. and WILTSCHKO, R., 1975 a: The Interaction of Stars and Magnetic Field in the Orientation System of Night Migrating Birds. I. Autumn Experiments with European Warblers (gen. *Sylvia*). Z. Tierpsychol. **37,** 337–355

WILTSCHKO, W. and WILTSCHKO, R., 1975 b: The Interaction of Stars and Magnetic Field in the Orientation System of Night Migrating Birds. II. Spring Experiments with European Robins, *Erithacus rubecula.* Z. Tierpsychol. **39,** 265–282

WIRTZ, P. and DAVENPORT, J., 1976: Increased Oxygen Consumption in Blennies (*Blennius pholis* L.) Exposed to Their Mirror Images. J. Fish Biol. **9,** 67–74

WOLFE, J. B., 1936: Effectiveness of Token-Rewards in Chimpanzees. Comp. Psychol. Monogr. **12,** 1–72

WÜRDINGER, I., 1970: Erzeugung, Ontogenetik und Funktion der Lautäußerungen bei vier Gänsearten. Z. Tierpsychol. **27,** 257–302

YERKES, R. M., 1940: Social Behavior of Chimpanzees. J. Comp. Psychol. **30,** 147–186

ZIMEN, E., 1972: Wölfe und Königspudel. Piper Verlag München

ZIMEN, E., 1976: On the Regulation of Pack Size in Wolves. Z. Tierpsychol. **40,** 300–341

ZUCKERMANN, S., 1932: The Social Life of Monkeys and Apes. London

Illustration List

English Index

A

B

C

D

E

Page numbers in normal impression indicate key-words followed by explanations in three languages. Page numbers in italic impression indicate only the translation of the key-word.

Index Français

D

E

L

M

N

Les chiffres inprimés en caractères normaux renvoient à des pages ou les notions éthologiques sont suivies d'une explication trilingue. Les chiffres en italique ne renvoient qu'à une traduction du terme.

Dr. Armin Heymer, D. Sc.
Chargé de Recherches au C.N.R.S

Geboren 1937, studierte Zoologie und Botanik und war 1956 Vogelwart auf der Insel Wangerooge/Nordsee. Seit 1959 regelmäßige Arbeitsaufenthalte am Laboratoire Arago in Banyuls-sur-Mer, 1965 Attaché de Recherches au C.N.R.S. und Mitarbeiter am Institut für Allgemeine Ökologie in Brunoy; 1967/68 Forschung am Institut für Höhlenforschung in Moulis/Pyrenäen. Promovierte bei Pierre-Paul Grassé und habilitierte sich 1970 an der Sorbonne in Ethologie. Im Jahre 1971 führte ihn ein Forschungsaufenthalt als Gast der Israelischen Akademie der Wissenschaften an die Universität Tel-Aviv. Heymers Arbeitsgebiet ist die vergleichende Verhaltensforschung, wobei sein besonderes Interesse dem Verhalten der Libellen und den Meeresfischen galt. Zahlreiche Reisen führten ihn zur Freilandarbeit in die Länder des Mittelmeerraumes, des Nahen Ostens und des Roten Meeres. In jüngster Zeit wandte er sich der Humanethologie zu. Sein besonderes Interesse gilt hier den Pygmäen.

Born in 1937, studied zoology and botany and was resident ornithologist at Wangerooge Island/North Sea. Since 1959 he regularly worked at the Laboratoire Arago/ Southern France. Attaché de Recherches au C.N.R.S. in 1965, and staff member at the Institute for General Ecology at Brunoy. In 1967/68 he worked at the Institute for Speleological Research at Moulis/Pyrenees. He received his Ph. D. with Pierre-Paul Grassé and habilitated in 1970 at the Sorbonne in Ethology. In 1971 he was a visiting scientist as a guest of the Israel Academy of Sciences at the University of Tel-Aviv. His special interest is the behaviour of dragonflies. However, he has worked with saltwater fish. His travels led him to field work in many Mediterranean countries, in the Near East and in the Red Sea. Most recently he turned to the study of human ethology. His special interest are the Pygmies.

Né en 1937. Etudes supérieures ès Sciences Naturelles, Zoologie et Botanique. Stage de recherches ornithologiques sur l'Ile de Wangerooge/Mer du Nord en 1956. Dès 1959, séjours fréquents et réguliers au Laboratoire Arago et au Centre d'Ecologie Méditerranéenne du Mas de la Serre à Banyuls-sur-Mer. En 1965, Attaché de Recherches au C.N.R.S., attaché au Laboratoire d'Ecologie Générale de Brunoy; 1967/68 séjour au Laboratoire Souterrain du C.N.R.S. à Moulis/Pyrénées. Thèse d'Etat ès Sciences Naturelles en 1970 à la Sorbonne et en 1971, invité de l'Académie des Sciences d'Israél à l'Université de Tel-Aviv. Son domaine de recherches est l'Ethologie. Ses investigations ont porté sur le comportement des Libellules, Insectes Paléoptères, mais également sur celui des Poissons du milieu marin. De nombreuses missions scientifiques ont été effectuées dans les pays du pourtour méditerranéen, du Proche Orient et de la Mer Rouge. Dernièrement, son intérêt s'est porté sur l'éthologie humaine, particulièrement sur les Pygmées.